Volume 3

Mechanisms of Inorganic and Organometallic Reactions

Volume 3

Mechanisms of Inorganic and Organometallic Reactions

Edited by

M. V. Twigg

Imperial Chemical Industries P.L.C.
Billingham, Cleveland, United Kingdom

PLENUM PRESS • *NEW YORK AND LONDON*

Library of Congress Cataloging in Publication Data

Main entry under title:

Mechanisms of inorganic and organometallic reactions.

Includes bibliographical references and indexes.
1. Chemical reactions. 2. Chemistry, Inorganic. 3. Organometallic compounds. I.
Twigg, M. V.
QD501.M426   1983   541.3'9   83-2140
ISBN 0-306-41960-2

© 1985 Plenum Press, New York
A Division of Plenum Publishing Corporation
233 Spring Street, New York, N.Y. 10013

Printed in the United States of America

Contributors

Dr. A. Bakác   Ames Laboratory and Department of Chemistry, Iowa State University, Ames, Iowa 50011, U.S.A.

Dr. J. Burgess   Chemistry Department, The University, Leicester LE1 7RH, U.K.

Dr. R. D. Cannon   Chemistry Department, University of East Anglia, University Plain, Norwich NR4 7TH, U.K.

Dr. R. J. Cross   Department of Chemistry, The University, Glasgow G12 8QQ, Scotland, U.K.

Dr. A. J. Deeming   Chemistry Department, University College London, 20 Gordon Street, London WC1H 0AJ, U.K.

Dr. R. van Eldik   Institute for Physical Chemistry, University of Frankfurt, Frankfurt am Main, Federal Republic of Germany

Dr. J. H. Espenson   Ames Laboratory and Department of Chemistry, Iowa State University, Ames, Iowa 50011, U.S.A.

Dr. M. Green   Chemistry Department, The University, York, North Yorkshire YO1 5DD, U.K.

Dr. D. N. Hague   Chemical Laboratory, The University, Canterbury, Kent CT2 7NH, U.K.

Dr. R. W. Hay   Department of Chemistry, University of Stirling, Stirling FK9 4LA, Scotland, U.K.

v

Dr. L. A. P. Kane-Maguire   Chemistry Department, Wollongong
   University, P.O. Box 1144, Wollongong, N.S.W. 2500, Australia

Dr. A. G. Lappin   Chemistry Department, University of Notre Dame,
   Notre Dame, Indiana 46556, U.S.A.

Dr. P. Moore   Department of Chemistry and Molecular Sciences,
   University of Warwick, Coventry CV4 7AL, U.K.

Dr. D. A. Sweigart   Department of Chemistry, Brown University,
   Providence, Rhode Island 02912, U.S.A.

Dr. C. White   Department of Chemistry, The University of Sheffield,
   Sheffield S3 7HF, U.K.

Dr. N. Winterton   ICI Plc, Mond Division, The Heath, Runcorn,
   Cheshire, WA7 4QE, U.K.

Preface

The purpose of this series is to provide a continuing critical review of the literature concerned with mechanistic aspects of inorganic and organometallic reactions in solution, with coverage being complete in each volume. The papers discussed are selected on the basis of relevance to the elucidation of reaction mechanisms and many include results of a nonkinetic nature when useful mechanistic information can be deduced. The period of literature covered by this volume is July 1982 through December 1983, and in some instances papers not available for inclusion in the previous volume are also included. Numerical results are usually reported in the units used by the original authors, except where data from different papers are compared and conversion to common units is necessary.

As in previous volumes material included covers the major areas of redox processes, reactions of the nonmetallic elements, reaction of inert and labile metal complexes and the reactions of organometallic compounds. While maintaining the space devoted to other areas, that given to the nonmetallic elements has been increased. In recognition of the increasing importance of the determination of volumes of activation in understanding the mechanisms of both inorganic and organometallic reactions a special reference section giving tabulated $\Delta V^{\ddagger}$ values has been included and this extensive compilation will be updated in future volumes.

Mention should be made of the devotion of the contributors, who in several instances produced their chapters in spite of unforeseen circumstances ranging from prolonged hospitalization to relocation to the other side of the world. The continuing support of those involved in the production of the series is also much appreciated, and particular thanks are due to Ken Derham in London and Jeanne Libby in New York. This series was established as a result of demand from members of the Inorganic Mechanisms Discussion Group (UK) and we thank them for their continuing support. Helpful suggestions and ideas from many readers are also gratefully acknowledged. Further comments and suggestions will of course be welcomed.

Contents

Chapter 3. *Metal–Ligand Redox Reactions*

A. Bakác and J. H. Espenson

Part 2: Substitution and Related Reactions

Chapter 4. Reactions of Compounds of the Nonmetallic Elements

N. Winterton

Chapter 5.   Substitution Reactions of Inert-Metal Complexes—
Coordination Numbers 4 and 5

R. J. Cross

Chapter 6.   Substitution Reactions of Inert-Metal Complexes—
Coordination Numbers 6 and Above: Chromium

P. Moore

*Chapter 7.   Substitution Reactions of Inert-Metal Complexes—
Coordination Numbers 6 and Above: Cobalt*

R. W. Hay

Chapter 8.   Substitution Reactions of Inert-Metal Complexes— Coordination Numbers 6 and Above: Other Inert Centers

J. Burgess

Chapter 9.   Substitution Reactions of Labile Metal Complexes

D. N. Hague

Part 3. Reactions of Organometallic Compounds

Chapter 10. Substitution and Insertion Reactions of Organometallic Compounds

D. A. Sweigart

Chapter 11. Metal–Alkyl Bond Formation and Fission; Oxidative Addition and Reductive Elimination

M. Green

Chapter 12.   *Reactivity of Coordinated Hydrocarbons*

L. A. P. Kane-Maguire

Chapter 13. Rearrangements, Intramolecular Exchanges, and Isomerization of Organometallic Compounds

A. J. Deeming

Chapter 14. Homogeneous Catalysis of Organic Reactions by Complexes of Metal Ions

C. White

Part 4. Compilations of Numerical Data

Chapter 15. Volumes of Activation for Inorganic and Organometallic Reactions: A Tabulated Compilation

R. van Eldik

Part 1
Electron Transfer Reactions

Chapter 1

Electron Transfer:
General and Theoretical

The highlight of 1983 has been the award of the Nobel Prize for Chemistry to Professor Henry Taube. The award recognizes Taube's fundamental contributions to the study of inorganic mechanisms, and especially to the mechanisms of electron transfer. Appropriately, Volume 30 of *Progress in Inorganic Chemistry* is devoted to reviews of areas in which Taube has worked.[1]

1.1. The Marcus–Hush Model

In this section we summarize developments in kinetics which are closely related to the standard Marcus formalism. That is, for a general electron transfer reaction (1),

$$A^+ + B \xrightarrow{\ k_{12}\ } A + B^+ \tag{1}$$

the rate constant is given by

$$k_{12} = Z \exp(-\Delta G^{\ddagger}_{12}/RT) \tag{2}$$

where

$$\Delta G^{\ddagger}_{12} = W_{12} + \tfrac{1}{4}\lambda_{12}(1 + \Delta G^{\ominus}_{12}/\lambda_{12})^2 \tag{3}$$

and the intrinsic free-energy barrier λ_{12} is related to corresponding quantities for the self-exchange reactions (4) and (5) by equation (6),

$$A^+ + A \xrightarrow{k_{11}} A + A^+ \tag{4}$$

$$B^+ + B \xrightarrow{k_{22}} B + B^+ \tag{5}$$

$$\lambda_{12} = \tfrac{1}{2}(\lambda_{11} + \lambda_{22}) \tag{6}$$

which leads to the well-known cross relation between rate constants k_{12}, k_{11}, and k_{22}.

Major new studies on the caged cobalt ammines have appeared. From Sargeson's laboratory come accounts of the synthesis of the "cobalt sepulchrates,"† $[Co(sep)]^{3+/2+}$, and of related complexes,[3,4] the structure of $[Co(sep)]^{2+}$,[2] and accounts of the electron transfer rates.[3,4,4a] Endicott, Brubaker, and co-workers report reactions of $[Co(sep)]^{3+/2+}$ with a number of other metal ions.[5] There is particular interest in the very wide range of self-exchange rates along the series $[Co(sep)]^{3+/2+}$, $[Co(NH_3)_6]^{3+/2+}$, and $[Co(en)_3]^{3+/2+}$. Most rates fit the Marcus equations (but see below, p. 7). All three Co(II) complexes are high-spin, and for each of the two oxidation states, Co–N bond lengths are similar in the hexammine and sepulchrate.[4] Infrared and Raman studies of the en complexes and sepulchrates do not disclose any great difference in the Co(III)–N force constants.[5] Overall ionic size is apparently not a dominant factor.[4] Differences in λ due to internal strain are still possible, however,[4] and detailed ground-state energy calculations do, to some extent, account for the rate differences.[5,6] In the couple $[Co(azacaptan)]^{3+/2+}$† with a CoS_3N_3 cage, the cobalt(II) ion is low-spin and the self-exchange is much faster.[3]

The great difference in self-exchange rates of the $[Co(bipy)_3]^{3+/2+}$ and $[Co(bipy)_3]^{2+/+}$ couples (18 and $10^9 \ M^{-1} \ s^{-1}$) has been explained quantitatively in terms of structural reorganization. Co–N bond length differences are $+0.20$ and -0.02 Å in the two couples, respectively, X-ray and EXAFS data being in good agreement.[7,8] The cobalt(I) complex can be described as high-spin d^8, but with appreciable electron donation to the ligands, shown by shortening of interring carbon–carbon bond distances.[8]‡

New crystallographic data on $[Ru(H_2O)_6]^{n+}$ give the Ru(III)–O and Ru(II)–O bond distances as 2.029(7) and 2.122(16) Å. With estimated values

---

† Sep = 1,3,6,8,10,13,16,19-octaazabicyclo[6,6,6]eicosane;
  sar = 3,6,10,13,16,19-hexaazabicyclo[6,6,6]eicosane; azacapten = 8-methyl 1,3,13,16-tetra-aza-6,10,19-trithiabicyclo[6,6,6]eicosane.

‡ Two further examples of comparative structural data for ions of the same metal in different oxidation states are Ce(III)–O and Ce(IV)–O, difference 0.07 ± 0.02 Å [EXAFS, aqueous solution], and the compound $Na_2[V(V)V(IV)O_3(S\text{-peida})_2]\text{-}[ClO_4]\cdot H_2O$ [$(S\text{-peida})]^2 =$ $(S)\text{-}[1\text{-}(2\text{-pyridyl})\text{ethyl}]\text{iminodiacetate}$, containing distinct V(V) and V(IV) atoms (X-ray data compared with related V(V) and V(IV) mononuclear complexes[111]).

of the breathing force-constants, Ludi and co-workers calculate self-exchange rates in good agreement with experiment. The results effectively rationalize the sequence of self-exchange rates $[Fe(H_2O)_6]^{3+/2+} <$ $[Ru(H_2O)_6]^{3+/2+} < [Ru(NH_3)_6]^{3+/2+}$.[110] The effect of noncancellation of work terms has been investigated in a careful study[5] of reaction (7):

$$[Fe(edta)]^- + [Ru(NH_3)_6]^{2+} \rightarrow [Fe(edta)]^{2-} + [Ru(NH_3)_6]^{3+} \quad (7)$$

Free energies of formation of the four possible ion pairs are calculated from the Debye–Hückel model, and when these are allowed for, the activation parameters of the cross reaction at zero ionic strength agree well with those calculated from the self-exchange rates.

The expected curvature in the plot of $\Delta G_{12}^{\ddagger}$ against $\Delta G_{12}^{\ominus}$, when the latter becomes strongly negative, has again been detected. In reactions of complexes $[(H_3N)_3Co(III)(OH)_2Co(III)(NH_3)_3OOCR]^{3+}$ with reductants generated by pulse radiolysis, rates tend to level off at high driving force, and the differences with different groups R decrease.[10]

The proposal[11] that λ values may be divided into contributions from the two couples that make up each reaction has been applied[2] to explain the notable breakdown of the cross relation in certain cases. For example, the self-exchange rate for $[Cu(phen)_2]^{2+/+}$, when calculated from cross reactions with $[Co(edta)]^-$ or with cytochrome-c(III/II), varies from 5×10^7 to $5 \times 10^1 \, M^{-1} s^{-1}$. The argument produces an intermediate value, $1.55 \times 10^4 \, M^{-1} s^{-1}$, and a new experimental determination is $\sim 10^5 \, M^{-1} s^{-1}$. Further examples of single-ion λ values [e.g., the free energy of the process $I\dot{\ }aq \rightarrow (I\dot{\ }aq)^*$, where the second species is the Franck–Condon excited state, having the coordination sphere appropriate to I^-aq] have been calculated by Delahay from comparisons of photoemission spectra and electrochemical data.[13]†

Other applications of Marcus theory include calculation of energy barriers in micelle reactions, with total micellar charges estimated via work terms,[14] assignment of outer-sphere mechanisms in reactions of organic radicals,[15] and calculation of the unknown redox potential of one reagent, either by curve fitting,[16] or by a linear extrapolation of rates to the diffusion-controlled limit.[17]

A limitation of the Marcus theory is that electron motion is coupled to only one vibrational frequency, expressed for example by the use of the single parameter λ_{12} in equation (3). The limitation has been widely recognized, especially in the case of very fast reactions, where the difference in frequency of inner-sphere vibration and solvent molecule motion becomes

† For a discussion of some complications which can affect these calculations see References 112–116.

significant. This has been confirmed experimentally in new measurements of intramolecular electron transfer rates, which correlate well with dielectric relaxation times of the different solvents. (The reactions studied were fluorescence decay of aromatic molecules such as 1-cyano-4-dimethyl-aminobenzene, in various primary alcohols).[17a] Calef and Wolynes have begun a theoretical treatment of the problem, first using a continuum description of the solvent,[18] then proceeding to a molecular approach.[19] With the continuum model, rates have been calculated for a number of intramolecular Ru(III/II) reactions. As yet there are no measured rates with which to compare them, however.

1.2. Quantum Mechanics of Electron Transfer

The book *Electrons in Chemical Reactions* by L. Salem places electron transfer in a broad context ranging from basic wave mechanics to ion–solvent interactions.[20] Sutin has reviewed theories of electron transfer with emphasis on nuclear, electronic, and frequency factors[21,22] and Wong and Schatz have given a useful unified account of the dynamics of mixed-valence complexes using the vibronic coupling model.[23]

A treatment of vibronic coupling by Fischer includes anharmonic effects, and like the Piepho–Krausz–Schatz (PKS) model, it avoids using the adiabatic potential surfaces, when the coupling is strong.[24]

1.2.1. Nonadiabaticity†

Kuznetsov has calculated transition probabilities for nonadiabatic reactions without assuming harmonic vibrations of the metal–ligand bonds.[25] Dahl and Ulstrup[26] have investigated methods of calculating transfer integrals using different sets of basis wave functions. Further *ab initio* calculations of the rate of the aqueous Fe(III/II) exchange have appeared. Logan and Newton[27] consider two modes of approach of the coordination octahedra, vertex to vertex and face to face. Again the latter is found to be the more favorable, with the faces rotated so that the two $(H_2O)_3$ units are staggered. A metal–metal distance as short as 5.3 Å is suggested to be the most probable. The main new feature of the calculation is the inclusion of the electrons of the water molecules. The transmission coefficient is now about 0.2, which though higher than previous estimates still indicates "a modest degree of nonadiabatic character."

---

† For a discussion of nonadiabaticity in reactions of organic molecules, see Reference 46.

Nonadiabaticity can, in principle, be revealed by deviations from the Marcus cross relation, if the three reactions involved are affected to different extents. In particular, it has been suggested that if reactions of a series of, say, oxidants A^+ with a common reductant B are all slightly nonadiabatic, then the self-exchange reaction which is included in the series (the case where $A = B$) will be favored, since in that case the donor and acceptor orbitals are strictly identical. Free-energy plots suggest that this may indeed be the case for $[Co(sep)]^{3+/2+}$ and $[Fe(H_2O)_6]^{3+/2+}$,[28] and for the iron couple, comparisons of self-exchange rates with electrochemical rates have shown a similar anomaly.[29] In all cases the effects are small, and other explanations cannot be ruled out. Indeed, Hupp and Weaver prefer to postulate a special mechanism for the $Fe^{3+/2+}$ reaction, such as bridging by water molecules.[29]

The very slow outer-sphere reactions of Eu^{3+} and Eu^{2+} have long been considered good candidates for nonadiabaticity, in view of the poor overlap of f orbitals. New data from europium cryptate complexes do not support this view.[30] From cross reactions, self-exchange rates of two cryptate couples $[Eu(2.2.1)]^{3+/2+}$ and $[Eu(2.2.2)]^{3+/2+}$† are calculated to be much higher than for $Eu_{aq}^{3+/2+}$, whereas for nonadiabatic reactions the rate is expected to fall with increasing metal–metal distance. Studies of cobalt ammines, including the encapsulated $[Co(sep)]^{3+/2+}$, $[Co(sar)]^{3+/2+}$, and $[Co(azacapten)]^{3+/2+}$, have led to a similar conclusion.[4a]

A test of nonadiabaticity proposed by Endicott and Ramaoami is the effect of anions in the outer coordination sphere. Cation–cation reactions are commonly catalyzed by anions, and the argument is that if the rate is significantly retarded by poor orbital overlap, then an anion X^- which is a good electron donor will tend to increase the coupling and so increase the rate. Data for the reaction $[Co(phen)_3]^{3+} + [Co(sep)]^{3+/2+}$, catalyzed by anions in the order $I^- < NO_2^- <$ ascorbate$^-$, are presented in support of this view.[31]

1.2.2. Nuclear Tunneling

The theory of tunneling transitions, not restricted to electron transfer, has been considered further by Trakhtenberg et al.,[32] and by Hush and co-workers.[33] In the low-temperature limit, the activation energy of electron transfer is expected to approach zero as nuclear tunneling becomes predominant. Two very different experiments provide new examples of activationless transfer. Goldanskiĭ, in Mössbauer-emission studies, uses the isotope ^{57}Co, in the compound cobalt(II) ferricyanide, to produce a transient

† (2.2.1) = 4,7,13,16,21-pentaoxa-1,10-diazabicyclo[8.8.5]tricosane;   (2.2.2) = 4,7,13,16,21,24-hexaoxa-1,10-diazabicyclo[8.8.8]hexacosane.

iron(II) in the nitrogen "hole" of the structure, which then leads to electron transfer:

$$^{57}Co(II)-N\equiv C-Fe(III) \xrightarrow{-\gamma} {}^{57}Fe(II)-N\equiv C-Fe(III)$$

$$\xrightarrow{k} {}^{57}Fe(III)-N\equiv C-Fe(II) \qquad (8)$$

In effect the reaction is the isomerization of Turnbull's blue to Prussian blue. Below 50 K, the reaction rate is nearly constant, $k = 1.4 \times 10^7 \text{ s}^{-1}$.[34] Gray and co-workers have synthesized an adduct of cytochrome-c(III) with $Ru(III)(NH_3)_5$ attached to the 33-histidine residue. On treatment with excited $[Ru(bipy)_3]^{2+}$, the ruthenium center is reduced and an intramolecular electron transfer follows on:

$$Fe(III)\cdots Ru(III) \xrightarrow{e^-} Fe(III)\cdots Ru(II) \xrightarrow{k} Fe(II)\cdots Ru(III) \qquad (9)$$

The rate constant $k = 20 \pm 5 \ M^{-1} \text{s}^{-1}$ is independent of temperature from 0 to 37°C[35] and even up to 80°C.[36] The authors propose "electron tunneling"—that is, in effect a low-temperature limit. In a comment on this work, Freed[36] has put forward qualitative arguments to the effect that intramolecular vibrational relaxation effects can also produce a high-temperature limiting rate. This is not a feature of any of the electron transfer theories in current use, and a more detailed discussion will be of interest.

1.2.3. Tunneling in the Inverted Region

The fact that the Marcus equation (see above, Eq. 3) predicts a positive increase in ΔG_{12} with an increasingly negative $\Delta G_{12}^{\ominus}$, beyond $\Delta G_{12}^{\ominus} = -\lambda_{12}$, is well known, as is the fact that the quantitative effect predicted has never been observed. What is usually seen is a leveling in rate at the diffusion-controlled limit, as in a recent electrochemical study.[37] Alternatively, products in long-lived electronically excited states may occur. The lifetimes are themselves sometimes evidence of slow electron transfer in the inverted region, since deactivation paths by electron transfer may be available thermodynamically, but not effective kinetically. In a recent study by Meyer and co-workers, deactivation rate constants k_{nr} of chemiluminescent states of $[OsL_3]^{2+}$ (L = various bipyridyls and phenanthrolines) have been shown to follow the energy-gap law

$$\log k_{nr} = a + b\Delta E \qquad (10)$$

where ΔE is the difference between ground and excited states and b is a negative constant.[38] The processes following laser excitation of $(NC)_5Fe(II)CNCo(III)L$ (L = hedta or N-benzyl-edta) involve excited states of both oxidation states of cobalt, that is, Co(II)-(^2E; $t_{2g}{}^6 e_g{}^1$) and Co(III)($^1A_{1g}$; $t_{2g}{}^5 e_g{}^1$).[39]

1.2.4. Isotope Effects

Predictions[40] of kinetic isotope effects in nonadiabatic reactions have been tested using the reactions of various strong oxidants of the $[M(bipy)_3]^{3+}$ type with $[Fe(^{18}OH_2)_6]^{2+}$ and $[Fe(OD_2)_6]^{2+}$. Agreement is good for the oxygen isotopomers, but the effect of deuteration is larger than was predicted simply on the basis of Fe–O vibration frequencies.[41] A similar discrepancy had been found previously in comparisons of $[Co(NH_3)_6]^{3+}$ and $[Co(ND_3)_6]^{3+}$ reactions.[40] In the inverted region, the inverse isotope effect is predicted, that is, an increase of rate with increase of ligand mass.[40a,40,41] Data for fluorescence quenching of naphthalene by CCl_4 in glasses at 77 K have been advanced in support of this effect.[42]

1.2.5. Bridging Groups and Long-Range Transfer

Bridged electron transfer continues to be of both theoretical and practical interest. Root and Ondrechen have calculated adiabatic energy curves for complexes of the type $A^+ \cdot X \cdot A$, where the energies of A^+ and A are varied, for example, by motion of coordinated ligands, and electron coupling occurs between the metals and the bridging group but not directly between the metals. It is suggested that these conditions apply to the Creutz–Taube ion (see below, p. 11). A feature of the results is that the line shape of the intervalence charge transfer band is affected by the simultaneous in-phase breathing mode of the two metal centers, as well as by the out-of-phase mode which controls the Marcus and PKS two-center models.[43]

Extended Hückel calculations have been used to argue that conjugated organic chains up to 20 Å long can permit adiabatic electron transfer.[44]

On the experimental side, Endicott and co-workers have shown that in chloride-bridged reactions of a series of couples LM(III/II), where L is a tetraazo macrocycle, reactivity is strongly influenced by the total number of electrons in the three-center system A^+–Cl–B, falling off in a series from A = B = Ni(II) to A = B = Co(II). Effective mixing of metal and bridge orbitals is also significant.[45]

There is mounting evidence of electron transfer between centers separated by long saturated bridges, such as aliphatic chains up to 8 Å long,[46] fused cyclohexane rings up to 15 Å,[47] and spirocyclic butane rings up to 17.3 Å.[48] (In the last case, rates were inferred from charge transfer spectra.) Transfer over long distances in proteins has been reported by several groups of workers,[35,49-53] and has mostly been discussed in terms of nonadiabatic processes. With flexible-chain complexes embedded in frozen matrices,[54,55] intermolecular transfer begins to compete with intramolecular, as the temperature is raised. The relatively familiar topic of long-range transfer of electrons trapped in glasses continues to generate theoretical[56] and experimental work. In reactions of a variety of donors and acceptors, the electron transfer mechanism is supported by agreement of the rates with the Marcus equation.[57] Studies of time-dependent fluorescence of excited reactants have again been used to infer transfer probabilities as a function of distance, that is, $k \propto \exp(-\alpha R)$, with transfer of electrons over 15 Å[58] or 25 Å being significant.[59]

1.2.6. Electron Transfer in the Gas Phase

Here we point out only a few references which bear particularly on the concerns of the solution chemist. Murrell has neatly illustrated the distinction between adiabatic and nonadiabatic processes, using the examples $Ar^+ + Cl_2 \rightarrow Ar + Cl_2^+$ and $Xe^+ + Cl_2 \rightarrow Xe + Cl_2^+$.[60] Experimental results for the $H_2^+ + Ar$ and $D_2^+ + Ar$ reactions have been reported.[61] Charge transfer is considered to occur at long range, and rates vary according to the vibrational state of the molecular ion, with a maximum at $v = 2$. For $v = 0$, moreover, rates vary with kinetic energy of the reactants, with two peaks reminiscent of the nonresonant transfer between atoms of different elements, to which the Massey adiabatic criterion applies.[62] Among quantum-mechanical treatments of atom–molecule and molecule–molecule electron transfer,[63,64] we note discussions of a number of symmetrical reactions such as $O_2 + O_2^+$,[65,66] in which particular attention was given to the validity of the Franck–Condon approximation—the idea that molecular bond lengths do not change appreciably in the time required for electron transfer.[66-69]† An especially detailed study[70] of the reaction $H_2^+ + H_2$ is designed as a model for electron transfer reactions in general, allowing for various vibrational states of reactants and products, for variation of H–H and $H_2 \cdots H_2$ distances, and for mutual deformation of the two molecules.

Also of interest are studies of charge transfer between excited states of He^+ and He, two-electron transfer in the reaction $CO_2^{2+} + CO_2$,[71] several low-energy reactions showing non-conservation of spin, for example, $H_2O^+(^2B) + NO_2(^2A) \rightarrow H_2O(^1A) + NO_2^+(^1\Sigma)$,[72] and electron attachment to relatively large molecules, such as the clusters $(Cl_2)_n$ and $(SO_2)_n$, $n = 1$-5.[73]

---

† For earlier references, see Reference 70.

1.3. Mixed-Valence Complexes

Here we review mainly experimental work, with emphasis on the determination of the extent of electron delocalization between metal ion centers (i.e., the placing of systems in the Robin and Day classification), measurements of rates of internal electron transfer, and the relationship between optical and thermal electron transfer processes.

1.3.1. Binuclear Systems

The chemistry and photophysics of d^5-d^6 binuclear metal complexes have been reviewed.[74] The prototype is the Creutz–Taube ion (**1**) ($n = 5$)

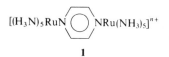

$$[(H_3N)_5RuN \bigcirc NRu(NH_3)_5]^{n+}$$

1

which was originally thought to contain localized Ru(III) and Ru(II) oxidation states. The classification as "localized" or "delocalized" must always be understood as being relative to the time scale of the particular experimental probe which is used, but even with this proviso, the status of the Creutz–Taube ion has proved exceptionally difficult to establish. The preliminary account of a major new collaborative effort on this problem has appeared.[75] X-ray data confirm that all three complexes, (III, III), (III, II), and (II, II), $n = 6, 5, 4$, have similar structures, though in bond distances the mixed-valence ion is closer to (II, II) than to the mean of (II, II) and (III, III). Mössbauer spectra do not imply localization as had been thought previously. Electronic and EPR spectra confirm the importance of metal-bridge interactions. The transferring electron is located in an orbital parallel to the π^* orbitals of the pyrazine ring. Infrared and Raman data are puzzling. Confirming previous reports, there is one band for $\delta_s(NH_3)$, intermediate between the corresponding bands for (II, II) and (III, III), as might be expected for delocalization, but certain pyrazine ring modes appear in both the infrared and the Raman, which suggests the absence of a center of symmetry. The conclusion so far is that in the Creutz–Taube ion, the valencies are certainly delocalized on the Mössbauer time scale, and possibly, if not probably, delocalized on the vibrational time scale as well. It is worth recalling here Taube's argument[76,77] from the kinetic behavior of related complexes, which implied localization on the time scale of the lifetime of an activated complex. This would place the internal electron transfer rate constant in the range 10^{12}–10^{13} s^{-1}. It remains true, as

Day pointed out in a recent review, that "the very existence of such a conflict of evidence means that the Creutz–Taube complex lies very near to the borderline between classes II and III, and is thus very far from being a typical case".[78]

The osmium analogue of (1) has now been reported and seems to fall unambiguously in class III, as judged from the comproportionation constant and infrared data.[79] Other related complexes are listed in Table 1.1. In the cases where K_{com} exceeds 10^{10}, the class III assignment is supported by the narrowness of the intervalence charge transfer band. In the others, the lines are broader and in some cases the expected solvent dependence is confirmed. Most authors have calculated valence delocalization coefficients and/or electron transfer rate constants by the Hush formulas, and have discussed these in relation to the structures of the bridging groups. In a comprehensive discussion, Richardson and Taube have used a molecular orbital description to consider metal–metal interaction via superexchange. They obtain rules for predicting the relative strengths of interactions in terms of the number of conjugated atoms and other structural features, such as extended saturated bridges and stacked aromatic rings.[80]

Long and Hendrickson have synthesized copper (II/I) complexes of the general type (2) with various three-carbon and four-carbon connecting chains R_1 and R_2. From temperature-dependent ESR line broadening, the rate constant for Cu(II)···Cu(I) exchange varies from below $2.2 \times 10^7\,\mathrm{s}^{-1}$

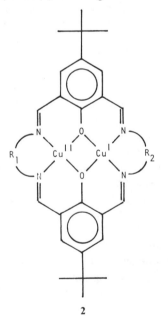

2

Table 1.1. Binuclear Mixed-Valence Complexes[a]

Complex A$^+$·X·A	ν_{IT} (10^3 cm^{-1})	ε (M^{-1} cm^{-1})	K_{com}	References
[(H$_3$N)$_5$Os(pyz)Os(NH$_3$)$_5$]$^{5+}$	b	b	7 × 10^{12}	79
[(bipy)$_2$ClOs(PPh$_2$CH$_2$PPh$_2$)OsCl(bipy)$_2$]$^{3+}$	6.0 / 10.8 / [13.3]^c	~100^d / ~150 / —	160^d	82
[(H$_3$N)$_5$RuNC⋯C(t-Bu)⋯CNRu(NH$_3$)$_5$]$^{4+}$	8.55	1.6 × 10^4	>10^{10}	118^e
[(H$_3$N)$_5$RuNC⋯CH⋯CNRu(NH$_3$)$_5$]$^{4+}$	7.8	>10^4	>10^{11}	118
[(H$_3$N)$_5$RuNC—C(CN)=CNRu(NH$_3$)$_5$]$^{4+}$	6.45	6.2 × 10^3	2.3 × 10^3	118
[(H$_3$N)$_5$RuNC—N=CNRu(NH$_3$)$_5$]$^{4+}$ ^{f}	9.1	2.82 × 10^3	340	119
[(H$_3$N)$_5$RuNCCH:CHCNRu(NH$_3$)$_5$]$^{5+}$ ^{g}	10.26	150	55	89
[(H$_3$N)$_5$Ru(pyrimidine)Ru(NH$_3$)$_5$]$^{5+}$	7.15	41	343	80
[(H$_3$N)$_5$RuNCC$_6$H$_4$CNRu(NH$_3$)$_5$]$^{5+}$ ^{h}	10.7	1100	23	80
[(H$_3$N)$_4$Ru(2,2'-bipyrimidine)Ru(NH$_3$)$_4$]$^{5+}$	—	—	1.5 × 10^3	120
[LRu(III)XRu(II)L]^i	—	—	19^j	121
[(H$_3$N)$_5$RuNC(CH$_2$Ph)$_2$—CNRu(NH$_3$)$_5$]$^{5+}$	10.10	1.6 × 10^{2j}	~10^6	148
[(H$_3$N)$_5$RuNCC(CH$_2$CH$_2$)$_3$CCNRu(NH$_3$)$_5$]$^{5+}$	12.5	10	~4	80
[(H$_3$N)$_5$RuS(CH$_2$)$_2$C(CH$_2$)$_2$C(CH$_2$)$_2$SRu-(NH$_3$)$_5$]$^{5+}$ ^{k}	12.4	9	—	48
[(H$_3$N)$_5$RuNC—(C$_5$H$_4$)Fe(C$_5$H$_4$)—CNRu-(NH$_3$)$_5$]$^{5+}$	—	—	10^d	117
[Cu(II)···Cu(I)]^l	10.5^m	401^m	4.3 × 10^6 ^{d}	81
[V$_2$O$_3$]$^{3+}$	10.1	≤110^n	—	122

^{a} ν_{IT} and ε refer to the intervalence transition usually assigned as A$^+$·X·A → (A·X·A$^+$).* K_{com} = [A$^+$·X·A]^2/[A$^+$·X·A$^+$][A·X·A]. Solvent is D$_2$O or H$_2$O at 25°C. Solvent dependence of ν_{IT} is reported in some papers but not in all.
^{b} Strong near-IR bands not characteristic of the separate Os(III) and Os(II) ions.
^{c} Band predicted but not observed.
^{d} Solvent MeCN or CD$_3$CN.
^{e} Cf. preliminary report, H. Krentzien and H. Taube, J. Am. Chem. Soc., **98**, 6379 (1976).
^{f} Data for [Ru(NH$_3$)$_4$py] and [Ru(NH)$_4$ (isonicotanamide)] terminal groups also reported.
^{g} Bridge is *trans*-dicyanoethylene. Data for tetracyanoethylene also reported.
^{h} Bridge is 1:4-dicyanobenzene. Data for other dicyanobenzenes, 4-cyanopyridine, several dicyanonapthalenes, and dicyanoanthracenes also reported.

^{i} L$^-$ = 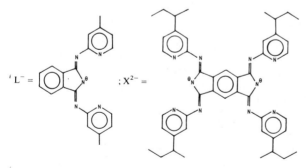 ; X^{2-} =

^{j} Solvent HCONMe$_2$.
^{k} System with four rings also studied.
^{l} See text and (2). Data here relate to R$_1$ = R$_2$ = (CH$_2$)$_3$. Several related complexes also reported.
^{m} Solvent acetone.
^{n} Shoulder.

to $1.6 \times 10^{10}\,\mathrm{s}^{-1}$. The authors note that in contrast to the widely discussed Ru(III/II) and bisferrocenyl Fe(III/II) series, electronic coupling is unlikely to vary much among the complexes studied, but that differences in geometry between the Cu(II) and Cu(I) centers will be sensitive to the ring size and may well have the dominant influence on the rates.[81] The complexity of intervalence charge transfer spectra is noted by Meyer and co-workers in the case of a weakly coupled Os(III)–Os(II) complex. Transfer of the electron from Os(II) can leave the resulting Os(III) in either of three electronic states, and the differences in energy between them can be estimated from spectra of the corresponding Os(III) monomer. Of the three predicted bands, two are actually observed.[82] The magnetic properties of mixed-valence species in which both of the component metal ions are paramagnetic are discussed by Girerd.[83] For class II compounds double-well potential curves are calculated for various values of the magnetic coupling constant. Activation energies for thermal electron transfer are predicted to increase with increasing antiferromagnetic coupling and this is thought to explain why some mixed-valence complexes are of class II in spite of the apparently intimate chemical bonding, for example, single-atom bridges in the case of $[(\mathrm{bipy}_2)\mathrm{Mn}(\mathrm{O})_2\mathrm{Mn}(\mathrm{bipy})_2]^+$. The class III complex $[\mathrm{Ni}(\mathrm{naphthyridine})_4\mathrm{Br}_2]^+$ is also discussed.

1.3.2. Polynuclear Systems

Mixed-valence trinuclear complexes have quite suddenly become a focus of attention. Launay and co-workers,[84] and Pope and co-workers,[85] report optical and ESR spectra of heteropolyions containing triangular units of metal ions. In the molybdates[84] the rate of electron transfer around a Mo(VI)Mo(VI)Mo(V) triangle is derived from ESR line broadening, and in each case the activation energies are substantially less than those calculated from the limiting Hush formula $E_a = E_{\mathrm{op}}/4$. This indicates appreciable delocalization of the extra electron, as is borne out by the ESR splitting patterns. In the ions[85] $\alpha - [\mathrm{P}_2(\mathrm{W}, \mathrm{V})_{18}\mathrm{O}_{62}]^{n-}$, a series has been synthesized including the mixed-metal and mixed-valence triangles W(VI)V(V)V(IV) and V(V)V(V)V(IV). Again there is rapid electron transfer at room temperature and partial delocalization of electrons at low temperature. Moreover, the trivanadium species has an activationless regime below 157 K. A theoretical model has been developed using adiabatic potential surfaces calculated for different values of a single resonance integral.[86,86a,87] Below a critical value, the lowest-energy surface has three wells, corresponding to the double-minimum potential in the binuclear case, and defining the three configurations of the mixed-valence unit, for example, Mo(VI)Mo(VI)Mo(V), Mo(VI)Mo(V)Mo(VI), and Mo(V)Mo(VI)Mo(VI), while above this value there is a single minimum for the symmetrical, class III configuration. In agreement with the model, among the molybdates, electron transfer is more rapid in ions of the Keggin structure than of the Lindqvist structure, and among the vanadotungstates it is more rapid in the trivanadium than in the divanadium ion.

Tetranuclear complexes involving two pairs of metal ions[88,89] include the species (3) in which the greatest stabilization (whether due to resonance or to solvation effects, or both) occurs in the half-reduced, Ru(III, III, II, II) species.[89]

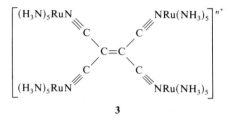

3

In the heteronuclear Ru(III)–Fe(II)–Ru(II) system listed in Table 1.1, Ru–Fe transitions but not Ru–Ru are seen in the spectrum. Electrochemical studies of a series of mixed-metal/mixed-valence complexes[90] have been used[91] to estimate mixed-valence stabilization effects. For example, in a series $M \cdots [Fe(CN)_5]^{n-}$, where M may be an organic group or a metallo-organic group, trends in $E^{\ominus}$ Fe(III/II) can be correlated with electrostatic interactions, but they also reveal small mixed-valence stabilization effects.

1.3.3. Other Mixed-Valence Materials

Prassides and Day[92] report a further study of the temperature dependence of the Sb(V/III) intervalence charge transfer spectrum in the salt $[MeNH_3]_4[Sb(V)Cl_6]_x[Sb(III)Cl_6]_x[Sn(IV)Cl_6]_{2-2x}$. The variation in band half-width with temperature accurately follows the semiclassical limiting equation, and the lattice vibration frequency obtained from the best fit to the curve is very close to the appropriately averaged frequencies of the breathing modes of the two $SbCl_6$ ions, that is, 290 ± 3 cm^{-1} and $[(327^2 + 267^2)/2]^{1/2} = 298$ cm^{-1}, respectively. The required difference of the Sb(III)–Cl and Sb(V)–Cl bond lengths is also similar to, though somewhat larger than, the typical measured values, that is, 0.19 and 0.13–0.14 Å. A further observation is that the intervalence band maximum shifts to the red on lowering the temperature. This is not predicted by the Hush or PKS models and the interesting suggestion is made that the effect is due to thermal contraction which shortens the metal–metal bond distance and hence lowers the reorganization energy.

Warren and co-workers have reported remarkable data on molten indium dihalides.[93,94] In the solid state these are typical class I compounds, $In(I)In(III)X_4$, but in molten "$InCl_2$"[94] the optical band for $In(III)\cdots In(I) \rightarrow In(II)\cdots In(II)$ is found at 17,000 cm^{-1}, and from ^{115}In NMR the nuclear relaxation rates above 400°C are interpreted in terms of the mechanism in equation (11):

$$In(III)\cdots In(I) \rightleftharpoons In(II)\cdots In(II) \tag{11}$$

The equilibrium constant is estimated as 3.5×10^{-7} at 550°C, and a combination of NMR and spectral data, using Hush's formulas, gives the activation energy for the forward reaction, 1.8 ± 0.1 eV. It should be noted, however, that only reaction (11) is considered as contributing to the nuclear relaxation rates: the self-exchange reactions In(III/II) and In(II/I) are not included.

Compounds with alternating metal-halide chains continue to be of great interest. New structural determinations[95-98] include a platinum-palladium mixed-metal compound,[90] and species involving binuclear metal units $PtPtXPtPtX\cdots$, with X = Cl or Br, and average oxidation state $Pt^{2.5}$.[100] Resonance Raman studies by Clark and co-workers[101-104] have identified the vibrational modes coupled to the intervalence transfer. They show that in general the process $Pt(IV)\cdots Pt(II) \rightarrow Pt(III)\cdots Pt(III)$ involves substantial movement of the bridging halide ion, together with rather smaller changes in the equatorial metal–ligand bond distances. These results are clearly related to the mechanism of the Pt(IV/II) exchange reaction in solution.

1.3.4. Electrical Conductivity of Solids

Again we select a few references which are closely related to reactions or complexes studied in solution. A study of spinel systems of the type $M_xFe_{3-x}O_4$, with M = Cu(II), Cd(II), Ti(IV) shows that with $x < 0.5$, these have the characteristics of Fe_3O_4, but with $x > 0.6$ they are semiconductors, implying localization of the Fe(II) oxidation state.[105] An important method of measuring conductivity in solids is to deposit the solid as a film on an electrode in solution. In a recent paper, Pickup and Murray review earlier studies of this type, and report new data on Os and Fe polypyridine polymers coated on platinum. As the potential is varied, conductivity sets in near the $E^{\ominus}$ value for the M(III/II) couple, and reaches a maximum when the overall composition is 50% M(III) and 50% M(II).[106]

One-dimensional platinum chain complexes with electrical conduction properties have been reviewed.[107] A novel nickel analogue of these materials is doubly mixed valent. The stoichiometry is interpreted as

[NiL]$_2$[NiL]$^+$[I$_3^-$], where L^{2-} = tetrabenzoporphyrinate. The conductivity is interpreted in terms of the Ni(III) and Ni(II) oxidation states in the NiNiNi... chains, together with rapid interconversion between metal-oxidized and ligand-oxidized forms [Ni(III)L]$^+$ and [Ni(II)(L$^{\overline{\cdot}}$)]$^+$, respectively.[108]

Chapter 2

Redox Reactions between Metal Complexes

2.1. Introduction

In the general area of electron transfer reactions between two metal complexes, several important review articles have appeared. These include a reexamination of the $[Co(III)/(II)(NH_3)_6]^{3+/2+}$ self-exchange reaction[1] and, of more biological interest, a summary of electron transfer reactions of iron–sulfur proteins[2] in *Advances in Inorganic and Bioinorganic Mechanisms*, as well as a volume of *Progress in Inorganic Chemistry* in appreciation of Henry Taube, including relevant articles by Creutz,[3] Endicott *et al.*,[4] Haim,[5] Meyer,[6] and Sutin.[7] Further articles appear in an ACS Symposium Series.[8] More theoretical aspects are treated in a *Faraday Discussion*[9] and a review by Sutin.[10]

On the scientific front, highlights have been the wealth of data published during the period on the electron transfer reactions of clathrochelate complexes following Sargeson's remarkable work,[11] and, in the bioinorganic field, the preparation and study of metalloprotein-bound ruthenium(II) derivatives.[12-14] Other important occurrences include a rash of papers dealing with redox chemistry of higher oxidation states of nickel and further advances in the acquisition and interpretation of activation volume data for electron transfer reactions. As with previous volumes, the text begins with intramolecular processes and then various topics are discussed. Data are presented in the text or in summary form in Tables 2.1 and 2.2.

2.2. Intramolecular Electron Transfer

2.2.1. Inert Systems

Aspects of this area have been reviewed by Creutz.[3] A number of new binuclear ruthenium and osmium complexes which can exist in mixed-valence form have been reported.[15-17] The ligand 2,2',3,3'-tetra-2-pyridyl-6,6'-biquinoxaline allows[18] attachment of up to four Ru(II)(bipy)$_2$ fragments. The molecule appears to act as two independent subunits which can exist in mixed-valence forms. Intervalence transfer (IT) bands detected for these mixed-valence subunits allow calculation of an electron transfer rate of $\sim 1 \times 10^{10}\ s^{-1}$ between the metal centers 6.8 Å apart.

Photoelectron spectroscopy has been used to probe the electronic structure of mixed-valence binuclear ruthenium complexes with a dithiaspirobridge.[19] These species show efficient electron transfer over substantial distances[20] and the electron is primarily transferred through the HOMO which is set up to provide a conducting pathway by "sideways π overlap" of p_z orbitals on carbon and sulfur.

Comparisons of the spectroscopic and electrochemical properties of $[(NH_3)_5RuNCFcCNRu(HN_3)_5]^{4+}$ and $[(NH_3)_5RuNCFcCN]^{2+}$, where Fc is ferrocene, leads to the conclusions that the two ruthenium chromophores in the trimetallic ion do not interact strongly.[21] This allows differences in the IT transition energies of the mixed-valence complexes to be ascribed to differences in charge distribution, in good agreement with theory.

Elegant use has been made[22] of a rigid 1,4-dicyanobicyclo[2.2.2]octane bridge to probe electron transfer between ruthenium(II) and cobalt(III). Generally, saturated alkane bridges are flexible and allow the ends to react in close proximity. However, this rigid system holding the metal centers 4 Å apart gives an intramolecular electron transfer rate constant of $<3 \times 10^{-6}\ s^{-1}$, more than four orders of magnitude slower than comparable outer-sphere reactions.

Picosecond spectroscopy[23] sheds some light on the electron transfer behavior of the binuclear ion $[(CN)_5Fe(II)CNCo(III)(hedta)]^{4-}$. Excited-state electron transfer takes place on a very short time scale giving iron(III) and cobalt(II) which undergoes a spin change and the back reaction is characterized by a 95-ps lifetime. The coordination of copper(I) to the olefinic chromophore of cobalt(III)-bound amino–alkene ligands results[23] in binuclear complexes (1) which undergo intramolecular electron transfer

$$[(NH_3)_5CoNH_2(CH_2)_nCHCH_2Cu]^{4+}$$

1

after Cu(d) $\rightarrow$ L(π^*) excitation. Electron transfer rates, $k \geq 10^7-10^8\ s^{-1}$, are more rapid than diffusion of the chain ends together discounting a mechanism of this sort and it is calculated that electron transfer is facile at distances up to 7-8 Å. However, the quantum yields do drop off as n increases.

Reactions of $[Co(NH_3)_6]^{3+}$ with pyridinyl carboxylate radicals, L^-, have been compared[25] with reactions of the carboxylate-bound radical in $[(NH_3)_5CoL^-]^{2+}$. Only the species $[(NH_3)_5CoO_2CCH_2OCOC_5H_4N^-]^+$ decays by an intramolecular route with a first-order rate constant of $4 \times 10^4\,s^{-1}$, while radicals of quaternized pyridine derivatives are oxidized by $[(NH_3)_5CoL]^{3+}$ complexes in solution. The coordinated imidazole in $[(NH_3)_5CoImH]^{3+}$ traps an electron more rapidly than the cobalt(III) center allowing detection[26] of an intermediate in pulse-radiolysis experiments and the determination of the intramolecular electron transfer rate as $2.5 \times 10^3\,s^{-1}$.

2.2.2. Labile Systems

A mixture of VO_2^+ and VO^{2+} in $HClO_4$ gives[27] a labile mixed-valence complex $[V_2O_3]^{3+}$ with a formation constant of $8.1\,M^{-1}$. The IT band is detected at $10,100\,cm^{-1}$. In methanol solution, a number of sulfur-bound copper(II) complexes, models for blue copper proteins, are reduced[28] by ferrocene in a mechanism which involves formation of a precursor ion pair followed by rate-determining electron transfer. The activation parameters suggest that the precursor complex undergoes desolvation in the transition state.

Selectivity in reduction of $[(NH_3)_3Co\mu(OH)_2\mu(O_2CR)Co(NH_3)_3]$, where $R = CX_3$, $X = H$ or F, is defined[29] by equation (1) in order to cancel

$$\log k_{red}^{CF_3} - \log k_{red}^{CH_3} = 0.95 E_{red}^0 + 1.0 \qquad (1)$$

ion pairing effects in outer-sphere electron transfer. This equation has been tested for very powerful reductants, such as nickel(I) macrocycles, Zn^+, Cd^+, and organic radicals using pulse radiolysis, and at high driving forces, selectivity is found to be independent of the reduction potential of the reductant, E_{red}^0. The results imply that the main contribution to the reorganizational term γ comes from rearrangements around the cobalt centers. Other examples of intramolecular electron transfer are discussed in the text, particularly in Section 2.15. The data are presented in Table 2.1.

2.3. Titanium(III)

Reductions of $[Ru(OAc)_4]^+$ and $[Ru(ptdn)_3]$ by titanium(III) are unusual[30] in that an $[H^+]$ independent term, reaction of $[Ti(H_2O)_6]^{3+}$, is

Table 2.1. Rate Constants for Intramolecular Electron Transfer at 25°C

Nonlabile systems	Medium (M)	k (s^{-1})	$\Delta H^{\ddagger}$ (kcal mol^{-1})	$\Delta S^{\ddagger}$ (cal K^{-1} mol^{-1})	References
$[(NH_3)_4(H_2O)CoNCC(CH_2CH_2)_3CCNRu(NH_3)_5]^{5+}$	0.1	$<3 \times 10^{-6}$			22
$[(NH_3)_5CoNH_2(CH_2)_nCHCH_2Cu]^{4+}$	0.1 (23°C)	$\geq 10^7\text{-}10^8$			24
$[(NH_3)_5CoO_2CCH_2O_2Cpy]^{2+}$	0.1	2.0×10^4			25
$[(NH_3)_5CoImH^+]^{2+}$	—	2.5×10^3			26
$[(NH_3)_5CoIm]^+$	—	3.2×10^3			26
$[cytc(III)\text{-}his33Ru(II)(NH_3)_5]$	pH 7.0 (23°C)	20	~0	1	12
$[cytc(III)\text{-}his33Ru(II)(NH_3)_5]$	pH 6.7 (22°C)	82	—	—	13
$[azurin(II)\text{-}his83Ru(II)(NH_3)_5]$	pH 7.0 (23°C)	1.9	~0	—	14
$[Hb(III)\text{-}^3Zn \text{ porphyrin}]$	pH 7.0	60	—	—	153

Labile systems	Medium (M)	$k(s^{-1})$ $K(M^{-1})$	$\Delta H^{\ddagger}$ ΔH°	$\Delta S^{\ddagger}$ ΔS°	References
$VO_2^+ + VO^{2+}$	8.7	8.1	-2.4	-8.5	27
$[Cu(Pr^nSCH_2IM)_2]^{2+} + [Cp_2Fe]$	0.1, MeOH	109	17.4	9	28
		26.5	-6.0	-13	
$[Cu(Bu^sSCH_2Im)]^{2+} + [Cp_2Fe]$	0.1, MeOH	402	18.4	15	28
		11.5	-4.8	-11	
$[Cu(PhCH_2SCH_2Im)_2]^{2+} + [Cp_2Fe]$	0.1, MeOH	149	19.6	17	28
		22.2	-2.9	-4	

Reaction	Conditions	k	$\Delta H^{\ddagger}$	$\Delta S^{\ddagger}$	Ref.
$[Cu(PhSCH_2Im)_2]^{2+} + [Cp_2Fe]$	0.1, MeOH	327	16.3	8	28
$[Cu(CH_2SCH_2Im)_2]^{2+} + [Cp_2Fe]$	0.1, MeOH	9.7	-4.3	-10	28
$[Ru_2(OX)_2(OAc)_2]^- + Ti^{3+}$	1.0	18, 32.2, 30, 8×10^2	18.6, -9.6, 13.9	10, -24	32
$[Co(NH_3)_5NCacac]^{2+} + Cr^{2+}$	1.0	7.3×10^2, 58	8.4, -7.1	-17, -14	34
$[NiLH]^{2+} + [Co(edta)]^{2-}$	0.1	4×10^{-2}, 480	—	—	77
$[Fe(bipy)_3]^{3+} + [Rh(2,4,6Me_3C_6H_2CN)_4]^+$	0.1, CH_3CN	2.7	—	—	99
$[Fe(phen)_3]^{3+} + [Rh(2,4,6Me_3C_6H_2CN)_4]^+$	0.1, CH_3CN	7.8×10^3, 20	—	—	99
$[Fe(phen)_3]^{3+} + [Rh(Bu^tCN)_4]^+$	0.1, CH_3CN	1.6×10^4, 40	—	—	99
$[Rh(bipy)_2]^{2+} + [Rh(bipy)_2]^{2+}$	—	9×10^{-3}	—	—	102
$[Mo(CN)_8]^{3-} + Cu^+$	50% Glycerol (93 K)	2.74×10^{-4}	—	—	106
$[Co(NH_3)_6]^{3+} + cytb_5(II)$	pH 7.4, 0.1	7.5×10^{-2}, 600	—	—	148
$[Pt(NH_3)_6]^{4+} + cytb_5(II)$	pH 7.4, 0.1	8.0×10^{-2}, 1.48×10^4	—	—	148
$[(NH_3)_5Co(\mu NH_2)Co(NH_3)_5]^{5+} + cytb_5(II)$	pH 7.4, 0.1	3.8, 1.66×10^4	—	—	148

observed. Normally, poor overlap between donor and acceptor orbitals prevents this pathway and the dominant mechanism requires OH^- as an outer-sphere electron mediator, but the acetate and pentane-2,4-dione ligands delocalize electron density to give a direct pathway. The catalytic effect of chromium(III) in titanium(III) reduction of ruthenium(III) has been reported.[31] A trinuclear intermediate is detected[32] in the reduction of the mixed-valence dimeric $[Ru_2(Oc)_2(OAc)_2]^-$ by $[Ti(H_2O)_6]^{3+}$. Electron transfer through the bridging oxalate has a value of $30\, s^{-1}$.

Acid dependencies in the reduction of $[(NH_3)_5CoL]^{2+}$ complexes, where LH is a keto-substituted carboxylic acid, by titanium(III) reveal[33] much about the mechanism. Remote keto substitution, in the β and γ positions, results in $[H^+]^{-1}$ terms in the rate law with rates comparable with those found for unsubstituted derivatives, while additional $[H^+]^{-2}$ terms with α-keto substituents and enhanced rates suggest attack by $TiOH^{2+}$ on the *gem*-diol hydrated form of the ligand. With phenyl-substituted keto acids, such as LH $= C_6H_5COCO_2H$, an acid-independent term reflects precursor-complex formation by reaction of $[Ti(H_2O)_6]^{3+}$.

2.4. Chromium(II)

The ligand 3-cyano-2,4-pentanedione, NCptdn$^-$, cyanide bound in $[(NH_3)_5CoNCptdn]^{2+}$ permits kinetic detection[34] of a chelated inner-sphere precursor on reduction by $[Cr(H_2O)_6]^{2+}$. A large negative $\Delta S^{\ddagger}$ for the electron transfer step may be indicative of nonadiabaticity although the activation parameters for the composite term Kk_{et} are in line with other chromium(II) reductions. Reduction of the oxygen-bound cobalt(III) complex, $[(en)_2CoptdnCN]^{2+}$ may also be inner sphere. In the related complex $[(bipy)_2Coptdn]^{2+}$, reduction by chromium(II) is very fast[35] and results in 81% formation of $[Cr(bipy)ptdn(H_2O)_2]^{2+}$ in which two different chelating ligands have been transferred. A dibridged intermediate is proposed and a caution is issued in assigning mechanisms in reactions where, like the bipy ligand, there are no apparent bridging ligands.

Nitrogen-bonded chromium(III) products are detected[36] in the reductions of a number of pentaamine cobalt(III) tetrazole complexes by chromium(II). The rate laws show inverse $[H^+]$ dependences and rate constants consistent with an inner-sphere mechanism in all cases. Product isolation confirms[37] remote attack by $[Cr(H_2O)_6]^{2+}$ on the 4-cyanobenzoic acid derivative $[(NH_3)_5CoNCC_6H_4CO_2H]^{3+}$ and its deprotonated form (pK_a 2.51). The $[V(H_2O)_6]^{2+}$ reaction is also consistent with a remote attack mechanism, while the 3-cyanobenzoic acid complex reacts by an outer-sphere pathway suggesting the possibility of a remote bridged outer-sphere mechanism rather than a through bridge mechanism for the 4-cyano derivatives. Bridged outer-sphere mechanisms are also proposed in the chromium(II) reductions of some pentaamine cobalt(III) complexes with sulfur-containing ligands.[38]

Comparisons[39] of the reduction of a series of alkyl- and aminocarboxylate cobalt(III) complexes which are isosteric but differ in overall charge, reveal that this charge has a marked effect where the reductant is U_{aq}^{3+} or Cr_{aq}^{2+}[40] and the mechanisms are predominantly inner sphere. For both these reductants this is a result of poor overlap between acceptor π orbitals on the bridging carbonyls of the cobalt(III) complexes and donor orbitals on the metal necessitating distortion of the aquocoordination environment. Further evidence for this comes from the observation[41] of a reverse kinetic salt effect in the Cr_{aq}^{2+} reduction of cobalt(III) dimers with the drop in water activity at high ionic strength.

The μ-carboxylato-cobalt(III) dimers of the type $[(NH_3)_3Co\mu(OH)_2\text{-}\mu(O_2CR)Co(NH_3)_3]^{3+}$ have no carboxyl carbonyl available for inner-sphere attack. Reductions by $[Ru(NH_3)_6]^{2+}$, Eu_{aq}^{2+}, Cr_{aq}^{2+}, V_{aq}^{2+}, and U_{aq}^{3+} with complexes with a variety of substituents R have been examined[42] and the rate-determining step is reduction of the first cobalt(III). A dominant outer-sphere mechanism is suggested by linear log–log plots for all reactions except those where carbonyl groups on the substituent are in conjugation. In this case the aquoreductants show some remote-attack inner-sphere character and rates enhanced over those predicted by an outer-sphere mechanism. An interesting observation is made in the reaction where R has a carbonyl group with Eu_{aq}^{2+} as reductant. Stoichiometry considerations suggest production of a cobalt(III)-bound radical which subsequently dimerizes. Similar studies have been carried out where R is an amide-substituted pyridine derivative.[43] In this case ligand radical formation might be favored but reductions take place with the aquo ions by remote attack through the amide function rather than through the pyridine ring.

2.5. Cage Complexes

Syntheses and characterization of a number of cobalt(III) and cobalt(II) cage complexes have been reported.[11] These species are unusual in that both oxidation states are inert to substitution and the closed structural framework precludes inner-sphere reactions. Facile resolution of the complexes into optical isomers provides a convenient method for investigating self-exchange rates which are much faster (see Table 2.3) than for other simple cobalt(III)/(II)–amine complexes.

Table 2.2. Rate Constants for Intermolecular Electron Transfer Reactions at 25°C[a]

Reaction	Medium (M)	k ($M^{-1}\,s^{-1}$)	$\Delta H^{\ddagger}$ (kcal mol^{-1})	$\Delta S^{\ddagger}$ (cal K^{-1} mol^{-1})	Ref.
Titanium(III):					
[Ru$_2$(OAc)$_4$]$^+$ + Ti^{3+}	1.0	2.3×10^2			30
+ TiOH^{2+}	1.0	3.4×10^3			30
[Ru(pd)$_3$] + Ti^{3+}	1.0	0.08			30
[Ru(pd)$_3$] + TiOH^{2+}	1.0	0.7			30
[Co(NH$_3$)$_5$(camphor-3-carboxylato)]$^{2+}$ + Ti(III)	1.0	4.7×10^{-3} s^{-1}, 1/[H$^+$]			33
[Co(NH$_3$)$_5$(levulinato)]$^{2+}$ + Ti(III)	1.0	7.3×10^{-3} s^{-1}, 1/[H$^+$]			33
[Co(NH$_3$)$_5$(γ-ketopimelato)]$^{2+}$ + Ti(III)	1.0	6.0×10^{-3} s^{-1}, 1/[H$^+$]			33
[Co(NH$_3$)$_5$(oxamato)]$^{2+}$ + Ti(III)	1.0	4.6×10^{-2} s^{-1}, 1/[H$^+$]			33
[Co(NH$_3$)$_5$(2-formylbenzoato)]$^{2+}$ + Ti(III)	1.7	1.7×10^{-3} s^{-1}, 1/[H$^+$]			33
[Co(NH$_3$)$_5$(4-formylbenzoato)]$^{2+}$ + Ti(III)	1.7	2.2×10^{-3} s^{-1}, 1/[H$^+$]			33
[Co(NH$_3$)$_5$(acetato)]$^{2+}$ + Ti(III)	1.0	4.9×10^{-3} s^{-1}, 1/[H$^+$]			33
[Co(NH$_3$)$_5$(formato)]$^{2+}$ + Ti(III)	1.0	7.0×10^{-3} s^{-1}, 1/[H$^+$]			33
[Co(NH$_3$)$_5$(glycolato)]$^{2+}$ + Ti(III)	1.0	3.1×10^{-1} s^{-1}, 1/[H$^+$]			33
[Co(NH$_3$)$_5$(CH$_3$COCO$_2$)]$^{2+}$ + TiOH^{2+}	1.0	5.2×10^2			33
[Co(NH$_3$)$_5$(C$_2$H$_5$COCO$_2$)]$^{2+}$ + TiOH^{2+}	1.0	3.5×10^2			33
[Co(NH$_3$)$_5$(CCl$_3$COCO$_2$)]$^{2+}$ + TiOH^{2+}	1.0	3.8×10^2			33
[Co(NH$_3$)$_5$(C(CH$_3$)$_2$COCO$_2$)]$^{2+}$ + TiOH^{2+}	1.0	6.6×10^1			33
[Co(NH$_3$)$_5$(HO$_2$CCOCO$_2$)]$^{2+}$ + TiOH^{2+}	1.0	4.0×10^2			33
[Co(NH$_3$)$_5$(HCOCO$_2$)]$^{2+}$ + TiOH^{2+}	1.0	8.6×10^1			33
[Co(NH$_3$)$_5$(phenylglyoxylato)]$^{2+}$ + Ti^{3+}	1.0	1.8×10^2			33
[Co(NH$_3$)$_5$(phenylglyoxylato)]$^{2+}$ + TiOH^2	1.0	8.9×10^3			33
[Co(NH$_3$)$_5$(4-OMephenylglyoxylato)]$^{2+}$ + Ti^{3+}	1.0	1.22×10^2			33
[Co(NH$_3$)$_5$(4-OMephenylglyoxylato)]$^{2+}$ + TiOH^{2+}	1.0	1.45×10^4			33
[Co(NH$_3$)$_5$(2,4,6-(OMe)$_3$phenylglyoxylato)]$^{2+}$ + Ti^{3+}	1.0	$>3.2 \times 10^2$			33
[Co(NH$_3$)$_5$(2,4,6-(OMe)$_3$phenylglyoxylato)]$^{2+}$ + TiOH^{2+}	1.0	4.3×10^2			33
[Co(NH$_3$)$_5$(salicylato)]$^{2+}$ + TiOH^{2+}	1.0	9.4×10^1			33
[Co(NH$_3$)$_5$(4-mesalicylato)]$^{2+}$ + TiOH^{2+}	1.0	1.17×10^2			33

Reaction				Ref.
[Co(NH₃)₅(5-mesalicylato)]²⁺ + TiOH²⁺	1.0	1.24×10^2		33
[Co(NH₃)₅(4-OHsalicylato)]²⁺ + TiOH²⁺	1.0	1.37×10^2		33
Vanadium(II):				
[Co(NH₃)₅(NCacac)]²⁺ + V²⁺	1.0	9.32	9.4	34
[Co(NH₃)₅(4-NCC₆H₄CO₂H)]³⁺ + V²⁺	1.0	11.1		37
[Co(NH₃)₅(4-NCCoH₄CO₂)]²⁺ + V²⁺	1.0	$0.21\ s^{-1},\ 1/[H^+]$		37
[(NH₃)₃Co(μOH)₂(μ-2-CHOC₆H₄CO₂)-Co(NH₃)₃]³⁺ + V²⁺	1.0	0.25	−23	41
[(NH₃)₃Co(μOH)₂(μ-4-CHOC₆H₄CO₂)-Co(NH₃)₃]³⁺ + V²⁺	1.0	0.13		41
[(NH₃)₃Co(μOH)₂(μ-HCO₂)Co(NH₃)₃]³⁺ + V²⁺	1.0	0.12		42
[(NH₃)₃Co(μOH)₂(μ-CH₃CO₂)-Co(NH₃)₃]³⁺ + V²⁺	1.0	0.053		42
[(NH₃)₃Co(μOH)₂(μ-CF₃CO₂)Co(NH₃)₃]³⁺ + V²⁺	1.0	0.41		42
[(NH₃)₃Co(μOH)₂(μ-C₆H₄CO₂)Co(NH₃)₃]³⁺ + V²⁺	1.0	0.067		42
[(NH₃)₃Co(μOH)₂(μ)-4-NO₂-C₆H₄CO₂)-Co(NH₃)₃]³⁺ + V²⁺	1.0	0.069		42
[(NH₃)₃Co(μOH)₂(μ-pyruvate)Co(NH₃)₃]³⁺ + V²⁺	1.0	0.30		42
[(NH₃)₃Co(μOH)₂(μ-phenylglyoxylato)-Co(NH₃)₃]³⁺ + V²⁺	1.0	0.83		42
[(NH₃)₃Co(μOH)₂(μ-4MeO-phenylglyoxylato)Co(NH₃)₃]³⁺ + V²⁺	1.0	1.07		42
[(NH₃)₃Co(μOH)₂(μ-2,4(MeO)₂-phenylglyoxylato)Co(NH₃)₃]³⁺ + V²⁺	2.0	1.00		42
[(NH₃)₃Co(μOH)₂(μ-2,4,6(MeO)₂-phenylglyoxylato)Co(NH₃)₃]³⁺ + V²⁺	1.0	0.88		42
[Co(NH₃)₅(phenylglyoxylato)]²⁺ + V²⁺	1.0	2.6		42
[Co(NH₃)₅(4-MeO-phenylglyoxylato)]²⁺ + V²⁺	1.0	3.5		42
[Co(NH₃)₅(2,4MeO)₂phenylglyoxylato)]²⁺ + V²	1.0	3.5		43
[Co(NH₃)₅(2,4,6-(MeO)₃phenylglyoxylato)]²⁺ + V²⁺	1.0	3.4		42
[Co(NH₃)₅(4-acetylbenzoato)]²⁺	1.0	0.83		42
[(NH₃)₃Co(μOH)₂(μ-2CO₂,3-CONH₂pyH)Co(NH₃)₃]⁴⁺ + V²⁺	1.0	0.69		43

(continued)

Table 2.2. (continued)

Reaction	Medium (M)	k ($M^{-1}\,s^{-1}$)	$\Delta H^{\ddagger}$ (kcal mol^{-1})	$\Delta S^{\ddagger}$ (cal K^{-1} mol^{-1})	Ref.
Vanadium(II) *(cont.)*:					
$[(NH_3)_3Co(\mu OH)_2(\mu\text{-}2CO_2,4\text{-}CONH_2pyH)\text{-}Co(NH_3)_3]^{4+} + V^{2+}$	1.0	9.9			43
$[(NH_3)_3Co(\mu OH)_2(\mu\text{-}2CO_2,5\text{-}CONH_2pyH)\text{-}Co(NH_3)_3]^{4+} + V^{2+}$	1.0	8.5			43
$[(NH_3)_3Co(\mu OH)_2(\mu\text{-}4CONH_2pyCH_2CO_2)\text{-}Co(NH_3)_3]^{4+} + V^{2+}$	1.0	1.82			43
$[(NH_3)_3Co(\mu OH)_2(\mu\text{-}3CONH_2pyCH_2CO_2)\text{-}Co(NH_3)_3]^{4+} + V^{2+}$	1.0	0.175			43
$[(NH_3)_3Co(\mu OH)_2(\mu\text{-}4CONHCH_2CO_2pyH)\text{-}Co(NH_3)_3]^{4+} + V^{2+}$	1.0	8.6×10^{-2}			43
$[(NH_3)_3Co(\mu OH)_2(\mu\text{-}4\text{-}COHN_2py\text{-}2\text{-}CH_2C_6H_4CO_2)\text{-}Co(NH_3)_3]^{4+} + V^{2+}$	1.0	0.138			43
$[Co(diamsarH_2)]^{5+} + V^{2+}$	0.2	0.40			48
$[Eu(2.2.1)]^{3+} + V^{2+}$	0.1	0.5			58
$[Eu(2.2.2)]^{3+} + V^{2+}$	0.1	1.4			58
Chromium(II):					
$[Co(en)_2(acacCN)]^{2+} + Cr^{2+}$	1.0	6.2×10^{-3}			34
$[Co(bipy)_2(pd)]^{2+} + Cr^{2+}$	1.0	$>3 \times 10^4$			35
$[Co(NH_3)_5N_4CH_2]^{3+}(N2) + Cr^{2+}$	1.0	19			36
$[Co(NH_3)_5N_4CH]^{2+}(N2) + Cr^{2+}$	1.0	$3.3\,s^{-1}$, $1/[H^+]$			36
$[Co(NH_3)_5N_4CHCH_3]^{3+}(N2) + Cr^{2+}$	1.0	0.15	4.2	-43	36
$[Co(NH_3)_5N_4CCH_3]^{2+}(N2) + Cr^{2+}$	1.0	$0.41\,s^{-1}$, $1/[H^+]$	5.7	-40	36
$[Co(NH_3)_5N_4CHCH_3](N1) + Cr^{2+}$	1.0 (23.9°C)	2.4			36
$[Co(NH_3)_5N_4CCH_3]^{2+}(N1) + Cr^{2+}$	1.0 (23.9°C)	$0.83\,s^{-1}$, $1/[H^+]$			36
$[Co(NH_3)_5N_4CHCN]^{3+}(N2) + Cr^{2+}$	1.0	2.1			36
$[Co(NH_3)_5N_4CHCONH_2]^{3+}(N2) + Cr^{2+}$	1.0	13.9			36

Complex					Ref.
[Co(NH$_3$)$_5$(4-NCC$_6$H$_4$CO$_2$H)]$^{3+}$ + Cr^{2+}	1.0	0.29	7.1	-37	37
[Co(NH$_3$)$_5$(4-NCC$_6$H$_4$CO$_2$)]$^{2+}$ + Cr^{2+}	1.0	0.99			37
[Co(NH$_3$)$_5$(3-NCC$_6$H$_4$CO$_2$H)]$^{3+}$ + Cr^{2+}	1.0	4.89×10^{-2}	8.3	-37	37
[Co(NH$_3$)$_5$(NH$_2$SO$_3$)]$^{2+}$ + Cr^{2+}	1.0	3.3	8.5	-27	38
[Co(NH$_3$)$_5$(NH$_2$SO$_2$NH$_2$)]$^{3+}$ + Cr^{2+}	1.0	0.197	9.3	-30	38
[Co(NH$_3$)$_5$(NHSO$_2$NH$_2$)]$^{2+}$ + Cr^{2+}	1.0	5.56×10^{-2}	11.5	-26	38
[Co(NH$_3$)$_5$(NH$_2$SO$_2$C$_6$H$_4$CH$_3$)]$^{3+}$ + Cr^{2+}	1.0	≥ 0.45	5	-46	38
[Co(NH$_3$)$_5$(NH$_2$SO$_2$C$_6$H$_4$CH$_3$)]$^{2+}$ + Cr^{2+}	1.0	6.57×10^{-2}	11.1	-27	38
[(NH$_3$)$_3$Co(μOH)$_2$(μ-2-CHOC$_6$H$_4$CO$_2$)-Co(NH$_3$)$_3$]$^{3+}$ + Cr^{2+}	1.0	37			41
[(NH$_3$)$_3$Co(μOH)$_2\mu$-4CHOC$_6$H$_4$CO$_2$)-Co(NH$_3$)$_3$]$^{3+}$ + Cr^{2+}	1.0	25			41
[Co(NH$_3$)$_5$(2-CHOC$_6$H$_4$CO$_2$)]$^{2+}$ + Cr^{2+}	1.0	73			41
[Co(NH$_3$)$_5$(3-CHOC$_6$H$_4$CO$_2$)]$^{2+}$ + Cr^{2+}	1.0	64			41
[(NH$_3$)$_3$Co(μOH)$_2$(μ-CH$_3$COCO$_2$)-Co(NH$_3$)$_3$]$^{3+}$ + Cr^{2+}	1.0	37			41
[Co(NH$_3$)$_5$(HCO$_2$)]$^{2+}$ + Cr^{2+}	1.0	6.8			41
[(NH$_3$)$_3$Co(μOH)$_2$(μ-CH$_3$CO$_2$)Co(NH$_3$)$_3$]$^{4+}$ + Cr^{2+}	1.0	1.4×10^{-3}			42
[(NH$_3$)$_3$Co(μOH)$_2$(μ-CF$_3$CO$_2$)Co(NH$_3$)$_3$]$^{4+}$ + Cr^{2+}	1.0	5.9×10^{-3}			42
[(NH$_3$)$_3$Co(μOH)$_2$(μ-C$_6$H$_4$CO$_2$)Co(NH$_3$)$_3$]$^{4+}$ + Cr^{2++}	1.0	1.7×10^{-3}			47
[(NH$_3$)$_3$Co(μOH)$_2$(μpyruvato)Co(NH$_3$)$_3$]$^{4+}$ + Cr^{2+}	1.0	4			42
[(NH$_3$)$_3$Co(μOH)$_2$(μ-phenylglyoxylate)-Co(NH$_3$)$_3$]$^{4+}$ + Cr^{2+}	1.0	1.6×10^4			42
[(NH$_3$)$_3$Co(μOH)$_2$(μ-phenylglyoxylato)-Co(NH$_3$)$_3$]$^{4+}$ + Cr^{2+}	1.0	1.1×10^4			42
[(NH$_3$)$_3$Co(μOH)$_2$(μ-2,4-(MeO)$_2$phenylglyoxylato)-Co(NH$_3$)$_3$]$^{4+}$ + Cr^{2+}	1.0	6×10^2			42
[Co(NH$_3$)$_5$(glyoxylato)]$^{2+}$ + Cr^{2+}	1.0	$>7 \times 10^3$			42
[Co(NH$_3$)$_5$(4-acetylbenzoato)]$^{2+}$ + Cr^{2+}	1.0	0.73			42
[Co(NH$_3$)$_5$(4-acetylbenzoato)]$^{2+}$ + Cr^{2+}	1.0	1.15 M^{-2} s^{-1}, [H$^+$]			42
[(NH$_3$)$_3$Co(μOH)$_2$(μ-2CO$_2$,3-CONH$_2$-pyH)-Co(NH$_3$)$_3$]$^{4+}$ + Cr^{2+}	1.0	8.7×10^{-2}			43

(continued)

Table 2.2. (continued)

Reaction	Medium (M)	k ($M^{-1}\,s^{-1}$)	$\Delta H^{\ddagger}$ (kcal mol^{-1})	$\Delta S^{\ddagger}$ (cal K^{-1} mol^{-1})	Ref.
Chromium(II) (*cont.*):					
[(NH$_3$)$_3$Co(μOH)$_2$(μ-2CO$_2$,4-CONH$_2$pyH)-Co(NH$_3$)$_3$]$^{4+}$ + Cr^{2+}	1.0 (20°C)	1.17×10^4			43
[(NH$_3$)$_3$CO(μOH)$_2$(μ-2CO$_2$,5-CONH$_2$pyH)-Co(NH$_3$)$_3$]$^{4+}$ + Cr^{2+}	1.0 (20°C)	1.16×10^3			43
[(NH$_3$)$_3$CO(μ-OH)$_2$(μ-4-CONH$_2$pyCH$_2$CO$_2$)-Co(NH$_3$)$_3$]$^{4+}$ + Cr^{2+}	1.0	16.8			43
[(NH$_3$)$_3$CO(μ-OH)$_2$(μ-3-CONH$_2$pyCH$_2$CO$_2$)-Co(NH$_3$)$_3$]$^{4+}$ + Cr^{2+}	1.0	4.3×10^{-1}			43
[(NH$_3$)$_3$CO(μ-OH)$_2$(μ-4-CONHCH$_2$CO$_2$pyH)-Co(NH$_3$)$_3$]$^{4+}$ + Cr^{2+}	1.0	1.4×10^{-2}			43
[(NH$_3$)$_3$CO(μ-4-CONH$_2$py-2-Co(NH$_3$)$_3$]$^{3+}$-CH$_2$C$_6$H$_4$CO$_2$) + Cr^{2+}	1.0	6.0×10^{-3}			43
[Co(sep)]$^{3+}$ + Cr^{2+}	0.5	8.7×10^{-4}			48
[Co(diamsarH$_2$)] + Cr^{2+}	0.5	8.1×10^{-3}			48
[Co(ammesarH)]$^{4+}$ + Cr^{2+}	0.5	4.9×10^{-4}			48
[Co(azamesar)]$^{3+}$ + Cr^{2+}	0.5	1.4×10^{-4}			48
Iron(II):					
[Fe(Me$_2$phen)$_3$]$^{3+}$ + Fe^{2+}	2.25 (20°C)	2.1×10^3	2.5	-35	62
[Fe(Mephen)$_3$]$^{3+}$ + Fe^{2+}	2.25 (20°C)	7.0×10^3	2.3	-33	62
[Fe(phen)$_3$]$^{3+}$ + Fe^{2+}	2.25 (20°C)	1.7×10^4	2.0	-33	62
[Fe(MeOphen)$_3$]$^{3+}$ + Fe^{2+}	2.25 (20°C)	1.9×10^4	1.9	-33	62
[Fe(Clphen)$_3$]$^{3+}$ + Fe^{2+}	2.25 (20°C)	8.4×10^4	1.3	-32	62
[Fe(NO$_2$phen)$_3$]$^{3+}$	2.25 (20°C)	4.6×10^5	0.6	-30	62
FeOH(edta)]$^{2-}$ + [V(edta)]$^-$	1.0	22.5			64
VO(edta)]$^{2-}$ + [Fe(Hedta)]$^-$	1.0	11.1			64
FeOH(cydta)]$^{2-}$ + [V(edta)]$^-$	1.0	74			64

		k	$\Delta H^{\ddagger}$	$\Delta S^{\ddagger}$	Ref.
$[VO(edta)]^{2-} + [Fe(Hcydta)]^-$	1.0	1.38			64
$[Cu(H_{-3}V_4)]^- + [Fe(CN)_6]^{4-}$	0.1	2.61×10^5			68
$[Cu(H_{-3}A_3G)]^- + [Fe(CN)_6]^{4-}$	0.1	7.8×10^6			68
$[Cu(H_{-3}A_4)]^- + [Fe(CN)_6]^{4-}$	0.1	9.8×10^6			68
$[Cu(H_{-3}AG_3)] + [Fe(CN)_6]^{4-}$	0.1	3.5×10^7			68
$[Cu(H_{-3}F_4)] + [Fe(CN)_6]^{4-}$	0.1	2.9×10^6			68
$[Cu(H_{-3}G_4)]^- + [Fe(CN)_6]^{4-}$	0.1	6.0×10^7			68
$[Cu(H_{-3}G_3A)]^- + [Fe(CN)_6]^{4-}$	0.1	8.0×10^7			68
$[Cu(H_{-3}VG_2a)]^- + [Fe(CN)_6]^{4-}$	0.1	5.1×10^7			68
$[Cu(H_{-3}G_3a)] + [Fe(CN)_6]^{4-}$	0.1	8.0×10^7			68
$[Cu(H_{-2}Aib_3)] + [Fe(CN)_6]^{4-}$	0.1	4.0×10^7			68
$[Cu(H_{-3}Aib_3a)]^- + [Fe(CN)_6]^{4-}$	0.1 (26.9°C)	1.21×10^6	8.3	-3	68
$[Cu(H_{-3}G_2AibG)] + [Fe(CN)_6]^{4-}$	0.1 (24.4°C)	9.9×10^6	5.6	-7.6	68
$[CuL(1)]^{3+} + Fe^{2+}$	1.0	1.5×10^6			71
		$2.6 \times 10^7\ M^{-2}\,s^{-1}\ [Cl^-]$			71
$[CuL(2)]^{3+} + Fe^{2+}$	1.0	2.4×10^6			71
		$3.6 \times 10^7\ M^{-2}\,s^{-1}\ [Cl^-]$			71
$CoCl^{2+} + Fe^{2+}$	1.0	1.0×10^4			71
$Ni((CH_3)_2bipy)_3]^{3+} + Fe^{2+}$	1.0	5.9×10^5	2.0	-25	72
$Ni(bipy)_3]^{3+} + Fe^{2+}$	1.0	6.7×10^6	1.6	-22	72
$NiL]^{2+} + Fe^{2+}$	1.0	1.42×10^2	4.3	-35	75
$Ni(dmg)_3]^{2-} + [Fe(CN)_6]^{4-}$	0.57 (35°C)	~ 0.06			82
$Ni(dmg)_3H]^- + [Fe(CN)_6]^{4-}$	0.57 (35°C)	0.20	21	5.5	82
$Ni(dmg)_3H]^- + [HFe(CN)_6]^{3-}$	0.57 (35°C)	6.2	14	9	82
$Rh_2(OAc)_4]^+ + Fe^{2+}$	1.0	1.07×10^5	3.9	-21.5	100
Cobalt(II):					
$[Co(NH_3)_5(NCacac)]^{2+} + [Co(sep)]^{2+}$	0.1	73			34
$[Co(NH_3)_5(acacCN)]^{2+} + [Co(sep)]^{2+}$	0.1	65			34
$[Co(diamsarH_2)]^{5+} + [Co(sep)]^{2+}$	0.2	145			48
$[Co(sep)]^{3+} + [Co(sar)]^{2+}$	0.2	20			48
$[Co(azamesar)]^{3+} + [Co(sar)]^{2+}$	0.2	10			48
$[Co(en)_3]^{3+} + [Co(sar)]^{2+}$	0.2	1.1			48

(continued)

Table 2.2. (continued)

Reaction	Medium (M)	k (M^{-1} s^{-1})	$\Delta H^{\ddagger}$ (kcal mol^{-1})	$\Delta S^{\ddagger}$ (cal K^{-1} mol^{-1})	Ref.
Cobalt(II) (*cont.*):					
$[Co(NH_3)_6]^{3+} + [Co(sar)]^{2+}$	0.2	2.3			48
$[Co(bipy)_3]^{3+} + [Co(diamsarH_2)]^{4+}$	0.2	33			48
$[Ru(NH_3)_6]^{3+} + [Co(sep)]^{2+}$	0.5	3.5×10^4			49
$[Co(bipy)_3]^{3+} + [Co(sep)]^{2+}$	0.2	1.0×10^4			49
$[Co(NH_3)_6]^{3+} + [Co(sep)]^{2+}$	0.2	0.15			49
$[Co(en)_3]^{3+} + [Co(sep)]^{2+}$	0.2	5.0×10^{-2}			49
$[Ru(NH_3)_4(phen)]^{3+} + [Co(sep)]^{2+}$	0.5	$>10^5$			49
$Co^{3+} + [CoL(2)]^{2+}$	3.0	10			70
$CoCl^{2+} + [CoL(2)]^{2+}$	3.0	1.6×10^6			70
$Co^{3+} + [CoL(3)]^{2+}$	3.0	2.2×10^2			70
$CoCl^{2+} + [CoL(3)]^{2+}$	3.0	1.5×10^7			70
$[CoL(3)(OH_2)Cl]^{2+} + [CoL(2)]^{2+}$	3.0	1.5×10^3			70
$[NiL(1)Cl]^{2+} + [CoL(2)]^{2+}$	1.0	3.1×10^6			70
$[NiL(1)]^{3+} + [CoL(1)]^{2+}$	1.5	2.3×10^4			71
$[NiL(1)]^{3+} + [CoL(2)]^{2+}$	1.5	7.6×10^2			71
$[NiL(1)]^{3+} + [CoL(3)]^{2+}$	1.5	4.6×10^4			71
$[NiL(1)]^{3+} + [CoL(4)]^{2+}$	1.5	1.3×10^5			71
$[NiL(2)]^{3+} + [CoL(1)]^{2+}$	1.5	2.0×10^5			71
$[NiL(2)]^{3+} + [CoL(2)]^{2+}$	1.5	2.8×10^3			71
$[NiL(2)]^{3+} + [CoL(3)]^{2+}$	1.5	3.1×10^4			71
$[NiL(2)]^{3+} + [CoL(4)]^{2+}$	1.5	9.5×10^3			71
$[NiL(2)]^{3+} + [CoL(5)]^{2+}$	1.5	1.5×10^4			71
$[NiL(3)]^{3+} + [CoL(2)]^{2+}$	3.0	2.1×10^4			71
$[NiL(6)]^{3+} + [CoL(2)]^{2+}$	1.0	2.0×10^2			71
$[NiL(6)]^{3+} + [CoL(3)]^{2+}$	1.0	1.4×10^4			71
$[NiL(6)]^{3+} + [CoL(4)]^{2+}$	1.0	3.1×10^4			71
$[NiL(1)Cl]^{2+} + [CoL(3)]^{2+}$	1.0	2.1×10^7			71

Reaction	Conditions	Rate constant			Ref.
$[NiL(2)Cl]^{2+} + [CoL(2)]^{2+}$	1.0	1.1×10^7			71
$[NiL(2)Cl]^{2+} + [CoL(3)]^{2+}$	1.0	1.0×10^7			71
$[NiL(3)Cl]^{2+} + [CoL(2)]^{2+}$	1.0	2.6×10^7			71
$[NiL(3)Cl]^{2+} + [CoL(3)]^{2+}$	1.0	3.4×10^7			71
$[CuL(1)]^{3+} + [CoL(2)]^{2+}$	1.0	1.2×10^6			71
$[CuL(1)]^{3+} + [CoL(2)]^{2+}$	1.0	$3.0 \times 10^7\,M^{-2}\,s^{-1}\,[Cl^-]$			71
$[CuL(1)]^{3+} + [CoL(3)]^{2+}$	1.0	4.0×10^7			71
$[CuL(1)]^{3+} + [CoL(3)]^{2+}$	1.0	$6.3 \times 10^8\,M^{-2}\,s^{-1}\,[Cl^-]$			71
$[CuL(2)]^{3+} + [CoL(2)]^{2+}$	1.0	2.7×10^6			71
$[NiL]^{2+} + [Co(phen)_3]^{2+}$	0.1	3.2×10^5			76
$[NiL]^+ + [Co(phen)_3]^{2+}$	0.1	8.3×10^2			76
$[NiLH]^{2+} + [Co(phen)_3]^{2+}$	0.1	4.1×10^4			76
$[NiL]^{2+} + [Co(edta)]^{2-}$	0.1	36			77
Metalloprotein reactions:					
$[(NC)_5FeCNCo(CN)_5]^{5-}$ + plastocyanin(I)	pH 7.0, 0.1	3.4×10^4	-2.9	-48	130
$[(NC)_5FeCNCo(CN)_5]^{5-}$ + azurin(I)	pH 5.9, 0.1	1.24×10^4	-6.0	-60	130
$[Fe(CN)_6]^{3-}$ + plastocyanin(I)	pH 7.0, 0.1	9.4×10^4	-3.3	-47	130
$[Co(dipic)]^-$ + plastocyanin(I)	pH 5.8, 0.1	2.87×10^2			134
$[Co(phen)_3]^{3+}$ + plastocyanin(I)	pH 5.8, 0.1	1.4×10^3			134
$[Co(bipy)_3]^{3+}$ + plastocyanin(I)	pH 5.8, 0.1	3.14×10^2			134
$[Ru(NH_3)_5py]^{3+}$ + plastocyanin(I)	pH 5.8, 0.1	4.2×10^4			134
$[Co(bipy)_2(O_2CMe)_2]^+$ + plastocyanin(I)	pH 5.8, 0.1	9.4			134
Plastocyanin(II) + $[Ru(NH_3)_5py]^{2+}$	pH 5.8, 0.1	3.8×10^5			135
Plastocyanin(II) + cytf(II)	pH 7.0, 0.1 (10°C)	2.1×10^7			135
Azurin(II) + cytc$_{551}$(II)	pH 7.0, 0.2	6.88×10^6			138
[Cr(III)azurin(II)] + cytc$_{551}$(II)	pH 7.0, 0.2	4.68×10^6			138
$[Co(dipic)_2]^-$ + plastocyanin(I)	pH 7.0, 0.2	4.57×10^2			140
$[Co(dipic)_2]^-$ + azurin(I)	pH 7.0, 0.2	1.41×10^2			140
Plastocyanin(II) + $[Fe(dipic)_2]^{2-}$	pH 7.0, 0.2	2.04×10^4			140
Azurin(II) + $[Fe(dipic)_2]^{2-}$	pH 7.0, 0.2	9.8×10^2			140
Stellacyanin(II) + $[Fe(dipic)_2]^{2-}$	pH 7.0, 0.2	6.8×10^4			140
$[Co(ox)_3]^{3-}$ + stellacyanin(I)	pH 7.0, 0.1	8.56×10^2			141

(continued)

Table 2.2. (continued)

Reaction	Medium (M)	k ($M^{-1}\,s^{-1}$)	$\Delta H^{\ddagger}$ (kcal mol^{-1})	$\Delta S^{\ddagger}$ (cal K^{-1} mol^{-1})	Ref.
Metalloprotein reactions (*cont.*):					
[Co(edta)]$^-$ + stellacyanin(I)	pH 7.0, 0.5	13.9			141
[Ru(en)$_3$]$^{3+}$ + stellacyanin(I)	pH 7.0, 0.1	3.4×10^4			141
Stellacyanin(II) + stellacyanin(I)	pH 7.7, 0.1	1×10^5			142
Azurin(II) + cytc(II)	pH 7.0 (20°C)	1×10^4			144
Azurin(II) + cytc$_{551}$(II)	pH 7.0 (20°C)	2×10^4			144
[Fe(CN)$_5$(4-NH$_2$py)]$^{2-}$ + Tz deoxyhemerthyrin	pH 8.2, 0.15	3.6×10^4			151
[Fe(CN)$_5$NH$_3$]$^{2-}$ + Tz deoxyhemerthyrin	pH 8.2, 0.15	2.4×10^4			151
[Fe(CN)$_6$]$^{3-}$ + Tz deoxyhemerthryin	pH 8.2, 0.15	1.0×10^5			151
[Fe(CN)$_5$PPH$_3$]$^{2-}$ + Tz deoxyhemerthyrin	pH 8.2, 0.15	7.3×10^5			151
[Fe(CN)$_4$(bipy)]$^-$ + Tz deoxyhemerthyrin	pH 8.2, 0.15	$\sim6.0 \times 10^6$			151
[Fe(CN)$_5$(4-NH$_2$py)]$^{2-}$ + P.g. semimethemerythrin	pH 8.2, 0.15	1.2×10^3			151
[Fe(CN)$_6$]$^{3-}$ + P.g. semimethemerythrin	pH 8.2, 0.15	4.5×10^2			151
[Fe(CN)$_5$PPh$_3$]$^{2-}$ + P.g. semimethemerythrin	pH 8.2, 0.15	4.4×10^4			151
[Fe(CN)$_4$(bipy)]$^-$ + P.g. semimethemerythrin	pH 8.2, 0.15	1.7×10^5			151
[Co(terpy)]$^{3+}$ + P.g. semimethemerythrin	pH 8.2, 0.15	5.1			151
Nickel(II):					
[Ni(H$_{-2}$G$_2$I)] + [Ni(H$_{-3}$G$_4$)]$^{2-}$	0.1	3.09×10^5			69
[Ni(H$_{-2}$GLG)] + [Ni(H$_{-3}$GF$_4$)]$^{2-}$	0.1	1.59×10^5			69
[Ni(H$_{-2}$DGEN)] + [Ni(H$_{-3}$F$_4$)]$^{2-}$	0.1	1.24×10^5			69
[Ni(H$_{-2}$GAG)] + [Ni(H$_{-3}$G$_4$)]$^{2-}$	0.1	1.2×10^5			69
[Ni(H$_{-2}$G$_2$V)] + [Ni(H$_{-3}$G$_4$)]$^{2-}$	0.1	2.64×10^5			69
[Ni(H$_{-2}$G$_2$A)] + [Ni(H$_{-3}$G$_4$)]$^{2-}$	0.1	1.44×10^5			69
[Ni(H$_{-2}$A$_3$)] + [Ni(H$_{-3}$G$_4$)]$^{2-}$	0.1	8.4×10^4			69
[Ni(H$_{-2}$G$_2$V)] + [Ni(H$_{-3}$Aib$_3$a)]$^-$	0.1	1.78×10^5			69
[Ni(H$_{-2}$GLG)] + [Ni(H$_{-3}$Aib$_3$a)]$^-$	0.1	1.04×10^5			69
[Ni(H$_{-2}$GAG)] + [Ni(H$_{-3}$Aib$_3$a)]$^-$	0.1	5.9×10^4			69

Reaction					Ref.
$[Ni(H_{-3}G_4a)] + [Ni(H_{-2}G_3)]^-$	0.1		2.8×10^4		69
$[Ni(H_{-2}G_2)]) + [Ni(H_{-3}G_5)]^{2-}$	0.1		8.1×10^4		69
$[Ni(H_{-2}GLG)] + [Ni(H_{-3}G_5)]^{2-}$	0.1		3.67×10^4		69
$[Ni(H_{-2}GAG)] + [Ni(H_{-3}G_5)]^{2-}$	0.1		2.22×10^4		69
$[Ni(H_{-2}G_2V)] + [Ni(H_{-3}G_5)]^{2-}$	0.1		8.2×10^4		69
$[Ni(H_{-2}Aib_3)] + [Ni(H_{-3}G_4)]^{2-}$	0.1		1.4×10^4		69
$[Ni(H_{-2}Aib_3)] + [Ni(H_{-3}G_5)]^{2-}$	0.1		6.3×10^3		69
$[Ni(H_{-2}Aib_3)] + [Ni(H_{-3}Aib_3a)]^-$	0.1		1.66×10^4		69
$[Ni(H_{-2}Aib_2)] + [Ni(H_{-3}Aib_3a)]^-$	0.1		1.6×10^4		69
$[Ni(H_{-3}G_4a)] + [Ni(H_{-3}Aib_3a)]^-$	0.1		9.1×10^3		69
$[Ni(H_{-3}Aib_3a)] + [Ni(H_{-2}Aib_3)]^-$	0.1		6.2×10^3		69
$[Ni(H_{-2}G_2V)] + [Ni(H_{-2}BG_2)]^-$	0.1		2.8×10^4		69
$[Ni(H_{-2}G_2V)] + [Ni(H_{-2}GAG)]^-$	0.1		3.5×10^4		69
$[Ni(H_{-3}Aib_2a)] + [Ni(H_{-3}G_5)]^{2-}$	0.1		8.0×10^4		69
$[Ni(H_{-2}G_2VO)] + [Ni(H_{-2}Aib_3)]$	0.1		7.6×10^3		69
$Co^{3+} + [NiL(2)]^{2+}$	3.0		1.4×10^2		70
$CoCl^{2+} + [NiL(2)]^{2+}$	3.0		7.9×10^5		70
$Co^{3+} + [NiL(1)]^{2+}$	3.0		86		70
$CoCl^{2+} + [NiL(1)]^{2+}$	3.0		5.3×10^7		70
$Co^{3+} + [NiL(3)]^{2+}$	3.0		50		70
$CoCl^{2+} + [NiL(3)]^{2+}$	3.0		1.6×10^5		70
$[NiL(2)]^{3+} + [NiL(1)]^{2+}$	3.0		4×10^6		70
$Co^{3+} + [NiL(4)]^{2+}$	3.0		947		71
$Co^{3+} + [NiL(6)]^{2+}$	3.0		750		71
$[Fe(phen)_3]^{3+} + [NiL(1)]^{2+}$	1.5		1.6×10^6		71
$[Ru(bipy)_3]^{3+} + [NiL(2)]^{2+}$	0.1		1.4×10^3		71
$[Ru(bipy)_3]^{3+} + [NiL(3)]^{2+}$	0.1		7.8×10^5		71
$[Ru(bipy)_3]^{3+} + [NiL(6)]^{2+}$	1.0		3.3×10^6		71
$[NiL(6)]^{3+} + [NiL(1)]^{2+}$	1.0		1.9×10^3		71
$[NiL(3)Cl]^{2+} + [NiL(1)]^{2+}$	1.0		1.2×10^8		71
$[CuL(1)]^{3+} + [NiL(1)]^{2+}$	1.0		7×10^6		71
$[CuL(1)]^{3+} + [NiL(1)]^{2+}$	1.0		$5.7 \times 10^7 \ M^{-2} \ s^{-1} \ [Cl^-]$		71
$[Ni(bipy)_3]^{3+} + [Ni((CH_3)_2bipy)_3]^{2+}$	1.0	8.0	2.1×10^4	−12	72

(continued)

Table 2.2. (continued)

Reaction	Medium (M)	k ($M^{-1}\,s^{-1}$)	$\Delta H^{\ddagger}$ (kcal mol^{-1})	$\Delta S^{\ddagger}$ (cal K^{-1} mol^{-1})	Ref.
Nickel(II) (cont.):					
[Ni(bipy)$_3$]$^{3+}$ + [Ni(phen)$_3$]$^{2+}$	1.0	1.1×10^3	8.9	−15	72
[Ni(phen)$_3$]$^{3+}$ + [Ni(bipy)$_3$]$^{2+}$	1.0	2.0×10^3	5.2	−26	72
[Ni(CH$_3$bipy)$_3$]$^{3+}$ + [NiH$_2$L]$^{2+}$	1.0	1.06×10^6	66	−9	72
[Ni(bipy)$_3$]$^{3+}$ + [NiH$_2$L]$^{2+}$	1.0	6.2×10^6	3.5	−16	72
[Ru(bipy)$_3$]$^{3+}$ + [NiH$_2$L]$^{2+}$	1.0	1.11×10^6			72
[CoOH]$^{2+}$ + [NiH$_2$L]$^{2+}$	0.5	1.75×10^4	6.2	−29.6	73
[Fe(dmbipy)$_3$]$^{3+}$ + [NiH$_2$L]$^{2+}$	1.0	1.25×10^3	4.8	−28	74
[Fe(dmphen)$_3$]$^{3+}$ + [NiH$_2$L]$^{2+}$	1.0	3.73×10^3	8.7	−13	74
[Fe(bipy)$_3$]$^{3+}$ + [NiH$_2$L]$^{2+}$	1.0	3.20×10^4	5.2	−20	74
[Fe(mphen)$_3$]$^{3+}$ + [NiH$_2$L]$^{2+}$	1.0	6.83×10^4	3.4	−25	74
[Fe(phen)$_3$]$^{3+}$ + [NiH$_2$L]$^{2+}$	1.0	9.16×10^4	4.0	−22	74
[Fe(Clphen)$_3$]$^{3+}$ + [NiH$_2$L]$^{2+}$	1.0	5.6×10^5	1.9	−26	74
[Ru(bipy)$_3$]$^{3+}$ + [NiH$_2$L]$^{2+}$	1.0	1.1×10^6			74
[NiL]$^{2+}$ + [NiL(1)]$^{2+}$	1.0	11.4	5.4	−35	74
Copper(I):					
[Ru(NH$_3$)$_5$py]$^{3+}$ + Cu$^+$	1.0	46.6			89
[Ru(NH$_3$)$_5$isn]$^{3+}$ + Cu$^+$	1.0	5.4×10^2			87
[Ru(NH$_3$)$_4$bipy]$^{3+}$ + Cu$^+$	1.0	3.8×10^3			89
[Ru(NH$_3$)$_4$(isn)$_2$]$^{3+}$ + Cu$^+$	1.0	4.4×10^4			89
[Co(acac)$_3$] + [Cu(phen)$_2$]$^+$	0.25	7.56×10^2	8.0	−19	90
[Co(edta)]$^-$ + [Cu(phen)$_2$]$^+$	0.25	4.48×10^2	7.5	−21	90
Copper(II):					
[Cu(H$_{-2}$Aib$_3$)] + [Cu(H$_{-3}$G$_2$AibG)]$^{2-}$	0.1	3.79×10^5			65
[Cu(H$_{-2}$Aib$_3$)] + [Cu(H$_{-3}$G$_4$)]$^{2-}$	0.1	6.51×10^4			65
[Cu(H$_{-2}$Aib$_3$)] + [Cu(H$_{-3}$V$_4$)]$^{2-}$	0.1	2.35×10^5			65
[Cu(H$_{-2}$Aib$_3$)] + [Cu(H$_{-3}$A$_3$G)]$^{2-}$	0.1	3.03×10^5			65

$[Cu(H_{-2}Aib_3)] + [Cu(H_{-3}Aib_3a)]^-$	0.1	5.8×10^6	65
$Co^{3+} + [CuL(1)]^{2+}$	3.0	2.8×10^3	70
$CoCl^2 + [CuL(1)]^{2+}$	3.0	$\sim 4.0 \times 10^3$	70
$Co^{3+} + [CuL(2)]^{2+}$	3.0	1.0×10^2	71
Ruthenium(II):			
$[Co(NH_3)_5CH_3CHCO_2]^{2+} + [Ru(NH_3)_6]^{2+}$	0.5	3.0×10^{-2}	39
$[Co(NH_3)_5NH_3CH_2CO_2]^{3+} + [Ru(NH_3)_6]^{2+}$	0.5	8.2×10^{-2}	39
$[Co(NH_3)_5((CH_3)_3CCH_2CO_2)]^{2+} + [Ru(NH_3)_6]^{2+}$	0.5	2.3×10^{-2}	39
$[Co(NH_3)_5((CH_3)_3NCH_2CO_2)]^{3+} + [Ru(NH_3)_6]^{2+}$	0.5	7.9×10^{-2}	39
$[Co(NH_3)_5((CH_3)_3CCO_2)]^{2+} + [Ru(NH_3)_6]^{2+}$	0.5	8.7×10^{-3}	39
$[Co(NH_3)_5((NH_3)C(CH_3)_2CO_2)]^{3+} + [Ru(NH_3)_6]^{2+}$	0.5	6.1×10^{-2}	39
$[Co(NH_3)_5(C_6H_5CH_2CO_2)]^{2+} + [Ru(NH_3)_6]^{2+}$	0.5	2.7×10^{-2}	39
$[Co(NH_3)_5(1-pyCH_2CO_2)]^{3+} + [Ru(NH_3)_6]^{2+}$	0.5	8.0×10^{-2}	39
$[Co(NH_3)_5(2-CH_3C_6H_4CO_2)]^{2+} + [Ru(NH_3)_6]^{2+}$	0.5	5.0×10^{-2}	39
$[Co(NH_3)_5(2-NH_3C_6H_4CO_2)]^{3+} + [Ru(NH_3)_6]^{2+}$	0.5	5.8×10^{-2}	39
$[(NH_3)Co(\mu OH)_2(\mu HCO_2)Co(NH_3)_3]^{3+} + [Ru(NH_3)_6]^{2+}$	0.5	1.77×10^{-1}	42
$[(NH_3)Co(\mu OH)_2(\mu\text{-}CH_3CO_2)Co(NH_3)_3]^{3+} + [Ru(NH_3)_6]^{2+}$	0.5	6.2×10^{-2}	42
$[(NH_3)Co(\mu OH)_2(\mu\text{-}CF_3CO_2)Co(NH_3)_3]^{3+} + [Ru(NH_3)_6]^{2+}$	0.15	7.2×10^{-1}	42
$[(NH_3)Co(\mu OH)_2(\mu\text{-}C_6H_4CO_2)Co(NH_3)_3]^{3+} + [Ru(NH_3)_6]^{2+}$	0.5	1.17×10^{-1}	42
$[(NH_3)Co(\mu OH)_2(\mu\text{-}4\text{-}NO_2\text{-}C_6H_4CO_2)Co(NH_3)_3]^{3+} + [Ru(NH_3)_6]^{2+}$	0.5	1.53×10^{-1}	42
$[(NH_3)Co(\mu OH)_2(\mu\text{-}pyruvato)Co(NH_3)_3]^{3+} + [Ru(NH_3)_6]^{2+}$	0.5	3.4×10^{-1}	42
$[(NH_3)Co(\mu OH)_2(\mu\text{-}phenylglyoxylato)Co(NH_3)_3]^{3+} + [Ru(NH_3)_6]^{2+}$	0.5	7.3×10^{-1}	42
$[(NH_3)Co(\mu OH)_2(\mu\text{-}4\text{-}MeO\text{-}phenylglyoxylato)\text{-}Co(NH_3)_3]^{3+} + [Ru(NH_3)_6]^{2+}$	0.5	6.0×10^{-1}	42

(continued)

Table 2.2. (continued)

Reaction	Medium (M)	k ($M^{-1}\,s^{-1}$)	$\Delta H^{\ddagger}$ (kcal mol^{-1})	$\Delta S^{\ddagger}$ (cal K^{-1} mol^{-1})	Ref.
Ruthenium(II) (cont.):					
$[(NH_3)Co(\mu OH)_2(\mu\text{-}2,4\text{-}(MeO)_2\text{-phenylglyoxylato})\text{-}Co(NH_3)_3]^{3+} + [Ru(NH_3)_6]^{2+}$	0.5	6.8×10^{-1}			42
$[(NH_3)Co(\mu OH)_2(\mu\text{-}2,4,6\text{-}(MeO)_3\text{-phenylglyoxylato})\text{-}Co(NH_3)_3]^{3+} + [Ru(NH_3)_6]^{2+}$	0.5	6.4×10^{-1}			42
$[(NH_3)Co(\mu OH)_2(\mu\text{-}4\text{-formylbenzoato})Co(NH_3)_3]^{3+} + [Ru(NH_3)_6]^{2+}$	0.5	1.82×10^{-1}			42
$[(NH_3)Co(\mu OH)_2(\mu\text{-}2\text{-formylbenzoato})Co(NH_3)_3]^{3+} + [Ru(NH_3)_6]^{2+}$	0.5	2.8×10^{-1}			42
$[Co(NH_3)_5(glyoxylato)]^{2+} + [Ru(NH_3)_6]^{2+}$	0.5	5.9×10^{-2}			42
$[Co(NH_3)_5(pyruvato)]^{2+} + [Ru(NH_3)_6]^{2+}$	0.5	1.17×10^{-1}			42
$[Co(NH_3)_5(4\text{-MeOphenylglyoxylato})]^{2+} + [Ru(NH_3)_6]^{2+}$	0.5	3.8×10^{-1}			42
$[Co(NH_3)_5(2,4\text{-}(MeO)_2phenylglyoxylato)]^{2+} + [Ru(NH_3)_6]^{2+}$	0.5	3.9×10^{-1}			42
$[Co(NH_3)_5(2,4,6\text{-}(MeO)_3phenylglyoxylato)]^{2+} + [Ru(NH_3)_6]^{2+}$	0.5	4.1×10^{-1}			42
$[Co(NH_3)_5(4\text{-formylbenzoato})]^{2+} + [Ru(NH_3)_6]^{2+}$	0.5	4.8×10^{-2}			42
$[Co(NH_3)_5(2\text{-formylbenzoato})]^{2+} + [Ru(NH_3)_6]^{2+}$	0.5	1.31×10^{-1}			42
$[(NH_3)Co(\mu OH)_2(\mu\text{-}2\text{-}CO_2,3\text{-}CONH_2pyH)\text{-}Co(NH_3)_3]^{4+} + [Ru(NH_3)_6]^{2+}$	0.5	1.10			42
$[(NH_3)Co(\mu OH)_2(\mu\text{-}2\text{-}CO_2,4\text{-}CONH_2pyH)\text{-}Co(NH_3)_3]^{4+} + [Ru(NH_3)_6]^{2+}$	0.5	0.63			43
$[(NH_3)Co(\mu OH)_2(\mu\text{-}2\text{-}CO_2,5\text{-}CONH_2pyH)\text{-}Co(NH_3)_3]^{4+} + [Ru(NH_3)_6]^{2+}$	0.5	0.35			43
$[(NH_3)Co(\mu OH)_2(\mu\text{-}4\text{-}CONH_2pyCH_2CO_2)\text{-}Co(NH_3)_3]^{2+} + [Ru(NH_3)_6]^{2+}$	0.5	0.94			43

Reaction					Ref.
[(NH₃)Co(μOH)₂(μ-3-CONH₂pyCH₂CO₂)-Co(NH₃)₃]⁴⁺ + [Ru(NH₃)₆]²⁺	0.5	0.60			43
[(NH₃)Co(μOH)₂(μ-4-CONHCH₂CO₂pyH)-Co(NH₃)₃]⁴⁺ + [Ru(NH₃)₆]²⁺	0.5	0.47			43
[(NH₃)Co(μOH)₂(μ-4-CONH₂py-2-CH₂C₆H₄CO₂)-Co(NH₃)₃]³⁺ + [Ru(NH₃)₆]²⁺	0.5	1.01			43
[Cu(H₋₃Aib₃a)] + [Ru(NH₃)₆]³⁺	0.1	1.5×10^7	1.5	−21	65
[Cu(H₋₃Aib₃a)] + [Ru(NH₃)₅py]²⁺	0.1	1.04×10^6	1.8	−25	65
[Cu(H₋₃Aib₃a)] + [Ru(NH₃)₅pic]²⁺	0.1	1.8×10^6			65
[Cu(H₋₃G₂AibG)]⁻ + [Ru(NH₃)₅py]²⁺	0.1	6.2×10^7			65
[Cu(H₋₂Aib₃)] + [Ru(NH₃)₅py]²⁺	0.1	1.5×10^8			65
[Cu(H₋₃A₃G)]⁻ + [Ru(NH₃)₅py]²⁺	0.1	1.01×10^8			65
[NiL(2)]³⁺ + [Ru(bipy)₃]²⁺	1.0	2.7×10^4			71
[Ni((CH₃)₂bipy)₃]³⁺ + [Ru(NO₂phen)₃]²⁺	1.0	4.1×10^6	2.4	−20	72
[Cu(dmp)₂]²⁺ + [Ru(NH₃)₅py]²⁺	0.1	5.5×10^5	5.2	−15	89
[Cu(dmp)₂]²⁺ + [Ru(NH₃)₅isn]²⁺	0.1	4.55×10^5	5.4	−14	89
[Cu(dmp)₂]²⁺ + [Ru(NH₃)₄bipy]²⁺	0.1	3.86×10^5	5.7	−14	89
[Co(NH₃)₅F]²⁺ + [Ru(en)₃]²⁺	0.2	22.8			94
[Co(NH₃)₅F]²⁺ + [Ru(NH₃)₆]²⁺	0.2	80			94
[Co(NH₃)₅F]²⁺ + [Ru(NH₃)₅OH₂]²⁺	0.2	3.3×10^{-2}			94
[Co(NH₃)₅Cl]²⁺ + [Ru(NH₃)₅OH₂]²⁺	0.2	1.8×10^{-1}			94
[Co(NH₃)₅Cl]²⁺ + [Ru(en)₃]²⁺	0.2	1.32×10^2			94
[Co(NH₃)₅Br]²⁺ + [Ru(NH₃)₅OH₂]²⁺	0.2	1.6×10^2			94
[Co(NH₃)₅Br]²⁺ + [Ru(en)₃]²⁺	0.2	2.34×10^2			94
[Co(NH₃)₅I]²⁺ + [Ru(NH₃)₅OH₂]²⁺	0.2	2.5×10^3			94
[Fe(edta)]⁻ + [Ru(NH₃)₆]²⁺	1.0	2.2×10^6	3.6	−17	97
Fe³⁺ + [(NH₃)₅RuORu(NH₃)₄ORu(NH₃)₅]⁶⁺	0.24 (20°C)	4.13			98
FeOH²⁺ + [(NH₃)₅RuORu(NH₃)₄ORu(NH₃)₅]⁶⁺	0.24 (20°C)	1.53			98
[*Ru(bipy)₃]²⁺:					
[Co(cpCO₂H)₃]⁺ + [*Ru(bipy)₃]²⁺	0.1	1.5×10^9			52
[Co(sep)]³⁺ + [*Ru(bipy)₃]²⁺	0.1	2.6×10^8 ᵇ			52
[Co(sep)]³⁺ + [*Ru(bipy)₃]²⁺	1.0 (H₂SO₄)	3.6×10^8 ᵇ			53

(continued)

Table 2.2. (continued)

Reaction	Medium (M)	k ($M^{-1}\,s^{-1}$)	$\Delta H^{\ddagger}$ (kcal mol^{-1})	$\Delta S^{\ddagger}$ (cal K^{-1} mol^{-1})	Ref.
[*Ru(bipy)$_3$]$^{2+}$ *(cont.):*					
[Co(sep)]$^{3+}$ + [*Ru(bipy)$_3$]$^{2+}$	0.5	1.5×10^8 ^{b}			54
[Co(meoxsar-H)]$^{2+}$ + [*Ru(bipy)$_3$]$^{2+}$	0.5	$<10^{-3}$			54
[Co(CO$_2$C$_2$H$_5$meoxsar-H)]$^{2+}$ + [*Ru(bipy)$_3$]$^{2+}$	0.5	$<10^{-3}$			54
[Co(sar)]$^{3+}$ + [*Ru(bipy)$_3$]$^{2+}$	0.5	3.0×10^6			54
[Co(NH$_2$mesar)]$^{3+}$ + [*Ru(bipy)$_3$]$^{2+}$	0.5	8.0×10^6			54
[Co(Clsar)]$^{3+}$ + [*Ru(bipy)$_3$]$^{2+}$	0.5	4.2×10^7			54
[Co(azacapten)]$^{3+}$ + [*Ru(bipy)$_3$]$^{2+}$	0.5	1.2×10^9			54
[Rh(phen)$_3$]$^{3+}$ + [*Ru(bipy)$_3$]$^{2+}$	0.5	4.8×10^8			103
[Os(NH$_3$)$_5$N$_3$]$^{2+}$ + [*Ru(bipy)$_3$]$^{2+}$	0.5 (H$_2$SO$_4$)	9×10^7			103
[Os(NH$_3$)$_5$Cl]$^{2+}$ + [*Ru(bipy)$_3$]$^{2+}$	0.5 (H$_2$SO$_4$)	$\leq 4 \times 10^7$			103
[Os(NH$_3$)$_6$]$^{3+}$ + [*Ru(bipy)$_3$]$^{2+}$	0.5 (H$_2$SO$_4$)	$\leq 1.4 \times 10^7$			103
[Os(NH$_3$)$_5$I]$^{2+}$ + [*Ru(bipy)$_3$]$^{2+}$	0.5 (H$_2$SO$_4$)	3.1×10^8			103
[Os(NH$_3$)$_5$OH$_2$]$^{3+}$ + [*Ru(bipy)$_3$]$^{2+}$	0.5 (H$_2$SO$_4$)	$\leq 7 \times 10^6$			103
Europium(II):					
[(NH$_3$)$_3$Co(μOH)$_2$(μ-2-CHOC$_6$H$_4$CO$_2$)-Co(NH$_3$)$_3$]$^{3+}$ + Eu^{2+}	1.0	0.058			41
[(NH$_3$)$_3$Co(μOH)$_2$(μ-4-CHOC$_6$H$_4$CO$_2$)-Co(NH$_3$)$_3$]$^{3+}$ + Eu^{2+}	1.0	0.70			41
[(NH$_3$)$_3$Co(μOH)$_2$(μ-HCO$_2$)Co(NH$_3$)$_3$]$^{3+}$ + Eu^{2+}	1.0	1.5×10^{-2}			42
[(NH$_3$)$_3$Co(μOH)$_2$(μ-CH$_3$CO$_2$)Co(NH$_3$)$_3$]$^{3+}$ + Eu^{2+}	1.0	8.6×10^{-3}			42
[(NH$_3$)$_3$Co(μOH)$_2$(μ-CF$_3$CO$_2$)Co(NH$_3$)$_3$]$^{3+}$ + Eu^{2+}	1.0	6.0×10^{-2}			42
[(NH$_3$)$_3$Co(μOH)$_2$(μ-C$_6$H$_4$CO$_2$)Co(NH$_3$)$_3$]$^{3+}$ + Eu^{2+}	1.0	1.0×10^{-2}			42
[(NH$_3$)$_3$Co(μOH)$_2$(μ-4-NO$_2$-C$_6$H$_4$CO$_2$)-Co(NH$_3$)$_3$]$^{3+}$ + Eu^{2+}	1.0	1.6×10^{-2}			42
[(NH$_3$)$_3$Co(μOH)$_2$(μ-pyruvato)Co(NH$_3$)$_3$]$^{3+}$ + Eu^{2+}	1.0	1.53			42

$[(NH_3)_3Co(\mu OH)_2(\mu\text{-phenylglyoxylato})\text{-}Co(NH_3)_3]^{3+} + Eu^{2+}$	1.0	86	42
$[(NH_3)_3Co(\mu OH)_2(\mu\text{-4-MeO-phenylglyoxylato})\text{-}Co(NH_3)_3]^{3+} + Eu^{2+}$	1.0	57	42
$[(NH_3)_3Co(\mu OH)_2(\mu\text{-2,4-(MeO)}_2\text{-phenylglyoxylato})\text{-}Co(NH_3)_3]^{3+} + Eu^{2+}$	1.0	10.1	42
$[(NH_3)_3Co(\mu OH)_2(\mu\text{-2,4,6-(MeO)}_3\text{-phenylglyoxylato})\text{-}Co(NH_3)_3]^{3+} + Eu^{2+}$	1.0	22	42
$[(NH_3)_3Co(\mu OH)_2(\mu\text{-4-formylbenzoato})\text{-}Co(NH_3)_3]^{3+} + Eu^{2+}$	1.0	6.9×10^{-2}	42
$[(NH_3)_3Co(\mu OH)_2(\mu\text{-2-formylbenzoato})\text{-}Co(NH_3)_3]^{3+} + Eu^{2+}$	1.0	5.4×10^{-2}	42
$[Co(NH_3)_5(glyoxylato)]^{2+} + Eu^{2+}$	1.0	80	42
$[Co(NH_3)_5\text{-(glyoxylato)}]^{2+} + Eu^{2+}$	1.0	1.2×10^4	42
$[Co(NH_3)_5(2,4\text{-(MeO}_2)\text{-phenylglyoxylato)}]^{2+} + Eu^{2+}$	1.0	7.4×10^2	42
$[Co(NH_3)_5\text{-(2,4,6-(MeO)}_3\text{-phenylglyoxylato)}]^{2+} + Eu^{2+}$	1.0	3.0×10^2	42
$[Co(NH_3)_5(4\text{-acetylbenzoato})]^{2+} + Eu^{2+}$	1.0	1.16	42
$[(NH_3)_3Co(\mu OH)_2(\mu\text{-2CO}_2\text{,3-CONH}_2pyH)\text{-}Co(NH_3)_3]^{4+} + Eu^{2+}$	2.5 (20°C)	0.58	43
$[(NH_3)_3Co(\mu OH)_2(\mu\text{-2CO}_2\text{,4-CONH}_2pyH)\text{-}Co(NH_3)_3]^{3+} + Eu^{2+}$	1.0 (20°C)	5.3×10^2	43
$[(NH_3)_3Co(\mu OH)_2(\mu\text{-2CO}_2\text{,5-CONH}_2pyH)\text{-}Co(NH_3)_3]^{3+} + Eu^{2+}$	1.0 (20°C)	1.7×10^2	43
$[(NH_3)_3Co(\mu OH)_2(\mu\text{-4-CONH}_2pyCH_2CO)\text{-}Co(NH_3)_3]^{3+} + Eu^{2+}$	1.0 (20°C)	86	43
$[(NH_3)_3Co(\mu OH)_2(\mu\text{-3-CONH}_2pyCH_2CO_2)\text{-}Co(NH_3)_3]^{3+} + Eu^{2+}$	1.0	0.103	43
$[(NH_3)_3Co(\mu OH)_2(\mu\text{-4-CONHCH}_2CO_2pyH)\text{-}Co(NH_3)_3]^{3+} + Eu^{2+}$	1.0	0.20	43
$[(NH_3)_3Co(\mu OH)_2(\mu\text{-4-CONH}_2py\text{-2-CH}_2C_6H_4CO_2)\text{-}Co(NH_3)_3]^{3+} + Eu^{2+}$	1.0 (20°C)	13.1	43

(continued)

Table 2.2. (*continued*)

Reaction	Medium (M)	k ($M^{-1}\,s^{-1}$)	$\Delta H^{\ddagger}$ (kcal mol^{-1})	$\Delta S^{\ddagger}$ (cal K^{-1} mol^{-1})	Ref.
Europium(II) (*cont.*):					
$[Co(sep)]^{3+} + Eu^{2+}$	0.2	0.12			48
$[Co(azamesar)]^{3+} + Eu^{2+}$	0.2	1.5×10^{-2}			48
$[Co(diamsarH_2)]^{5+} + Eu^{2+}$	0.2	0.13			48
$[Co(ammesarH)]^{4+} + Eu^{2+}$	0.2	0.03			48
$[Eu(2.2.1)]^{3+} + Eu^{2+}$	0.1	0.2			58
$[Eu(2.2.2)]^{3+} + Eu^{2+}$	0.1	1.4			58
$[Co(NH_3)_6]^{3+} + [Eu(2.2.1)]^{2+}$	0.1	5.5×10^{-2}			58
Uranium(III):					
$[Co(NH_3)_5(CH_3CH_2CO_2)]^{2+} + U^{3+}$	0.2 (22°C)	5.3×10^3			39
$[Co(NH_3)_5(CH_3CH_2CO_2H)]^{3+} + U^{3+}$	0.2 (22°C)	5.8×10^2			39
$[Co(NH_3)_5((CH_3O)_3CCH_2CO_2)]^{2+} + U^{3+}$	0.2 (22°C)	1.01×10^3			39
$[Co(NH_3)_5((CH_3)_3NCH_2CO_2)]^{3+} + U^{3+}$	0.2 (22°C)	66			39
$[Co(NH_3)_5((CH_3)_3CCO_2)]^{2+} + U^{3+}$	0.2 (22°C)	42			39
$[Co(NH_3)_5(NH_3C(CH_3)_2CO_2)]^{3+} + U^{3+}$	0.2 (22°C)	5.1			39
$[Co(NH_3)_5(C_6H_5CH_2CO_2)]^{2+} + U^{3+}$	0.2 (22°C)	3.1×10^3			39
$[Co(NH_3)_5(1\text{-}pyCH_2CO_2)]^{3+} + U^{3+}$	0.2 (22°C)	3.1×10^2			39
$[Co(NH_3)_5(2\text{-}CH_3C_6H_4CO_2)]^{2+} + U^{3+}$	0.2 (22°C)	3.4×10^2			39
$[Co(NH_3)_5(2\text{-}NH_3C_6H_4CO_2)]^{3+} + U^{3+}$	0.2 (22°C)	28			39
$[(NH_3)_3Co(\mu OH)_2(\mu HCO_2)CO(NH_3)]^{3+} + U^{3+}$	0.2	4.1			42
$[(NH_3)_3Co(\mu OH)_2(\mu CH_3CO_2)Co(NH_3)_3]^{3+} + U^{3+}$	0.2	1.26			42
$[(NH_3)_3Co(\mu OH)_2(\mu CF_3CO_2)Co(NH_3)_3]^{3+} + U^{3+}$	0.2	11.8			42
$[(NH_3)_3Co(\mu OH)_2(\mu C_6H_5CO_2)Co(NH_3)_3]^{3+} + U^{3+}$	0.2	2.8			42
$[(NH_3)_3Co(\mu OH)_2(\mu pyruvato)Co(NH_3)_3]^{3+} + U^{3+}$	0.2 (21°C)	1.5×10^5			42
$[(NH_3)_3Co(\mu OH)_2(\mu phenylglyoxylato)Co(NH_3)_3]^{3+} + U^{3+}$	0.2 (21°C)	1.2×10^5			42
$[(NH_3)_3Co(\mu OH)_2(\mu\text{-}4\text{-}MeO\text{-}phenylglyoxylato)\text{-}Co(NH_3)_3]^{3+} + U^{3+}$	0.2 (21°C)	1.7×10^5			42

Reaction		k			Ref.
$[(NH_3)_3Co(\mu OH)_2(\mu\text{-}2,4\text{-}(MeO)_2\text{phenylglyoxylato})\text{-}Co(NH_3)_3]^{3+} + U^{3+}$	0.2 (21°C)	1×10^5			42
$[(NH_3)_3Co(\mu OH)_2(\mu\text{-}2,4,6\text{-}(MeO)_3\text{phenylglyoxylato})\text{-}Co(NH_3)_3]^{3+} + U^{3+}$	0.2 (21°C)	8×10^4			42
$[(NH_3)_3Co(\mu OH)_2(\mu\text{-}4\text{-formylbenzoato})Co(NH_3)_3]^{3+} + U^{3+}$	0.2 (21°C)	4.5×10^2			42
$[(NH_3)_3Co(\mu OH)_2(\mu\text{-}2\text{-formylbenzoato})Co(NH_3)_3]^{3+} + U^{3+}$	0.2 (21°C)	7.3×10^4			42
$[(NH_3)_3Co(\mu OH)_2(\mu\text{-}2CO_2,3\text{-}CONH_2pyH)\text{-}Co(NH_3)_3]^{4+} + U^{3+}$	0.2 (20°C)	7.4×10^4			43
$[(NH_3)_3Co(\mu OH)_2(\mu\text{-}2CO_2,4\text{-}CONH_2pyH)\text{-}Co(NH_3)_3]^{4+} + U^{3+}$	0.2 (20°C)	2.5×10^6			43
$[(NH_3)_3Co(\mu OH)_2(\mu\text{-}2CO_2,5\text{-}CONH_2pyH)\text{-}Co(NH_3)_3]^{4+} + U^{3+}$	0.2 (20°C)	2.5×10^6			43
$[(NH_3)_3Co(\mu OH)_2(\mu\text{-}4CONH_2pyCH_2CO_2)\text{-}Co(NH_3)_3]^{4+} + U^{3+}$	0.2 (20°C)	2.5×10^6			43
$[(NH_3)_3Co(\mu OH)_2(\mu\text{-}3CONH_2pyCH_2CO_2)\text{-}Co(NH_3)_3]^{4+} + U^{3+}$	0.2 (20°C)	8.0×10^3			43
$[(NH_3)_3Co(\mu OH)_2(\mu\text{-}4CONHCH_2CO_2pyH)\text{-}Co(NH_3)_3]^{4+} + U^{3+}$	0.2 (20°C)	6.2×10^2			43
$[Co(sep)]^{3+} + U^{3+}$	0.2	15.8			48
Miscellaneous reactions:					
$[Co(NH_3)_5ImH]^{3+} + VO^{2+}$	0.5	1.3			26
$[NiL]^{2+} + VO^{2+}$	1.0	0.933	8.7	−30	75
$Ce(IV) + [Rh_2(OAc)_4]$	1.0	6.0×10^4	8.2	−8.7	100
$[Rh_2(OAc)_4]^+ + VO^{2+}$	1.0	1.98×10^2			100
$[Rh(bipy)_3]^{2+} + [Rh(II)(bipy)_2]^{2+}$		3.0×10^8			102
$VO_2^+ + [Mo(CN)_8]^{4-}$	1.0 (22.4°C)	$2.46 \times 10^6\ M^{-2}\,s^{-1}\,[H^+]$			105
		2.81×10^4			105
$[W(CN)_8]^{3-} + MnO_4^{2-}$	0.2	3.0×10^3			107

(continued)

Table 2.2. (continued)

Reaction	Medium (M)	k (M^{-1} s^{-1})	$\Delta H^{\ddagger}$ (kcal mol^{-1})	$\Delta S^{\ddagger}$ (cal K^{-1} mol^{-1})	Ref.
Miscellaneous reactions (*cont.*):					
$[Mo(CN)_6]^{3-} + MnO_4^{2-}$	0.2	8.5×10^4			107
$[Fe(CN)_6]^{3-} + MnO_4^{2-}$	0.2	3.8×10^2			107
$MnO_4^- + [H_2Os(CN)_6]^{2-}$	1.0	7×10^3			108
$MnO_4^- + [HOs(CN)_6]^{3-}$	1.0	1.8×10^4			108
$MnO_4^- + [Os(CN)_6]^{4-}$	1.0	3.3×10^4			108
$[IrCl_6]^{2-} + [Mo(H_2O)_6]^{3+}$	0.2	3.4×10^4			125
		2.9×10^4 s^{-1}, $1/[H^+]$			
$[Co(ox)_3]^{3-} + [Mo(H_2O)_6]^{3+}$	2.0	0.67			125
$VO^{2+} + [Mo(H_2O)_6]^{3+}$	2.0	6.5×10^{-3}			125
$MnO_4^- + Mo_2O_4^{2+}$	1.0	5.42×10^3			126
$HMnO_4 + Mo_2O_4^{2+}$	1.0	9.70×10^4			126

[a] $L(1) = [14]arcN_4$; $L(2) = Me_6[14]4,11\text{-dieneN}_4$; $L(3) = Me_4[14]\text{tetraeneN}_4$; $L(4) = Me_2pyO[14]\text{trieneN}_4$; $L(5) = [15]areN_4$; $L(6) = [14]4,11\text{-dieneN}_4$.
[b] Quenching rate constant.

Crystallographic studies reveal that Co(III)-N and Co(II)-N bond lengths in $[Co(sep)]^{3+/2+}$ are 1.99 and 2.16 Å, respectively, compared with 1.97 and 2.16 Å for the $[Co(NH_3)_6]^{3+/2+}$ case where the self-exchange rate is much lower. The close similarity in bond lengths and the similar spin-state rearrangement on electron transfer do not provide an explanation for the large differences in barriers to electron transfer.

The cage complexes provide an important new source of data in attempting to understand the mechanics of the cobalt(III)/(II) electron transfer barrier and a variety of complexes with reduction potentials varying from 0.06 to -0.04 $V^{(44)}$ have been prepared. In the azacapten complexes, magnetic measurements indicate that both cobalt(III) and cobalt(II) are in the low-spin configuration[45] with a significantly shorter Co(II)-N bond length, 2.07 Å, than in other cage complexes. The electron self-exchange is amenable to determination[46] by NMR line broadening of the capping methyl proton resonance and at 4500 $M^{-1} s^{-1}$ is larger than for the other cage complexes consistent with lower Frank–Condon barriers. A series of cage complexes without capping atoms have been synthesized[47] and self-exchange rates measured by loss of optical activity. The self-exchange rate for $[Co(amsartacn)]^{3+/2+}$ is 0.09 $M^{-1} s^{-1}$, larger than for its protonated analogue $[Co(amsartacnH)]^{4+/3+}$ but not significantly different from those cages where access along the C_3 axis is prevented by the ligand cap.[48] Agreement between rate constants measured for the cross reactions between the cage complexes in Table 2.2, and those calculated using the self-exchange rates and reduction potentials in Table 2.3 is excellent,[48] fully consistent with an outer-sphere mechanism.

Good agreement is also found between experimental and calculated data in cross reactions with a number of other oxidants and reductants where self-exchange data are accurately known. Since the reactions are necessarily outer sphere in nature, the rate data cast new light on aquosystems where self-exchange data are poorly defined. Estimates of 10^{-9}–10^{-10} $M^{-1} s^{-1}$ for $Cr_{aq}^{3+/2+}$, 10^{-4}–10^{-5} $M^{-1} s^{-1}$ for $Eu_{aq}^{3+/2+}$, and 10^{-4}–10^{-5} $M^{-1} s^{-1}$ for $U_{aq}^{4+/3+}$ at 25°C are derived, but perhaps the most significant revision lies in the $[Co(NH_3)_6]^{3+/2+}$ self-exchange, now thought to be around 10^{-7}–10^{-8} $M^{-1} s^{-1}$.[1]

The problems of explaining the five-orders-of-magnitude difference in self-exchange between $[Co(sep)]^{3+/2+}$ and $[Co(en)_3]^{3+/2+}$ remain. Sargeson and co-workers[48] are directing synthetic efforts to possible explanations involving reorganization of the hydration spheres around the cage, while Endicott and co-workers,[49] in a detailed molecular mechanics calculation, conclude that the differences can be explained by the published variations in Co-N bond lengths with a contribution from the interaction of the cage and not by any differences in Co-N force constants as had been suggested previously.[50,51] These calculations[49] are also consistent with the revised $[Co(NH_3)_6]^{3+/2+}$ self-exchange rate.[1]

A number of workers[52-54] have pointed out the potential usefulness of cobalt clathrochelate complexes as a relay in the $[^*Ru(bipy)_3]^{2+}$ water-splitting reaction. There is reasonable consistency in the measured rates of $[Co(III)(sep)]^{3+}$ reduction. Endicott and Ramasami argue[55] that anion effects in electron transfer reactions between positively charged ions are suggestive of a role for these ions in enhancing orbital overlap between the donor and acceptor and thus can be used as a probe for nonadiabatic behavior. Rate differences between CF_3^-, SO_3^- and ascorbate media of 50 for the reaction between $[Co(phen)_3]^{3+}$ and $[Co(sep)]^{2+}$ can be compared with a range of <3 in corresponding reduction by $[Ru(NH_3)_6]^{2+}$ which has greater adiabatic character.

Sepulchrate complexes of other metal ions have been prepared. Thus $[Pt(IV)(sep)]^{4+}$ has been isolated and can be reduced to give a long-lived platinum(III) transient in aqueous solution.[56] Photophysics of $[Cr(III)-(sep)]^{3+}$ have also been reported.[57] In a study[58] of the Eu(III)/(II) couple using the related cryptate ligand system, reductions of $[Eu(2.2.1)]^{3+}$ and $[Eu(2.2.2)]^{3+}$ by V^{2+} and Cr^{2+} are slow enough to be monitored polarographically. Self-exchange rates of $10\ M^{-1}s^{-1}$, $4 \times 10^{-2}\ M^{-1}s^{-1}$, and $5 \times 10^{-6}\ M^{-1}s^{-1}$ are evaluated for $[Eu(2.2.1)]^{3+/2+}$, $[Eu(2.2.2)]^{3+/2+}$, and $Eu_{aq}^{3+/2+}$, respectively. Using an argument that the normal cavity radius of cryptand(2.2.1), 1.1 Å, allows little expansion on reduction of Eu^{3+}, while that of cryptand(2.2.2), 1.4 Å, is more flexible the self-exchange rates are considered to reflect Frank–Condon barriers suggesting that nonadiabaticity is not responsible for the low reactivity of $Eu_{aq}^{3+/2+}$.

2.6. Iron(II)

Both inner-sphere and outer-sphere mechanisms are proposed in the oxidation of Fe_{aq}^{2+} by FeX^{2+}, where $X^- = Cl^-$, SCN^-, or N_3^-. By noting[59] that the outer-sphere mechanism is equivalent to the Fe_{aq}^{2+}-catalyzed dissociation of FeX^{2+}, the amount of outer-sphere character in these reactions has been estimated as 44% (Cl^-), 100% (SCN^-), and 0% (N_3^-), depending on the bridging ability of X^-. Relative reactivities for heterogeneous and homogeneous electron exchange of a number of aquo metal ions highlight[60] the odd behavior of $Fe_{aq}^{3+/2+}$ and it is again suggested that homogeneous self-exchange takes place by an inner-sphere pathway. The outer-sphere rate is predicted to be three orders of magnitude slower than observed. Inner-sphere and outer-sphere reorganizational energies have been estimated[61] from photochemical experiments.

Table 2.3. *Self-Exchange Rate Constants for Cobalt(III)/(II) Cage Complexes at 25°C*

Reaction	E^0 (V)	μ (M)	k_{11} ($M^{-1}\,s^{-1}$)	$\Delta H^{\ddagger}$ (kcal mol^{-1})	$\Delta S^{\ddagger}$ (cal K^{-1} mol^{-1})	References
[Co(sep)]$^{3+/2+}$	−0.29	0.2	5.1	9.6	−23	11
[Co(azamesar)]$^{3+/2+}$	−0.34	0.2	2.9			11
[Co(azacapten)]$^{3+/2+}$	+0.01	0.17–0.29	4500	7	−18	46
[Co(amsartacn)]$^{3+/2+}$	−0.31	0.2	0.09			47
[Co(amsartacnH)]$^{4+/3+}$	−0.17	0.2	0.04			47
[Co(sar)]$^{3+/2+}$	−0.45	0.2	2.1			48
[Co(diamsar)]$^{3+/2+}$	−0.31	0.2	0.50			48
[Co(diamsarH$_2$)]$^{5+/4+}$	+0.06	0.2	0.024			48

The Fe_{aq}^{2+} reduction of $[Fe(phen)_3]^{3+}$ would generally be predicted as outer sphere in nature, but an inner-sphere mechanism has been proposed[62] to explain a nonlinear Hammett plot obtained for substituted phenanthroline derivatives. Covalent hydration, that is, addition of H_2O to the 2-position of the phenanthroline, is suggested to provide a hydroxo bridge for an inner-sphere mechanism but the evidence is not convincing. Kinetic evidence has been presented[63] for an intermediate in the reduction of a variety of iron(III) complexes with donor ligands by $[Fe(3,4,7,8\text{-}Me_4phen)_3]^{2+}$ in acetonitrile. Both inner-sphere and outer-sphere pathways are found and the intermediate is thought to result from the former pathway.

A detailed study[64] of the reduction of $[Fe(edta)]^-$ by $[V(edta)]^-$ has appeared. The reaction can be studied between pH 5.4 and 7.1 and proceeds by an inner-sphere mechanism involving the edta complexes $[M(edta)]^-$ and their hydrolyzed derivatives $[M(edta)OH]^{2-}$. Between pH 2.5 and 4.0 the reverse reaction, oxidation of $[Fe(edta)]^{2-}$ by $[VO(edta)]^{2-}$, can also be studied, but at lower pH the thermodynamics favor iron(II) and vanadium(IV) and a pathway involving dissociation of $[Fe(edta)]^-$ is observed.

2.7. Copper(III), Nickel(III), and Nickel(IV)

Margerum and co-workers[65] have examined cross reactions of copper(III)- and copper(II)-peptide complexes and find, on application of Marcus theory, that the $[Cu(III)/(II)(peptide)]$ self-exchange rate is almost independent of peptide structure with a value of $2 \times 10^4\ M^{-1}\ s^{-1}$ at 25°C, 0.1 M ionic strength, in good agreement with the value obtained by NMR line broadening.[66] This value is large in view of the substantial differences in copper coordination and peptide backbone between oxidized and reduced forms.[67] Reductions of $[Cu(III)(peptide)]$ complexes by a series of $[Ru(NH_3)_5L]^{2+}$ reagents also conform to the Marcus expression and the large negative $\Delta S^{\ddagger}$ values for these outer-sphere reactions suggest extensive solvent reorganization, possibly a change in H_2O coordination at the copper center.

Reduction of $[Cu(III)(peptide)]$ complexes by $[Fe(CN)_6]^{4-}$ is markedly affected[68] by cation association with the reductant and kinetic data for the reaction with the neutral $[Cu(III)(H_{-2}Aib_3)]$ allows evaluation of stability constants for the ion pairs $LiFe(CN)_6^{3-}$, $NaFe(CN)_6^{3-}$, $KFe(CN)_6^{3-}$, and $CsFe(CN)_6^{3-}$ of 16, 13.2, 29, and 55 M^{-1}, respectively, at 25°C. Multiple ion association, postulated in other studies, is not detected. The reactions in 0.1 M $NaClO_4$ are around two orders of magnitude faster than predicted by Marcus theory, suggestive of an inner-sphere mechanism involving a cyanide bridge to the axial position of the copper(III) complex. The product

$[Fe(CN)_6]^{3-}$ binds axially to $[Cu(II)(H_{-2}Aib_3)]^-$ with a stability constant of 8.1 M^{-1} and the exchange rate calculated from EPR line broadening, $2 \times 10^8 M^{-1} s^{-1}$, is comparable with the value for exchange of $[IrCl_6]^{2-}$. Relatively small $\Delta S^{\ddagger}$ values contrast with those for ruthenium reduction and suggest that the copper center is solvated after the transition state.

Cross reactions of nickel(III)– and nickel(II)–peptide complexes conform readily to Marcus theory and are thought to be outer sphere but, in contrast to the copper system, give no unique self-exchange rate.[69] Instead values range from 550 $M^{-1} s^{-1}$ for $[Ni(III)/(II)(H_{-2}Aib_3)]^{0/-}$ through $1.3 \times 10^4 M^{-1} s^{-1}$ for tripeptide complexes to $1.2 \times 10^5 M^{-1} s^{-1}$ for tetrapeptides. The former value is consistent with NMR line-broadening experiments where an upper limit of 800 $M^{-1} s^{-1}$ is determined for the rate. Addition of halide or pseudohalide anions enhances the rate by axial binding to the nickel(III) complexes and allowing a bridged, inner-sphere pathway to operate. Ligands such as pyridine which cannot bridge inhibit the reaction.

Endicott and co-workers[70,71] have examined a number of outer-sphere and Cl^- bridged inner-sphere reactions of $[Cu(III)/(II)L]$, $[Ni(III)/(II)L]$, $[Co(III)/(II)L]$, and $[Fe(H_2O)_6]^{3+/2+}$, where L is a macrocyclic or square-planar chelating ligand. Using oxidations of the divalent complexes by $[Co(H_2O)_6]^{3+}$ to define outer sphere rates k_{os} and oxidations by $[Co(H_2O)_5Cl]^{2+}$ to define the Cl^- bridged inner-sphere rates, the inner-sphere advantage $\chi = (k_{is}/k_{os})_{\Delta G° \to 0}$ is computed by extrapolating both rates to $\Delta G° = 0$, assuming a Marcus-like dependence on the square root of the driving force for both reactions. It is found that the relationship (2) holds

$$\chi^{(A,B)} \sim (\chi^{(A,A)} \cdot \chi^{(B,B)})^{1/2} \qquad (2)$$

for reagents A and B and that there is a strong dependence on the metal. The ordering $\chi^{(Co,Co)} \geq \chi^{(Co,Ni)} > \chi^{(Ni,Ni)} > \chi^{(Co,Cu)} \gg \chi^{(Cu,Cu)}$ is best explained using a three-center bond model and reflects two important features. First, the inner-sphere pathway will be advantageous only if the binding strongly mixes donor and acceptor orbitals. This is not the case with Cu(III)/(II) ($d_{x^2-y^2}$) and Co(III)/(II) or Ni(III)/(II) (d_{z^2}). Second, more than four electrons in the three-center bond populates an antibonding orbital thereby destabilizing the bridging transition state.

Outer-sphere reductions of $[Ni(bipy)_3]^{3+}$, $[Ni(phen)_3]^{3+}$, and their substituted derivatives show[72] good agreement with Marcus theory and yield a self-exchange rate of $1.5 \times 10^3 M^{-1} s^{-1}$ for the nickel(III)/(II) couple at 25°C and 1.0 M ionic strength. Reductions by $[Fe(H_2O)_6]^{2+}$ are anomalous, thought to be due to differences in adiabicity between self-exchange and cross reactions. Exchange rates in a series of $[M(bipy)_3]^{n+/m+}$ complexes correlate with inner-sphere reorganization energies. The rate-determining step in the oxidation of the nickel(II) complex, $[Ni(II)LH_2]^{2+}$, to nickel(IV) by Co_{aq}^{3+} [73] and a series of substituted iron(III) polypyridine ions[74] is the formation of a nickel(III) intermediate. The reactions are outer sphere in nature and obey the Marcus relationship, which allows calculation of the $[Ni(III)/(II)LH_2]^{3+/2+}$ potential as 1.23 V. Reductions of $[Ni(IV)L]^{2+}$ by Fe_{aq}^{2+}, VO_{aq}^{2+}, and $[Ni([14]aneN_4)]^{2+}$ have been investigated[75] in aqueous acidic media. The rate-determining step is the outer-sphere electron transfer of an electron to form the nickel(III) intermediate which undergoes further rapid reduction, and Marcus considerations give a self-exchange rate for $[Ni(IV)/(III)L]^{2+/+}$ of $6 \times 10^4 \, M^{-1} \, s^{-1}$.

This self-exchange rate is in reasonable agreement with the value derived[76] from the $[Co(phen)_3]^{2+}$ reduction. Biphasic kinetics were noted, corresponding to a rapid one-electron reduction of $[Ni(IV)L]^{2+}$ followed by slower reaction of $[Ni(III)L]^+$. The kinetics of the nickel(III) reaction are complicated by protonation to give $[Ni(III)LH]^{2+}$ with a pK_a of 4.05. This protonation enhances the reactivity of the nickel(III) intermediate in accordance with the increase in reduction potential. In the corresponding reaction of $[Co(edta)]^{2-}$, similar biphasic behavior is observed[77] but $[Ni(III)LH]^{2+}$ is the only nickel(III) species which is directly reduced by $[Co(edta)]^{2-}$. Instead, $[Ni(III)L]^+$ decomposes by a disproportionation mechanism driven by nickel(IV) reduction. Considerable differences in the interactions of the similarly charged $[Ni(IV)L]^{2+}$ and $[Ni(III)LH]^{2+}$ with $[Co(edta)]^{2-}$ are detected. Limiting kinetics with the nickel(III) complex are ascribed to a hydrogen-bonded precursor which cannot form with the unprotonated nickel(IV) complex. Crystal-structure data[78] on $[Ni(IV)L]^{2+}$ and $[Ni(II)LH_2]^{2+}$ and a single-crystal EPR study[79] of $[Ni(III)L]^+$ give structural information on all three oxidation states and should allow some rationalization of the electron transfer data.

Other studies involving nickel(IV) complexes have been published. Schlemper and Murman[80] prefer a nickel(II)–diradical description of a formal nickel(IV)–oxime complex. Panda and co-workers have examined substitution properties of the inert $[Ni(dmg)_3]^{2-}$ ion[81] and the kinetics of reduction[82] by $[Fe(CN)_6]^{4-}$. The product is $[Ni(dmgH)_2]$ and the mechanism involves rate-determining outer-sphere one-electron transfer from $[Fe(CN)_6]^{4-}$ to $[Ni(dmg)_3]^{2-}$ and its protonated forms with formation of a nickel(III) intermediate which subsequently is reduced in a fast step.

A number of other studies of the stabilization and reactions of higher-oxidation-state metal ion complexes have been reported.[82–88]

2.8. Copper(II)/(I)

Substitution inert ruthenium(III) complexes have been used[89] to force an outer-sphere mechanism in the oxidation of Cu_{aq}^+. Good adherence to Marcus theory allows evaluation of a self-exchange rate for $Cu_{aq}^{2+/+}$ of

$1.9 \times 10^{-4} M^{-1} s^{-1}$, reflecting the structural alteration in this system. However, in the oxidation reactions of the corresponding ruthenium(II) complexes by $[Cu(dmp)_2]^{2+}$, where dmp is 2,9-dimethyl-1,10-phenanthroline, the rates are somewhat insensitive to a thermodynamic driving force. A value of $4 \times 10^4 M^{-1} s^{-1}$ is estimated for the self-exchange rate for the $[Cu(dmp)_2]^{2+/+}$ couple revealing a smaller reorganizational requirement.

Cobalt(III) complexes have been used to oxidize $[Cu(phen)_2]^+$ in aqueous and micellar solutions.[90] Lee and Anson have calculated[91] reorganizational energies for $[Cu(phen)_2]^{+/2+}$ from the cross reaction[92] with $[Co(edta)]^-$ and for $[Cu(phen)_2]^{2+/+}$ from the cross reaction[93] with cytc(II) and estimate a self-exchange rate of $1.55 \times 10^4 M^{-1} s^{-1}$. This value differs from previous estimates because it does not assume equal contributions from the couple to reorganizational energies in the self-exchange and cross reactions.

2.9. Ruthenium(II)

Reductions of a number of $[(NH_3)_5CoX]^{2+}$ complexes, where X is a halide by $[Ru(en)_3]^{2+}$, $[Ru(NH_3)_6]^{2+}$, and $[Ru(NH_3)_5H_2O]^{2+}$, have been reported.[94] For the first two reductants, an outer-sphere mechanism is proposed based on log–log plots since redox potential data for the oxidants are not available, although some advances in this area are being made.[95] For $[Ru(NH_3)_5H_2O]^{2+}$ however, reductions of $[Co(NH_3)_5F]^{2+}$ and $[Co(NH_3)_5Cl]^{2+}$ are inner sphere involving substitution at the ruthenium center. Crystal-structure data for $[Ru(H_2O)_6]^{3+}$ and $[Ru(H_2O)_6]^{2+}$ show[96] average Ru–O bond lengths of 2.029 and 2.122 Å, respectively, rationalizing the lower self-exchange rate, $60 M^{-1} s^{-1}$, for this couple compared with $[Ru(NH_3)_6]^{3+/2+}$ where inner-sphere rearrangement is much smaller. A detailed study[97] of the $[Ru(NH_3)_6]^{2+}$ reduction of $[Fe(edta)]^-$ includes ionic-strength dependencies which allow extrapolation to $\mu = 0$. Under these conditions, work-term contributions to the activation parameters are substantial and were estimated from a Debye–Hückel model.

The oxidation of ruthenium red, $[(NH_3)_5RuORu(NH_3)_4ORu(NH_3)_5]^{6+}$, by iron(III) in aqueous perchlorate media has been investigated.[98] Major pathways involve Fe_{aq}^{3+} and $FeOH_{aq}^{2+}$ as oxidants.

2.10. Cobalt(I) and Rhodium(I)/(II)

Square-planar rhodium(I) isocyanide complexes $[Rh(RNC)_4]^+$ are prone to oligomerization in acetonitrile solution.[9] Oxidations of the monomeric form by a series of $[Fe(phen)_3]^{3+}$ and $[Fe(bipy)_3]^{3+}$ derivatives and $[Co(bipy)_3]^{3+}$ have been examined and proceed through precursor complexes of the type [Fe(III), Rh(I)], [Fe(III)$_2$, Rh(I)], and [Fe(III), Rh(I)$_2$]. The rate constants for reaction of the monomer are in reasonable accord with Marcus theory and increased reactivity on oligomerization is correlated with the energy of the HOMO.

The rhodium(II) acetate dimer shows[100] good Marcus behavior in reactions with metal ion oxidants and reductants. This allows evaluation of a self-exchange rate around 16 $M^{-1}s^{-1}$ at 25°C and 1.0 M HClO$_4$ for $[Rh_2(OAc)_4]^{+/0}$. Dimeric species have also been noted[101] in a study of $[Rh(I)/(bipy)_2]^+$ in aqueous solutions. Pulse radiolysis of $[Rh(bipy)_3]^{3+}$ has been used[102] to generate $[Rh(bipy)_3]^{2+}$, which releases bipy to form $[Rh(bipy)_2(OH_2)_2]^{2+}$. Dimerization of $[Rh(bipy)_2(OH_2)_2]^{2+}$ is rapid and thermodynamically favored and the dimer $[Rh(bipy)_2]_2$ slowly disproportionates with a rate constant of $9 \times 10^{-3}\,s^{-1}$ to give Rh(I) and Rh(III). At higher pH, hydrolysis of $[Rh(bipy)_2(OH_2)_2]^{2+}$ occurs and the major pathway for disproportionation involves reaction of $[Rh(bipy)_3]^{2+}$ with $[Rh(bipy)_2(H_2O)(OH)]^+$.

Reductive quenching of $[*Ru(bipy)_3]^{2+}$ derivatives by $[Rh(4,4'-Me_2bipy)_3]^{3+}$ in aqueous solution is dependent[103] on the thermodynamic driving force for the reaction, and analysis of the data in terms of Marcus theory leads to a self-exchange rate for Rh(III)/(II) of $2 \times 10^9\,M^{-1}s^{-1}$ and $E^0 = 0.97$ V. Quenching by $[Os(III)(NH_3)_5X]^{2+}$ complexes is also in accord with Marcus predictions.

The self-exchange rate for $[Co(bipy)_3]^{2+/+}$ is $10^9\,M^{-1}s^{-1}$, eight orders of magnitude greater than for $[Co(bipy)_3]^{3+/2+}$. This difference in reactivity is due to nuclear configuration changes where differences in Co–N bond lengths are much smaller between Co(II) and Co(I) than between Co(III) and Co(II).[104]

2.11. Cyano Complexes

In perchloric acid solution, oxidation of $[Mo(CN)_8]^{4-}$ by VO_2^+ is biphasic, consisting of parallel cyanide-bridged inner-sphere and outer-sphere pathways.[105] Both pathways are catalyzed by $[H^+]$ which is presumed to coordinate to the molybdenum complex and ionic-strength dependencies indicate that $[HNa_2Mo(CN)_8]^-$ and $[HNaMo(CN)_8]^{2-}$ are the reactive species. Photoexcitation of the intervalence transfer band in the $[Mo(CN)_8]^{4-}$, Cu_{aq}^{2+} ion pair leads to electron transfer and allows determination of the dark reaction rate.[106] The oxidation of MnO_4^{2-} by $[Mo(CN)_8]^{3-}$ has also been examined, in alkaline media.[107] The reaction is outer sphere as are the corresponding reactions of $[W(CN)_8]^{3-}$ and

$[Fe(CN)_6]^{3-}$. Outer-sphere reactions are proposed[108] in the MnO_4^- oxidation of $[Os(CN)_6]^{4-}$ and its protonated forms leading to an estimate for the $[Os(CN)_6]^{3-/4-}$ exchange rate of $1 \times 10^7\ M^{-1}\,s^{-1}$.

Several mechanistic electrochemical studies of metal–cyanide complexes have appeared[109-113] as has a theoretical study of complexes of the type $[M(CN)_6]^{2-}$, where extensive π-back donation is proposed.[114]

2.12. Volumes of Activation

The activation volume $\Delta V^{\ddagger}$ for intramolecular electron transfer in the acid-catalyzed decomposition of $[(NH_3)_5CoOSO_2]^+$ is large and positive, $+34.4\ cm^3\,mol^{-1}$, and may be the result of solvation changes due to charge neutralization.[115] However, a mechanism involving initial cleavage of the *trans*(Co–N) bond cannot be ruled out. In contrast, electron transfer between iron(II) and cobalt(III) in the precursor complex *cis*-$[(en)_2CoCl_2Fe]^{3+}$ in DMSO solution is $-13.6\ cm^3\,mol^{-1}$ and might indicate atom transfer.[116] No conclusive statement on the number of bridging groups can be made from the reaction volume for precursor complexation.

The acid-catalyzed decomposition of μ-superoxo complexes of the type $[Co(III)_2(\mu O_2)(en)_4(NH_3)_2]^{4+}$ involves intramolecular electron transfer with $\Delta V^{\ddagger}$ around $20\ cm^3\,mol^{-1}$. Reductions by $[Fe(CN)_6]^{4-}$ of $[Co(edta)]^-$ and $[Co(NH_3)_5py]^{3+}$ are outer sphere but, in the latter case, detection of an outer-sphere precursor allows evaluation of $\Delta V^{\ddagger}$ for electron transfer of $23.9\ cm^3\,mol^{-1}$ comparable with other values.[117,118] Attempts to rationalize this suggest almost complete electron transfer in the transition state. Similar activation volumes for electron transfer are found[119] in reduction of $[(NH_3)_5CoOH_2]^{3+}$ by $[Fe(CN)_6]^{4-}$, but there is a marked discrepancy in reaction volumes for precursor formation with those of other cobalt(III) oxidants.[117,118] Such differences may be due in part to solvation effects which vary with ionic strength and the necessity of studying ionic strength dependencies of activation volumes for electron transfer reactions between complex ions has been pointed out.[120]

2.13. Stereoselectivity in Electron Transfer

Since the inclusion of a section on stereoselectivity in Volume 1 of this series, there have been a number of significant advances. Saito and co-workers[121] have correlated precursor-complex-formation constants and limiting electron transfer rates with stereoselectivity in the reaction of

Δ',Δ-[(en)$_2$Co(III)-μNH$_2$, μO$_2$-Co(III)(en)$_2$]$^{4+}$ with [Mo(V)$_2$O$_4$(R, S-pdta)]$^{2-}$. They conclude that both steps contribute to the stereoselective course of the reaction. It is noteworthy[122] also that Δ-[Co(III)(en)$_3$]$^{3+}$ associates preferentially with Λ-[Co(III)(edta)]$^-$ but that Δ-[Co(III)(en)$_3$]$^{3+}$ is formed preferentially[123] in the oxidation of [Co(II)(en)$_3$]$^{2+}$ by Δ-[Co(III)(edta)]$^-$. Again, stereoselective contributions from precursor complexation and electron transfer steps are established. Sargeson and co-workers[48] note that stereoselectivities as high as 10% are suggested in reactions of cobalt clathrochelates. These are in accordance with other values.

The chiral oxime–imine ligand $(+)$-(S)-Me$_2$LH$_2$ stereospecifically forms[77] a sexidentate complex with nickel(II), [Ni(II)(S)-Me$_2$LH$_2$]$^{2+}$, which can be oxidized to [Ni(IV)(S)-Me$_2$L]$^{2+}$ and [Ni(III)(S)-Me$_2$LH]$^{2+}$ producing 10–11% excess Δ-[Co(III)(edta)]$^-$ from [Co(II)(edta)]$^{2-}$ despite markedly different redox mechanisms. The nickel(III) complex forms a well-defined hydrogen-bonded precursor complex with [Co(II)(edta)]$^{2-}$, while nickel(IV) has no bridging proton. It is suggested that stereoselectivity is determined by the overall chirality of the reagent rather than the strength of the precursor interaction.

2.14. Miscellaneous Reactions

A review of the aquoions of molybdenum has been published.[124] Oxidations of [Mo(H$_2$O)$_6$]$^{3+}$ by [IrCl$_6$]$^{2-}$, [Co(ox)$_3$]$^{3-}$, and VO^{2+} have been examined[125] and show a variety of terms in their rate laws. For [IrCl$_6$]$^{2-}$, reactions are faster than substitution processes and an inverse [H$^+$] dependence is consistent with outer-sphere reactions, whereas the two other reactions are inner sphere and are complicated by oligomerization processes.

The permanganate ion oxidizes[126] the μ-oxo-molybdenum(V) dimer, [Mo$_2$O$_4$]$^{2+}$, to molybdenum(VI) and the rate-determining step involves one-electron transfer to form a mixed-valence species. An inner-sphere mechanism is suggested by the entropy of activation. The excited state of the molybdenum(II) cluster, [Mo$_6$Cl$_{14}$]$^{2-}$, is an excellent reductant which undergoes rapid reactions with no major structural change.[127] Electron transfer quenching by the heteropoly complexes [SiW$_{12}$O$_{40}$]$^{4-}$ and [PW$_{12}$O$_4$]$^{3-}$ and other reagents has been examined.

Oxidations of *cis*- and *trans*-[Pt(NH$_2$R)$_2$Cl$_2$] complexes by [AuCl$_4$]$^-$ in acetonitrile solution are catalyzed by Cl$^-$ suggesting an inner-sphere mechanism.[128] Increasing the bulk of the alkyl substituent R decreases the rate, particularly for the *cis* complexes for which an additional pathway is detected which is attributed to preequilibrium displacement of Cl^- by solvent in the platinum complex.

Disproportionation of the chromium(V) complex $[OCr(O_2CCOR_1-R_2)_2]$, where the alkyl substituents R_1 and R_2 are methyl or ethyl, to give chromium(VI) and chromium(III) is catalyzed[129] by cerium(III) in acetate media. The rate law suggests a mechanism in which dissociation of a ligand on chromium(V) precedes inner-sphere reduction by $[Ce(OAc)_2]^+$. Further partial chelate-ring opening in the chromium(IV) produced is required to explain the ultimate chromium(III) product.

2.15. Reactions of Metalloproteins

A novel development has been the publication of intramolecular electron transfer data on semisynthetic metalloproteins where a $(NH_3)_5Ru(II)$–chromophore has been attached to a histidine residue on the protein surface. Gray and co-workers[12] have prepared a cytochrome c(III) derivative with ruthenium(III) bound at histidine 33, 12.1 Å from the iron(III) center. The reagent $[*Ru(bipy)_3]^{2+}$ reacts faster $(6.6 \times 10^8 \ M^{-1} s^{-1})$ with the bound ruthenium than with the iron center $(1.2 \times 10^8 \ M^{-1} s^{-1})$ allowing observation of the Ru(II)–Fe(III) intramolecular electron transfer rate of $20 \pm 5 \ s^{-1}$. This rate is almost independent of temperature over a range of 0–37°C suggesting that reorganization at the heme c center is small and that electron tunneling may be important. Using the same approach[13] but with CO_2^- as a reductant, the rate is determined to be $82 \pm 20 \ s^{-1}$. Gray and co-workers[14] have also prepared *Pseudomonas aeruginosa* azurin with the ruthenium chromophore bound 11.8 Å from the copper center at histidine 83. In this case the rate is slower, $1.9 \ s^{-1}$, but again it is almost independent of temperature in the range -10 to 55°C. It is noteworthy that these unimolecular rates are comparable in magnitude to limiting rates detected in reactions where protein–complex association has been proposed.

Reactions of reduced plastocyanin and azurin with the negatively charged oxidants $[Fe(CN)_6]^{3-}$ and $[(NC)_5FeCNCo(CN)_5]^{5-}$ show[130] no evidence for strong protein–complex binding contrary to earlier reports.[13] This is in accord with electrostatic considerations. However, a number of positively charged redox-inactive complexes have been shown[132] to competitively inhibit the oxidation of parsley plastocyanin by $[Co(phen)_3]^{3+}$, and the results are consistent with a local charge model[138] for the interaction of the protein and the complexes.

The reaction of $[Ru(NH_3)_5py]^{2+}$ with oxidized parsley plastocyanin is markedly inhibited[134] by increasing acidity below pK_a 5.0 and by the addition of redox-inactive $[Pt(NH_3)_6]^{4+}$. This suggests that the reagent binds at the so-called "negative patch" around tyrosine 83 used by positively charged oxidants on the reduced protein. This site has also been shown[135] by pH, competitive inhibition, and chromium(III)-protein modification studies to be used by cytochrome f, a negatively charged protein thought to be the natural redox partner for plastocyanin. In contrast[136] with french bean, plastocyanin chromium(III) binding studies have an effect on the rate of photoreduction of P700, suggesting that this supposed natural partner and not cytochrome f occupies the negative patch. Other studies[137] reveal little effect of modification of the negative patch by ethylenediamine groups on the P700 interaction, suggesting that another site on plastocyanin is used.

Pecht and co-workers[138] have labeled *Pseudomonas aeruginosa* azurin with chromium(III) bound near histidine 35 and find, that compared with the native protein, the modified protein reacts more slowly with one of its physiological partners cyt c_{551} but that reaction with cytochrome oxidase, the other partner, is unaffected. Thus two electron transfer sites are available to the protein. Further studies of the azurin–cyt c_{551} interaction have been published.[139] It has been noted[140] that limiting kinetic behavior is also consistent with a dead-end-type mechanism. Application of relative Marcus theory to the oxidation of plastocyanin by $[Co(phen)_3]^{3+}$, where limiting kinetics are detected, and by $[Co(dipic)_2]^-$, where they are not, reveals no particular advantage from the binding as evidenced by the apparent self-exchange rates evaluated for the protein. Thus a dead-end mechanism is plausible and should not be ignored.

No evidence for protein–complex formation is noted[141] in reactions of stellacyanin with inorganic complexes though pH effects tentatively suggest that different protein sites are used for electron transfer with differently charged reagents. Stellacyanin alone, among the blue proteins, appears to conform to Marcus theory and the evaluated self-exchange rates are in excellent agreement with the value of $1 \times 10^5 \, M^{-1} s^{-1}$ measured by EPR experiments.[142] A number of relevant abstracts are to be found in the Proceedings of the First International Conference on Bioinorganic Chemistry.[143]

Hill and Walton have examined[144] the interaction of horse heart cytochrome c with azurin and cyt c_{551} using an electrochemical technique and a detailed theoretical study of cytochrome c has been published.[145] The interaction with cytochrome b_5[146] and redox potentials for this latter protein have also been studied.[147] The redox-inactive $[Cr(en)_3]^{3+}$ inhibits oxidation of cyt b_5 with $[Co(NH_3)_6]^{3+}$ and other positively charged reagents by blocking the heme edge site used by cyt c.[148] Electron transfer reactions of iron-sulfur proteins have been reviewed[2] and NMR studies[149] have been used to pinpoint reaction sites with inorganic complexes on the two iron ferredoxin from *Anabena variabilis* at residues 25 and 83.

A number of electron transfer studies have been carried out on proteins which have no biological electron transfer function. Thus redox potential measurements[150] have enabled a detailed examination[151] of the oxidation of deoxy and semi-met$_R$ forms of hemerythrin by iron(III)–cyano complexes. The results show good, if fortuitous, adherence to Marcus theory. Reduction of the two copper centers of bovine superoxide dismutase by $[Fe(CN)_6]^{4-}$ is a single kinetic process[152] and ionic-strength effects suggest that there is a direct interaction with the active site. Hoffman and co-workers[153] have substituted a zinc(II) porphyrin into a heme site in hemoglobin and have measured the rate of electron-transfer quenching of the triplet by a heme group 25 Å away as $60 \, s^{-1}$, substantial over such a long distance.

Chapter 3

Metal–Ligand Redox Reactions

3.1. Introduction

This chapter covers the literature during the period July 1982–December 1983. Its subject is broadly defined to include nonmetallic substrates of all types, irrespective of whether they are coordinated prior to or during reaction, or not at all. Like substrates are grouped together by their "central" or characteristic element. Many reactions between ligands and metals are noncomplementary, proceeding via free radicals—inorganic or organic. Studies limited to traditional determinations of reaction rates provide relatively few hard facts concerning such intermediates. Direct reports concerning free radicals are given in Sections 3.7 and 3.12.

Topics in this area which have been reviewed include reactions in aprotic solvents between Cu(I) complexes and molecular oxygen with an emphasis on the catalytic activation of O_2;[1] reactions examined by pulse radiolysis;[2] solvent effects on different reaction types, including redox processes;[3] redox and other reactions of metal complexes with sulfur-donor ligands;[4] and reactions relating to nitrogen fixation and nitrogenase.[5]

Topics specifically relating to free-radical reactivity include accounts of $[(H_2O)_5CrR]^{2+}$ complexes,[6,7] which can serve as "storage depots" for aliphatic free radicals. Superoxide radical anion and its chemical reactivity in numerous environments have also been reviewed.[8,9] An extensive compilation of kinetic data for aliphatic radicals has appeared in the Notre Dame series,[10] and the same group has published the UV-visible spectra of inorganic free radicals.[11]

Outer-sphere reactions of metal complexes with organic[12] and inorganic[13] substrates have been reviewed. The roles of free-radical intermediates (I^-, I_2^-, N_3^-, NO_2^-, etc.) and some calculated (Marcus equation) self-exchange rates and electrode potentials for selected free radicals are tabulated. The free-radical-induced chain reaction between $S_2O_8^{2-}$ and H_2 has been reviewed, including the inhibiting effects of such metal ions as Fe^{3+}, Ag^+, Cu^{2+}, Pb^{2+}, and CrO_4^{2-}.[14] Other compilations of kinetic data include two reviews of particular interest in water treatment and purification—one covering reactions of ozone particularly with organic reagents[15] and the other covering reactions of chlorine dioxide with inorganic and organic substrates.[16]

3.2. Nitrogen Compounds and Oxoanions

3.2.1. Nitrate

The autocatalytic reaction between iron(II) and nitrate ions exhibits bistability in flow reactors,[17] although oscillations are not sustained. Some key values (*M*, sec units at 25°C) are given in equations (1)–(6):

$$Fe^{2+} + NO_3^- + 2H^+ \rightarrow Fe^{3+} + NO_2 + H_2O \tag{1}$$

$$Rate_1 = 1.0 \times 10^{-7}[A][B] - 1.4 \times 10^{-5}[C][D][H^+]^{-2} \tag{2}$$

$$Fe^{2+} + NO_2 + H^+ \rightarrow Fe^{3+} + HNO_2 \tag{3}$$

$$Rate_3 = 3.1 \times 10^4[A][B] - 6.5 \times 10^{-4}[C][D][H^+]^{-1} \tag{4}$$

$$Fe^{2+} + HNO_2 + H^+ \rightarrow Fe^{3+} + NO + H_2O \tag{5}$$

$$Rate_5 = (7.8 \times 10^{-3} + 2.3 \times 10^{-1}[H^+])[A][B] - (5.6 \times 10^{-4}[H^+]^{-1}$$
$$+ 1.6 \times 10^{-2})[C][D](+\text{other terms}) \tag{6}$$

Vanadyl ion, VO^{2+}, is oxidized rapidly by NO_3^- in 5.8 *M* sodium nitrate at 80°C only after a long induction period,[18] during which neither NO nor NO_2^- is formed. Addition of VO_2^+ shortens the induction period, and NO_3^- catalyzes the oxidation of VO^{2+} by O_2. Nitrate ions are reduced to NH_4^+ by vitamin $B_{12's}$, the Co(I) state of B_{12}, at pH 1.5–2.5, at a rate $k[B_{12s}][NO_3^-][H^+]$ with $k = 2.1 \times 10^4 \, M^{-2} \, s^{-1}$ (25°C, $\mu = 0.11 \, M$).[19] An initial transformation to nitrite is proposed, although a clear distinction between the bimolecular paths $[Co(I)]^- + HNO_3$ *vs.* $[HCo(III)]^- + NO_3^-$ is not possible.

3.2.2. Hydroxylamine and Hydrazine

The chromate(V) metal–oxo complex $[L_2CrO]^-$ (L^{2-} = 2-ethyl-2-hydroxybutyrato) converts hydroxylamine to Cr–NO complexes[20] by prior loss of LH^-, coordination of NH_2OH, and a pair of two-electron reactions which occur in rapid succession, leading ultimately to the (formally) Cr(I)–NO^+ complexes, $[(LH)_2Cr(OH)(NO)]^-$ and $[LCr(OH)(H_2O)(NO)]$.

Hydroxylamine[21] and hydrazine [22] react with $[AuCl_4]^-$ in aqueous HCl according to equations (7)–(10):

$$Au(III) + 2NH_3OH^+ \rightarrow Au(I) + N_2 + 2H_2O + 4H^+ \qquad (7)$$

$$-d[Au(III)]/dt = 4.15 \times 10^{-4}[Au(III)][H_3NOH^+][H^+]^{-1} \quad (30°C) \quad (8)$$

$$Au(III) + 2N_2H_5^+ \rightarrow Au(I) + 2NH_4^+ + N_2 + 2H^+ \qquad (9)$$

$$-d[Au(III)]/dt = 2.39 \times 10^{-4}[Au(III)][N_2H_5^+]$$

$$\times [H^+]^{-1}/(1 + K_a[H^+]^{-1}) \quad (25°C) \quad (10)$$

In both cases it is proposed (albeit on evidence we consider to be scant) that the reactions proceed by single-electron transfer, forming Au(II) and radicals (e.g., $[AuCl_4]^- + NH_3OH^+ \rightarrow Au(II) + H_2NO\cdot + 2H^+$), followed by further rapid (but unverified) steps. The acid ionization of $HAuCl_4$ (K_a = 1.01 M) was invoked in the second of these reactions, studied over the range 0.20–2.00 M H^+, whereas in the first, the same authors treat it over a range of 0.1–1.00 M H^+ as a more one-sided equilibrium, with $HAuCl_4$ the predominant but unreactive form.

The initial rates of oxidation of $N_2H_5^+$ by Fe^{3+} in aqueous $HClO_4$ are reproducible only for a given stock solution of iron(III) perchlorate (otherwise scatter of 60–100% is noted).[23] In light of this problem, and also because the rate constants tabulated and the data depicted disagree numerically, it will not be considered further. The oxidation of NH_2OH by the peroxoacid H_3PO_5 is catalyzed by Fe^{3+}.[24]

The Mn(III) complex $[Mn(cdta)H_2O]^-$, with H_4cdta = *trans*-1,2-cyclohexanetetra-acetic acid, oxidizes hydrazine by a rate law in which the major term is $k[Mn(III)][N_2H_5^+][H^+]^{-1}$, whereas its oxidation of hydroxylamine is kinetically more complex and shows a quadratic dependence on hydroxylamine and a more complex dependence on $[H^+]$.[25] Still unresolved are the contradictory findings about the products, rates, and rate laws reported in two nearly simultaneous investigations of the oxidation of $N_2H_5^+$ by $[IrCl_6]^{2-}$.[26,27]

3.2.3. Nitrite

The oxidations of nitrite to nitrate by $[IrCl_6]^{2-}$, $[IrBr_6]^{2-}$, and $[Fe(bipy)_3]^{3+}$ all occur by way of outer-sphere electron transfer, forming NO_2 as an intermediate.[13] The respective second-order rate constants are 9.8, 2.8×10^1, and $3.3 \times 10^4\ M^{-1} s^{-1}$ (25°C). These data are used to infer (Marcus equation) an effective rate constant of $\sim 1 \times 10^{-2}\ M^{-1} s^{-1}$ for the NO_2^-/NO_2 self-exchange.

The cobalt(III) complexes $1,2,6\text{-}[Co(am)_3(NO_2)_3]$, with $(am)_3 = (NH_3)_3$ and diethylenetriamine, undergo intramolecular redox decomposition only after prior aquation of the unique nitrite (that trans to an amine). The first-order rate constants for Co^{2+} formation are given as $f([H^+])$ and T(35–50°C).[28] Several key questions are not addressed, however, such as the details of how Co^{2+} is formed, the species in which the transformed nitrite is released, or what (stable) nitrogen product is obtained; indeed, one wonders why the pathway apparently favored is that in which aquation of one nitrite precedes intramolecular electron transfer to one of the nitrites left behind.

3.2.4. Azide

The oxidation of azide ion by $[IrCl_6]^{2-}$ ($[IrCl_6]^{2-} + N_3^- \rightarrow [IrCl_6]^{3-} + 1.5N_2$) shows product inhibition, suggesting *reversible* oxidation of N_3^- to the radical, $N_3^{\cdot}$, present at steady state:

$$\frac{-d[IrCl_6^{2-}]}{dt} = \frac{k_1[IrCl_6^{2-}][N_3^-]}{1 + (k_{-1}/k_2)[IrCl_6^{3-}]} \qquad (11)$$

$$[IrCl_6]^{2-} + N_3^- \underset{k_{-1}}{\overset{k_1}{\rightleftharpoons}} [IrCl_6]^{3-} + N_3^{\cdot} \qquad (12)$$

$$N_3^{\cdot} \xrightarrow{k_2} N_2 + N^{\cdot}; \qquad 2N^{\cdot} \xrightarrow{fast} N_2 \qquad (13)$$

with $k_1 = 1.06 \times 10^{-1}\ M^{-1} s^{-1}$ and $k_{-1}/k_2 = 1.31 \times 10^3\ M^{-1}$ (25°C).[13] Less extensive data for $[IrBr_6]^{2-}$ and $[Fe(bipy)_3]^{3+}$ give $k_1 = 6.1 \times 10^{-2}$ and $10.8\ M^{-1} s^{-1}$, respectively, and similar mechanisms probably apply in that strong inhibition by $[Fe(bipy)_3]^{2+}$ is noted.

3.2.5. Organic Amines

The tris(dimethylglyoximato)nickelate(IV) ion, $[Ni(dmg)_3]^{2-}$, oxidizes NH_2OH[29] and PhN_2H_3[30] by Cu^{2+}-catalyzed pathways with a zero-order dependence on $[Ni(dmg)_3^{2-}]$, ascribed to rate-limiting internal electron transfer within $Cu(II)$–NH_2OH and $Cu(II)$–PhN_2H_3 precursor complexes. The authors cite the substitution lability of $Cu(II)$ as evidence "that $Cu(II)$–

phenylhydrazine precursor complexes are indeed formed in the redox system",[30] although we note that this really establishes only the *feasibility* that they are, not the *fact* of it. The reaction of PhN_2H_3 with $[Ni(dmg)_3]^{2-}$ also occurs directly.[31] The conclusion that the latter occurs by outer-sphere electron transfer is likely correct, but the reasons cited—that rate saturation does not occur at high $[PhN_2H_3]$ and that an adduct is not manifest spectrophotometrically—are insubstantial, since they attest only to the failure of any intermediate to attain a high concentration.

The oxidation of trialkylamine by $[Fe(CN)_6]^{3-}$ proceeds by parallel electron transfer from the free amine and hydrogen transfer from R_3NH^+, and is inhibited by added $[Fe(CN)_6]^{4-}$, consistent with a reversible first step.[32] In aqueous $HClO_4$, triethanolammonium ions are oxidized by V(V), the principal kinetic term being $k[VO_2^+][HN(C_2H_4OH)_3^+][H_3O^+]^2$ ($k = 2.36 \times 10^{-5} \ M^{-3} \ s^{-1}$ at 60°C).[33] The reaction is more rapid in sulfuric acid owing to an additonal term in $[HSO_4]^-$: $k'[VO_2^+][HN(C_2H_4OH)_3^+][H_3O^+]$-$[HSO_4^-]$ ($k' = 5.4 \times 10^{-5} \ M^{-3} \ s^{-1}$). The oxidation of primary amines by Tl(III) in HOAc involves α-C–H bond rupture.[34] Aliphatic amines are oxidized by a ditelluratocuprate(III) complex in alkaline solution; the reactivity order is $2° > 1° > 3°$.[35]

3.3. Oxygen, Peroxides, and Other Oxygen Compounds

3.3.1. Oxygen

The autooxidation of $[Cu(phen)_2]^+$ follows third-order kinetics[36] as in equation (14), with an inverse dependence of k_{obs} on $[Cu(phen)_2]^{2+}$.

$$- d[Cu(phen)_2^+]/dt = k_{obs}[O_2][Cu(phen)_2^+]^2 \qquad (14)$$

The following mechanism was proposed:.

$$[Cu(phen)_2]^+ + O_2 \rightleftharpoons [Cu(phen)_2]^{2+} + O_2^- \qquad k_1, k_{-1} \qquad (15)$$

$$[Cu(phen)_2]^+ + O_2^- + 2H^+ \longrightarrow [Cu(phen)_2]^{2+} + H_2O_2 \qquad k_2 \qquad (16)$$

$$[Cu(phen)_2]^+ + O_2 \rightleftharpoons [Cu(phen)_2O_2]^+ \qquad (17)$$

$$[Cu(phen)_2O_2]^+ + [Cu(phen)_2]^+ + 2H^+ \longrightarrow 2[Cu(phen)_2]^{2+} + H_2O_2 \qquad (18)$$

In the presence of superoxide dismutase the kinetic order in $[Cu(phen)_2]^+$ changes to 1 owing to the effective blocking of reactions -1 and 2 in equations (15) and (16) yielding $k_1 = 5 \times 10^4 \ M^{-1} \ s^{-1}$ and $k_2 = 2.95 \times 10^8 \ M^{-1} \ s^{-1}$.

The oxidation of CuCl by O_2 in pyridine occurs with a $4:1$ [Cu(I):O_2] stoichiometry and obeys the rate law[37] given in equation (19):

$$d[Cu(II)]/dt = k[CuCl]^3[O_2] \qquad (19)$$

It involves the reaction of CuCl with a μ-peroxo species containing two copper ions. No mechanistic conclusions were reached about the Cu(I)OAc–O_2 reaction, which proceeds with an overall third-order rate law shown in equation (20):

$$d[Cu(II)]/dt = k[Cu(OAc)]^2[O_2] \qquad (20)$$

The oxidation of halo(pyridine)copper(I) complexes by O_2 in CH_2Cl_2 and nitrobenzene also proceeds with an overall $4:1$ stoichiometry, producing tetranuclear oxocopper(II) complexes.[38] The kinetic behavior is governed by the involvement of Cu(I) in a monomer–dimer–tetramer equilibrium.

The oxidation of [Cu(bipy)$_2$]$^+$ with O_2 in CH_3NO_2 and CH_3CN in the absence of free bipy follows the rate-law[39] rate = $k[Cu(I)]^2[O_2]$ with $k =$ 135 and 3.0 $M^{-2}\,s^{-1}$ at 30°C in the respective solvents. The rate decreases with free bipy, owing to a rapid equilibrium between [Cu(bipy)$_2$]$^+$ and [Cu(bipy)]$^+$, with the former the less reactive.

The oxidation of [Co(sep]$^{2+}$ by O_2 occurs by an outer-sphere mechanism[40] with the formation of O_2^-:

$$[Co(sep)]^{2+} + O_2 \rightarrow [Co(sep)]^{3+} + O_2^- \qquad k = 43\ M^{-1}s^{-1} \qquad (21)$$

This has been confirmed[41] by a trapping reaction with Cu^{2+}. The overall rate constant for the oxygenation reaction is halved in the presence of Cu^{2+}, consistent with the scheme in equation (22):

$$O_2^- \begin{cases} \xrightarrow{\text{[Co(sep)]}^{2+},\,H^+} [Co(sep)]^{3+} + H_2O_2 \\ \xrightarrow{Cu^{2+},\,H^+} Cu^+ + O_2 \end{cases} \qquad (22)$$

The oxygenation of [Co(dmgH)$_2$] proceeds in three stages.[42] Initial formation of the μ-peroxocobaloxime(III) takes place rapidly:

$$[Co(dmgH)_2] + O_2 \rightleftharpoons [(dmgH)_2CoO_2] \qquad K = 1400\ M^{-1} \qquad (23)$$

$$[(dmgH)_2CoO_2] + [Co(dmgH)_2] \rightleftharpoons [(dmgH)_2CoO_2Co(dmgH)_2] \qquad (24)$$

The second-order decomposition of the μ-peroxo complex is proposed to be a disproportionation reaction, yielding [(dmgH)$_2$CoO$_2$Co(dmgH)$_2$]$^+$, [Co(dmgH)$_2$(OH)], and O_2. The dramatic inhibition of this step by free dmgH$^-$ was ascribed to the axial ligation of dmgH$^-$ to [(dmgH)$_2$CoO$_2$Co(dmgH)$_2$]. We note that this process alone cannot explain the quantitative data presented. The dependence of the reported second-order rate constant on $[Co(ClO_4)_2]_0$ was not discussed. The third step follows first-order kinetics and is presumably due to the decomposition of $[(dmgH)_2CoO_2Co(dmgH)_2]^+$, $k = 3.56 \times 10^{-2} s^{-1}$. The reaction of $[Co(opd)_2]^{2+}$ (opd = o-phenylenediamine) with O_2[43] takes place by a similar mechanism. The formation of $[Co(opd)_2O_2]^{2+}$, presumed to be kinetically controlled ($k = 47 M^{-1} s^{-1}$), is followed by the capture of a second $[Co(opd)_2]^{2+}$ and subsequent decomposition of $[(Co(opd)_2O_2]^{4+}$.

The reduction of O_2 at graphite electrodes is catalyzed by cofacial dicobalt and related porphyrins adsorbed on the electrode surface.[44] The positioning of the two cobalt centers in the molecule to permit formation of the proposed μ-peroxo complex and the protonation of the μ-peroxo bridge is crucial in the catalytic cycle by which O_2 is reduced to H_2O without release of H_2O_2. A mixed-valence dicobalt cofacial porphyrin, [Co(II), Co(III)], reacts rapidly with oxygen in benzonitrile and CH_2Cl_2 to form the μ-superoxo-dicobalt complex.[45] The [Co(II), Co(II)] analogue reacts with O_2 only very slowly. The initial formation of a superoxo intermediate, [Co(III), Co(III)]–O_2^-, was postulated to account for the high reactivity of the mixed-valence compound.

A dramatic salt effect was observed in the reactions of O_2 with [CpFe^I(arene)] complexes in THF.[46] Different products are formed depending on whether or not $NaPF_6$ is present in solution:

$$[CpFe(I)(\eta^6\text{-}C_6H_5CH_3)]$$

$$+\tfrac{1}{2}O_2 \xrightarrow[-80°C]{THF}
\begin{cases}
\xrightarrow{NaPF_6} [CpFe(II)(\eta^6\text{-}C_6H_5CH_3)]^+PF_6^- + \tfrac{1}{2}Na_2O_2 \\
\\
\longrightarrow [CpFe(II)(\eta^5\text{-}C_6H_5CH_2)] + \tfrac{1}{2}H_2O_2
\end{cases} \quad (25)$$

Electron transfer and the formation of O_2^- in the first step of the proposed mechanism is followed by deprotonation by O_2^- in the cage (no $NaPF_6$) or by ion exchange between the ion pairs $[(CpFe(II)arene)^+O_2^-]$ and $[Na^+PF_6^-]$, followed by precipitation of $[CpFe(II)(arene)]PF_6$ and dismutation of NaO_2.

Reactions of some lacunar and some novel binuclear iron(II) complexes with O_2[47] are only partly reversible and occur with the formation of free O_2^-, as shown by ESR spectroscopy. Reversible one-electron transfer occurs between O_2 and the pentacoordinated Fe(II) complex, followed by the coordination of a base (pyridine or 1-methylimidazole) yielding hexacoordinated Fe(III). The reaction of O_2 with Fe(II) complexes of bis(β-diimine)-containing macrocyclic ligands produces diketone ligands through the intermediacy of HO_2 and hydroperoxide species.[48]

Electrogenerated iron(II) tetrakis(N-methyl-4-pyridyl)porphyrin in aqueous solution reduces O_2 to H_2O, presumably through the intermediacy of H_2O_2[49] [equation (26)]:

$$[Fe(II)] + O_2 \rightarrow [Fe(III)O_2]^- \xrightarrow{H^+,[Fe(II)]} 2[Fe(III)] + H_2O_2 \quad (26)$$

This is completed either by the catlayzed disproportionation of H_2O_2 or a direct reduction of H_2O_2 by the ferrous porphyrin. A similar mechanism was suggested[50] for the reactions of some tridentate Schiff-base manganese(II) complexes with O_2 in aprotic solvents. The proposed formation of the "bare" peroxide (O_2^{2-}) and its subsequent disproportionation is not, however, an attractive possibility.

The kinetics and products of the reaction of phthalocyaninatoiron(II), FePc, with O_2 in DMSO[51] are strongly dependent on [FePc]. Under "high-concentration conditions" ($>1 \times 10^{-4}$ M FePc) the autocatalytic behavior of the system can be reversed by N_2 bubbling or dilution. $[(FePc)_2O]$ is the final product. Under "low-concentration conditions" ($\sim 10^{-5}$ M FePc) an irreversible degradation of the complex takes place. The data can be explained by the formation of an intermediate (presumably [FePcO]), which can either react reversibly with [FePc] to give the μ-oxo compound, or oxidize the solvent.

The autoxidation of binuclear Ru(II) complexes, $[(Ru(NH_3)_5)_2L]$, shows biphasic kinetics,[52] indicating two one-electron steps:

$$[Ru(II)LRu(II)] \xrightarrow{-e^-} [Ru(II)LRu(III)] \xrightarrow{-e^-} [Ru(III)LRu(III)] \quad (27)$$

Inhibition of the second phase by Ru(III) at higher pH's suggests outer-sphere electron transfer to form free superoxide.

$[Ru(II)(Me_2SO)_4Cl_2]$ is an effective catalyst for the oxidation of alkyl sulfides to sulfoxides by molecular oxygen in alcohols.[53] It was proposed that a different catalyst is generated in solution for each type of sulfide as in equation (28):

$$[Ru(Me_2SO)_4Cl_2] + SR_2 \rightleftharpoons [Ru(SR_2)_x(Me_2SO)_{4-x}Cl_2] \quad (28)$$

Kinetics and product analysis data, combined with some labeling studies, were used to deduce the following mechanism:

$$\text{"Ru(II)"} + O_2 \rightarrow \text{"Ru(IV)"} + O_2^{2-} \quad (29)$$

$$SR_2 + H_2O_2 \rightarrow R_2SO + H_2O \quad (30)$$

$$\text{"Ru(IV)"} + R_2CHOH \rightarrow \text{"Ru(II)"} + R_2CO + 2H^+ \quad (31)$$

We presume that the first step produces HO_2^- or H_2O_2, rather than O_2^{2-}.

In a qualitative and not well-documented study[54] of the reaction of $[Mn(II)(acac)]^+$ with O_2, $[Mn(H_2O)_6]^{3+}$ is reported as the manganese-containing product. The reaction photosensitized by use of Rose Bengal is also reported to produce $[Mn(H_2O)_6]^{3+}$. The proposed mechanism for the latter reaction is based on the formation of either $^1O_2$ or $[*Mn(acac)]^+$. The catalytic effect of $[Mn(acac)]^+$ on the autoxidation of SO_3^{2-} is explained by the occurrence of the reaction $Mn^{3+} + SO_3^{2-} \rightarrow SO_3^- + Mn^{2+}$, followed by a radical chain oxidation of SO_3^{2-}.

Reactions of tris(3,5-di-*tert*-butylcatecholato)manganate(IV) and the manganese(III) complex with O_2^- and O_2 in aprotic media[55] produce superoxo and peroxo adducts, ligand-centered radicals, and disproportionation products, as shown by magnetic, spectroscopic, and electrochemical studies.

3.3.2. Hydrogen Peroxide and Hydroperoxides

Light-induced homolysis of some organochromium complexes produces low concentrations of Cr^{2+} used to study the reaction of Cr^{2+} and H_2O_2.[56] The first and rate-determining step of the reaction shown in equation (32),

$$Cr^{2+} + H_2O_2 \rightarrow CrOH^{2+} + OH^{\cdot}(k_{Cr}) \qquad (32)$$

is followed by rapid production of a radical and its capture by Cr^{2+}:

$$OH^{\cdot} \xrightarrow{RH} R^{\cdot} \xrightarrow{Cr^{2+}} CrR^{2+} \qquad (33)$$

The formation of intensely colored CrR^{2+} complexes yields $k_{Cr} = (7.06 \pm 0.04) \times 10^4 \, M^{-1} \, s^{-1}$ (25°C). Identical results were obtained by the stopped-flow technique under different conditions.

The rate constant for the reaction of $[Cu(phen)_2]^+$ with H_2O_2,[36]

$$[Cu(phen)_2]^+ + H_2O_2 \rightarrow [Cu(phen)_2]^{2+} + OH^{\cdot} + OH^- \qquad (34)$$

is reported to be 937 $M^{-1} \, s^{-1}$, although we assign half that value ($k_{Cu} = 4.7 \times 10^2 \, M^{-1} \, s^{-1}$), because the second and much faster reaction in equation (35) must ensue:

$$[Cu(phen)_2]^+ + OH^{\cdot} \xrightarrow{H^+} [Cu(phen)_2]^{2+} + H_2O \qquad (35)$$

The oxidation of Cu(I) with H_2O_2 in aqueous HCl[57] shows second-order kinetics, but with a complicated dependence on Cl^-, consistent with the presence of different copper(I)-Cl^- species in solution. The kinetics of H_2O_2 oxidation of U(IV)[58] in aqueous HCl,

$$U(IV) + H_2O_2 \rightarrow U(VI) + 2H^+ \qquad (36)$$

yields $k_{obs} = 0.45 \, M^{-1} \, s^{-1}$ (1 M H^+ and 26°C). The order with respect to

H^+ is -1.3 in the acidity range 0.3–2.0 M, explained in terms of the reactions of U^{4+}, $[UCl(OH)]^{2+}$, and UO^{2+}.

H–D kinetic isotope effects of 22.1 and 16.7 were observed[59] in the oxidation of H_2O_2 by $[(bipy)_2(py)RuO]^{2+}$ and $[(bipy)_2(py)RuOH]^{2+}$. The mechanism is described in terms of proton-coupled electron transfer with strong electronic interactions between the Ru = O group and the OH group of H_2O_2. Alternatively, long-range proton tunneling may occur between the redox sites.

Diethylenetriaminepentaacetatocobalt(II), $[Co(dtpa)]^{3-}$, reacts with H_2O_2 with an approximate 2:1 stoichiometry, yielding $[Co(dtpa)]^{2-}$ at pH < 5, and a mixture of $[Co(dtpa)]^{2-}$ and a μ-peroxo dimer at pH > 5.[60] An acid-independent and an inverse acid-dependent term in the rate law are attributed to the reactions of H_2O_2 and HO_2^-. No experimental evidence was presented to support the proposed formation of $HO\cdot$ in the reaction. The H_2O_2 oxidation of the binuclear complex, formed initially from $[Co(dtpa)]^{3-}$ and HO_2^-, was proposed at pH $\gg$ 5. The activation parameters for the overall reaction at pH 5 are $\Delta H^{\ddagger} = 74\,kJ\,mol^{-1}$ and $\Delta S^{\ddagger} = -87\,J\,mol^{-1}\,K^{-1}$.

The oxidation of $[Co(cdta)]^{2-}$ ($H_4cdta = trans$-1,2-diaminocyclo-hexane-N,N,N',N'-tetraacetic acid) by H_2O_2 at pH 6–8[61] obeys the rate law in equation (37):

$$d[Co(cdta)^-]/dt = 2\{k_{H_2O_2} + k_{HO_2^-}K_a[H^+]^{-1}\}[Co(cdta)^{2-}][H_2O_2] \quad (37)$$

The ratio $k_{HO_2^-}/k_{H_2O_2} \sim 10^5$ is similar to that for related systems. The cobalt(II)- and manganese(II)-catalyzed reduction of H_2O_2 by tiron[62] was suggested to take place through the formation of ternary complexes.

H_2O_2 and some hydroperoxides (all 3 mM) oxidize $[Fe(bipy)_2]^{2+}$ (0.133 mM) with $k = 4.5 \times 10^{-4}$ (H_2O_2), 4.2×10^{-4} (cumene hydroperoxide), 5.0×10^{-4} (t-BuOOH), and 6.0×10^{-4} (EtOOH) s^{-1}.[63] The reaction of $[Co(dmgH)_2L]$ with $(CH_3)_3COOH$ produces $[CH_3Co(dmgH)_2H_2O]$ in H_2O[64] and $[(CH_3)_3COOCo(dmgH)_2(py)]$ in benzene.[64,65] The tert-butyloxy radical, produced in the rate-determining step in both solvents, shown in equation (38),

$$[Co(dmgH)_2L] + (CH_3)_3COOH \rightarrow [HOCo(dmgH)_2L] + (CH_3)_3CO\cdot \quad (38)$$

undergoes fast β-scission in H_2O, leading to the formation of CH_3 radicals, and ultimately $[CH_3Co(dmgH)_2H_2O]$. Hydrogen abstraction from $(CH_3)_3COOH$ by $(CH_3)_3CO\cdot$ in benzene is responsible for the formation of $[(CH_3)_3COOCo(dmgH)_2py]$. The first step in benzene has $k = (2.5 \pm 0.1) \times 10^3\,M^{-1}\,s^{-1}$. The reaction of $[Co(dmgH)_2L]$ with $(CH_3)_2CH(CH_3)_2COOH$ yields $[(CH_3)_2CHCo(dmgH)_2L]$ in both solvents, consistent with the fast β-scission of $(CH_3)_2CH(CH_3)_2CO$.

3.3.3. Other Oxygen Compounds

The quantum yield of $[Co(II)L(H_2O)_n]^{2+}$ ($L = Me_4[14]tetraeneN_4$) produced by the near-UV photolysis of $[Co(III)L(H_2O)_2]^{3+}$ increases appreciably in the presence of alcohols.[66] The results are interpreted within the context of the radical-pair model with the formation of OH radicals, which can be intercepted by the alcohol component of the solvent cage.

The Co(II)-catalyzed oxidation of H_2O by $[Ru(bipy)_3]^{3+}$ at pH 6.5–7.2 obeys the rate law[67] in equation (39),

$$\frac{-d[Ru(III)]}{dt} = a \frac{[Ru(III)]^2[Co(II)]}{[Ru(II)][H^+]^2} \tag{39}$$

where $a = (4 \pm 1) \times 10^{-10}\ M\ s^{-1}$. Rate-limiting formation of Co(IV) is proposed, followed by its decomposition to Co^{2+} and H_2O_2, and oxidation of H_2O_2 by $[Ru(bipy)_3]^{3+}$:

$$Co(OH)_2 + Ru(III) \rightleftharpoons Co(OH)_2^+ + Ru(II) \tag{40}$$

$$Co(OH)_2^+ + Ru(III) \xrightarrow{\text{slow}} CoO^{2+} + Ru(II) + H_2O \tag{41}$$

$$CoO^{2+} \xrightarrow{H_2O} Co^{2+} + H_2O_2 \tag{42}$$

$$2Ru(III) + H_2O_2 \longrightarrow 2Ru(II) + O_2 + 2H^+ \tag{43}$$

Fast redox reactions of $[Fe(N-N)_3]^{3+}$ (N–N = bipy or phen) in basic solutions produce $[Fe(N-N)_3]^{2+}$ and $[Fe(N-NO)_3]^{2+}$ as primary products.[68] Further reactions of $[Fe(N-NO)_3]^{2+}$ by parallel paths include dissociation to give free N–NO and catalysis by $[Fe(N-N)_2(OH)_2]^+$ leading to O_2. No O_2 is produced when dissociation of $[Fe(N-N)_3]^{3+}$ is suppressed by addition of excess ligand. Alternative mechanistic interpretations of these reactions were also presented.[69]

3.4. Halides, Halogens, and Their Oxoacids

3.4.1. Halogens

The oxidation of peroxovanadium(V) ion, OVO_2^+, by Cl_2 follows a rate law[70] showing kinetic retardation by V(V), $-d[OVO_2^+]/dt = k[Cl_2][OVO_2^+]/[VO_2^+]$ with $k = 2.2 \times 10^{-3}\ s^{-1}$ at 20°C. The rate-limiting step is the reaction of Cl_2 with H_2O_2, where the latter is formed in a rapid equilibrium $OVO_2^+ + H_2O = VO_2^+ + H_2O_2$.

3.4.2. Halides

The kinetics of oxidation of iodide ions have been examined for $[Fe(CN)_6]^{3-}$,[71,72] $[Pt(CN)_4Br(H_2O)/(OH)]^{1-/2-}$,[73] Cu_{aq}^{2+},[74] $[IrCl_6]^{2-}$,[13] $[IrBr_6]^{2-}$,[13] and $[Fe(bipy)_3]^{3+}$.[13] The reaction of ferricyanide ions has been reinvestigated during a study of the platinum-catalyzed reaction at a ring-disk electrode.[71] This study confirms the rate law in equation (44) and provides $k = 5.15 \times 10^{-4} M^{-2} s^{-1}$ ($E_a = 38.4 \text{ kJ mol}^{-1}$) and K = 0.9 (5.0°C, 1.00 M KNO$_3$):

$$\frac{-2d(I_3^-)}{dt} = \frac{k[Fe(CN)_6^{3-}][I^-]^2}{1 + K([Fe(CN)_6^{4-}]/[Fe(CN)_6^{3-}])} \tag{44}$$

The kinetics of oxidation of I^- by the platinum(IV) complex are consistent with parallel reactions of the aquo ($k = 2.0 \times 10^9 M^{-1} s^{-1}$) and hydroxo ($k = 2.8 \times 10^{-2} s^{-1}$) complexes (25°C, $\mu = 0.1 M$, $K_a \sim 0.025 M$).[73] Since the corresponding oxidations of CN^- and $S_2O_3^{2-}$ yield, respectively, BrCN and $BrS_2O_3^-$ as the immediate products, direct bromine transfer, formally as Br^+, is suggested to occur in this case as well:

$$[Pt(CN)_4Br(H_2O/OH)]^{1-/2-} + I^- \rightarrow IBr + [Pt(CN)_4]^{2-} + H_2O/OH^- \tag{45}$$

The rate of reaction between Cu^{2+} and I^- is[74] shown in equation (46),

$$-d[Cu(II)]/dt = k[CuCl_n^{2-n}]^2[I^-]^2 \tag{46}$$

with $k = 5.6 \times 10^4 M^{-3} s^{-1}$ (25°C, $\mu = 3.13 M$; $E_a = 54.8 \text{ kJ mol}^{-1}$, $\Delta S^{\ddagger} = 21.4 \text{ J mol}^{-1} K^{-1}$). The results suggest that reaction (47) is rate limiting:

$$2[CuCl_2I]^- \rightarrow 2[CuCl_2]^- + I_2 \tag{47}$$

The other one-electron reagents follow second-order rate laws,[13] suggesting formation of $I^{\cdot}$ and its subsequent reactions as in the example:

$$[IrCl_6]^{2-} + I^- \xrightarrow{k_1} [IrCl_6]^{3-} + I^{\cdot} \tag{48}$$

$$I^- + I^{\cdot} \underset{}{\overset{K}{\rightleftharpoons}} I_2^- \tag{49}$$

$$[IrCl_6]^{2-} + I_2^- \xrightarrow{k_2} [IrCl_6]^{3-} + I_2 \tag{50}$$

In these cases only the first step is of kinetic importance, the values (25°C, $\mu = 0.1 M$) of k_1 being 4.09×10^2 ($[IrCl_6]^{2-}$), 2.8×10^1 ($[IrBr_6]^{2-}$), and 9.5×10^4 ($[Fe(bipy)]_3^{3+}$) $M^{-1} s^{-1}$.[13] Initial-state and transition-state solvation effects in the $[IrCl_6]^{2-}$ reaction show the k_1 step to be really a composite of association, redox, and dissociation events:[75]

$$[IrCl_6]^{2-} + I^- \rightleftharpoons [IrCl_6^{2-}:I^-] \rightleftharpoons [IrCl_6^{3-}:I^{\cdot}] \rightleftharpoons [IrCl_6]^{3-} + I^{\cdot} \tag{51}$$

3.4.3. Hypohalites

Preliminary results[70] suggest direct oxidation of oxo(per-oxo)vanadium(V) ion by hypochlorous acid, $-d[OVO_2^+]/dt = 6.5 \times 10^{-2}[OVO_2^+][HOCl]$, although this equation does not hold accurately over the entire course of a kinetic run.

3.4.4. Halates

The reduction of IO_3^- to I^- by VO^{2+} in aqueous $HClO_4$ is catlayzed by OsO_4 and $Ru(III)$ (from $RuCl_3$, said to be $[Ru(H_2O)_6]^{3+}$).[76] In both cases the rate is independent of $[IO_3^-]$; in the former case, $-d[IO_3^-]/dt = k_{Os}[VO^{2+}][OsO_4][H^+]^{-1}$, with $k_{Os} = 4 \times 10^{-10}\,s^{-1}$ at 60°C, calculated by the present authors from the data given. Rate-limiting electron transfer occurs first, $VO(OH)^+ + OsO_4 \rightarrow VO_2^+ + OsO_3(OH)$, followed by several more rapid reactions. Catalysis by $Ru(III)$ shows a fractional-order dependence on $[VO^{2+}]$ suggesting extensive preassociation in a binuclear complex which undergoes internal electron transfer.

The oxygen-stable cobalt(II)–cdta complex (H_4cdta = trans-cyclo-hexane-1,2-diaminetetra-acetic acid) is oxidized by bromate ions. The $[H^+]$ dependence suggests a sequence of two steps[77]:

$$[Co(cdta)]^{2-} + H^+ + BrO_3^- \underset{k_{-1}}{\overset{k_1}{\rightleftharpoons}} [Co(cdta)^{2-} \cdot HBrO_3] \quad (52)$$

$$[Co(cdta)^{2-} \cdot HBrO_3] + H^+ \xrightarrow{k_2} [Co(cdta)^{2-} \cdot BrO_2^+] + H_2O \quad (53)$$

These would be followed by release of BrO_2 whose sequential reactions to Br_2 are, in comparison, much faster. We note that the first intermediate might instead be a bromate complex, $[(Hcdta)CoOBrO_2]^{2-}$, which could then release BrO_2 by further reaction with H^+ in a second step.

The substitution of $Mn(III)/(II)$ for $Ce(IV)/(III)$ in the Belousov-Zhabotinsky (B–Z) reaction results in oscillations when carried out in a stirred-flow reactor.[78] The computer modeling of the oscillation is, surprisingly, quite satisfactory when the rate constants for the several redox reactions of Mn^{3+} and Mn^{2+} with BrO_3^-, BrO_2, BrO_2^-, and $HBrO_2$, apparently unknown for manganese, are set at the values for cerium. The redox processes between $[Fe(phen)_3]^{3+/2+}$ and the same bromine species[79] are also related to the B–Z reaction. As usual, these reactions are coupled to the BrO_2 and $HBrO_2$ disproportionations. Other work on the B–Z reaction[80-84] and related[85,86] oscillating systems has appeared.

3.4.5. Perhalates

The oxidation of $[Fe(bipy)_3]^{2+}$ and three related complexes by per-bromate[87] occurs according to a rate law which suggests that ligand dissociation yields the redox-active form:

$$\frac{-d[FeL_3^{2+}]}{dt} = \frac{k_1 k_2 [FeL_3^{2+}][BrO_4^-]}{k_{-1}[L] + k_2[BrO_4^-]} \tag{54}$$

$$[FeL_3]^{2+} \underset{k_{-1}}{\overset{k_1}{\rightleftharpoons}} [FeL_2]^{2+} + L \tag{55}$$

$$[FeL_2]^{2+} + BrO_4^- \xrightarrow{k_2} [FeL_2]^{3+} + BrO_3^- + O^{\cdot -} \tag{56}$$

The inference of $O^{\cdot -}$ (or, if solvent is involved, $HO^{\cdot} + OH^-$) release was based on a nearly constant ratio k_{-1}/k_2 (10–50) for the four complexes ($E^0 = 0.88$–1.25 V); it is not clear to the present authors, however, whether intermediates seemingly less energetic than $O^{\cdot -}$ and $HO^{\cdot}$ (e.g., $[L_2FeOBrO_3]^{2+}$) should be discounted.

Several oxidations by periodate ions ($IO_4^-/H_4IO_6^-$) have been examined, including Cr^{3+} [88] and its edta complex, [89] $[Co(II)(nta)-(H_2O)_2]^-$, [90] and $[Mo(IV)(CN)_8]^{4-}$. [91] The lack of any kinetic inhibition by IO_3^- in the Cr^{3+} oxidation is taken to rule out a two-electron step for the rate-law term $k'[Cr(III)]^2[I(VII)]$, thus supporting a dinuclear intermediate, $[CrOH–IO_4H–OHCr]$. The remaining three reactions follow second-order kinetics, with the following variants: $k_1 = a[H^+]$ (for $[Mo(CN)_8]^{4-}$), $1/k = b + c[H^+] + d[H^+]^2$ (for $[Cr(III)(Hedta)H_2O]$), and $k = e + f[OAc^-]$ (for $[Co(II)(nta)(H_2O)_2]^-$). These, in turn, are interpreted in terms of inner-sphere processes when $I(VI)$ intermediates are involved and their rapid disproportionation is invoked: $2I(VI) \rightarrow IO_4^- + IO_3^-$.

3.5. Sulfur Compounds and Oxoacids

3.5.1. Thiocyanate

The reaction of SCN^- with $[Pt(CN)_4Br(OH)]^{2-}$ yields SO_4^{2-} and OCN^- [73] via BrSCN. It occurs by Br^+ transfer to SCN^- from $[Pt(CN)_4Br(OH_2)]^-$ ($k = 1.42 \times 10^8 \ M^{-1} s^{-1}$) and $[Pt(CN)_4Br(OH)]^{2-}$ ($k = 3.30 \times 10^{-3} \ M^{-1} s^{-1}$, 25°C, $\mu = 0.1 \ M$). One-electron oxidation to $SCN^{\cdot}$ is the initial step of the reaction between SCN^- and $[IrCl_6]^{2-}$ ($k = 4.55 \times 10^{-3} \ M^{-1} s^{-1}$) and $[Fe(bipy)_3]^{3+}$ ($k = 5.5 \ M^{-1} s^{-1}$). [13]

3.5.2. Dithionite

Several cyanide complexes of Co(III) porphyrins are reduced by $S_2O_4^{2-}$. [92] Thus $[NCCo(III)(P)OH_2]$ reacts directly with $S_2O_4^{2-}$ ($k = 0.05$–$0.35 \ M^{-1} s^{-1}$), whereas the bis complexes, $[Co(P)(CN)_2]^-$, react primarily with $SO_2^{\cdot -}$ ($k = 30$–$160 \ M^{-1} s^{-1}$). The former proceeds by an inner-sphere mechanism, whereas $SO_2^{\cdot -}$ reacts outer sphere. Other metalloproteins may react directly with $S_2O_4^{2-}$, but the parallel reaction of SO_2^- is always more favorable.[93] Only SO_2^- reacts with ferrioxamine B, $[Fe(Hdesf)]^+$, with $k = 2.7 \times 10^3$ (pH 5.8, 25°C).[94]

A dinitrohexaazacryptand complex of Co(III), $[Co(dinosar)]^{3+}$, is reduced by SO_2^- to $[Co(dinosar)]^{2+}$ ($k = 3.5 \times 10^5\ M^{-1}\,s^{-1}$ at 25°C).[95] The latter complex reacts further with SO_2^-, $k = 1.6 \times 10^4\ M^{-1}\,s^{-1}$ at 25°C, forming a Co(III) complex in which both nitro groups have been reduced to hydroxylamino groups.

3.5.3. Dimethyl Sulfoxide

Metal–oxo complexes may convert DMSO to $(CH_3)_2SO_2$ by direct oxygen transfer. The reaction of $[NiO(OH)_2]$ proceeds with a second-order rate constant $4 \times 10^5/(1 + 2 \times 10^4[DMSO])$ at 0.82 M H^+, 40°C.[96] Oxygen transfer has also been proposed for the Os(VIII) complex $[OsO_4(OH)_2]^{2-}$.[97]

3.5.4. Peroxodisulfate

The oxidation of $[Co(II)(cdta)]^{2-}$ (H$_4$cdta = *trans*-1,2-diaminecyclo-hexanetetra-acetic acid) by $S_2O_8^-$ has a second-order rate constant of $2.5 \times 10^{-4}\ M^{-1}\,s^{-1}$ (40°C, $\mu = 0.50\ M$; $\Delta H^{\ddagger} = 136$ kJ mol^{-1}, $\Delta S^{\ddagger} = 119$ J mol^{-1} K^{-1}).[61] The value of $\Delta S^{\ddagger}$ contrasts with negative values for related reactions (e.g., $[Co(edta)]^{2-}$). The iminodiacetate Co(II) complex $[Co(ida)_2]^{2-}$ is reported to react with $S_2O_8^{2-}$ by a sequence of two one-electron steps.[98]

The oxo(peroxo)vanadium(V) ion, OVO_2^+, is oxidized by $S_2O_8^{2-}$ in a reaction catalyzed by Ag^+ to yield ultimately VO^{2+}, O_2, and HSO_4^-. The probable mechanism[70] includes oxidation of OVO_2^+ by Ag^{2+}:

$$OVO_2^+ + Ag^{2+} \rightarrow Ag^+ + OVO_2^{2+}; \qquad OVO_2^{2+} \rightarrow VO^{2+} + O_2 \qquad (57)$$

The formation of OVO_2^{2+} is also indicated in the reaction of SO_4F^- with OVO_2^+ ($k = 28\ M^{-1}\,s^{-1}$ at 25°C) through the intermediacy of SO_4^-.[70]

A bimolecular reaction between Fe^{2+} and $S_2O_8^{2-}$ ($\Delta G^{\ddagger} = 14.53$ kcal mol^{-1} at 35°C) is responsible for its catalysis of the oxidation of $[Co(II)(edta)]^{2-}$ by $S_2O_8^{2-}$. The products of the rate-limiting step, Fe^{3+} and SO_4^-, rapidly oxidize the Co(II) complex.[99] The oxidation of $[Fe(phen)_3]^{2+}$ by $S_2O_8^{2-}$ occurs by direct and Ag^+-catalyzed pathways.[100] The respective second-order rate constants are 0.151 ($[Fe(phen)_3]^{2+}$) and 55 $M^{-1}\,s^{-1}$ (Ag^+), most probably at 313 K.

3.5.5. Sulfite

The oxidation of SO_3^{2-} to SO_4^{2-} by $[M(CN)_8]^{3-}$ (M = Mo, W) follows second-order kinetics,[101] but even with allowance for ionic-strength effects, the rate constant increases linearly with the concentration of alkali-metal cations, A^+ (0.05–0.8 M), the effectiveness increasing from Li^+ to Cs^+, in proportion to polarizability. The rate law valid under all conditions is given in equation (58):

$$\frac{-d[M(CN)_8^{3-}]}{dt} = \frac{C_1[A^+] + C_2[H^+]}{K_a + [H^+]}[M(CN)_8^{3-}][SO_3^{2-}] \qquad (58)$$

This is consistent with a unified picture of transition states (intermediates?) in which a monovalent cation, A^+ or H^+, bridges the reactants.

The Marcus equation correlates these two reactions and earlier data for $[Fe(CN)_6]^{3-}$ as well. The sulfite radical anion, SO_3^-, is proposed as a later intermediate, although this was not tested directly. Oxidations of SO_3^{2-} by other metal complexes believed to proceed via SO_3^-, follow second-order kinetics;[13] at 25°C, $k = 5.6 \times 10^4$ ($[IrCl_6]^{2-}$), 3.18×10^5 ($[IrBr_6]^{2-}$), and 2.1×10^8 ($[Fe(bipy)_3]^{3+}$) $M^{-1} s^{-1}$.

The free-radical SO_3^- is evidently avoided by the bromoplatinum(IV) complexes $[Pt(CN)_4Br(H_2O/OH^-)]^{1-/2-}$ which react with SO_3^{2-}/HSO_3^- by direct Br^+ transfer to yield the intermediate $BrSO_3^-$, which then rapidly hydrolyzes.[73] The reaction of vanadium(V) by SO_2-derived species shows a second-order dependence on [V(V)].[102] The principal pathway is shown in equation (59):

$$2[VO_2^+ \cdot SO_4^{2-}] + HSO_3^- \rightleftharpoons [(VO_2SO_4)_2HSO_3^{3-}]$$
$$\xrightarrow{2H^+} 2VOSO_4 + HSO_4^- + H_2O \qquad (59)$$

Intramolecular electron transfer to coordinated sulfite also occurs during the reaction of the O-sulfito complex $[Co(NH_3)_5OSO_2]^+$:

$$2[Co(NH_3)_5OSO_2]^+ \xrightarrow{H^+} 2Co^{2+} + SO_4^{2-} + SO_3^{2-} + 10\,NH_4^+ \qquad (60)$$

This has been shown by ^{17}O NMR to occur without Co–O bond breaking.[103] The intramolecular redox reaction follows first-order kinetics, $k = 1.4 \times 10^{-2} s^{-1}$ at 25°C. The associated value of $\Delta V^{\ddagger}$ is large and positive, $+34\ cm^3\ mol^{-1}$ at 25°C,[104] consistent with rate-limiting formation of SO_3^-. The radical then attacks a second complex, and is converted to SO_4^{2-} by oxygen atom abstraction.[104] The Pt(IV) complex $[Pt(NH_3)_5OSO_2]^{2+}$ reacts with free sulfite to yield an intermediate identified[105] as the S-bonded bis(sulfito) complex cis-$[Pt(NH_3)_4(SO_3)_2]\cdot 2H_2O$. In acidic solution it is converted ($k = 5.6 \times 10^{-4} s^{-1}$ at 25°C) to a planar Pt(II) complex and SO_3, which is hydrolyzed to sulfate:

$$[Pt(NH_3)_4(SO_3)(SO_3H)]^+ \rightarrow [Pt(NH_3)_3SO_3] + SO_3 + NH_4^+ \qquad (61)$$

3.5.6. Thiosulfate

Direct observation of the $BrS_2O_3^-$ produced in the reactions of $[Pt(CN)_4Br(H_2O/OH^-)]^{1-/2-}$ with $S_2O_3^{2-}$ ($k_{H_2O} = 4.6 \times 10^8\ M^{-1}\,s^{-1}$; $k_{OH} = 14.8\ M^{-1}\,s^{-1}$) substantiates the suggestion of direct Br^+ transfer. The intermediate decomposes by parallel reactions with bases,

$$BrS_2O_3^- + OH^- \xrightarrow{\ k\ } [HOS_2O_3^-] + Br^- \tag{62}$$

$$BrS_2O_3^- + S_2O_3^{2-} \xrightarrow{\ k'\ } S_4O_6^{2-} + Br^- \qquad k/k' = 0.031 \tag{63}$$

the first of which accounts for the small yield of SO_4^{2-} observed, since it reacts further with the Pt(IV) reagents.[73] In contrast, one-electron outer-sphere reagents form the intermediate radical anion $S_2O_3^-$, which is rapidly converted to the pseudohalide radical anion $(S_2O_3)_2^{3-}$, and then oxidized to the stable product $S_4O_6^{2-}$.[13] The first step, $S_2O_3^-$ formation, is a second-order reaction; at 25°C, $k = 1.74 \times 10^2$ ($[IrCl_6]^{2-}$), 1.75×10^1 ($[IrBr_6]^{2-}$), and 9.5×10^4 ($[Fe(bipy)_3]^{3+}$) $M^{-1}\,s^{-1}$.

3.5.7. Thiols

The oxidation of *l*-cysteine, mercaptoacetic acid, and so on, by 12-tungstocobaltate(III) ions, $[Co(III)O_4W_{12}O_{36}]^{5-}$, can be represented by the reaction $Co(III) + RSH \rightarrow Co(II) + \frac{1}{2}RSSR + H^+$. The rate laws are[106]

$$-d[Co(III)]/dt = \{k + k'[RSH][H^+]^{-1}\}[Co(III)][RSH] \tag{64}$$

No intermediates were detected. Parallel outer-sphere reactions were proposed, in the second of which an association of the two anions $[Co(III)O_4W_{12}O_{36}]^{5-}$ and RS^- precedes the rate-limiting electron transfer step:

$$Co(III) + RSH \rightarrow Co(II) + RS^. + H^+ \tag{65}$$

$$[Co(III)], RS^- + RSH \rightarrow Co(II) + R\dot{S}SRH \tag{66}$$

Complexes of iron(III) with mercaptocarboxylic acids (HORSH) undergo redox decomposition: $2Fe(OSR)^+ + 2H^+ \rightarrow 2Fe^{2+} + HORSSROH$. The kinetic data[107] suggest that the reactive species is $Fe(ORS)^-$, which yields a dimeric sulfur-centered radical. The latter is in turn oxidized by Fe(III) to the disulfide.

Intramolecular redox is also responsible for the decomposition of the 1:1 cysteine-$[Cu(tmpa)]^{2+}$ complex [tmpa = tris(2-pyridylmethyl)amine]. This is a multistage process with several intermediates assigned as chelates with S,O and S,N bonding as well as the monodentate S-bonded complex.[108]

Glutathione and some related thiols reduce Fe^{3+} to Fe^{2+} via blue intermediate(s) which were detected by fast kinetic techniques.[109]

3.6. Oxoanions of Phosphorus and Arsenic

3.6.1. Hypophosphite

In aqueous HCl, $[AuCl_4]^-$ oxidizes H_3PO_2 to H_3PO_3.[110] The kinetic data are represented by a second-order rate constant of $1.37 \times 10^{-2} \, M^{-1} \, s^{-1}$ (35°C, $E_a = 128$ kJ mol, $\Delta S^{\ddagger} = 136$ J mol^{-1} K^{-1}), which is independent of $[H^+]$ and $[Cl^-]$. The *pseudo*-first-order kinetic plot in experiments using a large stoichiometric excess of H_3PO_2 are *significantly* curved in the initial portions, however, which is not accounted for by the rate law. A mechanism proceeding by way of the reactive intermediates Au(II) and $HP(OH)_2$ radicals is proposed[110] on the basis that the reaction initiates polymerization of acrylamide. The oxidation of H_3PO_2 by Ce(IV) in aqueous $HClO_4$ is catalyzed by Ag^+; various complexes between Ce^{4+} and $H_2PO_2^-$ are the redox-active forms.[111]

3.6.2. Arsenite

Oxidation of As(III) by the silver(III) ion $[Ag(OH)_4]^-$ in alkaline solution occurs by parallel, second-order reactions of the three arsenite species: $H_2AsO_3^-$ ($k = 6.2 \times 10^3 \, M^{-1} \, s^{-1}$), $HAsO_3^{2-}$ ($k = 3.7 \times 10^4 \, M^{-1} \, s^{-1}$), and AsO_3^{3-} ($k = 1.8 \times 10^4 \, M^{-1} \, s^{-1}$). In all three cases a complementary, two-electron mechanism is proposed, with arsenate formation occurring by inner-sphere transfer of a bound OH group.[112]

3.6.3. Peroxodiphosphate

The Ag^+-catalyzed oxidation of Cr^{3+} by $P_2O_8^{4-}$ shows these reaction orders: Ag^+, first; Cr^{3+}, zero; and $P_2O_8^{2-}$, complex.[113]

3.7. Inorganic Radicals

3.7.1. Kinetics

Several studies on the reactivity of superoxide and hydroperoxide, O_2^- and HO_2, towards transition-metal complexes have appeared.[36,41,114–117] A summary of the kinetic data is given in Table 3.1. The data for oxidations by other radicals[118–123] are given in Table 3.2.

Table 3.1. *Kinetic Data for the Reactions of O_2^- and HO_2* [a]

M^n	k_f	k_r	References
Part A: $\quad O_2^- + M^n \underset{k_r}{\overset{k_f}{\rightleftharpoons}} O_2 + M^{n-1}$			
$[Co(sep)]^{3+}$	(~ 1)	43	40,41
$[Fe(III)(edta)]$ [b]	1.9×10^6 [c]		114
	$2 \times 10^6 [H^+]/(10^{-7.6} + [H^+])$	600	115
$[Fe(III)(Hedta)]$ [d]	7.6×10^5 [c]		114
$[Fe(III)(detapac)]$ [e]	$\leq 10^4$ [c]		114
$[Fe(III)(cyt\text{-}c)]$	2.6×10^5 [f]		116
$[Cu(phen)_2]^{2+}$	1.93×10^9	5×10^4	36
Cu^{2+}	8.1×10^9		5
Part B: $\quad HO_2 + M^n \underset{k_r}{\overset{k_f}{\rightleftharpoons}} O_2 + M^{n-1} + H^+$			
$[Ru(bipy)_3]^{3+}$	1.25×10^7		117
$[Co(sep)]^{3+}$	$(\sim 1 \times 10^{-5})$	43	41
Part C: $\quad O_2^- + M^n \underset{k_r}{\overset{k_f}{\rightleftharpoons}} H_2O_2 + M^{n+1}$			
$[Co(sep)]^{2+}$	4.6×10^7		41
$[Cu(phen)_2]^+$	2.95×10^8		36
$[Fe(II)(edta)]$	$\sim 10^6\text{-}10^7$		115

[a] Values given are measured quantities. Values calculated from thermodynamic or competition data
   are given in parentheses.
[b] edta = ethylenediaminetetraacetic acid.
[c] At pH 7.
[d] Hedta = *N*-(2-hydroxyethyl)ethylenediaminetriacetic acid.
[e] detapac = diethylenetriaminepentaacetic acid.
[f] At pH 7.8.

The one-electron oxidation of $Zr(IV)(O_2^{2-})$ by $Ce(IV)$ produces an
ESR-active complex, referred to as $Zr(IV)(O_2^-)$.[124] The latter reacts with
reducing agents according to

$$Zr(IV)(O_2^-) + red \rightarrow Zr(IV)(O_2^{2-}) + ox \qquad (67)$$

The rate constants at an unspecified temperature and $[H^+]$ have values
4×10^4 (Fe(II)), 2.1×10^4 (Ti(III)), 4×10^3 (resorcinol), 3×10^3 (guaiacol),
and 2.5×10^4 (ascorbic acid) $M^{-1} s^{-1}$.

O_2^- reacts with metalloporphyrins[125] with rate constants 5×10^5
[Fe(II)TAP; TAP = tetrakis(4-*N,N,N*-trimethylanilinium)porphyrin],
$\leq 10^5$ [Zn(II)TMpyP; TMpyP = tetrakis(4-*N*-methylpyridine)porphyrin],
3×10^6([Mn(III)TAP]), and $<10^5$([Cu(II)TAP]) $M^{-1}s^{-1}$; Mn(II)–pyro-
phosphate forms an adduct[126] with O_2^-, $k = 1.3 \times 10^7 M^{-1} s^{-1}$.
[Mo(V)O(tpp)Br](tpp = *meso*-tetraphenylporphyrin) reacts with O_2^- in

Table 3.2. Kinetic Data for Oxidations by Inorganic Radicals,
$$X^{\cdot} + M^n \rightarrow X^- + M^{n+1}$$

$X^{\cdot}$	M^n	$k\,(M^{-1}\,s^{-1})$	References
$\cdot ON(SO_3)_2^{2-}$	$[Fe(CN)_6]^{4-}$	0.67	118
$H\dot{O}N(SO_3)_2^-$	$[Fe(CN)_6]^{4-}$	>40	118
$\cdot OH$	$[(H_2O)_5CrCHMeOEt]^{2+}$	$1.6 \times 10^{9\ a}$	119
BrO_2	$[Fe(CN)_6]^{4-}$	1.9×10^9	120
BrO_2	$Mn(II)$	$\sim 1.5 \times 10^6$	120
ClO_2	$[Fe(phen)_3]^{2+}$	4.5×10^4	121
$[Mn(CO)_3L_2]$	$HSnBu_3{}^{\ b}$	$2 \times 10^{-3\ c}$	122
		0.110^d	122
		0.786^e	122
		10.7^f	122
Br_2^-	$[Co(tspc)]^{4-\ g}$	1×10^8	123

^{a} k represents the total rate constant for the loss of the organochromium, which occurs by
two pathways: $\cdot OH + [CrR]^{2+} \rightarrow ROH + Cr^{2+}$ and $\cdot OH + [CrR]^{2+} \rightarrow R\cdot + [CrOH]^{2+}$.
^{b} The reaction is $[Mn(CO)_3L_2] + HSnR_3 \rightarrow [Mn(CO)_3L_2H] + \cdot SnR_3$.
^{c} $L = P(i\text{-}Pr)_3$.
^{d} $L = P(O\text{-}i\text{-}Pr)_3$.
^{e} $L = P(i\text{-}Bu)_3$.
^{f} $L = PBu_3$.
^{g} tspc = tetrasulfophthalocyanine.

$CH_2Cl_2{}^{(127)}$ at $-72°C$ to form O_2 and $[Mo(IV)O(tpp)(O_2^-)]^-$. The latter undergoes $[Mo(IV)\text{-}(O_2^-)] \rightarrow [Mo(V)\text{-}(O_2^{2-})]$ interconversion at higher temperatures.

3.7.2. Detection of Radicals

Reactions of Fe(II) and Fe(III) porphyrins with O_2^- were studied in DMSO by electrochemical and spectroscopic techniques:[128]

$$[Fe(III)(TPP)] + O_2^- \rightarrow [Fe(II)(TPP)(O_2)] \rightarrow [Fe(II)(TPP)] + O_2 \quad (68)$$

$$[Fe(II)(TPP)] + O_2^- \rightarrow [Fe(III)(TPP)(O_2^{2-})] \quad (69)$$

Formation of $ClO_2{}^{(129)}$ in the systems $NaClO_3\text{-}Ti^{3+}$, $NaClO_2\text{-}Ti^{3+}$, and $NaClO_2\text{-}H_2SO_4$ in aqueous solutions was confirmed by ESR measurements. Free CN radicals are produced[130] in the course of the reaction (70),

$$[(PPh_3)_2N]_5[(CN)_5CoO_2Mo(O)(CN)_5] \xrightarrow{O_2} [(PPh_3)_2N]_2[O_2Mo(O)(CN)_4] \quad (70)$$

as demonstrated by a spin-trapping experiment with a nitrone. Also consistent with the involvement of $\cdot CN$ if the formation of $TCNE^-$ when CH_2Cl_2 is used as solvent.

3.8. Hydrogen

The generally accepted homolytic splitting of H_2 by $[Co(CN)_5]^{3-}$ was challenged because the formation of approximately equimolar quantities of $[Co(CN)_5D]^{3-}$ and $[Co(CN)_5H]^{3-}$ in the hydrogenation of $[Co(CN)_5]^{3-}$ in D_2O was taken as evidence for the heterolytic cleavage of H_2:[131]

$$H_2 + 2[Co(CN)_5]^{3-} \xrightarrow{\ D_2O\ } [Co_2(CN)_{10}H]^{7-}$$
$$\xrightarrow{\ D^+\ } [Co(CN)_5H]^{3-} + [Co(CN)_5D]^{3-} \qquad (71)$$

Halpern's reexamination[132] of the published data showed, however, that the $[Co(CN)_5D]^{3-}$ content varies during the reaction, and the high calculated[131] "average" amount is not representative of the hydrogenation reaction, but originates in the secondary exchange reaction (72):

$$[Co(CN)_5H]^{3-} + D_2O \rightarrow [Co(CN)_5D]^{3-} + HDO \qquad (72)$$

The proportion of $[Co(CN)_5D]^{3-}$ produced by the hydrogenation of $[Co(CN)_5]^{3-}$ in D_2O[133] is strongly pD dependent ($\sim 10\%$ at pD 11, $\sim 90\%$ at pD 13.7) owing to the pD dependence of the exchange reaction. The high proportion of $[Co(CN)_5H]^{3-}$ at low pD suggests that H^+ is not liberated in the hydrogenation reaction. This result can be explained[133] by both a homolytic fission and a heterolytic process without liberation of free H^+.

Reduction of $[Co(CN)_5OH]^{3-}$ by $[Co(CN)_5H]^{3-}$ in H_2-containing solutions,[134] as well as the aging of $[Co(CN)_5]^{3-}$, was observed only after the flow of the inert gas (Ar or N_2) was stopped. Thus both processes require catalysis by some unidentified, volatile compound. The equilibrium constant for the formation of $[Co(CN)_5H]^{3-}$,

$$2[Co(CN)_5]^{3-} + H_2 \rightleftharpoons 2[Co(CN)_5H]^{3-} \qquad (73)$$

$$[Co(CN)_5H]^{3-} + [Co(CN)_5OH]^{3-} \rightleftharpoons 2[Co(CN)_5]^{3-} + H_2O \qquad (74)$$

$$[Co(CN)_5]^{2-} + H^- \rightleftharpoons [Co(CN)_5H]^{3-} \qquad K \geq 10^{29}\ M^{-1} \qquad (75)$$

was calculated from the available thermodynamic data.

Isotope exchange between H_2O and D_2 is strongly catalyzed by $[Co(CN)_5H_2O]^{2-}$, but not $[Co(CN)_6]^{3-}$, although it was apparently not studied in inert atmosphere.[135] The following mechanism was proposed:

$$[Co(CN)_5H_2O]^{2-} + D_2 \rightarrow [Co(CN)_5D_2]^{2-} + H_2O \qquad (76)$$

$$[Co(CN)_5D_2]^{2-} \rightleftharpoons [Co(CN)_5D]^{3-} + D^+ \qquad (77)$$

$$2[Co(CN)_5D]^{3-} \rightleftharpoons 2[Co(CN)_5]^{3-} + D_2 \qquad (78)$$

$$[Co(CN)_5D]^{3-} + H_2O \rightleftharpoons [Co(CN)_5H]^{3-} + HDO \qquad (79)$$

V(II)–cys (cys = cysteine) at pH 7.5–8.5 reduces $H_2O^{(136)}$ according to

$$V(II)\text{–cys} + H_2O \rightarrow V(III)\text{–cys} + HO^- + \tfrac{1}{2}H_2 \qquad (80)$$

At 21°C, $k = 2.3 \times 10^{-3}\,s^{-1}$, independent of [cysteine] and pH. Both a direct reaction of V(II)–cys with H_2O and an H^+-catalyzed reaction of hydrolyzed V(II)–cys are consistent with the data.

The reaction of $[Mo_2Cl_8]^{4-}$ with HCl exhibits a first-order dependence on $[Mo_2Cl_8]^{4-}$ and 7.5-order dependence on HCl (5.78–11.85 M).[137] The reversible protonation of the quadruple Mo–Mo bond is followed by a rate-determining rearrangement to the hydrido-bridged product:

$$[Cl_4Mo \equiv MoCl_4]^{4-} + H^+ \rightleftharpoons [Cl_4Mo \equiv Mo(H^+)Cl_4]^{3-}$$
$$\rightarrow [Cl_3Mo(\mu\text{-Cl})_2(\mu\text{-H})MoCl_3]^{3-} \qquad (81)$$

$[Mo_2Cl_8H]^{3-}$ decomposes in HCl solutions ($<3\,M$ HCl) to yield H_2 and $[Mo_2(OH)_2Cl_x]^{(x-4)-}$. Interchange of terminal and bridging OH in the partly hydrolyzed complex in the suggested mechanism is followed by H_2 evolution.

Reductive elimination of H_2 from $[Pt_2H_2(\mu\text{-H})(\mu\text{-dppm})_2]^+$, induced by tertiary phosphine ligands (L), is an intramolecular process involving the intermediate $[Pt_2H_2(\mu\text{-H})L(\mu\text{-dppm})_2]^+$:[138]

$$[Pt_2H_2(\mu\text{-H})(\mu\text{-dppm})_2]^+ + L \rightarrow [Pt_2H_2(\mu\text{-H})L(\mu\text{-dppm})_2]^+$$
$$\rightarrow H_2 + [Pt_2H_2(\mu\text{-dppm})_2]^+ \qquad (82)$$

The activation parameters for the second step in dichloroethane are $\Delta H^\ddagger = 92 \pm 5\,kJ\,mol^{-1}$ and $\Delta S^\ddagger = 45 \pm 10\,JK^{-1}\,mol^{-1}$ (L = PPh_3). The H–D kinetic isotope effect of 3.5 is supportive of the proposed Pt–H bond cleavage in the transition state.

$[(PPh_3)Pd(OAc)_2]_2$ reacts with H_2 to produce $[(PPh_3)_2Pd_2H_2]$ through the intermediacy of $[(PPh_3)_2Pd_2]$.[139] The latter reacts with H_2O_2 to form $[(PPh)_2Pd_5]$, a very effective hydrogenation catalyst.

Absorption of H_2 by solutions of Pd(II) complexes in the presence of H^+ acceptors is proposed to occur by a heterolytic mechanism.[140]

H_2 and $[HCo(CO)_4]$ are both effective in reducing (ethoxycarbonyl)cobalt tetracarbonyl to ethyl formate.[141] The following mechanism was proposed:

$$[EtOCOCo(CO)_4] \underset{k_{-1}}{\overset{k_1}{\rightleftharpoons}} [EtOCOCo(CO)_3] + CO \qquad (83)$$

$$[EtOCOCo(CO)_3] + H_2 \xrightarrow{k_2} EtOCOH + [HCo(CO)_4] \qquad (84)$$

$$[EtOCOCo(CO)_3] + [HCo(CO)_4] \xrightarrow{k_3} EtOCOH + [Co_2(CO)_8] \qquad (85)$$

The kinetic parameters have values $k_1 = 1.73 \times 10^{-3}$, $k_2/k_{-1} = 6.7 \times 10^{-3}$, and $k_3/k_{-1} = 0.084$ (ether, 25°C) s^{-1}.

3.9. Alkyl Halides

The stoichiometry and kinetics of the reactions of $[Co(dmgH)_2L]$ and polyhalomethanes (RX) in acetone and benzene are in accord with a two-step mechanism[142]:

$$[Co(dmgH)_2L] + RX \rightarrow [XCo(dmgH)_2L] + R^{\cdot} \qquad (86)$$

$$[Co(dmgH)_2L] + R^{\cdot} \rightarrow [RCo(dmgH)_2L] \qquad k_{Co} \qquad (87)$$

The existence and reactivity of the free-radical intermediate for the case $R = CCl_3$ were examined by kinetic competition with 4-hydroxy-2,2,6,6-tetramethyl-piperidinyloxy (4-HTMPO). Numerical integration techniques yielded an estimate of the relative rate constants for the reaction of $^{\cdot}CCl_3$ with 4-HTMPO (k_{HTMPO}) vs. k_{Co} in benzene, $k_{HTMPO}/k_{Co} = 1.8 \pm 0.1$ (25°C).

Hydrogen-saturated alkaline solutions of cyanocobaltate with the CN^-:Co ratio of $<5:1$ react with aryl halides to produce σ-arylpentacyanocobaltates(III).[143] Hydrogenolysis and cyanation products were also observed. The reactive species is most likely $[Co(CN)_4]^{3-}$, since no organocobaltates were formed at CN^-:Co ratios $>5:1$.

The rapid reaction of $LiMe_2Cu$ with trityl halides in ether produces trityl radicals,[144] as shown by ESR. The reaction with a cyclizable alkyl halide, 6-iodo-1-heptene, produced the cyclic coupled product. Both results are strongly indicative of the intermediate radical formation by a single-electron transfer pathway.

Cobaloximes(I) and hydridocobaloximes can be arylated with aryl halides with electron-withdrawing substituents,[145] consistent with both electron transfer and S_N2 displacement mechanisms.

Electron transfer reduction of alkyl halides was discussed in terms of Marcus theory as a possible model for irreversible electron transfer.[146] Values of E^0 for the reaction $RX + e^- \rightarrow R + X^-$ were calculated for a series of alkyl halides in a number of solvents.

3.10. Quinols, Catechols, Diols, and Alcohols

3.10.1. Hydroquinone and Catechol

The oxidation of these substances ($H_2Q = 1,4$-dihydroxybenzene and $H_2Cat = 1,2$-dihydroxybenzene) by Ni(III)(cyclam) yields the corresponding quinone Q:[147] $2[NiL]^{3+} + H_2Q \rightarrow 2[NiL]^{2+} + Q + 2H^+$. With due allowance for acid ionizations, second-order rate constants at 25°C are 1.09×10^4 (H_2Q) and 7.0×10^2 (H_2Cat) for $[NiL]^{3+}$, but are much lower for the sulfate complex $[NiL(SO_4)]^+$. Related $[Ni(III)(N_4mac)]^{3+}$ complexes follow a similar pattern.[148] The reactions proceed by an outer-sphere electron transfer and a semiquinone radical. Marcus correlations are presented. A summary is given of much of the earlier kinetic data for oxidations of H_2Q and H_2Cat by metal complexes. Substituted quinoles also are oxidized by the Mn(III) complex $[Mn(cdta)]^-$ ($H_4cdta =$ *trans*-cyclohexane-1,2-diaminetetra-acetic acid) in a sequence of two one-electron steps, the first and rate limiting of which shows kinetic evidence of preassociation.[25] Semiquinone radicals were detected by use of ESR spectroscopy in a reaction which is, in essence, the reverse of these: reduction of quinone by $[Co(CN)_5]^{3-}$ when carried out at $-60°C$ in aprotic solvent.[149] Kinetics studies have also been carried out for oxidations of H_2Q and H_2Cat by alkaline ferricyanide[150,151] and by Hg^{2+} in aqueous $HClO_4$.[152]

3.10.2. Diols

Ethylene glycol, 2,3-butanediol, and pinacol are oxidized by $[IrCl_6]^{2-}$ at measurable rates in acetic acid–sodium acetate solutions to formaldehyde, acetaldehyde, and acetone, respectively.[153] Spectrophotometric evidence for prior complexation was obtained, and the protonation of the intermediate to an unreactive form invoked to explain the failure of the reaction to occur at higher acidities. The dependence of rate on diol concentration, and the observation that acrylamide is polymerized are cited as evidence to support the mechanism in equations (88) and (89):

$$[diol-Ir(IV)]^{2-} \xrightarrow{\ k\ } \text{free radical} + Ir(III) \tag{88}$$

$$\text{free radical} + [IrCl_6]^{2-} \xrightarrow{\ fast\ } 2R_2CO + Ir(III) \tag{89}$$

The oxidation of diols by V(V) is first order with respect to each reagent, and also occurs more rapidly in sulfuric than in perchloric acid, with a first-order dependence on $[HSO_4^-]$.[154] Oxidations by Ce(IV) are also reported.[155,156]

3.10.3. Alcohols

Allyl alcohol is oxidized by aqueous $[PdCl_4]^{2-}$ to β-hydroxypropanol, α-hydroxyacetone, and acrolein ($CH_2 = CHCHO$), or propanol. Deuterium labeling was used to show that acrolein does not result from dehydration of $HOCH_2CHCHO$, but is a direct oxidation product. Most probably the C=C bond as well as the OH group are coordinated to the metal.[157] The oxidation of a mixed alcohol–aldehyde system by Cr(VI) is a three-electron oxidation.[158] The oxidation of cyclic alcohols and α-hydroxy-cycloalkane carboxylic acids by Ce(IV) in aqueous $HClO_4$ has been reported.[159]

3.11. Ascorbic Acid

Ascorbic acid is oxidized by Ni(III) complexes of N_4 macrocycles (cyclam, etc.) by a sequence of two one-electron steps.[148] Both H_2A and HA^- react, and the rate constants for a series of $[NiL_2]^{3+}$ complexes are correlated by the Marcus cross relation. These reactions go by way of the free radical ($H_2A^{+\cdot}/HA^\cdot$), as does the oxidation by V(V) which induces the polymerization of acrylonitrile.[160,161] The oxidations by $[IrCl_6]^{2-}$ and $[IrBr_6]^{2-}$, which are also known to occur by parallel one-electron reactions of H_2A and HA^-, have been examined in (H, Li, Na)ClO_4 media at constant and variable ionic strength. Specific cation effects are important at higher electrolyte concentrations. The results reaffirm $LiClO_4$ as a substitute for $HClO_4$ in studies seeking to vary $[H^+]$ with minimal medium effects.[162]

3.12. Organic Radicals

3.12.1. Reactions

The kinetic data for the reactions of metal complexes with organic radicals[163-176] are summarized in Table 3.3. Arrhenius parameters were reported for H-abstraction reactions by $\cdot CF_3$ and $\cdot C_2F_5$ from CH_3GeCl_3, $(CH_3)_4Ge$, and $(C_2H_5)_4Ge$ in the gas phase.[177] UV-visible spectra of the immediate products of reactions of several C-centered radicals with B_{12r}[178] are indicative of the Co–C bond formation [R = $\cdot CH_2C(CH_3)_2OH$, $\cdot C(CH_3)_2OH$, $\cdot CH_2CHO$, and $\cdot CH(OH)CH_2OH$]. Reaction with $CO_2^{-\cdot}$ produces B_{12s} without any evidence for the intermediate $[Co(III)CO_2]$ formation. Cu(II) effectively oxidizes C-centered, but not O-centered α-diol radicals.[179]

3.12.2. Formation

Thermal homolysis of organometallic complexes results in the formation of C-centered radicals. For $[(H_2O)_5Cr(c-C_5H_9)]^{2+}$ $k_{hom} = 1.07 \times 10^{-4} \, s^{-1}$ at 25°C.[164] The values for $[CrC(CH_3)_2COOH]^{2+}$ and $[CrC(CH_3)_2CN]^{2+}$ are 4 and $10^4-10^6 \, s^{-1}$, respectively.[174] Activation parameters for the homolysis[180] of $[RCo(DO)(DOH)_{pn}]I$ are $\Delta H^\ddagger = 27.9 \pm 0.8 \, kcal \, mol^{-1}$, $\Delta S^\ddagger = 8 \pm 2 \, cal \, mol^{-1} \, K^{-1}$ (R = CH_2Ph), and $\Delta H^\ddagger = 32.2 \pm 2 \, kcal \, mol^{-1}$, $\Delta S^\ddagger = 18 \pm 6 \, cal \, mol^{-1} \, K^{-1}$ [R = $(CH_3)_3CCH_2$].

Table 3.3. Kinetic Data (log k) for the Reactions of Organic Radicals Which Proceed by Oxidation (O) or Reduction (R) of Metal Complexes, H-Atom Abstraction (H), or Metal-Carbon Bond Formation (M-R) in Aqueous Solution at 25°C

Radical	Metal complex	pH	$\log k$ $(M^{-1}\,s^{-1})$	Reaction type	References
CH_3	$[Co(II)P]^a$	8-13	9.15	M-R	163
$CH(CH_3)_2$	$[Co(II)P]^a$	8-13	9.30	M-R	163
$c\text{-}C_5H_9$	Cr^{2+}	3.3-4	7.90	M-R	164
CH_2OH	$[Co(II)P]^a$	8	9.04	M-R	163
$CH(CH_3)OH$	cyt-c_3	8	8.78	R	175
$C(CH_3)_2OH$	$[Co(II)P]^a$	8	9.08	M-R	163
	$[Fe(III)P']^b$	acidic^c	9.11	R	170
	$[Fe(III)P']^b$	7^c	8.57	R	170
	$[Co(NH_3)_5L_1]^{3+\,d}$	4.5-5.5	10.2	R	165
		1	9.62	R	165
	$[Co(NH_3)_5L_2]^{3+\,e}$	4.5-5.5	<8.45	R	165
		1	9.65	R	165
	$[Co(NH_3)_5L_3]^{3+\,f}$	4.5-5.5	9.45	R	165
		1	9.60	R	165
	$[Co(NH_3)_5L_4]^{3+\,g}$	4.5-5.5	9.46	R	165
		1	9.48	R	165
	$[Co(NH_3)_5F]^{2+}$	−0.3 to 1	6.34	R	166
	V^{2+}	−0.2 to 0.7	5.32	O	167
	$[Co(NH_3)_6]^{3+}$	1-3	5.61	R	168
	$[Co(ND_3)_6]^{3+}$	1-3	5.48	R	168
	$[Co(en)_3]^{3+}$	1-3	5.23	R	168
	$[Co(tn)_3]^{3+}$	1-3	6.28	R	168
	$[Co(chxn)_3]^{3+}$	1-3	<4	R	168
$CH_2C(CH_3)_2OH$	$[Co(II)(tspc)]_2^{8-\,h}$	6	9.65	M-R	169
$CH(CH_3)OC_2H_5$	$[Co(NH_3)_5F]^{2+}$	1	6.04	R	166
	V^{2+}	−0.5 to 0.7	4.77	O	167
$C(CH_3)_2O^-$	$[Co(II)P]^a$	13	8.84	R	163
CO_2^-	$[Co(II)P]^a$	8	8.23	R	163
	$[Co(II)P]^a$	13	8.41	R	163
	$[Co(II)(tspc)]^{4-\,h}$	3-11	8.18	R	169
CF_3CHCl	$[Fe(II)P']^b$	Acidic	10.15	M-R	170
		Alkaline	9.84	M-R	170
	$[Fe(III)P']^b$	—	≤6	—	170
$L_1^{-\,d}$	$[Co(NH_3)_6]^{3+}$	4.5-5.5	8.04	R	165
L_1H^d	$[Co(NH_3)_6]^{3+}$	1	6.38	R	165
$L_3^{-\,f}$	$[Co(NH_3)_6]^{3+}$	4.5-5.5	8.00	R	165
L_3H^f	$[Co(NH_3)_6]^{3+}$	1	6.28	R	165
$L_4^{-\,g}$	$[Co(NH_3)_6]^{3+}$	4.5-5.5	7.65	R	165
L_4H^g	$[Co(NH_3)_6]^{3+}$	1	6.32	R	165
$MV^{+\,i}$	$Cr_2O_7^{2-}$	—	9.70	R	171
	$[Fe(CN)_6]^{3-}$	—	9.94	R	171
	cyt-C_3	8	$8.65^{\,j}$	R	175
	cyt-C_3	9	9.20	R	175
	$ZnP'^{+\,a}$	7	9.11	R	176

(continued)

Table 3.3. *(continued)*

Radical	Metal complex	pH	log k ($M^{-1}\,s^{-1}$)	Reaction type	References
$CH_2 = CH(CH_2)_3CH_2$	Bu_3GeH	—	4.97^k	H	172
	Bu_3GeH	—	4.89^l	H	172
ɔzH ^{m}	$[Co(NH_3)_5Cl]^{2+}$	0	0.279	R	173
	$[Co(NH_3)_5Br]^{2+}$	0	1.50	R	173
	$[Co(NH_3)_5I]^{2+}$	0	2.04	R	173
2-CH_3-pzH	$[Co(NH_3)_5Cl]^{2+}$	0	0.991	R	173
	$[Co(NH_3)_5Br]^{2+}$	0	2.34	R	173
	$[Co(NH_3)_5I]^{2+}$	0	2.84	R	173
,5-$(CH_3)_2$pzH	$[Co(NH_3)_5Br]^{2+}$	0	2.66	R	173
	$[Co(NH_3)_5I]^{2+}$	0	3.32	R	173
,6-$(CH_3)_2$pzH	$[Co(NH_3)_5Br]^{2+}$	0	2.67	R	173
	$[Co(NH_3)_5I]^{2+}$	0	3.41	R	173
uinH ^{n}	$[Co(NH_3)_5Cl]^{2+}$	0	−0.161	R	173
	$[Co(NH_3)_5Br]^{2+}$	0	0.919	R	173
,3-$(CH_3)_2$-quinH	$[Co(NH_3)_5Cl]^{2+}$	0	1.20	R	173
	$[Co(NH_3)_5Br]^{2+}$	0	2.60	R	173
hH ^{o}	Fe^{3+}	0	3.26	R	173
	Tl(III)	0	2.67	R	173
	V^{2+}	0	4.50	O	173
N-(CH_3)PhH	V^{2+}	0	4.70	O	173
$CH_2CH_2CN/$ $CH(CH_3)CN$	Cr^{2+}	—	8.36	M–R	174
$C(CH_3)_2CN/$ $CH_2CH(CH_3)CN$	Cr^{2+}	1–2	8.28	M–R	174
$C(CH_3)_2COOH/$ $CH_2CH(CH_3)COOH$	Cr^{2+}	1–2	8.06	M–R	174

P = tetrakis(4-sulfonatophenyl)porphyrin.
P′ = ferrideuterioporphyrin(IX) dimethyl ester.
In 2-propanol-H_2O.

^{g} $L_4 = $

^{h} tspc = tetrasulfophthalocyanine.
^{i} MV^+ = methyl viologen radical cation.
^{j} The reverse reaction has $k = 7.8 \times 10^4\,M^{-1}\,s^{-1}$.
^{k} In n-octane.
^{l} In benzene.
^{m} pz = pyrazine.
^{n} quin = quinoxaline.
^{o} Ph = phenazine.

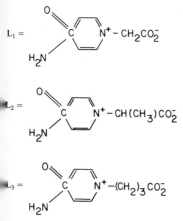

$L_1 = $

$L_2 = $

$L_3 = $

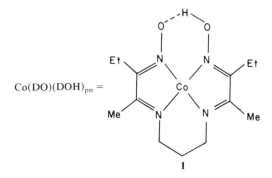

$Co(DO)(DOH)_{pn} =$

1

Rate constants were reported for the formation of substituted pyrazinium radicals[173] by the reduction of pyrazines with V^{2+} or by oxidation of hydropyrazines with Co(III) complexes and UO_2^{2+}.

3.13. Carboxylic Acids, Carboxylates, and Carbonyl Compounds

3.13.1. Carboxylic Acids

The oxidation of methylmalonic and ethylmalonic acids (LH_2) by Ce(IV) in acidic aqueous solution,[181] equation (90),

$$RCH(COOH)_2 + 4Ce(IV) + H_2O$$

$$\rightarrow RC(O)COOH + CO_2 + 4Ce(III) + 4H^+ \qquad (90)$$

is consistent with the formation of two intermediate complexes, $CeLH_2^{4+}$ and $CeLH^{3+}$, which undergo rate-determining electron transfer to produce a free radical, $LH\cdot$. Further oxidation leads to the final product. Dimethylmalonic and diethylmalonic acids do not react with Ce(IV) like unsubstituted and monosubstituted malonic acids, illustrating the importance of the methylene (or methine) hydrogens in the $LH\cdot$ formation step. Ce(IV) oxidizes α-hydroxycycloalkane-carboxylic acids in acidic perchlorate[182] to the corresponding cyclic ketones. Complex formation between Ce(IV) and the hydroxy acids, followed by rate-determining decarboxylation to a reactive intermediate is proposed.

3.13.2. Carboxylates

Oxalate complexes of Mn(III)[183] decompose to Mn(II) and CO_2. The rate-determining step is a unimolecular formation of Mn(II) and CO_2^-. The radical scavenger acrylic acid has little effect on the rate, indicating that the dimerization of $CO_2^{\doteq}$ is more important than its reaction with Mn(III). In air the decomposition of Mn(III)oxalate induces the autoxidation of oxalate through the formation of $O_2COO^{\doteq}$ and its reduction by Mn(II):

$$Mn(II) + O_2COO^{\doteq} + 2H^+ \rightarrow Mn(III) + CO_2 + H_2O_2 \qquad (91)$$

Cr(IV), produced *in situ* from Cr(VI) and V(IV), oxidizes Co(III) complexes[184] of mandelic, lactic, and benzilic acids with the rate law in equation (92):

$$\text{rate} = \frac{k[\text{Cr(VI)}][\text{V(IV)}]^2[\text{Co(III)L}]^0}{[\text{V(V)}]} \qquad (92)$$

The rate-determining formation of Cr(IV) is followed by the rapid oxidation of the hydroxyl center, the C-C bond scission, and electron transfer to Co(III) as in Scheme 1.

Scheme 1

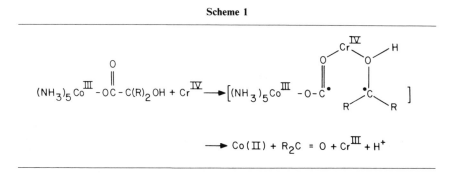

Alkaline $KMnO_4$ oxidizes sodium acetate to oxalate.[185] At 91.2°C, $k = 2.34 \times 10^{-5} + 1.88 \times 10^{-4} [\text{NaOH}]^2 \; M^{-1} \, \text{min}^{-1}$. The oxidation of coordinated oxalate in $[(NH_3)_5Co(OCOCOOH)]^{2+}$ by Ce(IV)[186] was proposed to take place via a precursor complex, in which reduction of Ce(IV) to Ce(III) and Co(III) to CO(II) occurs by a C–C bond scission.

$$CH_3COO^- + 6MnO_4^- + 7OH^- \rightarrow C_2O_4^{2-} + 6MnO_4^{2-} + 5H_2O \qquad (93)$$

$$[(NH_3)_5Co(OCOCOOH)]^{2+} + Ce(IV) \longrightarrow Co^{2+} + Ce(III)$$
$$+ 2CO_2 + 5NH_4^+ \qquad (94)$$

3.13.3. Carbonyl Compounds

A deuterium kinetic isotope effect of 6.1 was observed in the oxidation of acetaldehyde by permanganate in aqueous $HClO_4$,[187] implying a hydride transfer from the aldehyde hydrate to $KMnO_4$. Substituted benzaldehydes are oxidized by $KMnO_4$[188] at pH 6.5 in the following order:

$$p\text{-}NO_2 > m\text{-}NO_2 > H > p\text{-}CH_3 > m\text{-}CH_3 > o\text{-}NO_2 > m\text{-}Cl > o\text{-}Cl > o\text{-}CH_3$$

Aliphatic esters are oxidized by chromic acid in acetic acid–H_2O mixtures.[189] The reaction shows an unusually strong H^+ dependence: An increase in $[H_2SO_4]$ from 2 to 3.5 M causes a 38-fold increase in rate. The rate acceleration by oxalic acid was explained by a three-electron mechanism.

$$3CH_3C(O)OCH_2R + 2HCrO_3^+ \rightarrow 3CH_3COOH + 3RCHO + 2Cr(III) \tag{95}$$

3.14. Alkanes, Alkenes, Alkynes, and Arenes

3.14.1. Alkanes

Permanganate oxidizes n-butane, n-pentane, and methyl-c-hexane in acidic aqueous solution[190] (H_2SO_4 or $HClO_4$) according to the rate law:

$$-d[RH]/dt = k_0[HMnO_4] + k_{-1}[MnO_4^-] \tag{96}$$

For CH_3-c-C_6H_{11} at 25°C, $k_0 = 2.8\ M^{-1}\,s^{-1}$ and $k_{-1} = 9 \times 10^{-4}\ M^{-1}\,s^{-1}$. A late-transition state involving an oxygen bridge was proposed. Oxidation of alkanes by $KMnO_4$ in aqueous H_2SO_4[191] follows the reactivity order $3° > 2° > 1°$. A kinetic deuterium isotope effect of 3.8 was observed in the oxidation of c-C_6H_{12}.

$[Fe(CN)_6]^{3-}$ oxidizes Ph_2CH_2 and Ph_3CH in aqueous acetic acid containing $HClO_4$[192] to Ph_2CO and Ph_3COH, respectively, at a rate independent of $[HClO_4]$ and added $[Fe(CN)_6]^{4-}$. Hydrogen-atom abstraction and subsequent oxidation of the radical by $[Fe(CN)_6]^{3-}$ were proposed.

A kinetic deuterium isotope effect of 2.8,[193] and the inhibiting effect of Ce(III) were taken as evidence that oxidation of hexamethylbenzene by Ce(IV) acetate in acetic acid proceeds by electron transfer with some contribution from the proton transfer step.

3.14.2. Alkenes

$[(cod)RhO_2]_2$ (cod = 1,5-cyclooctadiene) decomposes at 140°C in C_6D_6 to cyclooctanone quantitatively,[194] even in the presence of free olefins. Pyrolysis under $^{18}O_2$ produces cyclooctanone-^{16}O (90%) and cyclooctanone-^{18}O (10%). The two findings strongly support the oxygen transfer reaction between two coordinated species.

Oxidation of methyl(E)-cinnamate with quarternary ammonium permanganates in CH_2Cl_2[195] is fastest for the ions which minimize the interionic distance in the quarternary ammonium ion pair. An inverse secondary deuteriom isotope effect was observed. The reaction is initiated by an electrophilic attack, followed by a slower conversion of a π-complex to a cyclic manganate(V) diester (**2**).

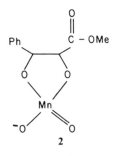

2

Alkenes, such as (E)- and (Z)-β-methylstyrene, are converted stereospecifically by chromyl nitrate to the corresponding epoxides with high selectivites in aprotic media.[196] Cosolvents (DMF, acetone, pyridine) are required for effective epoxidation. Products derived from a Cr(IV) intermediate are absent suggesting that the active species is oxochromium(V), formed *in situ* by the prior one-electron oxidation of solvent.

3.14.3. Arenes

Oxidation of p-methoxytoluene by 12-tungstocobaltate(III) ion, $[Co(III)W_{12}O_{40}]^{5-}$ (or $Co(III)W^{5-}$) in HOAc–OAc$^-$–H$_2$O yields net two-electron oxidation to $ArCH_2OAc$ (36%) and $ArCH_2OH$ (47%), although the kinetic data,[197] particularly the inverse dependence of rate upon $[Co(II)W^{6-}]$, suggest an electron transfer mechanism:

$$[Co(III)W]^{5-} + ArCH_3 \underset{k_{-1}}{\overset{k_1}{\rightleftharpoons}} [Co(III)W]^{6-} + ArCH_3^{+\cdot} \qquad (97)$$

$$ArCH_3^{+\cdot} + OAc^- \ (or\ H_2O) \overset{k_2}{\longrightarrow} products \qquad (98)$$

The composite rate constant shows a large kinetic isotope effect for $ArCD_3$ ($k_H/k_D = 4.5$–7.3), strong salt effects, specific cation accelerations (Na$^+$) and inhibitions (Bu$_4$N$^+$), and a marked dependence on solvent ratio and dielectric constant. Marcus-equation estimates of k_1 and k_{-1}, the latter diffusion controlled, yield $k_2 = 3.6 \times 10^{-6}$ (OAc$^-$) and 80 (H$_2$O) $M^{-1}s^{-1}$.

Positional selectivities[198] in the Ce(IV) and Co(III) side-chain oxidations of alkyl aromatics in CH_3COOH at 60°C were interpreted by suggesting that Ce(IV) reacts by electron transfer, and Co(III) by H-atom transfer.

Part 2

Substitution and Related Reactions

Chapter 4

Reactions of Compounds of the Nonmetallic Elements

4.1. Introduction

During the period covered useful reviews have appeared on mechanistic aspects of Group I–III elements, including boron,[1] substitution reactions at silicon by Corriu and Guerin,[2] organosilicon reaction mechanisms by Cartledge,[3] and the synthesis of five-coordinate compounds of the nonmetals as analogues for the transition states of associative nucleophilic displacements.[4] Useful mechanistic and other information on reactions involving boron (Chapter 5), silicon (Chapter 9), germanium (Chapter 10), and arsenic and antimony (Chapter 13) may be found in an important nine-volume treatise on organometallic chemistry.[5]

4.2. Boron

^{11}B and variable temperature NMR studies on aqueous solutions of $KB_5O_8 \cdot 4H_2O$ and $K_2B_4O_7 \cdot 4H_2O$ at various pH's have been interpreted in terms of equilibria involving $B(OH)_4^-$, $B_3O_3(OH)_4^-$, and $B_5O_6(OH)_4^-$.[6] In neither solution can $B_4O_5(OH)_2^-$ be detected, despite its being the most abundant polyborate species in the solution of the tetraborate salt, according to earlier Raman spectroscopic studies. Emsley and Lucas[7] suggest that the ease with which $B_5O_6(OH)_4^-$ may be formed at room temperature from boric acid in aqueous potassium fluoride is associated with the enhanced nucleophilicity of oxygen, resulting from hydrogen bonding with fluoride.

$B(OH)_3(OOH)^-$ has been detected by Raman spectroscopy in equilibrium (1) with $B(OH)_4^-$ and H_2O_2 in solutions of alkaline lithium metaborate:[8]

$$B(OH)_4^- + H_2O_2 \overset{K}{\rightleftharpoons} B(OH)_3(OOH)^- + H_2O \qquad (1)$$

$$K = {\sim}20 \, \text{liter mol}^{-1}$$

4-methylcatechol in aqueous boric acid gives a $1:1$ complex with $K_1 = 9.8 \times 10^6$, whereas, for the $2:1$ complex, K_2 is only 1.9. Nevertheless, the $2:1$ complex precipitates from solution, presumably because of its lower solubility.[9] Complexation of boric acid and tetrahydroxyborate ion with sorbitol, mannitol, D-glucose, glycerol, and ethylene glycol has been studied by potentiometric techniques.[10]

Emri and Györi[11,12] continue their mechanistic studies of the hydrolysis of various (pyrrolyl-1)borates, including $BH_n(NC_4H_4)_{4-n}^-$ ($n = 1, 2, 3$), $BH_n(NC_4H_4)_{3-n}(CN)^-$, and $B(C_6H_5)_n(NC_4H_4)_{3-n}(CN)^-$ ($n = 1, 2$). Their earlier work concluded that hydrolysis of $B(C_6H_5)_n(NC_4H_4)_{4-n}^-$ ($n = 0, 1, 2, 3$) involved rate-determining proton attack on the nitrogen-bound substituent, whereas in the pH range 3.2–13.1 the rates of hydrolysis of B–H and B–NC_4H_4 in $BH_n(NC_4H_4)_{4-n}^-$ were almost identical. The overall process has the stoichiometry shown in equation (2):

$$BH_n(NC_4H_4)_{4-n}^- + H^+ + 3H_2O \rightarrow B(OH)_3 + nH_2 + (4-n)HNC_4H_4 \qquad (2)$$

Formation of 1,2- and 1,3-$[d_2]$-pyrrole in D_2O suggests processes involving both C and N protonation. The cyano group significantly increases the hydrolytic stability of the B–N bond in $BH_n(NC_4H_4)_{3-n}(CN)^-$ and $B(C_6H_5)_n(NC_4H_4)_{3-n}CN^-$. In aqueous acid, $B(OH)_3$ and $B(C_6H_5)_n(OH)_{3-n}$ are formed from these compounds, whereas, in alkaline solution, cyanide loss is the favored process for the phenyl derivatives and concomitant cyanide and B–H bond fission for the hydridoborates. Kinetic studies reveal that both $B(C_6H_5)_n(NC_4H_4)_{3-n}(CN)^-$ ($n = 1, 2$) and $BH(NC_4H_4)_2CN^-$ hydrolyze with a mixed mechanism with the observed hydrolysis rate constant having the form $k_1[H^+] + k_2$ in which the acid-independent term relates to B–CN cleavage, with the activation parameters favoring an S_N1 process. For $BH_2(NC_4H_4)CN^-$ the observed hydrolysis rate constant approaches a limiting value in high acid, ascribed by Emri and Györi to the generation of a C-protonated intermediate.

Electrophilically induced substitution of R in $[R_2B(C_3H_3N_2)]_2$ (R=H, SMe, Et) by Br (using $^{10}BBr_3$) to give $[Br_2B(C_3H_3N_2)]_2$[13] proceeds without ^{10}B incorporation, indicating that B–N bond fission is not involved. Partly alkylated intermediates may be detected by ^{11}B NMR. Similar labeling techniques have been used to study substituent exchange processes of 1,3,2-diathiaborolanes and related heterocycles with BBr_3.[14]

Boulton and Prado[15] have concluded from the similarity in the activation free energies for the exchange process (3) as determined from the coalescence of the benzyl methylene proton NMR signals in (1) and (2) and those for the enantiotopomerization of similar compounds studied earlier by Russian workers, that it was unnecessary to postulate a process involving a planar tetracoordinate boron atom. Boulton and Prado favored planar three-coordinate intermediates instead.

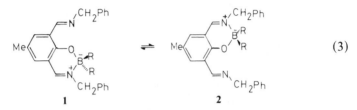 (3)

Matteson *et al.* have studied the homologation of boronic esters with dichloromethyl lithium.[16,17] The stereoselectivity of the reaction involving pinanediol boronic esters is decreased due to epimerization of the homologated product by co-produced lithium chloride.[18,19] Assembling the proposed intermediate by the reaction of the dichloromethyl boronic ester with an alkyl lithium does not give the same enantioselectivity as with the previous procedure. This was thought to be associated with kinetic selectivity in the attack by the nucleophiles R^- and $CHCl_2^-$.

Brown and co-workers have continued their extensive studies of the mechanisms of hydroboration. Some of this work has recently been reviewed.[1] Kinetic studies have been reported of the reaction of 9-borabicyclo[3.3.1]nonane dimer (3) with alkenes,[20,21] alkynes,[22] aldehydes, and ketones,[23] in various solvents and with various Lewis bases[20,24] and Brønsted acids.[25,26] Earlier studies suggested that the reaction of (3) with alkenes in a variety of solvents proceeded via the initial dissociation as in equation (4) to the monomer (4) followed by the hydro-

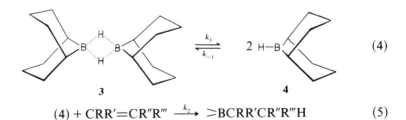 (4)

$$(4) + CRR'{=}CR''R''' \xrightarrow{k_2} {>}BCRR'CR''R'''H \qquad (5)$$

boration step, reaction (5), with k_1 (at 298 K) increasing from $1.5 \times 10^{-4} \, \text{s}^{-1}$ (CCl_4, cyclohexane) to $2.0 \times 10^{-4} \, \text{s}^{-1}$ (benzene) to $2.8 \times 10^{-4} \, \text{s}^{-1}$ (Et_2O) and $1.4 \times 10^{-3} \, \text{s}^{-1}$ (THF). It has been found that hydroboration in hindered ethers such as 2,5-dimethyltetrahydrofuran[20] occurred at rates similar to those found in noncomplexing solvents. More detailed studies of reactions (4) and (5) in THF revealed that the expected first-order reaction seen on addition of alkene to (3) in THF was preceded by a more rapid process ascribed to the reaction between the alkene and an equilibrium concentration of the THF complex, (4)·THF.

Similar behavior in the presence of dimethylsulphide was seen, for which the equilibrium constant for formation of (4)·SMe$_2$ was estimated to be 2.1×10^{-3} liter mol^{-1}. Dissociation of solvent from (4)·THF or (4)·SMe$_2$ would seem to be much more rapid than the dissociation of (3) to give (4). The diborane (3) undergoes a similar symmetrical cleavage with amines, unhindered amines[24] showing second-order kinetics, first order in [(3)] and [amine], ascribed to a direct attack of the amine on the borane. For sterically more hindered amines, kinetics first order in [(3)] only are observed, with rates corresponding to dimer dissociation. For a series of pyridines in which (3) is essentially completely dissociated to (3)·NR$_3$, reaction with added alkene was found to be slow in the presence of sterically nonhindered pyridines and more rapid the greater the steric hindrance. Brown and Chandrasekharan have been able to rationalize some of these solvent effects and some unexpected catalytic effects of BBr$_3$[27] noted in hydroborations. These are thought to arise from a combination of basicity factors (giving good conversions to monomeric borohydride complexes) and steric and electronic effects on kinetics (resulting in ready dissociation of coordinated base before reaction with olefins). The view that THF is a better leaving group from $H_3B \cdot THF$ than BH_3 is from B_2H_6 is supported from MO (molecular orbital) calculations by Clark *et al.*[28] in which $H_3B \cdot OH_2$ was used as a model for the ether complex. However, these authors tend towards the view that, in unhindered boranes, displacement of the Lewis bases occurs with participation of the olefin, being S_N2-like with a late-transition state, rather than S_N1-like. Chadha and Ray have used MNDO MO calculations in theoretical studies of the reaction paths involved in the diboration of ethylene[29] and acetylene.[30]

Similar kinetic behavior to that seen with olefins takes place in the reactions of aldehydes and ketones with (3),[23] with the more reactive compounds showing rate-determining dissociation of (3) and the less reactive ketones showing intermediate or $\frac{3}{2}$-order kinetics. Reductions by (3) are less susceptible to steric effects than those involving BH_4^-, with the largest effect being noted in ketones substituted with bulky groups on either side of $\diagup\!\!=\!\!O$. For both p-XC$_6$H$_4$CHO and p-XC$_6$H$_4$COCH$_3$ reactivity decreases in the order $X = OCH_3 > CH_3 > H > Cl$ are consistent with coordination of carbonyl oxygen to boron during the reduction process. A theoretical study of BH_4^- addition to formaldehyde supports the view that a $[2 + 2]$ concerted four-center process involving synchronous B–H breaking and C–H and B–O bond making is not involved.[31] A transition state is preferred in which BH_3 shifts from C–H to O, with the calculated activation energy of the process markedly reduced by involvement of a mole of H_2O hydrogen bonded to O.

Spielvogel *et al.*[32] have investigated the structural and electronic analogies between $CH_3C(O)X$ and $BH_3C(O)X^-$, particularly towards hydride transfer and have predicted that reactivity should increase in the order $H_3BC(O)Cl^- < H_3BC(O)H^- < H_3BC(O)R^- < H_3BC(O)OR^- < H_3BC(O)NR_2^- < H_3BC(O)O^{2-}$. The reactivity of $BH_3C(O)Cl^-$ was confirmed from the marked catalytic effect of Cl^- on the conversion of B_2H_6 and CO to $[CH_3BO]_3$ in $CHCl_3$.

Protonolysis of (3) in both THF and CCl_4 with alcohols and phenols[25] is a process with first-order dependence on $[(3)]$ for hindered alcohols, with a rate constant essentially that of the dimer dissociation process (4), whereas with less hindered alcohols, mixed kinetics consistent with an additional pathway involving direct attack of the alcohol on (3) were noted. The electronic effects of substituents in the reactivity of p-$XC_6H_4CH_2OH$ are consistent with complexation of alcohol with boron before elimination of H_2. Protonolyses of boranes such as Et_3B by carboxylic acids are known to be especially facile. This has been ascribed to coordination of RCO_2H prior to proton transfer. Brown and Hébert[26] have shown from qualitative rate measurements for $R' = Me$, Et, *i*-Pr, Hept, CF_3, the reactivity decreases along the series $Et_3B > Et_2BOCOR' > EtB(OCOR')_2$, with better complexing solvents (ethylene glycol, DMF) slowing the reaction compared with acetic anhydride or nitrobenzene. Other studies by Brown *et al.*[33-37] and Wood and Rickborn[38] are concerned with isomerization and the stereospecificity and regiospecificity of hydroboration mechanisms. Regioselectivity in the hydroboration of amines can be controlled by B–N coordination.[39]

The borane radical anions $BH_3^{\cdot-}$ and $BH_2CN^{\cdot-}$ (but not $B(Ot\text{-}Bu)_3^{\cdot-}$),[40] may be formed either by reaction with photochemically generated *t*-BuȮ or $(Me_3Si)_2\dot{N}$ in ether/alcohol mixtures or directly by photolysis in liquid ammonia or $NH_3/MeOCH_2CH_2OMe$ mixtures. After interruption of photolysis, both $BH_3^{\cdot-}$ and e_{amm}^- decayed with approximately first-order kinetics, $t_{1/2}$ being ~40 and 10 ms, respectively. This reaction is cation dependent, since replacing the $(n\text{-}Bu)_4N^+$ countercation by Li^+, Na^+, or K^+ gave only e_{amm}^-. No $BH_3^{\cdot-}$ or e_{amm}^- is formed in the presence of *t*-BuCl, with the detection of *t*-Bu$^\cdot$ indicating the occurrence of the halogen-abstraction reaction to give BH_3Cl^-. Symons *et al.* have tentatively suggested that γ radiolysis of solid $NaBH_4$ at 77 K leads to $BH_4^{\cdot}$.[41]

Free B_3H_7 has never been isolated though the complexes $B_3H_7X^-$ (X = Cl, NCO, H) are well known. Irreversible electrochemical one-electron oxidations of these compounds in acetonitrile solution results in $B_3H_7(NCCH_3)$ via a proposed B_3H_7 intermediate.[42] $B_3H_7^{\dot-}$ has been detected in a variety of solvents by Roberts *et al.*[43] using hydrogen-atom abstraction from $B_3H_8^-$ by t-BuȮ or $(Me_3Si)_2\dot{N}$. MNDO calculations have suggested that $B_3H_7^{\dot-}$ is fluxional though the actual activation barrier to the equivalencing of the boron atoms appears to be larger than that calculated. $B_3H_7^{\dot-}$ appears to be less reactive than $BH_3^{\dot-}$ with no evidence of bromine abstraction from either t-BuBr or i-PrBr. Amine adducts of boryl radicals $H_2\dot{\bar{B}}\overset{+}{N}R_3$ have also been obtained by reaction with BuȮ.[44] ESR measurements suggest that, while $BH_3^{\dot-}$ and $BH_2CN^{\dot-}$ are effectively planar, $H_2\dot{\bar{B}}\overset{+}{N}R_3$ (isoelectronic with $Et_3\overset{+}{N}\dot{C}H_2$) is pyramidal, in contrast to $Et_3\overset{+}{\bar{P}}\dot{B}H_2$ generated similarly, which is thought to be planar. The amino and phosphinoboryl radicals are both much less reactive than $BH_3^{\dot-}$ towards alkyl bromides.

In the borane clusters $B_nH_n^{2-}$ (n = 8–12), n = 11 is fluxional even at 183 K, n = 8 is nonrigid at 303 K, and n = 9, 10, 12 is stereochemically rigid under normal conditions (with the nine-vertex polyhedral borane $B_9H_9^{2-}$ maintaining its tricapped trigonal prismatic geometry even up to 523 K). Replacement of H^- by a single SMe_2 in the 1- and 4-position and disubstitution to give 1,5, 4,5 and 1,8(9) isomers of $B_9H_7(SMe_2)_2$ profoundly affects the rearrangement barriers. Variable-temperature NMR studies are interpreted[45] in terms of intramolecular rearrangements via simple distortions which are normal cage vibrations which result in interchange of two capping five-coordinate boron atoms with two prismatic (six-coordinated) vertex borons via a $D_{3h} \leftrightarrows C_{4v} \leftrightarrows D_{3h}$ sequence based on Lipscomb's diamond–square–diamond mechanism. ^{1}H, ^{11}B, and ^{31}P NMR allow the topological transformation to be followed in which a Ph_2P bridge between B(5) and B(6) in (**5a**) is transformed into one between B(6) and B(9) in (**5b**) for the deprotonated form of the strongly acidic arachno-$B_{10}H_{13}PPh_2$.[46]

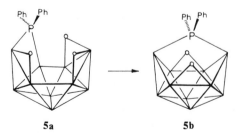

(6)

5a   **5b**

Pentaborane-(9) intramolecular rearrangements have been studied by ^{2}H NMR.[47] Oh and Onak[48] have set up complex mechanistic schemes to account for the rearrangement of 5-methyl-closo-2,4-$C_2B_5H_6$ to other isomers at 468 K, with the methyl group displaying preferences for the positions in the order $3 > 1$, $7 > 5$, 6. The fit of data appears to be better for a scheme based upon the diamond–square–diamond mechanism, though intramolecular methyl transfer was not ruled out.

The barriers to rotation about B–C, B–N, B–O, B–S, B–Se bonds have been studied by Wilson et al.[49] and about B–N bonds by Cragg.[50] Activation barriers to rotation, $\Delta G^\ddagger$, about the B–C and B–Se bond in (mesityl)$_2$BSeR were found, respectively, to be (R = Me) 36.0 and 76.6 kJ mol^{-1}, and (R = Ph) 38.8 and 71.8 kJ mol^{-1}.[49] Comparisons with the author's earlier work indicate the B–C rotation barrier increases along the series X = Se < N, S < 0 ascribed to secondary effects arising from changes in the B–X bond length. The B–X rotation barrier is governed by electronic effects and increases along in series $0 < S \sim Se < N$, the same as the order of increasing back donation from X to B. Both steric and electronic effects control the influence of the substituent X in the barrier to the isomerization of PhB(X)NR$_1$R$_2$, where R$_1$ = R$_2$ = Me, R$_1$ = R$_2$ = i-Pr, and X = F, Cl, Br, OMe, SMe. When R$_1$ and R$_2$ are small, barriers to B–N rotation are governed by the inductive and mesomeric effects of X, whereas when R$_1$ and R$_2$ are bulky steric effects dominate.[50]

The relative tendencies of various groups on boron to migrate to carbon have been qualitatively estimated[51] from product ratios in the reaction of PhB(X)NR$_2$ with phenylisocyanate, reaction (7):

$$PhB(X)NR_2 + PhNCO \begin{cases} \longrightarrow PhB(X)NPhC(O)NR_2 \\ \longrightarrow PhB(NR_2)N(Ph)C(O)X \end{cases} \quad (7)$$

Migratory aptitude of NR$_2$ appears to be governed by steric effects, whereas electronic effects are thought to be much more important in controlling X migration (which increases along the series Ph, halide, OR < R$_2$N < RNH < RS).

Synthetic and structural work has concluded that 1,3,2-dioxaborinium salts, thought to contain a three-coordinate boron center, in fact are four coordinate with a bound counterion.[52] Evidence has been presented for a B–C double bond[53] and Paetzold et al. have studied the chemistry of monomeric boron imides containing B–N double bonds.[54-56]

4.3   Carbon

While the hydration of CO_2 is known to be slow, the reaction with OH$^-$, important at alkaline pH's, is much more rapid. This reaction has been studied by the *in situ* generation of CO_2 from pulse radiolysis of N_2O/O_2 saturated aqueous sodium formate[57] in which $\dot{O}H$ abstracts $\dot{H}$ from HCO_2^-, with the resulting CO_2^- being oxidized by O_2 (possibly via the short-lived radical $[O_2CO_2]^-$). The reaction between $CO_2 + OH^-$ was followed conductimetrically, with a second-order rate constant at 294 K of 6900 ± 700 liter $mol^{-1} s^{-1}$ being obtained, in agreement with earlier studies. Coordination of OH^- to Co(III), Rh(III), and Ir(III) reduces this rate constant (at 298 K) to 199 ± 20, 335 ± 35, and 411 ± 50 liter $mol^{-1} s^{-1}$, respectively, from data obtained by van Eldik *et al.*[58]

Alkyl carbonates, $ROCO_2^-$, act as anionic inhibitors of the BCA-catalyzed (bovine carbonic anhydrase) CO_2 hydration, the binding constant for $ROCO_2^-$ being of similar order to that for $CH_3CO_2^-$. Pocker and Deits[59,60] suggest that while $ROCO_2^-$ and $HOCO_2^-$ (the natural substrate for BCA) bind with the enzyme in a similar fashion, a critical proton transfer mediated by Zn(II) is denied $ROCO_2^-$, accounting for the $>10^5$ drop in turnover number compared with $HOCO_2^-$. A series of related papers was presented at a conference held in Florence.[61-64] The reaction of amines with CO_2, equation (8), has been investigated by two groups.[65,66] Stopped-flow kinetics using indicators[65] established that in all cases the reaction was first order in $[CO_2]$ and in [amine], with proton transfers assumed to be diffusion controlled. Observed rate constants at 293 K (corrected if necessary for the CO_2-OH^- reaction) increased along the series glycylglycinate (1826 liter $mol^{-1} s^{-1}$) < 2-aminoethanol < benzylamine < glycinate < 3-aminopropan-1-ol < 2-phenethylamine < 3-phenyl-1-propylamine < 4-phenylbutylamine (9142 liter $mol^{-1} s^{-1}$) and E_a varying between 39.5 ± 2.4 and 51.0 ± 2.2 kJ mol^{-1} and $\log_{10}A$ between 10.7 and 13.0. The data are consistent with mechanism (8) if the zwitterion is formed irreversibly:

$$CO_2 + R\overset{+}{N}H_2 \rightleftharpoons R\overset{+}{N}H_2\overset{-}{C}O_2 \xrightarrow{B = RNH_2} RNHCO_2^- + BH^+ \qquad (8)$$

Castro *et al.*[67] have found the reaction of CS_2 and amines in ethanolic lithium chloride more complicated, with the involvement of solvent in the proton transfer which is thought to occur during nucleophilic attack at carbon. The rate of the reaction of amine oxides and CS_2 in dilute aqueous acetonitrile, which leads to quantitative formation of amine and a range of sulfur- and carbon-containing products, is first order in the concentration of each reagent. For a series of p-$XC_6H_4N(O)Me_2$, $\log k_2$ correlated well with the Hammet σ value with a small negative ρ. The authors propose[68] nucleophilic addition of amine oxide to CS_2 to give $R_3\overset{+}{N}OC(S)\overset{-}{S}$, which then reacts to give products.

The uncatalyzed and general acid-catalyzed hydrolyses of alkyl xanthates, $ROCS_2^-$, and monothiocarbonates, $ROCOS^-$, have been studied in aqueous solution,[69] with the effect of electron-withdrawing groups in R being characteristic of type n [reaction (9)] rather than type e [reaction (10)] addition–elimination mechanisms:

$$\text{type } n: \quad B + HNu + \text{>C=X} \rightleftharpoons BH^+ + Nu-\overset{|}{\underset{|}{C}}-X^- \tag{9}$$

$$\text{type } e: \quad Nu^- + \text{>C=X} + BH^+ \rightleftharpoons Nu-\overset{|}{\underset{|}{C}}-XH + B \tag{10}$$

It is possible to predict values of k_{-a} ($4.85 \times 10^{-5}\,s^{-1}$) for reaction (11) and of k_{-b} ($6.67 \times 10^{-1}\,s^{-1}$) for reaction (12) with known values of k_a and

$$HO^- + COS \underset{k_{-a}}{\overset{k_a}{\rightleftharpoons}} HOC(O)S^- \qquad K_A = k_a/k_{-a} \tag{11}$$

$$HO^- + CS_2 \underset{k_{-b}}{\overset{k_b}{\rightleftharpoons}} HOC(S)S^- \qquad K_B = k_b/k_{-b} \tag{12}$$

k_b allowing estimates (albeit of low precision) to be made of K_A (2.9×10^5 liter mol^{-1}) and K_B (1.6×10^4 liter mol^{-1}).

In superacid media (HF/SbF_5) at 273 K, alkanes may be oxidized to carbenium ions and H_2.[70] A two-step mechanism is proposed in which the formation of the protonated intermediate from various isomeric pentanes is rate determining, whereas for methylcyclopentane its decomposition is rate determining.

^{13}C-labeled t-butyl carbocation $(CH_3)_3{}^{13}C^+$, obtained at 195 K in $FSO_3H/SbF_5/SO_2FCl$ from labeled t-BuCl, exhibits complete scrambling of the label after 20 h at 343 K. No deprotonation occurs, and the scrambling is thought to involve a delocalized protonated methylcyclopropane intermediate or transition state.[71]

Diazomethane may be C protonated in $HOSO_2F/SO_2ClF$ at 153 K and C and N protonated in the more strongly acidic medium $HOSO_2F/SbF_5$ at the same temperature,[72] reaction (13):

$$CH_2N_2 \begin{cases} \xrightarrow{HOSO_2F} CH_2=\overset{+}{N}=NH + CH_3\overset{+}{N}_2 \\ \xrightarrow[HOSO_2F]{SbF_5} \\ \xrightarrow[SO_2ClF]{} CH_3\overset{+}{N}_2 \xrightarrow{188\,K} CH_3OSO_2F \end{cases} \tag{13}$$

The protonation is irreversible and $CH_2=N^+=NH$ was found to be the thermodynamically less stable isomer and was formed as a result of kinetic control. Studies of the decomposition of $CH_3N_2^+$ in a range of solutions allowed a tentative order of nucleophilicity to be drawn up, namely, $FSO_3^- > SO_2 > FSO_3^-/SbF_5 > SO_2ClF$.

4.4. Silicon

NMR studies of aqueous alkaline solutions of ^{29}Si-enriched silica have allowed 12 silicate species to be identified in a single solution and the structures for a further six to be tentatively assigned. These include linear and cyclic oligomers, substituted cyclic oligomers, polycyclic compounds, and bridged oligomers.[73,74] From solutions containing SiO_2, Al_2O_3, and [NEt_4]OH, from which the synthetic zeolite ZSM-5 crystallizes, a double five-ringed silicate (**6**) has been characterized using ^{29}Si NMR, attenuated total reflectance FT–IR, and mass spectrometry.[75] Ray and Plaisted,[76] using a derivatization–GLC method, have analyzed the distribution of anions in solutions of Li^+, Na^+, K^+, Rb^+, Cs^+, Me_4N^+, n-Pr_4N^+, and n-Bu_4N^+ silicate and have suggested that the degree of polymerization increases with increasing cation radius, the ratio of SiO_2 to base and concentration, and decreases with temperature.

Martin has examined further the mechanisms of the nucleophile-catalyzed racemization at the four-coordinate silicon of (**7**), which has been shown to be first, and never second, order in nucleophile and to involve one, and not two, nucleophile molecules. The kinetics of the inversion at the five-coordinate silicon in (**8**), where Nu = p-XC_6H_4CHO (X = H, 3-CF_3,

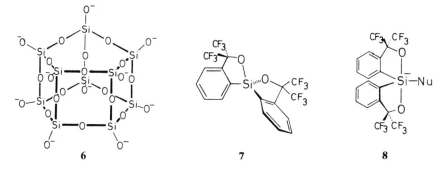

6   7   8

4-OMe) has confirmed[77] the previously proposed five-step pseudorotation route. The highest-energy structure along this pathway is likely to be (**9**) and the acceleration of reaction by electron-withdrawing substituents X rules out a dissociative process via (**10**). Inversion rates measured using spin-polarization transfer techniques allowed an activation free energy of 42 kJ mol^{-1} to be measured for (**8**) (Nu=NMe_2), which is in rough agreement with estimates resulting from structure-activity correlations.

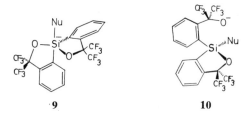

·9   10

The importance of apicophilicity of nucleophile, leaving group, or other substituent in five-coordinate silicon intermediates involved in substitution in five-coordinate silicon intermediates involved in substitution or racemization processes is well known. ^{19}F NMR has been used by Corriu et al.[78] to draw up a scale of apicophilicity for X in the trigonal bipyramidal (11), Z=Me, the order being X=H,OR,NR$_2$ < F < Cl < OC(O)Ph, which is the same sequence derived earlier by Corriu from studies on the racemization of $R_1R_2R_3\overset{*}{S}iX$ or substitutions involving inversion of configuration. The major trigonal bipyramidal conformer is thus not determined by the electronegativity of F. A direct method, using ^{29}Si chemical shifts, has supplemented earlier ^{1}H NMR studies of the equilibrium between ring-opened and ring-closed forms of compounds of the type (11).[79] The tendency for pentacoordination is least when X, Y, and Z are H or OR; most when substituents are Br, Cl, and OC(O)Me; and intermediate for SR and F. K_{eq} (303 K) for a variety of halide derivatives lies in the range 1.0–8.1 with a single ^{29}Si resonance being observed at all temperatures, indicative of a rapid equilibration on the NMR time scale.

11

The activation parameters for racemization of tris(tropolonato)silicon(IV) (12), the first-order kinetics without ligand substitution, and the catalysis by CCl$_3$CO$_2$H are all consistent with Si–O bond rupture to give a five-coordinate intermediate[80] rather than the non–dissociative twist mechanism favored by other workers. It is believed that the racemization of (12) and of other cationic silicon(IV) tris-β-ketonate complexes follows a common mechanism.

Racemization at chiral silicon and nucleophilic substitution may occur either by the intermediacy of five-coordinate silicon favored by Corriu or by pathways involving ion pairs, as proposed by Chojnowski. While Corriu has suggested that the latter process is a special case for labile bromosilanes in the presence of good nucleophiles, Bassindale and Stout[81] have suggested that the latter mechanism merits more serious consideration after interpreting ^{1}H, ^{13}C, and ^{29}Si NMR data in terms of the formation of equilibrium amounts of [Me$_3$SiDMF]X from Me$_3$SiX (X = I, Br, Cl, OSO$_2$CCl$_3$) in DMF. The reaction of (arythio)trimethylsilane with benzoyl chloride in CH$_2$Cl$_2$ obeys second-order kinetics yielding Me$_3$SiCl and PhC(O)SAr. The reaction was accelerated in polar solvents and showed a large positive ρ value for substitution in the arylthio group and a negative ρ for substitution in the benzoyl group.[82] Rate-determining heterolysis of Si–S in a five-coordinate Si intermediate was proposed.

In gas-phase reactions followed using the flowing afterglow technique, Me_3SiCl and MeI were found to react with anions (H_2N^-, OH^-, F^-, HS^-, NC^-) at each encounter, providing the anions were sufficiently basic.[83] As the basicity of the nucleophile decreased, so did the rate of its reaction with MeI. In contrast, Me_3SiCl reacted at each encounter with these nucleophiles until the net reaction ceased, when the displacement became endothermic. In the thermoneutral reaction between $^{37}Cl^-$ and Me_3SiCl, the displacement of unlabeled Cl^- was rapid with the proposed five-coordinate intermediate collisionally stabilized by the carrier gas and having a lifetime $>10^{-7}$ s.

The exchange equilibria established between alcohols (R'OH) and a variety of alkoxytriethylsilanes (e.g., Et_3SiOR)[84] and dialkoxydimethyl-silanes [e.g., $Me_2Si(OR)_2$][85] have equilibrium constants K between 1 and 1.5 when both R and R' are primary radicals, whereas, when R = Et and R' is a secondary radical, K's are $\ll 1$. The reactions are catalyzed by I_2 or IBr.

Corriu *et al.*[86] have shown that the stereochemistry of nucleophilic substitution at silicon is not altered by the presence of $GePh_3$ as a substituent. The extents of retention and inversion seen in reactions of $PhMe\overset{*}{S}i(X)GePh_3$ (**13**) and $PhMe\overset{*}{S}i(F)GePh_3$ with different reducing agents were similar to those found earlier for similar derivatives not containing Ge. In related studies,[87] Si–Ge bond cleavage was seen, leading to products corresponding to formation of Ph_3Ge^- and transient radicals. Stereoselective SiGe cleavage in (**13**) (X = H, OR, i.e., for a poor leaving group) by α-naphthyllithium occurs with retention of configuration to give Ph_3GeLi, good leaving groups (X = Cl, F) favoring substitution at Si–X instead, whereas with ethyllithium a competition between Si–F and Si–Ge cleavage is seen. Nucleophilic cleavage of transition-metal silicon bonds in (**13**) occurs according to the general rules governing stereochemistry of S_N(Si), that is, when X is polariz-able (e.g., $[Co(CO)_4]$), it behaves as a good leaving group and S_N occurs with inversion. Nonpolarizable groups (e.g., $[(\pi\text{-}C_5H_5)Fe(CO)_2]$) are bad leaving groups and are replaced with retention. Contrary to predictions, however, $[(\pi\text{-}C_5H_4Me)Mn(CO)H]SiR_3$ gives inversion, possibly accounted for by the involvement of Mn–H–Si interactions. For octahedral $R_3\overset{*}{S}i$ complexes of Mn, Re, and W, cleavage always occurred with low retention of configuration, regardless of the nature of the ligands on the transition metal or the incoming nucleophile, providing the first instances in which the stereochemistry of the displacement at silicon is independent of both leaving group and nucleophile.[88]

Bromine reacts with trialkylsilanes in CCl_4 by a molecular mechanism to give R_3SiBr and HBr with kinetics which are first order in each reactant, with higher-order kinetics being seen at $[R_3SiH] > 0.1$ mol liter^{-1}.[89] Electron-releasing substituents in $(p\text{-}XC_6H_4)_3SiH$ accelerate the reaction, indicating a buildup of positive charge in the transition state confirmed by a correlation of the observed rate constants in CCl_4, octane, and $CHCl_3$–CCl_4 mixtures with the solvent parameter $E_T(30)$. The steric effects in these reactions were considered unimportant,[89] whereas Cartledge[90] has attempted to correlate structural changes with reactivity trends in a representative series of silane reaction types and to isolate steric effects from other reactivity variables. Three conventional measures of steric effects gave unsatisfactory correlations when applied to eight series of reactions at silicon. Using the rate data for the acid-catalyzed R_3SiH hydrolysis, Cartledge has defined a new set of substituent constants applicable to silicon. If the results are confined to alkyl groups, the derived parameters give reasonable $(r > 0.98)$ correlations when applied to the reactions R_3SiH/OH^-, R_3SiH/O_3 in CCl_4, R_3SiOPh/OH^-, R_3SiOPh/H^+, R_3SiF/H_2O, $R_3SiH/\text{hex-1-ene}$, though not for R_3SiH/Br_2.

An interactive computer program CAMEO (Computer-Assisted Mechanistic Evaluation of Organic Reactions), used to predict the products of organic reactions using mechanistic information, has been refined[91] to include data relating to nucleophilic and electrophilic processes involving silicon, taking into account the high affinity of Si for oxy and halide ions, the directing effects for electrophilic addition to allyl and vinyl silanes, stabilization of carbanions and of S_N2 transition states by adjacent Si, stereochemistry of β-eliminations, and steric effects in the formation and removal of silyl protecting groups.

Eaborn and his group have continued their studies of highly sterically hindered organosilicon compounds some of which have been reviewed.[92] Solvolysis in MeOH of $(Me_3Si)_3CSiCl_3$ (14) and $(Me_3Si)_3CSiR_2X$ (15) ($R = Me$; $X = I$, $OClO_3$, OSO_2CF_3), to give the monosubstituted products $(Me_3Si)_3CSiCl_2(OMe)$ and (15) ($R = Me$; $X = OMe$), are unimolecular processes unaffected by MeO^- up to 0.5 mol liter^{-1}.[92,93] Prolonged reaction of (14) with $MeO^-/MeOH$ leads to an E2 fragmentation of the monosubstituted product. More recent work on (15) shows that nucleophilic attack by inorganic anions[93,94] competes with solvolysis, with the ratio of nucleophilic displacement to solvolysis increasing in the order $NCO^- < NCS^- < CN^- < F^- < N_3^-$ (though conversion to products after 24-h reflux are 33%, 62%, 70%, and 15% for NCO^-, NCS^-, CN^-, and F^-, respectively). The rate of formation of (15) ($R = Me$; $X = N_3$) is first order in $[N_3^-]$. For (15) ($R = Me$; $X = OH$), reaction in $MeO^-/MeOH$ leads to a 1,3 C to O silyl migration (**14**) whose facility is ascribed[95] to the relief of ring strain and the stabilization of the intermediate carbanion by three silicons:

$$(Me_3Si)_3CSiMe_2OH \underset{B^+H}{\overset{B}{\rightleftharpoons}} (Me_3Si)_3CSiMe_2O^-$$

$$(Me_3Si)_2\bar{C}(SiMe_2OSiMe_3) \underset{B}{\overset{B^+H}{\rightleftharpoons}} (Me_3Si)_2CH(SiMe_2OSiMe_3)$$

(14)

The first-order rate constant for the base cleavage of R–Si in $RSiMe_2OMe$ rises sharply to a plateau level on addition of increasing amounts of H_2O to the reaction medium. Rate-determining R^- loss from a silanoate $RSiMe_2O^-$ intermediate is proposed. This leads to a transient silanone, $Me_2Si=O$, which is rapidly trapped by solvent.[96] Silanones have also been suggested to explain the catalysis by MeO^- of the methanolysis of $(Me_3Si)_3CSiPh(OH)I$ to give $(Me_3Si)_3CSiPh(OH)OMe$ within 5 min at room temperature, whereas $(Me_3Si)_3CSiPh(OMe)I$ is unaffected by MeO^- even after 2-h reflux.

Electrophilic attack of $AgNO_3$ on (**15**) (R = Ph; X = I) in MeOH or CH_2Cl_2 leads to a 1,3 Me_3Si migration to give $(Me_3Si)_2(MePh_2Si)CSiMe_2X$ (X = MeO, NO_3), via a cationic intermediate possibly stabilized by Me bridging from a neighboring $SiMe_3$.[92] Methanolysis of (**15**) (R = Me; X = $OClO_3$, I) is believed to be accelerated by the anchimeric assistance of a Me from an adjacent $SiMe_3$ and similar interactions have been further probed in a study of solvolysis rates of $(Me_3Si)_2C(SiMe_2X)(SiMe_2Y)$ (**16**) which show that X facilitates the loss of Y = $OClO_3, OSO_2CF_3, Cl, I$ in the order   X = OMe > OSO_2CF_3 > $OClO_3$ > F > Cl,I > Me.[97] The MeO group in (**16**) (X = MeO; Y = Cl) accelerates loss of Cl^- by a factor of $\sim 10^6$ compared with the corresponding derivative, (**16**) (X = H; Y = Cl). $(Me_3Si)_2C(SiMe_2Y)_2$ reacts more readily than $(Me_3Si)_3C(SiMe_2Y)$ by a factor of >35 for Y = $OClO_3$, 48 for Y = I, and >110 for Y = OSO_2CF_3.

Using the bulk of the Me_3Si group to suppress competing reactions, Eaborn has observed the oxidatively assisted loss of I from (**15**), (R = Me, Ph; X = I) induced by methanolic *m*-chloroperbenzoic acid (MCPBA).[98] For R = Me, $(Me_3Si)_3CSiMe_2OH$ is formed at a rate qualitatively dependent on MCPBA concentration, with LiCl not only accelerating the reaction but diverting the product to the corresponding chloro derivative. For R = Ph, exclusive formation of the rearranged product $(Me_3Si)_2(MePh_2Si)C(SiMe_2OH)$ was seen. The better kinetic behavior of *t*-Bu_3SiI allowed a first-order dependence of rate on [MCPBA] to be estimated from initial rates. While the overall stoichiometry of these reactions is not yet clear, silicocation intermediates have been proposed, by analogy with reactions of alkyl iodides. Oxidative cleavage of Si–C bonds by MCPBA has also been reported.[99] Stereospecific and regioselective insertions of O into Si–Si bonds of the four geometric isomers of [*t*-BuMeSi]$_4$ by MCPBA in CCl_4 have been studied kinetically,[100] the reactions being first order in each reagent.

(15) (R = Ph; X = H) reacts with one mole of ICl in CCl_4 to give (15) (R = Ph; X = I), this reacting further with excess ICl to give the rearranged product $(Me_3Si)_2(MePh_2Si)C(SiMe_2Cl)$. Polar transition states have been proposed, though the use of the more polar MeOH as a solvent for the reaction of (15), (R = Me; X = H) and ICl does not lead to the expected rate acceleration.[101] In fact, the formation of (15) (R = Me; X = Cl and MeO) in MeOH occurs more slowly than in CCl_4.

While Lambert and Schulz have established from conductivity and ^1H and ^{13}C NMR that $(i\text{-PrS})_3SiH$ may be reversibly deprotonated in CH_2Cl_2 to give the fully ionized salt of the silicon-centered cation $(i\text{-PrS})_3Si^+$ uncomplexed by solvent,[102] organosilicon cations R_3Si^+ analogous to R_3C^+ have never been directly observed in solution. In weakly coordinating solvents (CH_2Cl_2, CH_2Br_2) Lewis acids may complex with R_3SiX and the extent of X^- removal to give R_3Si^+ probed by ^{29}Si NMR chemical shifts. $R_3SiBr/AlBr_3$ results in the most highly deshielded ^{29}Si resonance yet reported, associated with the formation of the complex (17) $[\overset{\delta+}{\equiv}Si-Br \rightarrow \overset{\delta-}{AlBr_3}]$ which may be in equilibrium with very small concentrations of ion pairs R_3Si^+, $AlBr_4^-$. ^{27}Al NMR linewidths decrease from 1250 Hz for $AlBr_3/CH_2Br_2$ to 950 Hz in the presence of the silane which may be compared with the 20-Hz linewidth for $AlBr_4^-$ in CH_2Br_2.[103]

Reactions of unsaturated organosilanes with electrophiles are often surprisingly rapid, ascribed to the formation of carbocations β to silicon. Jones et al. have found no evidence from studies of the acid cleavage of exo- and endo-2,2-dimethyl-3-neopentyl-2-silanorborn-5-enes (18) that rapidly equilibrating or delocalized cations such as (19) are formed.[104] The absence of exo–endo isomerization or deuterium incorporation into (18) in reactions in the presence of D_2SO_4 suggests that silyl-group loss is faster than proton loss. Equilibration of (19a) and (19b) via (20) was not seen, consistent with the observations of Olah et al.[105] that loss of the silyl group from β-silylcarbocations would be rapid even at temperatures as low as 133 K.

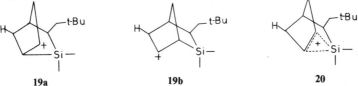

19a   19b   20

Chatgilialoglu et al.[106] have measured the rate of hydrogen-atom abstraction by $t\text{-BuO}$ from a series of alkyl and aryl silicon hydrides at 300 K, with second-order rate constants in the range 0.6×10^7 to

2×10^7 liter mol^{-1} s^{-1} (compared with 3.6×10^6 liter mol^{-1} s^{-1} for Ph$_3$CH). Reactions with analogous germanium and tin hydrides were more rapid. The arylsilyl radicals formed add readily to their precursors with Ph$_3\dot{S}i$ reacting with Ph$_3$SiH with $k = (2.10 \pm 0.04) \times 10^5$ liter mol^{-1} s^{-1} at 300 K.

The kinetics of the high-temperature thermal decomposition of silanes and alkyl silanes have also been discussed.[107] The absolute rate constants for the halogen-atom abstraction reaction of Et$_3\dot{S}i$ with 21 organic halides, RX, at 200 K, which span a range of $>10^5$, show[108] that for a particular R, rate constants increase in the order F < Cl < Br < I and for a particular X in the order phenyl < cyclopropyl < primary alkyl < secondary alkyl < tertiary alkyl < benzyl < allyl. Polychlorinated alkanes show an unusually high reactivity compared with monochlorinated substrates and this difference is reflected in the difference in the Arrhenius pre-exponential factor, which is larger for CCl$_4$ compared with monochloro compounds by an amount too large to be accounted for by statistical factors. Charge transfer interactions of the type shown would allow a gain of two extra rotational degrees of freedom; the greater the importance of the polar interactions the smaller would be the restriction on the orientation of Et$_3\dot{S}i$ with respect to the C–Cl bond being broken:

$$[Et_3\dot{S}i\cdots Cl-CR_3] \leftrightarrow [Et_3\overset{+}{S}i\cdots\bar{C}l\cdots\dot{C}R_3]$$
$$\updownarrow$$
$$Et_3SiCl + \dot{C}R_3 \leftarrow [Et_3\overset{+}{S}i\cdots\bar{C}l\cdots\dot{C}R_3]$$

Using data reported by Sommer, Chatgilialoglu *et al.*[108] have estimated the rate constant for inversion in (1-napthyl)PhMe$\dot{S}i$, derived from an optically active precursor, to be $\sim 6.8 \times 10^9$ s^{-1}, with an associated activation barrier ~ 23 kJ mol^{-1}.

Rates of addition of Et$_3\dot{S}i$ to olefins to give Et$_3$SiCR$_2\dot{C}R_2$ span a similarly large range,[109] with chlorinated olefins being more reactive than their alkylated counterparts. Second-order rate constants at 300 K ranged from 1.1×10^9 liter mol^{-1} s^{-1} for acrylonitrile to 9.4×10^5 for cyclohexene. The addition is regiospecific for vinylidene chloride which in cyclopropane gave only Et$_3$SiCH$_2\dot{C}Cl_2$ with $k_2 = (2.7 \pm 0.3) \times 10^8$ liter mol^{-1} s^{-1}. With other types of multiple bonds, reactivities decreased in the order isocyanide > nitrone > nitro > isocyanate > nitrile. The kinetics of the gas phase addition of Cl$_3\dot{S}i$ to olefins has also been studied.[110–112]

Polar effects were found to be particularly important in the reaction of Et$_3\dot{S}i$ with carbonyl compounds to give Et$_3$SiO$\dot{C}$RR′, with second-order rate constants at 300 K covering the range 2.5×10^9 liter mol^{-1} s^{-1} for duroquinone to 3.5×10^4 liter mol^{-1} s^{-1} for ethyl formate.[113]

Much work of a synthetic nature with mechanistic implications has been published and it is inappropriate that it is discussed in detail here.

However, mention should be made of the use of a trimethylsilyl methyl acetal of dimethylketene as an initiator for addition polymerization[114] and characterization or detection of compounds with Si=Si or Si=C bonds or containing divalent silicon or interchanges between these species. The chemistry of reactive intermediates in organosilicon chemistry has been reviewed by Barton.[115]

4.5. Germanium

The studies of Ge–Si bond cleavage in optically active $MePh\overset{*}{S}i(X)GePh_3$, (13), have been reviewed under silicon[87] and, more generally, the stereochemistry of germanium compounds has been surveyed by Gielen.[116] The activation parameters for the sigmatropic rearrangement for $(C_5Me_5)GeMe_2X$ show that the dynamics are, as also noted for the silicon analogues, little affected by the nature of X(= F,Cl,I,OMe,SMe).[117] R_4Ge (R = Ph, Et) is oxidized by ozone in CCl_4 in a process which is first order in each reagent with a second-order rate constant of 0.12 liter mol^{-1} s^{-1} at 293 K. It is proposed that for R = Et the reaction proceeds via an intermediate trioxide ($Et_3GeOOOEt$) which rearranges to give Et_3GeOOH and CH_3CHO. Et_4Ge reacts more rapidly than Et_4Si but more slowly than Et_4Sn.[118]

The effect of a strong magnetic field on the reaction products formed from $PhCH_2Cl$ and Et_3GeM (M = Na,K)[119] has been interpreted in terms of the differing extents of formation of cage ($PhCH_2GeEt_3$) and escape products ($PhCH_2GeEt_3$, $(PhCH_2)_2$, and $(Et_3Ge)_2$) following the initial electron transfer. Evidence has been presented for the generation of $Me_2Ge=GeMe_2$ [120] and for the singlet ground state of the germylene, Me_2Ge.[121]

4.6. Nitrogen

4.6.1. Nitric Acid, Nitration, and Nitrate Esters

The kinetics of ion-pair formation in concentrated aqueous $NaNO_3$ and of the protolysis of 2.0–14.7 mol $liter^{-1}$ nitric acid have been studied by Raman line broadening. Rate constants for the individual processes (15) and an equilibrium constant for the dissociation have been estimated, the latter having a value of 8.9 liter mol^{-1} at 298 K:

$$HNO_3 + H_2O \rightleftharpoons H_3O^+,NO_3^- \rightleftharpoons H_3O^+ + NO_3^- \tag{15}$$

$$K = \frac{[H_3O^+][NO_3^-]}{[HNO_3] + [H_3O^+,NO_3^-]}$$

The pK_a of HNO_3 was estimated to be -0.9 ± 0.3 at 298 K.[122] HNO_3 and NO_2^+ give separate ^{14}N NMR signals in 81–96.7% H_2SO_4, with only HNO_3 present in $<85\%$ H_2SO_4 and only NO_2^+ present in $>93\%$ H_2SO_4.[123] Line-shape analysis allows estimates to be made for the forward and back reaction rates for equation (16) with k_1 varying from 182 s^{-1} in 86.2% H_2SO_4 to 3236 s^{-1} in 92.6% H_2SO_4 and k_{-1} from 1288 to 81 s^{-1} in the same acids.

$$HNO_3 \underset{k_{-1}}{\overset{k_1}{\rightleftharpoons}} NO_2^+ \qquad (16)$$

In 88.6% H_2SO_4, the NMR signals coalesce on warming and allow $\Delta H_1^{\ddagger}$ (59.8 ± 9.6 kJ mol^{-1}), $\Delta S_1^{\ddagger}$ (0 ± 8 gibbs mol^{-1}), $\Delta H_{-1}^{\ddagger}$ (70.3 ± 10.0 kJ mol^{-1}), or $\Delta S_{-1}^{\ddagger}$ (8 ± 8 gibbs mol^{-1}) to be estimated.[123] A linear relationship has been established between the M_C activity coefficient function and a corrected observed second-order rate constant for nitration of aromatics in concentrated H_2SO_4.[124] Ipsonitration of 4-[^{15}N]-$O_2NC_6H_4OH$ with unlabeled HNO_3 has been found to lead to 11% incorporation of the label in the 2-position[125] of the $2,4$-$(O_2N)_2C_6H_3OH$ product, accounted for by migration of NO_2 in a radical pair (21) for which CIDNP evidence was reported [equation (17)]:

$$ArOH + NO^+ \rightarrow ArO^{\bullet} + NO + H^+; \qquad NO + NO_2^+ \rightarrow NO^+ + NO_2$$

$$\qquad (17)$$

The rates and isotope effects for the rearrangement of ipsonitrated intermediates from isotopically labeled N,N-4-trimethylaniline and N,N-3,4-tetramethylaniline have been studied.[126] The absence of ^{15}N incorporation in the 4-nitrophenol resulting from the nitrodenitrosation of 4-nitrosophenol by $H^{15}NO_2$[127] rules out the ipso attack mechanism proposed by earlier workers. Ipsonitrated intermediates formed from N,N-2,4,6-pentamethylaniline in 70% HNO_3 at 273 K have been isolated and characterized.[128] Kinetics of ipso attack suggest that two mechanisms are operative.[129]

Maya and Stedman have studied the decomposition of HN_3 in HNO_3[130] and propose the approximate stoichiometry (370 K, 9.18 mol-liter^{-1} HNO_3) in equation (18):

$$0.486HNO_3 + HN_3 \rightarrow 1.15N_2 + 0.27NO + 0.46N_2O + 0.73H_2O \quad (18)$$

Simple first-order dependence on $[HN_3]$ was seen with the observed first-order rate constant k_1 increasing with $[HNO_3]$. Plots of $\log k_1$ vs. H_0 (6.1-10.7 mol liter^{-1} HNO_3) have slope = 1.71, ascribed to changes in the concentration of the active N(V) species, probably NO_2^+, which gave N_3NO_2; the N_3NO_2 either fragmented to N_2 and 2NO or dissociated into N_3 and NO_2, with N_2O arising from sequence (19):

$$2NO_2 \rightarrow N_2O_4; \qquad N_2O_4 + HN_3 \rightarrow N_3NO + HNO_3;$$

$$N_3NO \rightarrow N_2 + N_2O \qquad (19)$$

Protonated HN_3 is not thought to be an intermediate. Using ^{15}N and ^{13}C NMR Olah et al. has detected the formation of the highly reactive $[NHRN_2]^+$ from RN_3 in FSO_3H/SbF_5 or (R = H) from $NaN_3/AlCl_3/HCl$.[131]

The solvolysis kinetics data for secondary alkyl nitrates have been used to question[132] current methods of calculating activation heat capacities $\Delta C_p^{\ddagger}$, a parameter of interest because it has been argued that a major contribution to its sign and magnitude arises from solvation changes during activation.

4.6.2. Dinitrogen Tetroxide

The kinetics and mechanisms of gas-phase reactions involving N_2O_4 have been reported for Me_3CNO (oxygen-atom transfer to give Me_3CNO_2),[133] alkenes (study of addition vs. hydrogen-atom abstraction),[134] and H_2NNMe_2 (giving HNO_2 and $Me_2NN=NNMe_2$).[135]

4.6.3. Nitrous Acid, Nitrite Esters, Nitrosation, and Related Topics

Nitrous acid and peroxomonophosphoric acid react in aqueous acid with a 1:1 stoichiometry, giving nitrate and phosphate. The acid dependency suggests involvement of H_3PO_5 and $H_2PO_5^-$ and HNO_2, with NO_2^- being kinetically relatively unimportant.[136,137] The specific rate constants for the processes (20) and (21) are 4407 and 4.43 liter mol^{-1} s^{-1}, respectively:

$$H_3PO_5 + HNO_2 \rightarrow HNO_3 + H_3PO_4 \qquad (20)$$

$$H_2PO_5^- + HNO_2 \rightarrow HNO_3 + H_2PO_4^- \qquad (21)$$

The reaction of HNO_2 and NH_4^+ and of HNO_2 and urea manifest oscillatory gas-evolution behavior dependent on experimental conditions.[138]

A reinvestigation[139] of the kinetics of the NO_2^-/SO_3^{2-} reaction to give hydroxylamine N,N-disulfonate, $HON(SO_3)_2^{2-}$, confirms the rate law previously established except that, in the pH range studied and with the buffer systems employed, no bisulfite independent path could be detected. In addition, evidence for $ONSO_3^-$ as an intermediate was obtained and this either hydrolyzed to give N_2O or reacted further with SO_3^{2-} to give $HON(SO_3)_2^{2-}$.

The kinetic order in $[HONHSO_3^-]$ for the reaction of NO_2^- with hydroxylamine N-sulfonate changes from zero to unity as $[HONHSO_3^-]$ increases. The reaction was first order in $[NO_2^-]$ and the rate dependent on $[H^+]$ possibly to the second order, though this was difficult to establish. A steady-state treatment for NO^+ in its nitrosation of $HONHSO_3^-$ to give H_2O, HSO_4^-, and H^+ gave a rate law which was fitted by the kinetic data.[140] It has been suggested that hydroxylammonium O-sulfonate $[\overset{+}{N}H_3OSO_3^-]$ reacts with Fe(II) to give $\overset{\cdot}{N}H_3$, which may abstract hydrogen atoms from organic species to give nucleophilic radicals which may be used synthetically to functionalize heteroaromatic bases.[141] In aqueous solution, Fremy's salt, $\overset{\cdot}{O}N(SO_3)_2^{2-}$ (NDS), is reduced either by $[Fe(CN)_6]^{4-}$ or H_2O_2, with a stoichiometry (22) for the latter reaction:

$$2\overset{\cdot}{O}N(SO_3)_2^{2-} + H_2O_2 \rightarrow O_2 + HON(SO_3)_2^{2-} \tag{22}$$

and kinetics in accord with the rate law and Scheme 1 for the $[H_2O_2]$-dependent route involving rate determining HO_2^- attack on N of NDS and unimolecular heterolysis of NDS giving $\overset{\cdot}{O}NSO_3$ and SO_3^{2-} for the $[H_2O_2]$-independent route.[142] Fremy's salt has also been shown to be an effective nitrosating agent for tertiary amines under mildly basic conditions.[143]

Scheme 1

$$\text{Rate} = \{k_1[NDS][H_2O_2]/(([H^+]/K_A) + 1) + k_0[NDS])\}$$

$$H_2O_2 \overset{K_a}{\rightleftharpoons} H^+ + HO_2^-; \qquad HO_2^- + \overset{\cdot}{O}N(SO_3)_2^{2-} \rightarrow \overset{\cdot}{O}N(SO_3)OOH^- + SO_3^{2-}$$

$$\overset{\cdot}{O}N(SO_3)OOH^- \rightarrow HO_2^{\cdot} + ONSO_3^-$$

$$HO_2^{\cdot} + \overset{\cdot}{O}N(SO_3)_2^{2-} \rightarrow O_2 + HON(SO_3)_2^{2-}$$

$$ONSO_3^- \rightarrow HONSO_3 \rightarrow HON(SO_3)_2^{2-}$$

The first-order rate constant for the nitrosation of alcohols at modest acidity (<0.1 mol liter^{-1}) reaches a limiting value at high [ROH].[144,145] This leveling off, seen also with other substrates, has been previously ascribed to the rate-limiting formation of NO^+, though the fact derived from this work that different alcohols at the same acidity gave markedly different limiting rates led to the proposal of an alternative, medium effect supported by additional observations in mixtures of THF and EtOH. Equilibrium concentrations of alkyl nitrite are formed by nitrosation of ROH (R = Me,Et,n-Pr,i-Pr), diols, and triols, with plots of observed first-order rate constants $vs.$ [ROH] giving rate constants for the forward and back reactions, both of which are catalyzed by H^+ and Cl^- or Br^-, interpreted in terms of rate-determining O nitrosation by $H_2NO_2^+$ (or NO^+), NOCl, or NOBr. The overall equilibrium constant in the absence of halide ions decreased in the order MeOH > EtOH > i-PrOH > t-BuOH, this being associated almost totally with the change in the rate of the forward reaction, suggesting the importance of steric effects.[145] Hydrolysis of nitrite esters RONO under neutral conditions is relatively slow, whereas exchange with R'OH is rapid.[146] Photolysis of methyl nitrite in an argon matrix[147] produces a hydrogen-bonded complex between H_2CO and HNO which react further to give two rotameric forms of the previously unknown C-nitrosomethanol. A correlation has been reported[148] for the computed values of the perturbation energies for the reaction of $H_2NO_2^+$ and various anions (F^-, Cl^-, Br^-, NO_2^-, CN^-, SCN^-, HS^-) and the experimentally determined equilibrium constant for the reaction.

The chemistry of N-nitrosamines has recently been reviewed.[149] In dilute aqueous H_2SO_4, the denitrosation of DL-N-acetyl-N^1-nitrosotryptophan occurred irreversibly and quantitatively with the kinetics first order in both [nitrosamine] and [H^+] and was unaffected by added Br^-, SCN^-, N_3^-, or I^-. At pH 6, catalysis by these anions was observed which became independent of X^- at high anion concentrations. These observations were interpreted in terms of two possible sites of protonation, O or C-3, and a change in the rate-determining step as the concentration of the nucleophile was changed.[150] Further mechanistic aspects of nitrosation and denitrosation processes involving organic substrates have been reported.[151-156] Nitrosations of a series of substituted thioureas, under low-acidity conditions favoring HONO or N_2O_3, lead to attack at nitrogen rather than at the normally reactive sulfur.[157] Diazotization at 298 K, pH 1.4, of $MeSCH_2CH_2CH(NH_2)CO_2H$, $MeSCH_2CH(NH_2)CO_2H$, and $MeCH(NH_2)CO_2H$ follows the rate-law: rate = k[HNO_2][substrate], the values of k being, respectively, 0.127, 0.109, and 0.0013 liter mol^{-1} s^{-1}. Initial S nitrosation followed by intramolecular S to N NO transfer[158] was proposed to account for the greater reactivity of the sulfur-containing compounds.

While earlier evidence for the intermediate formed in the reaction of hydroxylamine and nitrous acid has been interpreted in terms of either a symmetrical or nonsymmetrical structure, recent labeling studies are claimed definitively to point to a symmetrical intermediate with earlier evidence suggesting otherwise arising from experimental artifacts.[159] Reaction of $Na^{15}NO_2$ with excess $NH_2OH \cdot HCl$ solutions of varying acidity gave no significant amounts of NO, negligible $^{15}N_2O$, and abundance ratios for mass numbers 30 and 31 in the mass spectrum suggesting that N_2O containing equal amounts of $^{14}N^{15}NO$ and $^{15}N^{14}NO$ was formed over the entire acidity range. Bonner *et al.* further suggest that even in the pH range 6-9 the nitrosating species was HNO_2 and not N_2O_3.

Further details have been published of the reaction of doubly ^{15}N labeled NH_2NH_2 and excess HNO_2 under acid conditions[160] in which the isotopic distributions of ^{14}N and ^{15}N in the products N_2 and N_2O are not in accord with previously proposed mechanisms. No entirely satisfactory explanation of the data has been proposed with the least implausible being isotopic scrambling via a cyclic HN_3 species (see Volume 2 of this book, p. 87) or during the decomposition of N_3NO.

Nitrite may be oxidized to nitrate by O_2 in $NaNO_3/KNO_3$ melts at 773-873 K without the formation of intermediates, the forward rate being proportional to $[NO_2^-]$ and $[O_2]$. The rate data are thought not to be subject to the mass transfer limitations of earlier studies.[161]

4.6.4. Hydroxylamine and Nitroamine

In the borderline conditions which favor either NH_2OH scavenging of HNO_2- or HNO_2-catalyzed autoxidation of NH_2OH by HNO_3, shown in equations (23a) and (23b), the NH_2OH/HNO_3 system is very sensitive even

$$HNO_2 + NH_3OH^+ \rightarrow N_2O + H_2O + H_3O^+ \qquad (23a)$$

$$NH_3OH^+ + 2HNO_3 \rightarrow 3HNO_2 + H_3O^+ \qquad (23b)$$

to small changes of any reaction parameter, including the nature and dimensions of the reaction vessel.[162] Unusually, the initially homogeneous solution may develop two sharply delineated layers, an upper layer of HNO_2 in HNO_3 and a lower layer of NH_2OH in HNO_3, the boundary between them consisting of a zone which moves downward with time and in which HNO_2 diffuses downwards and NH_2OH upwards.

Oxidation of NH_2OH by iodate or periodate in acetic or perchloric acid displays oligo-oscillatory behavior depending on conditions, the phenomena being interpreted in terms of a set of reactions which include the intermediate formation of NH_2OHI^+ from NH_2OH and I_2.[163] Bistability has been seen in the autocatalytic reaction of Fe(II) and HNO_3 in a continuous-stirred tank reactor (CSTR) and has been interpreted in terms of mechanisms proposed to account for the batch clock reaction involving these reagents.[164] (See Ref. 138 for oscillatory behavior in the NH_3/HNO_2 system.)

Hughes *et al.* have continued their studies of nitroamine. They have confirmed an acid-catalyzed pathway in the decomposition of NH_2NO_2 by kinetic studies in HCl, H_2SO_4, and $HClO_4$[165] and by ^{18}O labeling studies in $HClO_4$.[166] The acidity dependence is indicative of the involvement of H_2O as a nucleophile in a rate-determining step. In $HClO_4$ up to 8 mol liter^{-1}, the N_2O produced by the acid-catalyzed pathway contained ^{18}O incorporated from the solvent H_2O. In neutral or slightly acidic solutions such incorporation was not observed and, as the acidity was increased, the amount of ^{18}O incorporation also increased. These results have required a modification of earlier mechanistic proposals, with H_2O attack on the amino N of the protonated aci-form of nitroamine in equation (24) now being favored:

$$NH_2NO_2 \rightleftharpoons NH{=}N(O)OH;$$

$$NH{=}N(O)OH + H^+ \rightleftharpoons [\overset{+}{N}H_2{=}N(O)OH] \qquad (24)$$

$$H_2O + [\overset{+}{N}H_2{=}N(O)OH] \rightarrow [\overset{+}{N}H_3OH] + HNO_2 \rightarrow N_2O + H^+ + 2H_2O$$

The NH_2OH so produced is isotopically labeled and HNO_2 exchanges O's with solvent in an acid-dependent process so that the extent of solvent label incorporation into N_2O will vary between 50 and 100%, depending on acidity. Protonation of $NH{=}N(O)OH$ will favor nucleophilic attack by H_2O without adversely affecting loss of neutral HNO_2 (in contrast to the effect of acid on hydrolysis of $NHOSO_3^-$).

Nitroamine and nitrous acid react with the stoichiometry in equation (25) in a process catalyzed by $X = Cl^-$, SCN^-. The observed rates fit a law

$$NH_2NO_2 + HNO_2 \rightarrow N_2 + HNO_3 + H_2O \qquad (25)$$

corresponding to the rate-determining attack of NOX on the deprotonated nitroamine, $NHNO_2^-$, with the specific rate constants for this process at 273 K being 1.3×10^4 liter mol^{-1} s^{-1} ($X = SCN$; $\sim 10^3$ times less reactive than for NH_2OH or NH_2OMe) and $\sim 2 \times 10^7$ liter mol^{-1} s^{-1} ($X = Cl$; similar to values for NH_2OH and NH_2OMe).[167] Scheme 2 was proposed. A change of stoichiometry was observed for $X = Br$ ascribed to a rapid reaction of NH_2NO_2 and Br_2 formed from NOBr.

Scheme 2

$$H^+ + HNO_2 + X^- \rightleftharpoons NOX + H_2O$$

$$NOX + NHNO_2^- \rightarrow ONNHNO_2 + X^-$$

$$ONNHNO_2 \rightarrow HONN{=}NO_2 \rightarrow HO^- + N_2 + NO_2^+$$

NH$_2$OH is oxidized by peroxomonophosphoric acid to H$_2$N$_2$O$_2$, HNO$_2$, or HNO$_3$ depending on the ratio [H$_3$PO$_5$]:[$\overset{+}{N}$H$_3$OH]. The process is catalyzed by Fe(II) or I$^-$.[168] In excess $\overset{+}{N}$H$_3$OH, H$_2$N$_2$O$_2$ was formed, in an acid-independent process, at a rate which depended linearly on [H$_3$PO$_5$] and [$\overset{+}{N}$H$_3$OH]. The derived rate expression, including the effect of I$^-$, is

$$-d[H_3PO_5]/dt = [H_3PO_5](k_1[\overset{+}{N}H_3OH] + k_3[I^-] + k_4[I^-][\overset{+}{N}H_3OH]).$$

4.6.5. Haloamines

NHCl$_2$ may be prepared essentially free of NH$_2$Cl and NH$_3$ by the disproportionation of NH$_2$Cl on an ion exchange resin in the H$^+$ form.[169] The decomposition of NHCl$_2$ is autocatalyzed by NCl$_3$ and HOCl, though the effect of the latter may be suppressed by NH$_4^+$. The reaction between NHCl$_2$ and HOCl is general base catalyzed [third-order rate constants decrease in the order OH$^-$ (3.3 × 10^9 liter^2 mol^{-2} s^{-1} at 298 K) > CO$_3^{2-}$ > OCl$^-$ > HPO$_4^{2-}$ (1.6 × 10^4 liter^2 mol^{-2} s^{-1})] giving NCl$_3$, OH$^-$, and BH$^+$, with B assisting proton removal from NHCl$_2$ as the nitrogen attacks the Cl of HOCl. NCl$_3$ reacted with NHCl$_2$ to give a range of products with reaction (26) thought to be of most importance. Ammonia reacts with NHCl$_2$ to give NH$_2$Cl by two processes, one first order in [NHCl$_2$] and in [NH$_4^+$]

$$NHCl_2 + NCl_3 + 3OH^- \rightarrow N_2 + 3Cl^- + 2HOCl + H_2O \qquad (26)$$

($k_2 = 3 \times 10^{-5}$ liter mol^{-1} s^{-1}) and the second first order in [NHCl$_2$] and first order in [OH$^-$] with $k_2 = 150$ liter mol^{-1} s^{-1}.

Cl transfer from NH$_2$Cl to amines, amino acids, and peptides[170,171] proceeds via nucleophilic attack of the unprotonated nitrogenous compound on the chlorine of the protonated amine. The apparent rate constants are quite small ($k = 0.11$ (MeNH$_2$), 0.51 (β-alanine), 5.9 (glycylglycine) liter mol^{-1} s^{-1}) because the large separation of pK_a for NH$_2$Cl and amine prevents the reactive forms of the two species from being abundant simultaneously. Specific second-order constants k_{Cl} increase with amine nucleophilicity from (3.3 ± 0.2) × 10^7 liter mol^{-1} s^{-1} for glycylglycine to a limiting value of (2.4 ± 0.1) × 10^8 liter mol^{-1} s^{-1}, suggesting the formation of an intermediate whose rate of decomposition is independent of amine nucleophilicity[171] (27):

$$[H_3NCl]^+ + RNH_2 \rightleftharpoons [H_3NClNH_2R]^+ \rightleftharpoons NH_3 + [RNH_2Cl]^+ \qquad (27)$$

It was concluded that $\overset{+}{N}$H$_3$Cl is a more effective chlorinating agent than HOCl and is only 10^1–10^3 less effective than Cl$_2$.

Oxidation of various amino acids by sodium N-chloro-p-toluene sulfonamide has been studied kinetically in both acidic and basic media. In alkali at 308 K, arginine and histidine react at a rate first order in both [substrate] and [oxidant] and with an inverse fractional order dependence on [OH⁻],[172,173] interpreted in terms of parallel reactions involving $R\bar{N}Cl$ and RNHCl and the amino acid to give the monochloroamino acid. This reacted with more oxidant to give the corresponding nitrile. In HCl, alanine and phenylalanine[174] show a similar stoichiometry in reactions with sodium N-chlorobenzenesulfonamide (28), with kinetic regimes dependent on

$$RCH(NH_2)CO_2H + 2PhSO_2NClNa \rightarrow RCN + 2PhSO_2NH_2$$

$$+ 2NaCl + CO_2 \qquad (28)$$

[HCl]. At <0.1 mol liter^{-1}, the reaction is independent of [amino acid] with the components of the rate law $-d[PhSO_2NClNa]/dt = [PhSO_2NClNa]\{k_1[H^+][Cl^-] + k_4[Cl^-] + k_8\}$ being associated with the following processes (29)-(31). The oxidants generated react rapidly with the amino acid. At >0.2 mol liter^{-1} HCl, rates are independent of [H⁺] and [Cl⁻], have a first-order dependence on [PhSO₂NClNa], and a fractional-order dependence on [amino acid], associated with reactions involving protonated substrates.

$$PhSO_2NHCl + H^+ + Cl^- \underset{}{\overset{k_1}{\rightleftharpoons}} PhSO_2NH_2 + Cl_2 \qquad (29)$$

$$PhSO_2NHCl + Cl^- \underset{}{\overset{k_4}{\rightleftharpoons}} PhSO_2NH^- + Cl_2 \qquad (30)$$

$$PhSO_2NHCl + H_2O \underset{}{\overset{k_8}{\rightleftharpoons}} PhSO_2NH^- + H_2OCl^+ \qquad (31)$$

Margerum et al. have more carefully studied the latter stages of these reactions when the N-chloroamino acids fragment to give NH_3, CO_2, Cl^-, and carbonyl compounds via imino intermediates.[175] These reactions were found to be independent of acidity (pH 5-9), the rates being first order in [N-chloroamino acid], and the rate constants at 298 K varying from 4.2×10^{-6} s^{-1} for glycine to 9.0×10^{-2} s^{-1} for 1-amino-1-carboxycyclohexane. The reactivity range and the large positive $\Delta S^{\ddagger}$ values were thought to be consistent with a concerted fragmentation process controlled by conformational, steric, and electronic effects of R on the ease with which the intermediate ($O_2CRR'CNHCl$) could adopt the antiperiplaar structure about the C–N bond.

Oxidations of α-amino acids by sodium N-bromobenzenesulfonamide[176] and the decomposition of N-brominated alanine[177] have both been studied mechanistically. The reaction of HOBr and OBr⁻ with NH_3 at $\geq$pH 9.9 gave only NH_2Br, whereas at $\leq$pH 8.2 $NHBr_2$ was the ultimate product.[178] The kinetics at 293 K were consistent with the rate law $-d[total bromine]/dt = k_a[HOBr][NH_3] + k_b[OBr^-][NH_3]$ having $k_a = 7.5 \times 10^7$ liter mol^{-1} s^{-1} and $k_b = 7.6 \times 10^4$ liter mol^{-1} s^{-1}. Contrary to earlier reports, BrO_2^- was found not react with NH_3 to give NH_2Br. The reactions of aqueous bromine with amino compounds were more rapid than with NH_3, with k_a for glycine $= 3.8 \times 10^8$ liter $mol^{-1} s^{-1}$ and $k_b = 2.1 \times 10^5$ liter $mol^{-1} s^{-1}$. For $MeNH_2$ the reaction was too rapid to measure.

Nitrogen triiodide ammoniate has been reported to react in liquid NH_3 to give an unstable species thought to be NH_2I and, in dilute solution, NH_2I decomposed according to the stoichiometry (32) in a first-order

$$3NH_2I \cdot xNH_3 \rightarrow 3[NH_4]I + N_2 + (x - 2)NH_3 \qquad (32)$$

process having $k = (7.22 \pm 0.17) \times 10^{-6} s^{-1}$ at 236 K, $E_a = 60 \pm 2$ kJ mol^{-1}, and log $A = 8.2$.[179]

4.6.6. Other Systems

First-order kinetics have been observed for the decomposition of alkyl hyponitrites ($RON=NOR$) in isooctane. Alcohols, aldehydes, and ketones are formed though the stoichiometry has not been established. Half-lives increase in the order $R = PhCH_2$ (3.0 ± 0.1 min) $< PhCHMe <$ $PhCH_2CH_2 < Me <$ cyclohexyl $< t$-butyl $< i$-propyl.[180]

A theoretical study has proposed that cis–trans isomerization of $MeN(O)N(O)Me$, which is symmetry forbidden by a twisting mechanism, occurs via dissociation and recombination of $MeNO$ monomers.[181]

$(CF_3S)_2NN(SCF_3)_2$ gives $2(CF_3S)_2\dot{N}$ in a homolytic process with $K_{eq} = (4 \pm 2) \times 10^{-6}$ mol liter^{-1} and a forward rate constant (at 298 K and independent of the galvinoxyl added to suppress the back reaction) of $1.9 s^{-1}$. The reverse rate constant was then computed to be $(5 \pm 2) \times 10^5$ liter $mol^{-1} s^{-1}$.[182] Aminyl radicals, $(RCH_2)_2\dot{N}$, are formed by hydrogen-atom abstraction from $(RCH_2)NH$ by t-BuO. $RCH_2NH\dot{C}HR$ is not co-produced by H abstraction from the starting amine by the aminyl radical.[183] The nature of the aminyl radical formed in bromine/N-bromosuccinimide brominations continues to be the subject of controversy.[184]

$NH_4NO_3 \cdot 3NH_3$ is formed from the reaction on warming of O_3 and NH_3 condensed together at 77 K. At >103 K, spectroscopic studies suggest two intermediates are formed, a charge transfer complex $\overset{+}{N}H_3 \cdot \overset{-}{O}_3$ and possibly $NH_2\overset{+}{O} \cdot \overset{-}{O}_2H$, which are transformed at higher temperatures to the observed product. The system NH_2OH/O_3 was also studied.[185]

The prevention of the back reaction (usually diffusional controlled) in light-activated redox processes would allow such systems to be considered for the quantum storage of light energy. One approach has involved the sensitized photoreduction of NO_3^- to NO_3^{2-} in micellar solution with NO_3^- being rapidly converted to NO_2 and OH^- thus preventing the back reaction.[186]

4.7. Phosphorus

The hydrolyses of cyclic triphosphates and tetraphosphates have been studied in the presence of a series of metal halides (M = Li, Na, K, Mg, Ca, Ni, Cu, Al). The reactions were all first order in substrate and subject to catalysis by acid and base. Alkali metals retard the acid hydrolysis and accelerate reaction in basic media, in the order $Li^+ > Na^+ > K^+$. Divalent cations retard the hydrolysis at pH < 2.7 and accelerate it at pH > 2.7.[187] Kura[188] has examined chromatographically the products from the acid hydrolyses of cyclohexaphosphate and cyclooctaphosphate and has confirmed first-order kinetic behavior, with the rate constant declining for $[PO_3]_n^{n-}$ with $n = 3$-6 and then increasing to $n = 8$. Ring opening of $[PO_3]_n^{n-}$ for $n = 3$ by ethylenediamine has also been investigated.[189] Urea accelerates the hydrolysis of pyrophosphate $P_2O_7^{4-}$ possibly via hydrogen-bond formation, though the inhibition of triphosphate hydrolysis by urea remains unexplained.[190]

Because of its importance in the synthesis of phosphoanhydride linkages in nucleoside polyphosphates, the stereochemistry of the coupling process (33) has been investigated using chiral, specifically ^{18}O-labeled, substrates, for example, (22),[191] with loss of chirality at the $[^{18}O]$-phos-

$$(PhO)_2\overset{\overset{\displaystyle ^{18}O^-}{|}}{\underset{\overset{\displaystyle ||}{O}}{P}}-O-\overset{\overset{\displaystyle *}{|}}{\underset{\overset{\displaystyle ||}{S}}{P}}-OAdo + ROPO_3^{2-} \xrightarrow{\text{pyridine}} R\overset{\overset{\displaystyle O}{||}}{\underset{\overset{\displaystyle |}{O^-}}{P}}-O-\overset{\overset{\displaystyle S}{||}}{\underset{\overset{\displaystyle |}{O^-}}{P}}-OAdo$$

22

$$+ (PhO)_2PO_2^- \qquad (33)$$

(Ado = adenosyl-; R = H, HOP(O)$_2^-$, HOP(O)$_2$OP(O)$_2^-$)

phorothioate center being associated with involvement of pyridine in the formation of a rapidly equilibrating intermediate, $[C_5H_5NP(^{18}O)\text{-}(S)OAdo]$.

The intermediacy of cyclodiphosphates [e.g., (23)] has been proposed[192] to explain the consequences of the reaction of the bis^{18}O-labeled species (24) with CNBr in H_2O. ^{31}P NMR chemical shifts suggest that sulfur may be replaced by the oxygen bridging α- and β-phosphorus, according to the Scheme 3.

In a combination of kinetic stucies and ^{18}O-labeling studies, Meyerson *et al.*[193] have examined the nonenzymatic hydrolysis of adenosine-5′-triphosphate (25) which, in principle, can involve one of four processes, depending on whether attack occurs at the α-, β- or γ-phosphorus. In the acid region, 93% of reaction occurs at the terminal (γ) phosphorus and 7% on β-phosphorus, with the latter giving ADP and orthophosphate and

Scheme 3

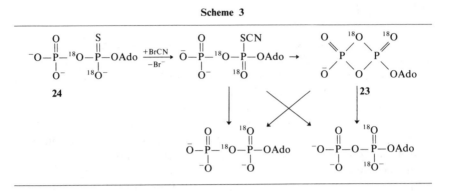

not AMP and diphosphate.[193] Similarly, ADP is hydrolyzed primarily by attack at terminal (β)-phosphorus. No exchange was noted between ATP, ADP, or PO_4^{3-} and water.

$$[\text{Ado-O-P}_\alpha(O)_2\text{-OP}_\beta(O)_2\text{-OP}_\gamma(O)_2O]^{4-}$$

25

A bicarbonate-dependent enzyme-catalyzed positional isotope exchange has been observed for (**25**) labeled either by ^{18}O or ^{17}O in the β-γ-bridge position,[194] in which the label is transferred to the β-nonbridge position. Labeled (**25**) was produced as a mixture of nonlabeled, bridge-labeled, and nonbridge-labeled compounds characterized by ^{31}P NMR isotopic chemical shifts. Cullis[195] has recently reported a means of obtaining specifically the β-γ ^{18}O-bridged compound.

Thiaminopyrophosphate L^{2-} (the coenzyme form of vitamin B_1) undergoes a spontaneous pH-dependent hydrolysis first order in the substrate. The kinetics[196] have been analyzed in terms of the three most important species in solution, namely H_2L, HL^-, and L^{2-}, with activation parameters pointing to an S_N1 process involving the intermediacy of PO_3^- for L^{2-} and internal proton transfers via cyclic six-membered rings for H_2L and HL^-, also leading to PO_3^-.

Reynolds, *et al.* consider ^{17}O NMR to be a more sensitive tool than ^{31}P NMR for studying the interactions of protons and phosphoryl oxygens[197] and imidophosphate derivatives.[198] Assignments of all resonances were made for ATP and ADP and pK_a values obtained using ^{17}O NMR for a variety of compounds, the values obtained being in reasonable agreement with those obtained by ^{31}P NMR. ^{17}O and ^{15}N NMR on enriched $O_2P(O)NHP(O)O_2^{4-}$ and $(EtO)_2P(O)NHP(O)(OEt)_2$ showed ^1H-^{15}N couplings of 70 Hz pointing to the imido tautomeric structure in both cases. An imido analogue of ATP, similarly enriched, showed similar couplings which were retained in the presence of stoichiometric quantities of Mg^{2+}. ^{17}O NMR showed that protonation of $O_2P(O)NHP(O)O_2^{4-}$ occurred exclusively on oxygen and that monoprotonation of the ATP-imido analogue occurred exclusively on the oxygens of terminal P. The tautomerism of oxocyclotriphosphazadienes has also been investigated.[199]

The previously unknown symmetrical monothiopyrophosphate $O_3PSPO_3^{4-}$ has been synthesized by cleavage of $(MeO)_2P(O)SP(O)(OMe)_2$. The lithium salt hydrolyzes without isomerization to equimolar amounts of PO_4^{3-} and PSO_3^{3-} with a half-life of 4.23 h at pH 10.05 and 0.92 h at pH 9.00, in a reaction slowed by Mg^{2+}.[200]

The acid-catalyzed hydrolysis of a series of dimethyl N-(alkylphenyl)phosphoramidates $(MeO)_2P(O)NHAr$ to give $(MeO)_2P(O)-OH$ and $\overset{+}{N}H_3Ar$ gave activation parameters indicative of an A2 mechanism with log k_{obs} vs pK_a for the leaving group giving two linear correlations, one for orthosubstituted Ar and the other for meta- and parasubstituted Ar, suggesting the importance of steric effects for orthosubstitution.[201] Activation-parameter data for the alkaline solvolysis of diamidophosphates, $(ArO)P(O)(NHR)_2$, have led Mollin et al.[202] to prefer an S_N2 mechanism involving pentacoordinate intermediates rather than the E_1CB mechanism proposed by earlier workers.

Evidence from ^{18}O-labeling studies has been obtained which argues against hexacoordinate phosphorus intermediates in the alkaline hydrolysis of phosphate esters. Treatment of $EtO\overset{\frown}{P}(O)OCH_2CH_2\overset{\urcorner}{O}$ in dry dioxane with 5.0 mol liter^{-1} NaOH in 72% $H_2^{18}O$ followed by controlled reacidification gave only $EtOP(O)(OH)OCH_2CH_2OH$ and $EtOP(^{18}O)(OH)OCH_2CH_2OH$, whereas a hexacoordinate intermediate would require incorporation of two moles of label.[203] At pH 2–7 both exo and endo cleavage is seen with the former giving monolabeling and the latter leading to the incorporation of up to two atoms of label.

The reaction between methanol and 2-methoxy-2-oxo-4,5-dimethyl-1,3,2-dioxaphosphol-4-ene (26) in $CHCl_3$ is first order in (26) and zero order in MeOH, leading to solvolytic ring opening of the PO_2C_2 ring. The reaction is retarded by an equimolar quantity of LiX with the half-life for the reaction increasing from 5.6 to 11.4, 13.2, and 31.8 min for X = F, X = Cl, and X = Br, respectively. ^1H NMR in acetone is in accord with the complexation of Li^+ by P=O and the shift in the reaction pathway to one involving dealkylation of the exocyclic methoxy group.[204]

Abell and Kirby[205] have shown that bis(2-carboxyphenyl)phosphate is hydrolyzed via its dianion at a rate 10^{10} times greater than diphenylphosphate, with intramolecular nucleophilic catalysis by one ionized carboxyl group and (unexpectedly inefficient) intramolecular general-acid catalysis of the breakdown of the pentacoordinate monocarboxyphenyl ester intermediate by the other.

Cyclic anhydrides have been proposed as intermediates in the anomalous reactions of N-(amino(methyl)phosphinyl)-L-phenylalanine derivatives in alcohols to account for the formation of $MeP(O)(X)$-$NHCH(CH_2Ph)C(O)Y$, (**27**), $X = OR$, and $Y = OH$ from $MeP(O)(NH_2)$-$NHCH(CH_2Ph)CO_2^-$ and (**27**), $X = OR$, and $Y = NH_2$ from $MeP(O)$-$(NH_2)NHCH(CH_2Ph)CO_2Me$.[206,207]

The neutral and acid-catalyzed hydrolyses of a series of phosphoric carboxylic imides $X_2P(O)NRC(O)R'$ (**28**) ($X = EtO$, MeO, Et; $R = H$, Me; $R' = Me$, Ph) have been compared with the behavior of related carboxylic acid and phosphoric amides, particularly in acidic conditions where the latter two systems differ greatly. In acid the carboxylic amides react slowly by attack of solvent on the oxygen-protonated form of the substrate and the phosphoric amide reacts rapidly by direct $S_N2(P)$ displacement via an N-protonated substrate. The acid hydrolysis of (**28**) lies between these two extremes with the protonation of carbonyl oxygen stabilizing the resonance structure which renders the phosphorus atom more electrophilic and the P–N bond less labile with the consequence that the reactivity of the P–N and C–N linkages become comparable.[208]

The arylphosphorane (**29**) displays a single aliphatic ^{13}C NMR resonance at room temperature due to a rapid pseudorotation which ^1H NMR shows does not equilibrate methylene protons. A pseudorotation via (**30**) would lead to such equivalencing but the process is not favored at ≤ 453 K. However, the methylene proton signal is coalesced by the addition of $MeOSO_2CF_3$[209] via the formation of a phosphonium intermediate (**31**).

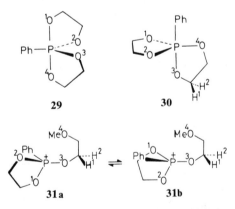

Hydrolysis of the phosphoranes $\overline{OC(CF_3)_2C(CF_3)_2O}$–PABC (A = Ph, OR; B = Ph, OR; C = Cl, OR) containing the perfluoropinacolyl ring gives acyclic products under neutral, acidic, or basic conditions. The equilibrium between acyclic and cyclic isomers of $Ph_2P(O)OC(CF_3)_2C(CF_3)_2OH$ in solvents such as Me_2CO, CH_2Cl_2, and MePh leads to equilibration of CF_3 groups at temperatures >333 K. This exchange may be promoted at 298 K by bases such as pyridine, suggesting deprotonation followed by ring closure.[210]

The activation parameters for the pseudorotation of (32), which equilibrates the oxygen donors of the chelate between apical and equatorial positions, suggest that the geometrical preferences for the six-membered ring are determined by the R group.[211]

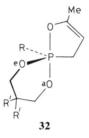

32

^{31}P NMR line-shape analysis has yielded measures of the barriers for pseudorotation in aminophosphoranes F_4PNHMe, $F_4PNMeBz$, $Me(CF_3)_2P(F)NMe_2$, $Me(CF_3)_2PF(NHMe)$, $F(CF_3)_3PNMe_2$, and $F(CF_3)_3$-PNHMe, where the $\Delta G^{\ddagger}_{298}$ values for the process are consistently larger for the derivatives containing the primary amido group compared with those for a secondary compound. These differences were ascribed in part to H bonding between N–H and the axial fluoride.[212,213]

Bimolecular nucleophilic substitution at phosphorus has been studied by Corriu using concepts based on the nature of the nucleophile and leaving group developed for silicon. Displacement of Cl^- from derivatives of the type $\overline{OCR_2CR_2O}P(O)Cl$ (33), $\overline{OCR_2CR_2CR_2O}P(O)Cl$ (34), or $(EtO)_2P(O)Cl$ (35) by a variety of alcohols, phenols, and amines all obey second-order kinetics, with the rate constants k_2 for (34) and (35) being highly dependent on the nucleophile (covering a range of $\sim 10^4$), whereas for the more highly strained (33) the reaction was generally more rapid, with k_2 being less sensitive to the nature of the nucleophile (covering a range of ~ 50). The nucleophile influences more the rates of those reactions which lead to inversion (with apical entry of nucleophile and apical departure of leaving group in the trigonal bipyramidal intermediate) than it does the rates of those which lead to retention (with nucleophile and leaving group in an equatorial–apical relationship in the intermediate), a situation similar to that seen for substitutions at silicon.[214]

Contrary to expectations based on an associative mechanism, the reaction of t-BuP(O)Cl$_2$ with amines such as t-BuNH$_2$ gave only t-BuP(O)-(NHt-Bu)$_2$ and unchanged starting material without detectable amounts of t-BuP(O)Cl(NHt-Bu)(36).[215] It was shown that when independently prepared, (36) reacted with amines in a process that was unusually insensitive to steric effects, which the author interpreted in terms of an elimination-addition process (34). Less bulky alkylphosphonic dichlorides (RP(O)Cl$_2$; R = Me, Et, i-Pr) reacted as expected by an associative process.

$$t\text{-BuP(O)Cl(NH}t\text{-Bu)} + R^1NH_2 \rightarrow [R^1NH_3]^+Cl^- + t\text{-BuP(O)}{=}Nt\text{-Bu}$$
$$\mathbf{36} \qquad\qquad\qquad\qquad\qquad\qquad\qquad \downarrow$$
$$t\text{-BuP(O)(NH}t\text{-Bu)(NHR)} \qquad (34)$$

The mechanism of phosphoryl-group transfer between heterocyclic nitrogens has been investigated by two groups of workers.[216,217] It was established that a process involving a PO$_3^-$ intermediate with a significant lifetime could be excluded and the distinction between a concerted process with one transition state and a preassociation mechanism with two was resolved by the use of Brønsted plots for a series of pyridines which both groups found to be linear, consistent with a single transition state which is symmetrical over a wide range of basicities.

Dillon et al. have examined the structure and fluxionality of a range of pseudohalogen derivatives of P(V) using ^{31}P NMR,[218-221] including PF$_{6-n}$(CN)$_n^-$ ($n = 1$-4) and PF$_3$Cl$_{3-n}$(CN)$_n^-$ ($n = 1$-3),[218] in which a sharp division between stereochemical rigidity and fluxionality is seen when the number of F's is increased from 3 to 4. LiN$_3$ reacts with PF$_n$Cl$_{6-n}^-$ ($n = 1$-3) and AgNCS with PF$_n$Cl$_{6-n}^-$ ($n = 2,3$) to displace Cl$^-$ only. PCl$_6^-$ reacts rapidly with LiNCS to give P(NCS)$_6^-$ and rather more slowly with LiCN only as far as the two isomers of PCl$_3$(CN)$_3^-$.[219] PF$_6^-$ and P(N$_3$)$_6^-$ exchange ligands only slowly in CH$_2$Cl$_2$ at room temperature, with PF(N$_3$)$_5^-$ detected after four days and PF$_2$(N$_3$)$_4^-$ after two weeks. In contrast, distribution reactions between PF$_6^-$ and P(NCS)$_6^-$ are more rapid with the fluxional species PF$_5$(NCS)$^-$ and cis-PF$_2$(NCS)$_4^-$, being seen after only one day.[220] In the reaction of RPCl$_5^-$ (R = Me,Ph), complete substitution by CN$^-$ occurs for R = Me, but reaction proceeds only as far as RPCl$_2$(CN)$_3^-$ for R = Ph.[221]

(EtO)$_2$Ṗ$=$O radicals, obtained from t-BuȮ and (EtO)$_2$P(H)$=$O,[222] abstract halogen from alkyl halides, with competition studies being used to establish relative rates. Absolute second-order rate constants for reaction with butyl halides are ~10^4 times slower for the phosphonyl radical compared with Et$_3$Ṡi.[223] Phosphinyl-radical intermediates are implicated from ESR and ^{31}P CIDNP experiments when ketoximato-phosphorus(III) intermediates (formed from ketoximes and X$_2$PCl at low temperatures) rearrange unimolecularly to the corresponding N-phosphinylated imine as in equation (35):[224]

$$R_2C=NOH + X_2PCl \rightarrow R_2C=NOPX_2 \rightarrow R_2C=NP(O)X_2 \quad (35)$$

Catalytic oxidation of H_2O by peroxodiphosphate has been studied under acid conditions where competing hydrolysis to peroxomonophosphate is avoided. The stoichiometry (36) has been established and the rate law found to be first order in $[P_2O_8^{4-}]$ and $[Ag^+]$ from initial rate measurements, $H_2P_2O_8^{2-}$ and $HP_2O_8^{3-}$ being the most important species.[225] Oxidations of aliphatic aldehydes with peroxomonophosphate have been studied mechanistically.[226]

$$K_4P_2O_8 + H_2O \xrightarrow{Ag^+} 2K_2HPO_4 + \tfrac{1}{2}O_2 \quad (36)$$

Nucleophilic substitutions at phosphorus(III) have been little studied though it is known that R^- or MeO^- displace leaving groups such as OR^-, NR_2^-, and Ph^- with inversion at P, though when the leaving group is Cl^- the reaction is stereochemically unselective. Dahl[227] has shown that displacement of Me_2N from $P(NMe_2)_3$ by i-PrOH in $CDCl_3$ is very slow in the absence of amine salts $[R_2NH_2]Cl$ present as impurities. Kinetic studies in the presence of catalytic quantities of $[Me_2NH_2]Cl$ have allowed a reactivity order for a series of amines to be established, though they do not clarify the precise nature of the substitution-step proper. P(III) compounds such as $(Me_2N)_3P$, $(t\text{-BuCH}_2O)_3P$, Ph_3P, or $(PhO)_3P$ may desulfurize, deselenize, or deoxygenate $\rangle P(O)SSP(O)\langle$, $\rangle P(O)SeSeP(O)\langle$, or $\rangle P(O)S\text{-}Cl$ in facile processes.[228] The salt (37) is identified as an intermediate at 173 K and reacts further by paths a (17%) or b (83%) shown in Scheme 4.

Scheme 4

$(EtO)_2P(O)SSP(O)(OEt)_2 + Ph_3P \rightleftharpoons [(EtO)_2P(O)S\overset{+}{P}Ph_3][(EtO)_2PO\overset{-}{S}]$

$$\xleftarrow{\quad a \quad} \; 37 \; \xrightarrow{\quad b \quad}$$

$(EtO)_2P(O)OP(S)(OEt)_2 + Ph_3PS \qquad (EtO)_2P(S)O\overset{+}{P}(S)(OEt)_2 + Ph_3PO$

In a related reaction, Ph_3P is oxidized by peroxodisulfate to give Ph_3PO and sulfate with kinetics first order in each reagent. Radicals are not involved. The rate-determining step is thought to involve nucleophilic displacement of SO_4^{2-} from $S_2O_8^{2-}$ giving a $[Ph_3\overset{+}{P}OSO_3^-]$ intermediate which hydrolyzes rapidly.[229] The chiral phosphine (R)-$(-)$-MePrPhP* reacted with retention of configuration with O-mesityl sulfonyl hydroxylamine, consistent with bimolecular substitution at nitrogen, whereas chloramine reacted with racemization, consistent with the formation of a trigonal bipyramidal intermediate.[230]

The alkaline hydrolysis of $[Et_nPPh_{4-n}]^+I^-$ is a third-order process proceeding according to (37). The third-order rate constants were greater in THF/H_2O at 313 K than they were in MeOH/H_2O at 323 K, with the ratio decreasing as n, the number of ethyl groups, increases,[231] ascribed to the greater electron-releasing power of Et which affects solvation of the reactants.

$$R_4P^+OH^- \rightleftharpoons R_4POH \underset{H^+}{\overset{OH^-}{\rightleftharpoons}} R_4PO^- \rightarrow R_3P{=}O + R^- \xrightarrow{H_2O} RH + OH^-$$

$$(37)$$

Synthetic work has been directed towards the isolation or detection of species which may be implicated as intermediates in mechanistic schemes, including monomeric aryldioxophosphoranes $ArPO_2$.[232-234]

4.8. Arsenic and Antimony

Kinetic studies of the hydrolysis of trialkyl arsenites, $(RO)_3As$, have confirmed earlier work that the removal of the first OR is slower than that of the second or third. In solutions in which $[H_2O] \gg [(RO)_3As]$, first-order plots departed from linearity and indicated the buildup of either $(HO)As(OR)_2$ or $(HO)_2As(OR)$ as intermediates.[235] In 13.0–20.0 mol liter^{-1} H_2O in CH_3CN, the rate law $-d[As(OR)_3]/dt = k[H_2O][(RO)_3As]$ was derived with k (298 K, 30% aqueous CH_3CN) dropping from 4.4 liter mol^{-1} s^{-1} (R = Me) to 0.29 (R = Et) and to 0.0086 (R = i-Pr). The large negative $\Delta S^{\ddagger}$ values were consistent with earlier oxygen-exchange studies and pointed to a highly associative mechanism with the formation of an intermediate.

The kinetics of the reduction of $[Ag(OH)]_4^-$ by arsenite have been reported[236] and the oxidation of Sb_2O_3 in alkaline H_2O_2 studied by Russian workers.[237] The bromide exchange between $SbBr_3$ and t-BuBr[238] and $PhCH_2Br$[239] in nitrobenzene or 1,2,4-$C_6H_3Cl_3$, monitored using ^{82}Br labels, exhibits first-order dependence on the alkyl halide and first-order dependence on $SbBr_3$ at low concentrations and second-order dependence on $SbBr_3$ at high concentrations, in the case of $PhCH_2Br$, but 1.5-order dependence on $SbBr_3$ at all concentrations studied, in the case of t-BuBr. The authors propose that both reactions proceed via rate-determining formation of $[R^+][SbBr_4^-]$ and/or $[R^+][Sb_2Cl_7^-]$.

Stereochemical changes in $[Sb(OR)_5]_2$ have been studied by NMR.[240]

4.9. Oxygen

Accurate kinetic data for the reactions of benzyl, cyclohexyl, t-butyl, 1-(1-diethylamino)ethyl, and 2-(2-hydroxy)propyl radicals with oxygen in the liquid phase have been reported by Scaiano et $al.$[241] who find that, under conditions where competing decay processes can be minimized and the reaction can be assumed to be irreversible, rate constants for the $Ph\dot{C}H_2/O_2$ reaction at 300 K fall from $(2.78 \pm 0.32) \times 10^9$ liter $mol^{-1} s^{-1}$ in hexane to $(1.04 \pm 0.05) \times 10^9$ liter $mol^{-1} s^{-1}$ in hexadecane, ascribed to the effect of increased solvent viscosity.

The peroxy radicals, t-Bu$\dot{O}_2$ and tetralylperoxy, have been spin trapped using phenyl-N-($tert$-butyl)nitrone and methyl-N-durylnitrone.[242] Singlet oxygen is formed when gaseous Cl_2 is mixed with concentrated aqueous H_2O_2, the percentage of $^1O_2$ decreasing with pH, dropping to zero at pH 5. MNDO calculations,[243] taken with earlier evidence, suggest that the reactions of Cl_2 at either the terminal O or the H of HO_2^- should be the most important processes, the latter leading to HCl, Cl^-, and $^1O_2$ directly, with the former proceeding via a reaction with base of the intermediate $HO_2Cl_2^-$. The lifetime of $^1O_2$ ($^1\Delta_g$) increases in the order MeOH (9 μs) < pyridine < cyclohexane $\sim$ 1,4-dioxan < acetone < CH_2Cl_2 < $CHCl_3$ (62 μs) at room temperature.[244] The reactions of singlet oxygen with conjugated dienes,[245] enol ethers,[246,247] enol esters,[248] and olefins[249] involve [2 + 2] cycloadditions to give dioxetanes, [4 + 2] cycloadditions to give endoperoxides, and the "ene" reaction to give hydroperoxides via concerted or stepwise processes possibly involving biradical, perepoxide, or zwitterionic intermediates. Hurst and Schuster[249] found that the rates of the ene reaction for 12 olefins spanned a 10^5-fold range, with the variation determined almost totally by changes in the activation entropy, associated with the stereochemical requirements in forming the transition state from a weakly bound encounter complex. Gorman et $al.$[246] favor solvation entropy arguments and reject direct product formation from encounter complexes to account for their observations on enol ethers.

The reaction between the hydrated electron and O_3 in bicarbonate-containing solution at pH 9 gives O_3^- with a rate constant of $(3.60 \pm 0.2) \times 10^{10}$ liter $mol^{-1} s^{-1}$ with routes via $O^{\cdot-}$ (formed by deprotonation of $\dot{O}H$), O_2^- or O being ruled out. Peroxy radical[250] [formed from e^-(aq) and O_2] reacted with O_3 to give O_3^- and O_2 stoichiometrically, with a rate constant at pH 10.3 of $(1.52 \pm 0.05) \times 10^9$ liter $mol^{-1} s^{-1}$. From competition studies involving $\dot{H}$ with O_3 and O_2, the rate constant for the $\dot{H}/O_3$ reaction at pH 2 was estimated to be $(3.65 \pm 0.4) \times 10^{10}$ liter $mol^{-1} s^{-1}$.

The kinetic and mechanistic aspects of the reaction of ozone with olefins[251-256] and acetylene[257] in the liquid phase have been extensively studied. Details of the Criegee mechanism have been probed using deuterium labeling,[253] by stereochemical studies of the cross ozonide product arising from reactions involving propene,[254] by IR matrix isolation studies using $^{18}O_3$,[255] and by ESR investigations of radicals formed in minor reaction pathways.[256] A hydrotrioxide is the suggested intermediate in the reaction of O_3 and ćumene,[258] carboxylic acids, alcohols, and carbonyl compounds.[259] The reaction of O_3 and alcohols is catalyzed by H_2O_2, the effect being ascribed to the formation of $\dot{O}H$ from the reaction of H_2O_2 and O_3.[260] Spin-trapping experiments have shown that both $\dot{H}$ and $\dot{O}H$ are generated in the cavitation bubbles produced by ultrasound in argon-saturated H_2O. The presence of air suppressed the formation of $\dot{H}$ but not $H\dot{O}$.[261]

The nature and reactions of the superoxide, $O_2^{\pm}$, have been studied by a number of groups. Sawyer *et al.*[262] suggest that solid $[Me_4N][O_2^-]$ contains significant amounts of the dimer O_4^{2-}. This dissociates readily in water and the solution is found to be surprisingly stable if 10% acetonitrile is present.[262,263] Electron-spin echo studies on frozen aqueous solutions of $O_2^{\pm}$ at $4.2\,K$[264] suggest that each $O_2^{\pm}$ is surrounded by four water molecules. $O_2^{\pm}$ disproportionates in excess acid giving H_2O_2 and O_2, with reactions involving mixtures of $^{16}O_2^{\pm}$ and $^{18}O_2^{\pm}$ ruling out H_2O_4 as an intermediate.[262]

The kinetics of $O_2^{\pm}$ protonation in EtOH and H_2O have been studied.[265] The dismutation of $O_2^{\pm}$ to H_2O_2 and O_2 is catalyzed by Fe(II) and Fe(III) complexes.[266] A value of the pK_a for $H\dot{O}_2$ of 4.8 ± 0.05 is proposed by different workers who studied the formation and decay kinetics of $O_2^{\pm}$ and $H\dot{O}_2$ generated by pulse radiolysis,[267] and obtained a value for $2k$ ($H\dot{O}_2$ + $H\dot{O}_2$) of $(3.7 \pm 0.2) \times 10^6$ liter $mol^{-1}\,s^{-1}$ in the pH range 1.5–8 and for $2k$ ($O_2^{\pm} + O_2^{\pm}$) < 10 liter $mol^{-1}\,s^{-1}$. $O_2^{\pm}$ may also be generated from alkaline H_2O_2.[268] The kinetics and mechanisms of reactions of $O_2^{\pm}$ and ethyl acetate,[269] ascorbic acid, and dehydroascorbic acid[270,271] have been reported.

Using purified alkali and chelating agents to control catalytic processes, Galbács and Csányi have estimated a rate constant for the noncatalyzed alkali-induced decomposition of H_2O_2 at pH 11.6 and 308 K of 3×10^{-6} liter $mol^{-1}\,s^{-1}$.[272] On the other hand, Špalek *et al.* believe that the base-dependent kinetics are not due to the participation of OH^- but result from heterogeneous processes occurring at the surfaces of colloidal particles of metal oxide or hydroxide or at glass surfaces.[273] The decomposition of $Na_2CO_3 \cdot 1.5H_2O_2$ in the presence of water probably involves metal ion catalysis and may be inhibited by sodium silicate.[274] A chain decomposition of H_2O_2 is induced by catalytic quantities of the reduced form of 5-methylphenazinium cation (MPH) and metal ions (e.g., Fe^{2+}, Fe^{3+}). The kinetics of the reaction (zero order in $[H_2O_2]$; first order each in the initial concentrations of (MPH) and Fe(III)) show that the rate-determining step involves a redox reaction between MPH and an Fe(III) citrate complex.[275]

H_2O_2 and the reduced form of methyl viologen react to give $O_2^{\dot-}$ or $H\dot{O}_2$.[276] H–D kinetic isotope effects of 16 and 20 are reported in the oxidation of H_2O_2 by Ru(IV) and Ru(III) complexes.[277] H_2O_2 has been studied as one of a series of monooxygen transfer agents ROOH (R = H, $R^1C(O)$, alkyl) using a range of substrates of differing charge, polarizability, and basicity. The kinetics[278] suggest that the ability of the RO group to support a negative charge in the transition state is of equal importance for the nucleophiles studied.

An attempt has been made to rationalize oxygen transfer reactions from inorganic and organic peroxides.[279] Alkaline H_2O_2 cleaves alkynes with the ratio of the rates of addition of HO_2^- and OH^- towards PhC≡CC(O)Me being 1400.[280] Two studies have been reported of the kinetics of the reaction of HO_2^- with methyl aryl sulfates[281,282] in methanol which estimated the pK_a for H_2O_2 in MeOH to be 15.8 ± 0.2 at 298 K.

^{17}O NMR line-shape studies of aqueous solutions containing HCl, NaOH, and various inorganic and organic salts have allowed proton life-times τ in the water molecule to be determined.[283] The classical structure breaking salts cause the proton exchange rate to decrease. A minimum in the rate of proton exchange is seen in 8.0 mol liter^{-1} LiCl at pH 5.2 as a consequence of the superposition of two opposing effects, one favoring proton transfer (polarizing effect of coordination to Li^+) and one retarding (effect of the cation or dynamic effects such as rotation or translation) such a transfer.

4.10. Sulfur

The kinetics of oxidation of SO_2 by H_2O_2 have been studied in the pH range 0–4,[284] and the following rate law verified:

$$\frac{d[S(VI)]}{dt} = \frac{k_1 K_{al}[H_2O_2][S(IV)](k_2[H^+] + k_3[HA])}{(k_{-1} + k_2[H^+] + k_3[HA])(K_{al} + [H^+])}$$

consistent with the mechanism in Scheme 5 not involving radicals, with k_1 estimated to be $(2.6 \pm 0.5) \times 10^6$ liter mol^{-1} s^{-1} at 288 K, $k_2/k_{-1} = 16 \pm 4$ liter mol^{-1} and k_2/k_3 (A = OAc$^-$) = $(5 \pm 1) \times 10^2$.

Scheme 5

$$SO_2 + (aq) \xrightarrow{fast} H^+ + HSO_3^-, \quad K_{al}$$

$$HSO_3^- + H_2O_2 \underset{k_{-1}}{\overset{k_1}{\rightleftharpoons}} \bar{O}(O)SOOH + H_2O$$

$$H^+ + \bar{O}(O)SOOH \xrightarrow{k_2} H_2SO_4; \quad \bar{O}(O)SOOH + HA \xrightarrow{k_3} H_2SO_4 + A^-$$

Oxidation of aqueous SO_2 by O_2 in the presence of transition-metal catalysts, with and without added antioxidants and chelating agents, has been studied as a function of pH and ionic strength.[285-287] Detailed kinetic studies have allowed different mechanistic schemes for the involvement of Mn(II) or Fe(II) catalysts to be proposed. For Fe(II) the complex rate expression revealed a pronounced negative dependence on $[H^+]$, an order on catalyst concentration which was ~1 but which increased with pH up to pH 7, a zero-order oxygen dependency at high $[O_2]$, and a positive (though <0.5) order with respect to $[HSO_3^-]$. The reaction was subject to a negative ionic-strength effect, was inhibited by added chelates and SO_4^{2-}, and slowed by antioxidants. The volume of activation, $\Delta V^{\ddagger}$, was shown to be 9.9 ± 2.0 cm^3 mol^{-1}. A complex mechanistic scheme was proposed in which a series of iron(II) complexes initiated a series of propagation processes, some involving free radicals, as shown in Scheme 6, and some involving Fe(II)–Fe(III) redox chemistry.

Scheme 6

$$SO_4^{\cdot -} \text{ (formed in initiation step)} + HSO_3^- \rightarrow HSO_4^- + SO_3^{\cdot -}$$

$$SO_3^{\cdot -} + O_2 \rightarrow SO_5^{\cdot -}$$

$$SO_5^{\cdot -} + HSO_3^- \rightarrow HSO_4^- + SO_4^{\cdot -}$$

$SO_2^{\cdot -}$ is the product of the primary electron transfer step in the reduction of SO_2 in DMF containing $NEt_4^+ClO_4^-$. Laman *et al.*[288] have detected $SO_2^{\cdot -}$ and the monosolvated product $S_2O_4^{\cdot -}$ with ESR and other spectroscopic data suggesting the latter is a charge transfer complex of $SO_2^{\cdot -}$ and SO_2 formed with a forward second-order rate constant of 10^5 liter mol^{-1} s^{-1} and an equilibrium constant for the reaction (38), $K = 611 \pm 160$ liter mol^{-1}.

$$SO_2^{\cdot -} + SO_2 \rightleftharpoons S_2O_4^{\cdot -}, \qquad K = 611 \text{ liter mol}^{-1} \qquad (38)$$

Using solutions containing NEt_4^+, Parker[289] sees only a reversible dimerization process for $SO_2^{\cdot -}$ and suggests K for reaction (38) $\leq$ ~200 liter mol^{-1} at 293 K. However, use of NBu_4^+ increases K for (38) to 1640 liter mol^{-1}, with the $S_2O_4^{\cdot -}$ product reacting further by a kinetically second-order process to give $S_2O_4^{2-}$ and SO_2, either by an electron transfer mechanism or by a nucleophilic displacement, the latter being favored on the basis of E_a data. Parker ascribes these differences to the more pronounced ion pairing involving NEt_4^+ compared with NBu_4^+. Reactivities of the radical anions $SO_2^{\cdot -}$, $So_3^{\cdot -}$, and $SO_4^{\cdot -}$ in aqueous solution have also been studied by ESR.[290]

The reactions in aqueous solution of SO_2 and HSO_3^- with $[Pt(NH_3)_5OH]^{3+}$ to give an O-bonded sulfite complex and of SO_3^{2-} with $[CO(NH_3)_5OH]^{2+}$ to give an S-bonded complex have been studied kinetically by Keshy and Harris[291] and Spitzer and van Eldik,[292] respectively.

Hopkins et al.[293] have examined substituent effects on the rates of SO_3-group transfer between isoquinoline-N-sulfonate and a range of pyridines to distinguish between concerted and stepwise preassociation mechanisms. The linear Brønsted plot establishes the concerted process since the stepwise process should show a break in such a plot at the pK_a of the pyridine derivative equal to that of isoquinoline. Hopkins and Williams[294] have also shown that SO_3 is not an intermediate in the hydrolysis of isoquinoline-N-sulfonate in 80% CH_3CN/H_2O.

Nucleophilic attack at sulfonyl (and sulfinyl) sulfur is thought to occur via a one-step concerted process or via a two-step process involving a sulfurane. Perkins and Martin[295] have reported the synthesis of the first example of an observable analogue to the intermediates (38) and (39) postulated to lie along the pathway of associative attack:

38   X = lone pair
39   X = O

Exchange of $^{36}Cl^-$ between LiCl and $X-C_6H_4SO_2Cl$ in sulfolane displays[296] the characteristics indicative of an S_N2 process involving a trigonal bipyramidal transition state with limited bond making. Acid hydrolysis of alkyl sulfates, $ROSO_3^-$, is known to occur via S–O bond cleavage, with the lower alkyl groups having little affect on the straightforward second-order kinetics. Longer alkyl chains ($R = C_{12}H_{25}$, $C_{12}H_{25}OCH_2CH_2OCH_2CH_2$) lead to a change in acid dependence. Values of the second-order rate constant are found to be dependent on the substrate concentrations, reaching a maximum above the critical micelle concentration and showing a sharp drop thereafter. The nature of the countercations affects $k_2(obs)$, in the order $NH_4^+ > Li^+ > Na^+ \gg Mg^{2+}$,[297,298] the effects being ascribed to the relative binding of these cations and protons in the micelle Stern layer.

Cleavage of trisulfides by R_3P may occur either via attack on central or terminal sulfur with ^{35}S labeling showing that both processes may occur depending upon the nature of the substrate, the nucleophile, and the solvent. The selectivity for central sulfur attack favored by $(p-XC_6H_4)_3P$ is little affected by p-X substituents (X = H, Me, MeO, Cl) where 91–99% central sulfur removal was seen. While displacements employing R_3P are insensitive to solvent, those using $(Me_2N)_3P$ showed an increased tendency to remove central sulfur going from Et_2O to CH_3CN. The process was further probed using an optical label at the $C\alpha$ to S, with central sulfur removal expected not to affect configuration, though terminal extrusion should give inversion at one C leading to a mesodisulfide. $(Me_2N)_3P$ gave 33% central sulfur removal in Et_2O and 82% in CH_3CN.[299]

The trisulfide $HOCH_2CH_2S_3CH_2CH_2OH$ undergoes a one-electron reduction on reaction with CO_2^-, e_{aq}^-, or $Me_2\dot{C}OH$ to give $HOCH_2CH_2SH$ and $HOCH_2CH_2S_2^-$, the latter dimerizing to give $HOCH_2CH_2S_4CH_2CH_2OH$ with a second-order rate constant[300] of $(1.4 \pm 0.3) \times 10^9$ liter mol^{-1} s^{-1}.

The activation parameters for the thiolate–disulfide interchange process involving oxidized glutathione and thiols $HOCH_2CH_2SH$, $HSCH_2CH(OH)CH_2SH$, and $HSCH_2CH(OH)CH(OH)CH_2SH$ in aqueous solution[301] are consistent with the three-step mechanism of Scheme 7.

Scheme 7

$$R'SH \rightleftharpoons H^+ + R'S^-$$

$$R'S^- + RSSR'' \rightleftharpoons [R'S \cdots SR \cdots SR'']^{\ddagger} \rightleftharpoons R'SSR + SR''^-$$

$$R''S^- + H^+ \rightleftharpoons R''SH$$

The equilibrium constants for the ring-opening S–S(O) bond fission in cyclic aryl thiosulfenates by SO_3^{2-}, t-BuS^-, or CN^- are much greater than those for similar reactions at the S–S(O)$_2$ bond of the related cyclic aryl thiosulfinates. The major factor responsible for this difference seems to be the large increase, compared with the S(IV) compound, with which the RSO^- group in the ring-opened product displaces the nucleophile[302] to reform the S–S bond. Reactions involving MeO^- are anomalous, showing a second-order dependence on $[MeO^-]$[303] and are thought to involve intermediates the reactivity of which are consistent with Martin's arguments relating to the stabilization of hypervalent sulfur species by apical alkoxy ligands.

The sulfonyl compounds $ArSO_2X$ are reduced to thiols by Ph_3P and a catalytic amount of I_2, with reactivity being in the order X = $Cl^- > SAr^- > H^- > OH^- > OR^- > O^{2-} > SO_2Ar^- \gg R^-$, Ar^-.[304] Similar reductions may also be achieved by polyphosphoric acid derivatives and iodide in organic solvents via the formation of mixed anhydrides containing P–O–S bonds from which the phosphate residue may be displaced by iodide.[305] Sulfuric acid and sodium sulfate yield elemental sulfur and H_2S on similar treatment.[306] SO_2 may be detected in the absence of iodide and SO_3 trapped by reaction with mesitylene leading the authors to propose Scheme 8.

Scheme 8

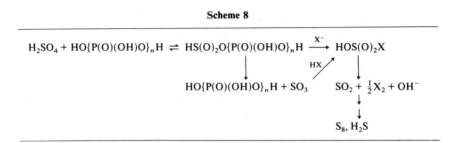

Parenthetically, the published values of pK_{a_2} for H_2S cover a range of eight orders of magnitude, 22 of them larger than K_w. Raman spectra on H_2S in 8.9 mol liter^{-1} NaOH allow the detection and estimation of HS$^-$, which was present at concentrations which would argue against any $pK_a < K_w$. The data allowed a value of the $pK_a = 17 \pm 1$ to be estimated, the highest value yet reported, though under circumstances of low [water] and high ionic strength.[307]

Unsymmetrical thiosulfinates RS(O)SR' may be catalytically transformed to the sulfinate RS(O)OR'' by R''OH in the presence of I_2, Br_2, or HCl (XY), probably via the intermediacy of RS(O)Y and R'SX[308] with the yield substantially increased by the addition of H_2O_2, presumably by the consumption of R'SX to give R'SSR'. The reactions of thiols and SO_2 are catalyzed by $BF_3 \cdot OEt_2$, giving RS_nR ($n = 2, 3, 4$) depending on the conditions, possibly via RSS(O)OH, RSS(OH)$_2$SR, or RSS(O)SR, with the composition of the products in the presence of $BF_3 \cdot OEt_2$ being similar to that from the reaction of RSH and SO_2 alone, at least during the early stages of the reaction.[309]

Oxidation using Ru[^{17}O]$_4$ of diastereotopically labeled cyclic sulfite diesters [e.g., (**40**)] to give [^{17}O]-labeled cyclic sulfate esters occurs with retention of configuration, as demonstrated by the ^{17}O NMR chemical shifts of the two sulfate esters differentially shifted by [Eu(fod)$_3$].[310]

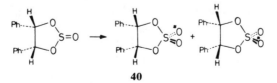

40

$PhSCH_2CO_2H$ is oxidized by peroxydiphosphate to give $PhS(O)CH_2CO_2H$ with kinetics similar to those reported earlier for $P_2O_8^{4-}$ and $S_2O_8^{2-}$ oxidations of alkyl aryl sulfides (first order in each reactant), consistent with rate-determining nucleophilic attack of S on peroxidic oxygen. A first-order dependence on H^+ was noted for $P_2O_8^{4-}$ with no acid-independent pathway. The effect on rates of parasubstituents in (p-$XC_6H_4)SCH_2CO_2H$ demonstrates the benefit of electron-releasing substituents to a positive sulfur center formed in the rate-determining step[311] (Scheme 9). Ph_2S oxidation by $P_2O_8^{4-}$ or $S_2O_8^{2-}$, while giving similar products, showed kinetics which were first order in oxidant and zero order in substrate,[312] the data being consistent with rate-determining acid hydrolysis to the peroxydianions of the peroxymonoanions HSO_5^- and $H_2PO_5^-$, followed by rapid reaction to give Ph_2SO.

Scheme 9

$$H_2P_2O_8^{2-} + H^+ \rightleftharpoons H_3P_2O_8^-$$

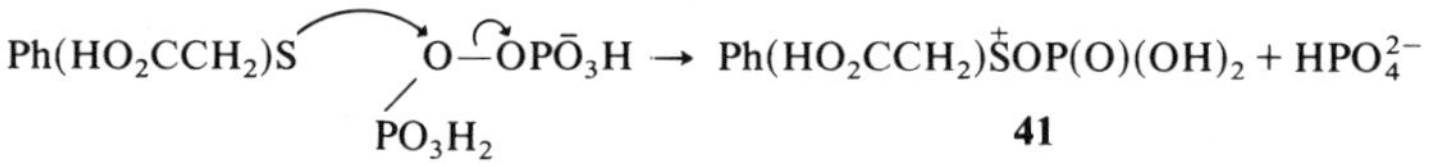

$$Ph(HO_2CCH_2)S \overset{\frown}{\quad} O\!-\!\overset{..}{O}P\bar{O}_3H \rightarrow Ph(HO_2CCH_2)\overset{+}{S}OP(O)(OH)_2 + HPO_4^{2-}$$

$$\underset{PO_3H_2}{\diagup} \qquad\qquad \mathbf{41}$$

$$(\mathbf{41}) + H_2O \rightarrow PhS(O)CH_2CO_2H + H_2PO_4^- + 2H^+$$

The uncatalyzed reaction of aliphatic diamines with $S_2O_8^{2-}$ displays kinetics and products consistent with the intermediacy of $H_2N(CH_2)_nNHOSO_3^-$, which subsequently rapidly hydrolyzes.[313] Reactions of $S_2O_8^{2-}$ with cyclic ethers occur via radical-chain processes[314] involving O–O bond homolysis and $\dot{H}$ abstraction from the organic substrate. Kinetics of the oxidation of ascorbic acid by $S_2O_8^{2-}$ in the pH range 3.4–4.6, in the absence of metal ion catalysis, gave the empirical rate law $-d[S_2O_8^{2-}]/dt = k([H_2A] + [HA^-])[S_2O_8^{2-}]$ which is unaffected by the addition of a radical trap suggesting parallel two-electron transfer processes from H_2A and HA^- to dehydroascorbic acid.[315] The role of transition-metal-ion catalysis on these[316-318] and other reactions of $S_2O_8^{2-}$ has been studied.

van Noort et al.[319] have shown that peroxyacylnitrates $RC(O)O_2NO_2$ ($R = Me$, n-C_5H_{11}) rapidly convert thioethers (RSR') into sulfoxides ($RS(O)R'$). The rates, unaffected by added radical scavengers, were essentially independent of solvent though subject to marked steric hindrance. Relative-rate studies for p-XC_6H_4SMe in methanol gave a Hammett ρ of -1.7 indicating that the sulfur displayed nucleophilic character. The ^{18}O label in $RC(O)O_2N(^{18}O)_2$ was not transferred to S. The results, particularly the absence of solvent effects, were rationalized on the basis of the formation of a sulfurane intermediate $RR'S(ONO_2)(OCOR)$.

Three studies have been reported of the oxidation of S–S compounds by m-chloroperbenzoic acid, particularly directed to an investigation of the relative importance of the two pathways for oxidation of thiosulfinic S-esters $RS(O)SR'$, via the formation of $RS(O)S(O)R'$ or $RS(O)_2SR'$.[320-322] Reactions are complex being subject to regioselectivity governed by electronic and steric effects, further complicated by the lability of some of the intermediate species. For MeSSPh, oxidation by MCPBA takes place at the most electron-rich site to give $MeS(O)SPh$ and this unsymmetrical ester is reported[320] to react further to give a distribution of products different from that derived from the isomeric $PhS(O)SMe$, which was taken to indicate the absence of the common intermediate $PhS(O)S(O)Me$. Using symmetrical S-alkyl alkanethiosulfinates at 233–253 K in $CDCl_3$, other workers[321] have shown that MCPBA oxidation led to the two diastereoisomers of $RS(O)S(O)R$, $RS(O)OH$, $RS(O)_2OS(O)_2R$, $RS(O)_2SR$, and $R'CH{=}S{=}O$, with NMR evidence at 233 K suggesting $MeS(O)S(O)Me$ to be the initial product from $MeS(O)SMe$. At 253 K, NMR shows the presence of $MeS(O)_2SMe$, $MeS(O)OH$, and $MeS(O)OS(O)Me$. Work on t-$BuCH_2S(O)SPh$ by the same authors[322] has also been reported.

Persulfoxide intermediates, $R_2\overset{+}{S}O\overset{-}{O}$, have been proposed for the photosensitized oxidation of sulfides[323-325] and the oxygen-atom transfer of $R_2\overset{+}{S}O\overset{-}{O}$ to ^2H- and ^{18}O-labeled DMSO[326,327] shown to involve a linear sulfurane, $Me_2\overset{+}{S}OO\overset{-}{S}(CD_3)_2{}^{18}O$. Kinetic studies show that the formation of $R\dot{S}O_2$ by the reaction $Et_3\dot{S}i + RSO_2Cl \rightarrow Et_3SiCl + R\dot{S}O_2$ is diffusional controlled for both R = Me and Ph at 300 K, whereas for RSO_2F the second-order rate constants are lower by $\sim 10^2$, and, significantly, are similar for $R = CH_3$ and $4\text{-}MeC_6H_4$.[328] The radicals generated decay with clean second-order kinetics with $2k_2$ being in the range $(4.5 \pm 1.5) \times 10^9$ liter mol^{-1} s^{-1} at 223 K for all the radicals studied.

4.11. Selenium and Tellurium

Nucleophiles, Nu^-, may react with bis(alkylthio)selenides, RSSeSR, at either S or Se, to give either RSNu and $RSSe^-$ (k_S) or RSSeNu and RS^- (k_{Se}). Kinetic data[329] for $Nu = RS^-$ and R = t-Bu show that $k_{Se} \gg k_S$ but when R = n-alkyl, $k_S \gg k_{Se}$. For R = i-Pr, $k_S \sim k_{Se}$ revealing the greater sensitivity to steric effects of displacements at sulfur compared with those at selenium. It was argued that the better leaving ability of $RSSe^-$ compared with RS^- and the greater ease of nucleophilic attack on Se compared with S, cancel in such a way as to make steric bulk the determining factor. The

$\Delta S^{\ddagger}$ value for the identity exchange reaction for the $Nu^- = {}^{35}SR^-$ and $R = t\text{-Bu}$ was found to be $-3\,JK^{-1}\,mol^{-1}$ and raises the possibility that reaction at selenium with nucleophiles may not take place by a simple S_N2 process and may involve electron transfer.

The reaction between $(4\text{-}XC_6H_4)SeCl$ and a series of ring-substituted (Z)- and (E)-1-phenylpropenes proceeds in CH_2Cl_2 at a rate first order in each reagent with electronic effects[330] suggesting that a positive charge is built up in the transition state on both α-C of the propene and on Se of ArSeCl. Nucleophilic displacement at divalent Se is proposed, with C–Se bond making lagging behind Se–Cl bond breaking in the transition state.

$TeCl_4$ adds to olefins in $CDCl_3/CH_3CN$ to give mixtures of syn–anti addition products, whereas 2-naphthyl-$TeCl_3$ gives exclusively antiaddition.[331] The latter is consistent with an ionic mechanism via a telluronium intermediate, whereas the former is more complex, though with the ionic mechanism ruled out.

The reaction of bis(morpholinoselenocarbonyl)triselenide with iodine is a first-order process from a $1:1$ charge transfer precursor complex and gives

In excess iodine, the reaction is more complex.[332] Mahdi and Miller[333] have shown that $[p\text{-EtOC}_6H_4Te]_2$ and I_2 react to give $p\text{-EtOC}_6H_4TeI$, at a rate which is first order in each reagent, via a proposed square Te_2I_2 transition state. With excess I_2, a I_3^- complex is formed, as in equation (39), ($K_{eq} = 1570\,\text{liter mol}^{-1}$ at 298 K) which oxidizes to a tri-iodo Te(IV) complex in a unimolecular process with $k = 6.2 \times 10^{-4}\,s^{-1}$:

$$p\text{-EtOC}_6H_4TeI + I_2 \underset{}{\overset{K_{eq}}{\rightleftharpoons}} [p\text{-EtOC}_6H_4Te(I_3)] \rightarrow p\text{-EtOC}_6H_4TeI_3 \qquad (39)$$

$PhSe\cdot$, generated by flash photolysis from PhSeSePh, reacts with a variety of vinyl monomers, $CH_2{=}CHY$, in a reversible process to give $PhSeCH_2\dot{C}HY$, at a rate some 10–50 times slower than $Ph\dot{S}$.[334]

The redox reaction between TeO_3^{2-} and BrO_3^- has been shown to be a second-order process.[335] The hydrolysis of TeF_6 in aqueous solution at 363 K has been followed by ^{125}Te NMR and a reaction scheme proposed.[336]

4.12. Halogens

4.12.1. Fluorine

It has been shown that the enhanced reactivity of fluoride ion in solutions of KF in nonhydroxylic solvents containing crown ethers is not as great as expected on the basis of a "naked" F^-. Miller and Clark[337] suggest from ^{19}F NMR linewidth studies that, in a variety of organic solvents, F^- and crown-ether complexed cations exist as tight ion pairs or higher aggregates. ^{35}Cl NMR has been used by Sugawara *et al.*[338] to study crown-ether KCl interactions. ^{19}F NMR has also been used to probe the strong H bonds between F^- and aliphatic diols and acids,[339] though it could not establish (along with ^{17}O and ^{31}P NMR, pH, and conductance evidence) that $HP(OH)OH \cdot F^-$, which *ab initio* calculations suggest may be formed from F^- and phosphorous acid with the release of 61 kJ mol^{-1}, is an abundant component in solution.[340]

In a preliminary report without experimental detail[341] the half-life for the reaction of F_2 and H_2O in water is estimated to be 7×10^{-6} s.

4.12.2. Chlorine

The kinetics of the reaction in aqueous MeOH and in MeCN of perchloryl fluoride and iodide ion to give F^-, ClO_3^-, and I_2[342] follow a second-order rate law, first order in each reagent. The rate decreases as the proportion of H_2O in MeOH increases and is 450 times faster in MeCN than in 90% aqueous MeOH, the two solvents having similar dielectric constants. Vigalok *et al.* do not distinguish between three two-step processes: the first involving the displacement of F^- by I^- to give ClO_3I, the second the displacement of ClO_3^- by I^- to give FI and, third, outer-sphere electron transfer to give $ClO_3F^{\pm}$ and $\dot{I}$; all giving intermediates which react further to give ClO_3^-, F^-, and I_2.

Methanolysis of methyl perchlorate in nitromethane (containing 10% benzene) gives Me_2O and $HClO_4$. The kinetics (corrected for a solvolysis reaction which gives toluene) obey the rate law $d[Me_2O]/dt = k_2[MeOClO_3][MeOH] + k_3[MeOClO_3][MeOH]^2$, where $k_2 = 5.03 \times 10^{-5}$ liter mol^{-1} s^{-1} and $k_3 = 5.70 \times 10^{-3}$ liter^2 mol^{-2} s^{-1}.[343] Kevill *et al.* suggest the k_2 term is associated with a reaction with no base catalysis (or base catalysis involving CH_3NO_2), the k_3 term with either an S_N2 process in which the transition state is solvated by a MeOH molecule or one which is general-base catalyzed. General-base catalysis was found to be less important in CH_3NO_2 compared with benzene since the addition of $Nn\text{-}Bu_4^+ClO_4^-$ produces only a modest acceleration in nitromethane and not the very marked rate increase seen in benzene.

Oxidation of V(IV) by chloramine-T at various acidities was shown to be first order in oxidant and to follow Michaelis–Menten kinetics.[344] Chlorination of toluene by the same oxidant has also been studied.[345]

4.12.3. Bromine

Most of the work reported is covered in Section 4.12.5 on oscillating reactions. Anomalous kinetic isotope effects, which lead to values of the relationships between the activation parameters A_D/A_H and $(E_a)_D - (E_a)_H$ being greater than those predicted from semiclassical transition-state theory, have had two separate rationalizations: (1) quantum-mechanical tunneling in a single-step process; and (2) a three-step internal return process in which the partitioning of the intermediate is modified by isotopic substitution. The oxidation of HCO_2^- or DCO_2^- by Br_2 in aqueous acid to give H^+ or D^+, CO_2, and $2Br^-$, gives anomalous values of A_D/A_H and $(E_a)_D - (E_a)_H$ which cannot be accounted for by either explanation, Brusa and Colussi[346] proposing instead a mechanism involving tunneling in the decomposition of a charge transfer complex in equilibrium with reactants.

A variable order with respect to pinacol concentration in its oxidation by acidic bromate catalyzed by Mn(II) has been interpreted[347] in terms of equilibria involving 1:1 and 1:2 complexes between Mn(II) and pinacol in which only the 1:1 complex is involved in oxidation with BrO_3^-. Bromate oxidation of $Ce(III)$[348] and $Co(II)$ in the presence of cyclo-hexanediaminetetraacetate[349] have been studied. Electron transfer between BrO_4^- and tris complexes of Fe(II) and bipyridyl and o-phenanthroline requires initial dissociation of the chelate followed by rapid electron transfer.[350] Oxidation of V(IV) by N-bromosuccinimide in aqueous acetic/perchloric acids is first order in oxidant and independent of acid.[344] The mechanisms of oxidation of alcohols in alkaline N-bromoacetamide have been studied.[351,352] Controversy abounds concerning the nature of radical intermediates in N-bromosuccinimide brominations.[184,353,354]

4.12.4. Iodine

Electron transfer between V(IV) and I(V) (as IO_3^-) catalyzed by Ru(VIII) and Os(VIII)[355] and the inner-sphere oxidation of diaqua (nitrilotriacetato)cobalt(II) by IO_4^- [356] have both been studied kinetically. The radiolysis of IO_3^- and the photolysis of IO^- in aqueous solution have also been studied.[357,358] N-iodosuccinimide and V(IV) react in a process first order in oxidant and inverse first order in acid [in contrast to the reaction involving V(IV) and NBS].[344]

Oxygen exchange between $PhI^{18}O$ and H_2O in neutral or basic aqueous solution is very slow, though such exchange occurs readily in methanol in which polymeric PhIO dissolves to give $PhI(OMe)_2$ and which may then be hydrolyzed to $PhI^{18}O$ by $H_2^{18}O$.[359]

The kinetics of the iodine-atom transfer reaction $Ar + Ar'I \rightarrow ArI + Ar'$[360] favor a two-step process involving an intermediate $[Ar\overset{+}{I}Ar']$ (generated independently by electrolytic reduction of $[Ar\overset{+}{I}Ar']$) over the single-step process with $[Ar\overset{+}{I}Ar']^{\ddagger}$ as a transition state, proposed by other workers.

Iodine is reported to react with various cyclic polyethers (in excess) in $CHCl_3$ in a kinetically first-order process.[361]

4.12.5. Oscillating Reactions

A bumper crop of papers has appeared dealing with this topic and it is outside the scope of this chapter to do anything other than to provide the briefest of summaries. Readers are referred to recent surveys by Gurel and Gurel,[362] Zhabotinskii,[363] Geiseler,[364] and Epstein *et al.*[365] A theoretical model for oscillating chemical systems has been proposed[366] based on phase exchange and pulsating supersaturation. Tyson has applied single perturbation theory to the detailed mechanism of the Belousov–Zhabotinskii (B–Z) reaction, has generated several models, and has suggested that the Oregonator is the best of these.[367] Other theoretical approaches based on new experimental observations have also been reported.[368,369] Theoretical studies of coupled chemical oscillation[370] suggested that regimes exist (in addition to those in which a single stationary or oscillatory state is stable) which exhibit stability of both a stationary and oscillatory state ("hard excitation"), a stable aperiodic oscillation ("chaos"), and two different stable oscillatory states ("biorhythmicity"). Chaos has recently been noted[371] and biorhythmicity seen in experiments in a continuous-stirred tank reactor (CSTR) involving ClO_2^-, BrO_3^-, and I^-.[372] A further phenomenon was seen in which one oscillator was "entrained" by another to create a single oscillator with a complex wave form ("compound oscillation"). While bromate oscillators are the most thoroughly studied, a preliminary classification has now been made of the less well-investigated chlorite oscillators, involving ClO_2^-, I^-, and an oxidizing substrate (IO_3^-, $Cr_2O_7^{2-}$, MnO_4^-, BrO_3^-) or ClO_2^-, I_2, and a reducing substrate ($[Fe(CN)_6]^{4-}$, SO_3^{2-}, $S_2O_3^{2-}$).[373] Further details have been reported of the first iodide-free chlorite oscillator—the $ClO_2^-/S_2O_3^{2-}$ system.[374,375] Studies of $ClO_2^-/I^-/IO_3^-$/arsenate in a CSTR reveal three different steady states ("tristability") as well as oscillations, depending on the conditions.[373]

Understanding of the B–Z oscillator system has been inhibited by the lack of knowledge of the organic intermediates involved and this has motivated attempts to devise a wholly inorganic BrO_3^- oscillator. Epstein *et al.* showed that the metal ion in $BrO_3^-/Br^-/M^{n+}$ oscillators, for example, Mn(II),[376,377] (see also Geiseler[378]), can be replaced by ClO_2^- and that Br^- flows (which replace the organic substrate in the B–Z system) can be replaced by N_2H_4, Sn(II), I^-, SO_3^{2-}, and AsO_3^{3-}. NH_2OH and $[Fe(CN)]_6^{4-}$ do not give oscillatory phenomena, prompting the suggestion that oscillatory behavior may be associated with two-electron reductants. Adamčíková and

Ševčík reported that sodium hypophosphite (NaH_2PO_2) generated temporal chemical oscillations with BrO_3^- in the absence of organic reducing agents.[379] Bar-Eli and Geiseler have reported the $BrO_3^-/Br^-/Ce(III)$ oscillator.[380,381] The BrO_3^-/I^- reaction has been shown to generate oscillations in a CSTR with two schemes being proposed, one involving autocatalytic formation of $HBrO_2$ and the other involving $IBrO_2$.[382]

Other mechanistic aspects of catalyzed and uncatalyzed bromate oscillator systems related to the B–Z and similar reactions have been studied.[383-393] Spacial and temporal oscillations in the same system have been reported with different activation energies.[394] Evidence of the intermediacy of BrO_2 in the B–Z reaction involving aliphatic alcohols has been put forward[395,396] (BrO_2 has been proposed from kinetic studies on the Br_2-O_2 system[397]). Oscillating chemiluminescence has been reported.[398] The B–Z reaction with other substrates has been studied, including 2,4-pentanedione,[399] tartaric acid,[400,401] an ester of 3-oxobutanoic acid,[402] phenol,[403] phosphonoacetic acid,[404] pyrogallol,[405] 1,4-cyclohexanedione,[406] formic acid,[407] gallic acid,[408] ethylacetoacetate,[409] α-ketoglutaric acid,[410] and mixed organic substrates.[411]

Edelson has carried out a sensitivity analysis on the proposed mechanisms of the Briggs–Rauschler (B–R) oscillating system ($H^+/IO_3^-/H_2O_2/Mn^{2+}/malonic\ acid$)[412] and a "double-barreled" oscillator devised involving hypochlorite but based on the B–R system.[413] The subsystems of the B–R oscillator responsible for iodine formation ($H_2O_2/IO_3^-/H^+/Mn^{2+}$) have been studied using *trans*-2-butenoic acid to consume iodine.[414]

The iodate/arsenous acid system has been studied in thin unstirred films in which the chemical changes are discussed in terms of two component processes (40) and (41):

$$IO_3^- + 5I^- + 6H^+ \rightarrow 3I_2 + 3H_2O \tag{40}$$

$$H_3AsO_3 + I_2 + H_2O \rightarrow 2I^- + H_3AsO_4 + 2H^+ \tag{41}$$

The single propagating wave, developed with starch, divides the mixture into two—that ahead of the wave being in a kinetic state associated with the initial reaction mixture and that behind the wave front corresponding to the state of thermodynamic equilibrium. In a CSTR the system exhibits bistability.[415,416]

Chapter 5

Substitution Reactions of Inert-Metal Complexes— Coordination Numbers 4 and 5

5.1. Introduction

Once again studies of square-planar palladium(II) and platinum(II) complexes dominate this section, most of them extending generally accepted ideas on nucleophilic ligand replacement reactions. Isomerization reactions remain a fruitful area for study, and new interpretations and mechanistic pathways have emerged. Unexplained isomerization phenomena are also common, however, indicating perhaps that this field will continue to demand attention in the future. Investigations of chelation reactions have been extended in the period covered by this survey by the publication of a number of papers on the formation of large-ring chelates spanning *trans* coordination sites. Structural dependences on ring size and ligand-donor atoms have been emphasized, though as yet kinetic investigations on such systems are rare.

The number of dynamic NMR spectroscopic studies devoted to mechanistic aspects of Pd and Pt complexes continues to grow, and a review on ^{195}Pt NMR spectroscopy has appeared.[1]

Several other recent reviews contain material relevant to this section. An article by Blandamer and Burgess[2] on the thermodynamics, kinetics, and mechanisms of solvation, solvolysis, and substitution in nonaqueous solvents contains a contribution on the controversial dissociative mechanism for isomerization of square-planar molecules. This is outlined in Section 5.5. A review of ligand substitution reactions at low-valency transition-metal centers[3] contains sections on five-coordinate metal carbonyl complexes and on ML_4 complexes (mainly tetrahedral configurations with L being a tertiary phosphine), as well as on acid- and base-catalyzed reactions. A review by Constable[4] surveying the reactions of nucleophiles with complexes of chelating heterocyclic imines contains a sizable section on square-planar palladium and platinum derivatives. Most discussion centers on $[Pt(bipy)_2]^{2+}$ and $[Pt(phen)_2]^{2+}$ (bipy = 2,2'-bipyridine; phen = 1,10-phenanthroline). The metal center, ligand, or both are susceptible to nucleophilic attack and the mechanisms involved are critically assessed.

A review by Hartley and Davies[5] deals with complexes of Pd(II) and Pt(II) with weak ligands (sulfoxides, DMF, RCN, acetone, ROH, H_2O, and OH^-) which can readily be replaced, for example, by olefins. An article by Mondal and Blake[6] on the thermochemistry of oxidative addition reactions contains much information on square-planar complexes of iridium, palladium, and platinum, including tables of bond dissociation energies, reaction enthalpies and entropies, equilibrium constants, and heats of solution and sublimation. Included also are discussions of other addition reactions, where additional ligands coordinate above the square plane, and relating directly to some intermediates or transition states of associative nucleophilic ligand displacements at square-planar molecules. A review[7] of structures and properties of metal-complex stacks concentrates on metal-ligand (as well as metal–metal) interactions along the z-axis of square-planar species.

Finally, the appearance of more reviews of platinum-containing anti-cancer agents reflects the enormous amount of work devoted to this subject.[8] Features of chemical interest in them include the effects of leaving group (in substitutions), solubility and electronic charge, and the nature of the binding of platinum compounds to nucleic acids.

5.2. Substitution at Square-Planar Complexes

5.2.1. General

A systematic study of the effects of solvent on the leaving group in substitution reactions at palladium(II) has been carried out.[9] Rate and equilibrium constants were reported for reactions (1) and (2) in methanol,

$$[Pd(dien)X]NO_3 + Y \rightleftharpoons [Pd(dien)Y]X(NO_3) \tag{1}$$

$$[Pd(biL)X_2] + en \rightleftharpoons [Pd(biL)(en)]X_2 \tag{2}$$

dimethylformamide (DMF), and dimethylsulfoxide (DMSO) [dien = $H_2NC_2H_4NHC_2H_4NH_2$; biL = phen, 1,2-bis(diphenylphosphino)ethane (DPPE) or 1,2-diaminoethane (en); X = Cl, Br, I, N_3, SCN, or NO_2; Y = propylamine or thiourea (tu)]. The usual two-term rate law [equation (3)] was obeyed in all cases, though the contribution of the solvolytic path

$$k_{obs} = k_1 + k_2[Y] \tag{3}$$

k_1 could not always be distinguished. All the k_2 values were influenced by the solvent nature, but the dependence was not large, differing by less than a factor of 100 for a given reaction in different solvents. The lability sequence of the leaving groups X was notably dependent on the solvent, as well as on the system being investigated, and it was concluded that no lability sequence should be assessed unless specific reference to the reaction medium was made.[9] It was not possible, however, to establish whether the solvent dependence was a ground-state or transition-state effect.

A study of salt effects on the kinetics of the reaction of *cis*-[PtCl$_2$L$_2$] (L = 4-cyanopyridine) with thiourea has enabled an analysis to be made of initial-state and transition-state contributions, however.[10] The reaction was followed in aqueous solutions of KCl, GdCl$_3$, and Et$_4$NCl and here, too, although the usual rate law was adhered to [equation (3), Y = tu] the strength of the nucleophile made the k_1 term negligible. The effects of KCl and GdCl$_3$ were opposite those of added Et$_4$NCl, in keeping with their hydration characteristics. The initial-state chemical potential changes in all cases were found to be larger than transition-state changes, and account for the differences in reactivity trends observed. KCl and GdCl$_3$ additions lead to destabilization of both initial and transition states, whereas Et$_4$NCl addition destabilizes the initial state but stabilizes the transition state.

Rate and equilibrium constants for reactions (4) (L is tertiary phosphine or arsine; am is a heterocyclic nitrogen base)[11] in methanol have been

$$[PtLCl_3]^- + am \rightleftharpoons trans-[PtL(am)Cl_2] + Cl^- \tag{4}$$

compared to previously obtained values when L was a sulfur donor. The lability of the chloride was found to be very dependent on the donor atom (P, As, or S), but much less sensitive to the nature of the substituent groups attached to these atoms.

The kinetics of reaction (5), where RR'SO represents a set of 11 sulfoxides, allowed a comparison to be made of the effects of the substituent

$$[PtCl_4]^{2-} + RR'SO \rightarrow [PtCl_3(RR'SO)]^- + Cl^- \tag{5}$$

organic groups on the entering ligand,[12] and extended previous studies on amines and thioethers. Equation (3) (Y = RR'SO) was again adhered to, with small k_1 values. Although the k_2 values showed a marked dependence on the steric bulk of R and R', they were not sensitive to inductive effects. Using data from the literature, Annibale et al. compared the sensitivity of reactions like (4) and (5) to the inductive effects of the substituents of the entering ligands, and pointed out that reactions fall into one of two classes: those markedly sensitive to the inductive effects, and those very insensitive. The nature of the complexes under attack dictates the behavior pattern adopted.[12] Inductive-effect sensitivity was taken as an indication that the bond-breaking (second) energy maximum was rate limiting, whereas insensitivity of the type found for reaction (5) was taken to indicate that the bond-forming maximum was the rate-determining step. Annibale et al. also commented on the low nucleophilicity and relatively high trans effect of sulfoxides,[12] a combination which, it has been pointed out,[13] resembles ethylene. They speculated that the initial bond could be formed between platinum and oxygen, the changeover to the Pt–S link of the final products taking place between the bond-forming transition state and the five-coordinate intermediate, to account for the moderate trans effect. Also mentioned was the intriguing possibility that the trans effect of sulfoxides might arise from an S=O olefinic bond in the five-coordinate intermediate.

Kinetics of cyanide exchange at $[M(CN)_4]^{n-}$ (M = Ni, Pd, Pt, and Au) have been followed by ^{13}C NMR spectroscopy, using ^{13}C enriched samples in D_2O solvent.[14] Table 5.1 gives the rate constants and activation parameters. Only the second-order dependence could be detected, leading to a simple rate law in equation (6). The exceptionally high value for the nickel complex (too high to be measured by ^{13}C NMR spectroscopy) was assigned to the known stability of the five-coordinate species $[Ni(CN)_5]^{3-}$, which may be similar to the intermediate or transition state for this associative process.

$$\text{Rate} = k_2[M(CN)_4^{n-}][CN^-] \tag{6}$$

Table 5.1. Rate Constants and Activation Parameters for $CN^- + [M(CN)_4]^{n-}$ (Reference 14)

$[M(CN)_4]^{n-}$	$k_2/(\text{mol}^{-1}\,\text{s}^{-1}$ at 24°C)	$\Delta H^{\ddagger}/(\text{kJ mol}^{-1})$	$\Delta S^{\ddagger}/(\text{J K}^{-1}\,\text{mol}^{-1})$
$[Ni(CN)_4]^{2-}$	5×10^5	—	—
$[Pd(CN)_4]^{2-}$	120	17 ± 2	-178 ± 7
$[Pt(CN)_4]^{2-}$	26	26 ± 3	-148 ± 8
$[Au(CN)_4]^{-}$	3900	28 ± 1	-100 ± 3

5.2.2. Anation Reactions

A comment by Beattie[15] on reaction (7) reconciles two previous conflicting views. Original work[16] described the kinetics as adhering to a

$$[Pt(dien)(OH_2)]^{2+} + Cl^- \rightleftharpoons [Pt(dien)Cl]^+ + H_2O \tag{7}$$

two-term rate law [equation (3), Y = Cl], unusual in that a solvent attack step is not applicable to a solvent complex. It was thus interpreted as being due to the operation of a dissociative pathway. Later reinvestigation failed to detect the k_1 term, however, and dismissed the original report as experimental error.[17] Beattie points out[15] that the original data are consistent with the measurements made when reaction (7) did not reach equilibrium. The rate constant for the approach to equilibrium is given by equation (8) (k_f is the forward reaction and k_r the reverse step). A calculation of the equilibrium constant for (7) by equating k_f/k_r to k_1/k_2 is in good agreement with that available from literature data.

$$k_{obs} = k_f + k_r[Cl^-] \tag{8}$$

Platinum and palladium complexes of the tridentate ligands $Et_2NC_2H_4N(R)C_2H_4NEt_2$ have been considered so sterically cluttered above and below the coordination plane that they have been described as "pseudo-octahedral," and an earlier kinetic study of H_2O substitution of the palladium complex was interpreted in terms of an ion-pairing mechanism.[18] A further study of anation reactions of sterically hindered substituted dien complexes of palladium and platinum throws more light on their behavior[19] and casts doubt on the earlier interpretation. Anation of the complexes $[Pt(Me_5dien)(OH_2)]^{2+}$ $[Me_5dien = Me_2NC_2H_4N(Me)C_2H_4NMe_2]$ and $[Pd(triL)(OH_2)]^{2+}$ $[triL = Me_5dien$ or $Et_2NC_2H_4N(Me)C_2H_4NEt_2]$ by Cl^-, Br^-, I^-, SCN^-, $S_2O_3^{2-}$, and tu were examined in water. Despite the steric hindrance, all the substitutions proceeded via a nucleophile-dependent path [equation (3), $k_1 = 0$]. The kinetic behavior depends markedly on the number and size of the substituents on the nitrogen atoms, and both the reactivity towards a given reagent and the order of effectiveness of the entering group varies along the series of complexes. For the Me_5dien compounds, the k_2 values followed the usual nucleophilicity order of $Y = Cl^- < Br^- < I^- < SCN^- < S_2O_3^{2-}$ at both Pd and Pt. Different orders were found with the more bulky chelates, however, and it was concluded that the greater steric shielding prevented some of the entering groups from exerting their whole potential nucleophilicity.[19] While the ion-pairing mechanism previously proposed[18] is not ruled out by these studies, it becomes difficult to explain the reactivity variations in terms of it.

In a related work,[20] the anation of $[Pd(triL)(OH_2)]^{2+}$ [triL = Me_5dien, $EtHNC_2H_4N(Et)C_2H_2NEtH$, or $Me_2NC_2H_4NHC_2H_4NMe_2$] was followed as a function of Y, T, and P. The second-order rate constants decreased with increasing triL bulk, and were accompanied by an increase in $\Delta H^{\ddagger}$. The values of $\Delta S^{\ddagger}$ and $\Delta V^{\ddagger}$ were consistent with associative mechanisms in each case. The nucleophilicity order was Y = $Cl^- < Br^- < I^- < N_3^-$, in contrast to an earlier report of $Cl^- < N_3^- < Br^- < I^-$ for $[Pt(dien)(OH_2)]^{2+}$. The authors point to precedents for the higher reactivity of azide with palladium, but the rate of anation of the unsubstituted $[Pd(dien)(OH_2)]^{2+}$ is too fast to allow a direct comparison.

5.2.3. Chelate Complexes

An NMR spectroscopic study of complex formation with long-chain diphosphines, $Ph_2P(CH_2)_nPPh_2$, revealed that maximum stability and minimum ring strain for monomeric trans-chelating complexes were found at a chelate ring size of 15 (trans-$[PtCl_2(Ph_2P(CH_2)_{12}PPh_2)]$). Cis-chelated monomers had optimum ring sizes of 14 (n = 11) and 19 (n = 16), but a cis-dimeric configuration was preferred by most ligand chain lengths.[21] Arsenic analogues behaved similarly.[22] The bulk of the ligands was found not to be of importance in determining any preference for trans chelation. In all cases, formation from $K_2[PtCl_4]$ favored a cis substitution pattern as kinetic products, whereas use of $K[PtCl_3(C_2H_4)]$ led to trans complexes initially. When the chain length was too short to allow trans chelation, diphosphine-bridged dimers resulted.[23]

In the formation of related palladium complexes $[PdX_2(Ph_2As-(CH_2)_nAsPh_2)]$ (X = Cl, Br, I, or SCN; n = 6–12, or 16),[24] the geometry was always trans when n > 9, and mainly trans when n < 9, reflecting the common predisposition of Pd towards trans complexes with P and As donors. The shorter-chain diarsines produced polymeric materials, whereas the larger ones led to trans-chelating monomers.

Substitutions of Cl^- by pyridine-4-azo-4'-(N,N-dimethyl)aniline (1), have been followed by stopped-flow spectrophotometry in methanol for trans-[PtRCl(biL)] (R = H or Me; biL = 2),[25] and compared to those for trans-$[PtRClL_2]$ (L = 3 or PEt_3). Somewhat surprisingly, the steric effects of the bulky ligands (3) and the bulky trans-chelating ligand (2) were found to be small, not having a pronounced effect on rates or mechanism. Reactions proceeded by the usual two-term rate law and, interestingly, the k_2 path predominated to the extent that the k_1 contribution was difficult to determine accurately. Clearly, any steric blocking effects are insufficient to hinder attack by the entering nucleophile, even when the ligand spans one of the axial sites, and there was no evidence for a dissociative pathway. A suggested reason for the small k_1 contribution is that the aromatic groups of (**2**) and (**3**) prevent efficient solvation.

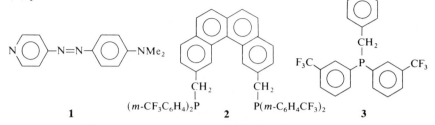

Finally, a ring-closure reaction reported by Breet and van Eldik[26] may be of practical value to chemists active in this area. An attempted modification of the usual synthesis of [PdCl(Me₃dien)]Cl [Me₃dien = MeHNC₂H₄N(Me)C₂H₄NMeH] led instead to [PdCl₂(Me₃dien)]HCl, where Me₃dien appears to behave as a bidentate ligand. In water it can be made to undergo the ring-closure reaction by HCl release, readily forming the desired product.

5.3.   Cis and Trans Effects

A kinetic study of the replacement of Cl⁻ from [PtCl(en)L]⁺ (L = PMe₃, PEt₃, PBu₃ and P(OMe)₃] by a range of nucleophiles Y (Br⁻, I⁻, N₃⁻, or SCN⁻) in water offers some insight into cis effects.[27] As usual with strong nucleophiles, the two-term rate law [equation (3)] was dominated by the k_2 term, and k_1 was determined independently using Y = OH⁻, which does not have an OH⁻-dependent term. Comparison of the k_2 values, and those from previous work where L is DMSO or thioether, with those for [PtCl(en)NH₃]⁺, used as a reference substrate, revealed differences which represent *cis* effects. The nucleophilic discrimination of the phosphine ligands resembled that of DMSO, and all were greater than those of thioethers. The conclusion was reached that the main cause of substituent *cis* effects was steric in origin. The effects do not appear to be as great as those previously found with *trans*-[PtCl₂L₂],[28] and the charge on the complex may account for the difference.

A paper by Mochida and Bailar[29] questioned the operation of a secondary *trans* effect at [Pt(dien)I]⁺. A report in 1968[30] described the methylation of that complex by MeI, forming [Pt(Me-dien)I]⁺. It was concluded that the nitrogen trans to iodide had been methylated, and the authors attributed this rather unexpected result to a secondary *trans* effect along the I–Pt–N–H chain. The new work suggests that the earlier paper misidentified the [Pt(Me-dien)I]⁺, probably because of an impure sample of commercial Me-dien used for comparison. It appears that methylation occurred at a terminal nitrogen (Scheme 1), *cis* to the iodide, and the *trans* effect of iodide is not great enough to change the relative reactivities of the NH bonds.[29]

Scheme 1

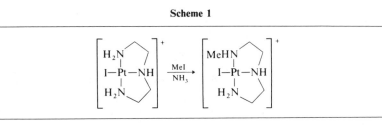

^{119}Sn NMR spectroscopic measurements on a series of $SnCl_3^-$ complexes of platinum assigned a *trans* influence series of $H^- > R_3P > R_3As > SnCl_3^- >$ olefin $> Cl^-$, based on $^1J(^{195}Pt-^{119}Sn)$.[31] [195]Pt NMR spectroscopic investigations on the reactions of aqueous NO_2^- with chloro and bromo complexes of Pt(II) and Pt(IV) found both the $\delta(Pt)$ and $J(^{195}Pt-^{15}N)$ changes on ligand substitution to be dominated by the nature of the trans ligands.[32]

Microcalorimetric measurements at elevated temperatures of the thermolysis of several *cis*-[PtX$_2$L$_2$] [L = NH$_3$, amine, or pyridine(py); X = Cl, Br, or I] allowed derivation of standard enthalpies of formation and other thermodynamic data.[33] Calculations of standard enthalpies of formation of diamine (dicarboxylate)platinum(II) and [Pt(acac)$_2$] led to estimates of mean Pt-O bond dissociation enthalpies.[34]

5.4. Five-Coordinate Complexes

A series of complexes [PdX(Me$_6$tren)]Y [X = Y = Br, I, or SCN; X = Cl, Br, I, or SCN; Y = PF$_6$ or BF$_4$; Me$_6$tren = N(C$_2$H$_4$NMe$_2$)$_3$] has been examined by ^{1}H NMR spectroscopy in various solvents. In protic and aprotic solvents, there is a rapid (NMR time scale) intramolecular rearrangement (Scheme 2) involving four- and five-coordinate complexes.[35] Although Me$_6$tren is known to favor five-coordinate complexes in the solid state, the palladium complexes are the first examples which are diamagnetic, allowing this solution investigation. At low temperatures the rate of equilibration is significantly retarded and the position is shifted to the right. In water the free arm of tridentate Me$_6$tren is readily protonated.

Scheme 2

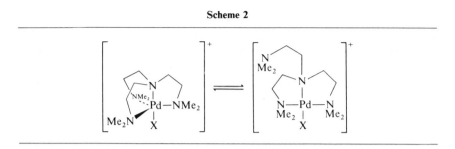

X-ray crystallographic studies on a tetradentate N_2S_2-macrocyclic complex of palladium, [Pd(tetraL)]Cl$_2$·2H$_2$O, reveal the existence of two five-coordinate isomers, (4) and (5), one with a Pd–Cl distance of 3.20 Å, and the other with a Pd–Cl distance of 3.68 Å.[36] This supports previous solution studies, which indicate that as Cl⁻ approaches [Pd(tetraL)]²⁺, the square folds back about the N–Pd–N axis to form a trigonal bipyramid.[37]

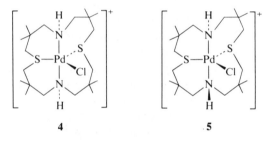

4	**5**

A variable-temperature and variable-pressure ^{13}C NMR spectroscopic study in CH_3CN has been made of solvent exchange at the five-coordinate nickel(II) complex, **6**.[38] The activation volume of $2.3 \pm 1.3 \, cm^3 \, mol^{-1}$ suggests a dissociative mechanism. UV-visible spectra of **6** are compatible with a $5 \rightleftharpoons 4$ coordinate equilibrium, so it is reasonable that the solvent exchange in the five-coordinate species proceeds dissociatively via this four-coordinate complex.

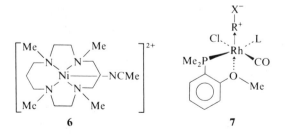

6	**7**

Finally, an interaction between the oxygen of coordinated P(o-C$_6$H$_4$OMe)Me$_2$ or As(o-C$_6$H$_4$OMe)Me$_2$ (L) with Pt(II) in *cis*-[PtMe$_2$L$_2$] and with Rh(I) in *trans*-[RhCl(CO)L$_2$] has been postulated to account for the enhanced oxidative additions of organic halides to these metal ions. Structure (**7**) indicates how the interaction works.[39] This chelating interaction forming a pseudo-five-coordinate complex enhances the reactivity of the metal to the extent that less reactive organic halides can be employed in oxidative addition reactions.

5.5. Isomerization Reactions

The relationship between five-coordinate complexes (or intermediates) and isomerization has been illuminated by a study of Pt(II) phosphole (**8**) complexes.[40] Equilibria (9) [L = (**8**) with R = Me, n-Bu, t-Bu, Ph, or CH$_2$Ph; X = Cl, Br, or I] were examined by ^{31}P and ^{195}Pt NMR spectroscopy

$$[L_2PtX_2] + L \rightleftharpoons [L_3PtX_2] \tag{9}$$

and conductivity measurements. At 25°, phosphole exchange is rapid (NMR time scale) and occurs by both intramolecular and intermolecular mechanisms in CHCl$_3$ or CH$_2$Cl$_2$. The formation of the five-coordinate complexes is enthalpy favored but entropy disfavored. Small ligands give greater stability, but lower the barrier to intramolecular rearrangements. The five-coordinate complexes are more rigid in methanol than in less polar solvents, but there is no conductivity evidence for the formation of [PtXL$_3$]X. Neither is there any detectable anion scrambling in mixtures of [L$_3$PtClBr] and [L$_3$PtBr$_2$].

8

[L$_2$PtX$_2$] are *cis* in solution at ambient temperatures and, curiously, the only system which isomerizes in the presence of excess L is that which forms the least stable but most rigid five-coordinate complex [L = (**8**); R = t-Bu; X = Br]. The reaction coordinate shown in Figure 5.1 rationalizes this, as only when there is little difference in energy between the *cis* and *trans* forms will isomerization occur. Thus the majority of these systems leading to stable [L$_3$PtX$_2$] need not lead to isomerization, despite a low-energy barrier to pseudorotation. MacDougall *et al.* suggested that in com-

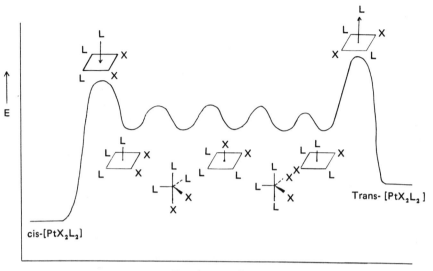

Figure 5.1. Reaction coordinate for the addition of a phosphine (L) to [PtX$_2$L$_2$].

plexes of this type, conditions favoring isomerization by pseudorotation would be less likely than routes proceeding via consecutive-anion or neutral-ligand displacement.[40] Analogous palladium systems revealed less stable and less rigid five-coordinate compounds. In these cases, however, addition of excess L to *cis*-[PdX$_2$L$_2$] led to isomerization in each case.

With examples of isomerization via pseudorotation rare, a kinetic investigation which reveals another is the more interesting.[41] The isomers (9) and (10) (Scheme 3, where O⌢N represents the amino acids glycine, sarcosine, or *N,N*-dimethylglycine and DMSO is S-bonded dimethylsulfoxide) have been isolated and their isomerizations followed. At equilibria, configuration (10) [labeled *cis(N, S)* by the authors] predominated, markedly for glycine but less so with increasing N methylation. Isomerization and substitution kinetics were followed by ^{1}H NMR spectroscopy in D$_2$O and use was also made of ^{36}Cl and d^6-DMSO to follow exchange reactions. Scheme 3 shows the reactions responsible for isomerization. Isomerization via chloride-catalyzed consecutive displacement involves k_c, k_{-c}, k_t, and k_{-t}. Direct isomerizations catalyzed by Cl$^-$ or DMSO are represented by k_x, k_{-x}, and k_L, k_{-L}, respectively. The constants k_c, k_{-c}, k_t, and k_{-t} were determined from ligand displacement reactions, and compared to isomerization rates. The results revealed that for chloride-catalyzed isomerizations, a direct pseudo-rotation-type mechanism dominates. The

Scheme 3

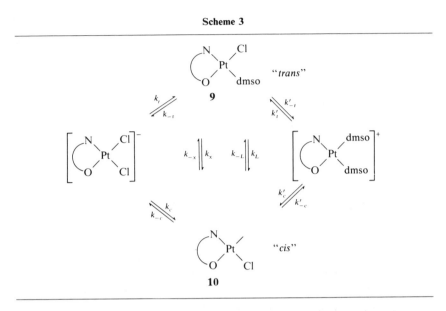

9

10

neutral DMSO also catalyzes the interconversions but in this case a consecutive displacement mechanism involving k'_c, k'_{-c}, k'_t, and k'_{-t} best fits the evidence, and k_L and k_{-L} are probably negligible.

Although both Cl$^-$ and DMSO exchange take place without isomerization, and probably follow the accepted pathway, the authors point out that a route involving rapid pseudo-rotation-type steps via more than one five-coordinate intermediate cannot be ruled out and need not be distinguished kinetically. They favor a "turnstile" type of fluxionality[42] for the intermediate rearrangements, because of the presence of the chelating ligands. While there is no evidence for this type of rearrangement in complexes of this type, it must be kept in mind that details for the usual "pseudorotation" interconversions between trigonal bipyramidal and square pyramidal intermediates are lacking also. Complexes related to (**9**) and (**10**) with DMSO replaced by olefins, also undergo isomerization reactions.[43] The relationship between olefin and DMSO complexes has already been referred to,[13] and it is entirely plausible that similar mechanisms operate for these complexes also.

The complex [PtCl$_2$(PPh$_3$)$_2$] is commonly encountered in synthetic studies, but the very low solubility of both its *cis* and *trans* forms makes it difficult to study. Davies and Uma[44] have applied cyclic voltammetry to its investigation, and find that the isomers can be distinguished by their Pt(II) → Pt(0) reduction potentials in dilute benzene–acetonitrile solution.

In the triphenylphosphine-catalyzed isomerization, the presence of $[PtCl(PPh_3)_3]Cl$ could also be detected, suggesting a consecutive displacement mechanism, as in equation (10). No isomerization took place from

$$trans\text{-}[PtCl_2(PPh_3)_2] \xrightarrow{PPh_3} [PtCl(PPh_3)_3]Cl \xrightarrow{-PPh_3} cis\text{-}[PtCl_2(PPh_3)_2$$

(10)

either isomer without added catalyst, but slow isomerization became apparent when pure solutions were examined by cyclic voltammetry. The most likely explanation is that reduction leads to $[Pt(PPh_3)_2]$, which can lose PPh_3. This, in turn, catalyzes the isomerization in the usual way.

Calculations based on leaving-group solvation have been performed to test the validity of the controversial dissociative route for isomerizations of $cis\text{-}[PtRCl(PEt_3)_2]$ in alcohols. Considerations of the solvent effects on rate constants give remarkable agreement with experimental values, and lend support to Scheme 4 being the rate-limiting step.[2,45] The figures suggest that the Pt–Cl bond is almost broken at the transition state, chloride having developed a charge of about 0.9. The calculations involved many assumptions, however, and experimental work aimed at allowing a full initial-state/transition-state analysis is promised.[45]

Scheme 4

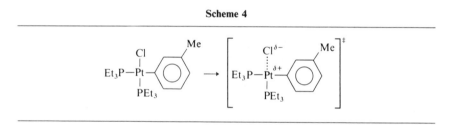

The thermal equilibration of *cis*- and *trans*-$[PtH_2L_2]$ $[L = PMe_3$ or $PEt_3]$ has been examined.[46] The equilibrium positions are solvent dependent (as expected, polar solvents favor *cis* complexes) and ligand dependent. PMe_3 leads to more of the *cis* isomer than PEt_3. The isomerization route was not determined, but it was noted that the compounds catalyze scrambling between H_2 and D_2. An oxidative-addition/reductive-elimination route could accomplish this and could also be involved in the isomerization.

A ^{31}P, ^{195}Pt, and ^{199}Hg NMR investigation of *cis*-$[Pt(GePh_3)\text{-}(HgGePh_3)(PPh_3)_2]$ between −40°C and 90°C reveals the molecule to be fluxional. Effective site exchange takes place by an intramolecular route, and does not involve catalysis by the free ligand.[47] The motion is described as a digonal twist with a transition state probably close to tetrahedral, though no paramagnetic tetrahedral complex takes part in the equilibrium. Whatever the intimate nature of the fluxionality turns out to be, it seems likely that such nonrigidity could contribute to isomerization in square-planar complexes. The NMR parameters also revealed that the group Ph_3GeHg has a smaller *trans* influence than Ph_3Ge.

The isolation of two isomeric forms of $[Rh_2(\mu\text{-}PBu_2^t)_2(CO)_4]$ revealed a new form of isomerism and isomerization of d^8 metal molecules.[48] The geometry of each form was determined by X-ray crystallography. The red isomer contains a metal–metal bond (Rh–Rh is 2.76 Å)—one Rh is planar, and the other resembles tetrahedral coordination. The orange-yellow isomer is symmetrically planar and the Rh–Rh bond is 3.72 Å, too large for a bonding interaction. The two forms equilibrate in solution (Scheme 5), a process which can be followed by UV-visible or ^{31}P NMR spectroscopy. The equilibrium is temperature dependent, low temperatures favoring the red, metal–metal bonded isomer. Although the detailed mechanism is not yet known, it is interesting to speculate whether such a process could contribute to *cis–trans* isomerization in other dinuclear planar complexes, such as $[Pt_2(\mu\text{-}PR_2)_2R_2'L_2]$ and its halide-bridged analogues.

Scheme 5

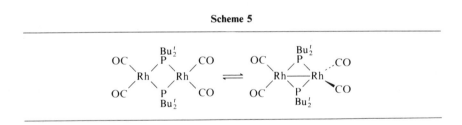

Despite the above, studies of isomerization processes at planar rhodium complexes are almost unknown, and this may reflect in part the fact that few isomer pairs are known for this element. The isolation of a series of iodo complexes *cis-* and *trans-*$[RhI(CO)(PR_3)_2]$ (R = aryl) may help this situation slightly, though the steric crowding due to the iodide makes these materials very prone to PR_3 dissociation.[49]

Cis and *trans* isomers of arylazo oximato platinum complexes have been isolated.[50] *Trans* to *cis* isomerization can be achieved simply by heating in hydrocarbon or alcohol solvents, and the reverse step is performed by treating the *cis* complex with hydrogen chloride, followed by neutralization (Scheme 6). Hydrogen chloride opens the arylazo oximate ring, but the intimate nature of the site exchange has not yet been elucidated.

Scheme 6

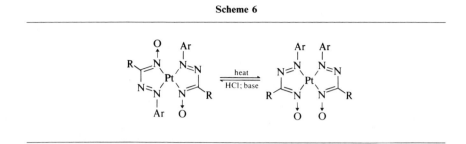

A report on the conversion of platinum sulfinate-*S* complexes to sulfin-ate-*O,O'* complexes involves isomerization following Cl⁻ abstraction.[51] Once again the geometry change mechanism is not identified, but this system resembles those where three-coordinate intermediates are likely (Scheme 7).

Scheme 7

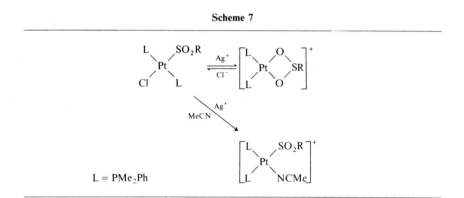

Finally, a comment on the solid-state ^{13}C NMR of [PdCl$_2$L$_2$] (L = PMe$_2$Ph or PMePh$_2$) is relevant to this section.[52] About equal amounts of the *cis* and *trans* isomers were detected for each compound, although the composition of their solutions are variable and solvent dependent. When isomerization is rapid in solution it is usual for the less soluble isomer to crystallize, and a previous crystal structure revealed a cis geometry for [PdCl$_2$(PMe$_2$Ph)$_2$]. Nevertheless Bodenhausen *et al.* speculate that in the present study mixed crystals with *cis* and *trans* molecules occupying alternate lattice sites may be growing.

5.6. Miscellaneous

The kinetics of cis-$[Pt(NH_3)_2(OH_2)_2]^{2+}$ dimerization to $[Pt_2(\mu\text{-}OH)_2(NH_3)_4]^{2+}$ have been examined in aqueous solution by UV spectrophotometry at pH range 4–5.[53] The product is formed mainly from cis-$[Pt(NH_3)_2(OH_2)(OH)]^+$, with only a small contribution from cis-$[Pt(NH_3)_2(OH_2)_2]^{2+}$ itself. Cis-$[Pt(NH_3)_2(OH_2)_2]^{2+}$ reacts with various alkylcobalamins via the sequence in Scheme 8.[54] A fast pre-equilibrium K between the two reactants forms an adduct in which cobalt is still in the "base-on" form. The rate-determining step exchanges a coordinated water of platinum with N(3) of the 5,6-dimethylbenzimidazole (dbzm) ligand.

Scheme 8

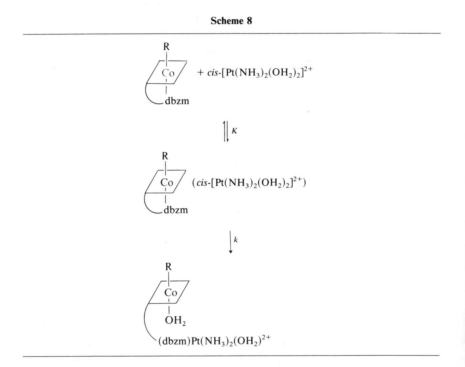

α-Amino acid esters react with aqueous $[Pd(bipy)(OH_2)_2]^{2+}$ by equilibria (11), and then hydrolyses (12) and (13).[55] The ester hydrolyses are

$$[Pd(bipy)(OH_2)_2]^{2+} + [NH_3CHRCOOR']^+$$
$$\rightleftharpoons [Pd(bipy)(NH_2CHRCOOR')]^{2+}$$
$$+ H_3O^+ + H_2O \qquad \mathbf{11} \qquad (11)$$

$$(11) + H_2O \xrightarrow{k_{H_2O}} [Pd(bipy)(NH_2CHRCOOH)]^+ + R'OH + H^+ \qquad (12)$$

$$(11) + OH^- \xrightarrow{k_{OH^-}} [Pd(bipy)(NH_2CHRCOOH)]^+ + R'OH \qquad (13)$$

substantially enhanced [except for L-cysteinates and methyl L-histidinates, which can form Pd(II) complexes not involving their alkoxycarbonyl groups], and activation parameters support the idea that the reactions involve hydroxide-ion attack on the chelated ester complexes (11) similar to the en complexes previously reported.

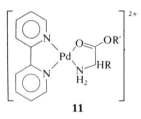

11

A spectroscopic study of aqueous solutions of $[Rh(bipy)_2]^+$ revealed a complex set of equilibria, underlining the problems of investigating the chemistry of such species.[56] Figure 5.2 shows the dominant species present, where the dividing lines approximate to $K = 1$ for the various equilibria.

Sulfur inversions in *trans*-$[PdX_2L_2]$ (X = Cl or Br; L = (12), (13), or (14) have been investigated by vt NMR spectroscopy up to pressures of 220 MPa.[57] Activation volumes $\Delta V^{\ddagger}$ are very small in each case, indicating no dissociative or associative character, or solvent (CH_2Cl_2 or $CHCl_3$)

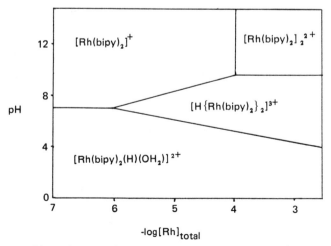

Figure 5.2.   Dominant species in aqueous "$[Rh(bipy)_2]^+$".

participation. The major contribution to the entropy is the reorganizational effect on attaining ring planarity at the transition state.

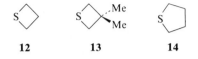

12   **13**   **14**

A comparison of 15 molecular structures $[MX(PR_3)_3]^x$ [M = Rh(I) or Pt(II); X = H$^-$, F$^-$, CO$_3^{2-}$, or Cl$^-$; R = Me, Et, Pr^i, or Ph] draws attention to the torsion angles of the M–P links.[58] The distribution of conformations is nonrandom, and indicates that two routes should operate for conformational change. Both involve a cog-wheel interaction of the intermeshed PR$_3$ groups, and illustrate how structural information can be transmitted across such molecules.

Chapter 6

Substitution Reactions of Inert-Metal Complexes— Coordination Numbers 6 and Above: Chromium

6.1. Introduction

This chapter covers the period between the end of the previous report[1] and the literature available up to December 1983. A review (220 references) of the coordination chemistry of chromium(III) has appeared.[2]

6.2. Aquation and Solvolysis of Chromium(III) Complexes

6.2.1. Unidentate Leaving Groups

6.2.1.1. Aquo Complexes

Studies of alkylpentaaquochromium(III) complexes of the type $[(H_2O)_5CrR]^{2+}$ continue to be actively investigated.[1-9] Recent examples include studies of complexes with $R = CMe_2X$ ($X = OH$,[3,4] CO_2H, and $CN^{(4)}$, $CHCl_2$ and CH_2OMe,[5] cyclopentyl,[6] and $CH_2OH^{(7)}$). A related study of the cleavage of the Cr–C bond in $[Cr(edta)(CR_1R_2OH)]^{2-}$ ions has also appeared.[8] Homolysis of $[(H_2O)_5CrCMe_2OH]^{2+}$ gives the CMe_2OH

radical and $[Cr(H_2O)_6]^{2+}$. The radical either recombines with the Cr^{2+}, or can react with added substrates. Reactions of $\cdot CMe_2OH$ with a series of cobalt(III) complexes have been investigated in this way. For example, at pH 5-9, $[(H_3N)_5Co(NH_2CMe_2OH)]^{2+}$ is the dominant initial product of the reaction with $[Co(NH_3)_6]^{3+}$, which rapidly decomposes. No reaction of the radical with chromium(III) complexes was observed.[3] Pulse radiolysis of the tertiary organochromium(III) complexes $[(H_2O)_5CrCMe_2X]^{2+}$ also involves homolytic Cr–C bond cleavage with rate constants of 0.15, 4, and $\sim 10^5\ s^{-1}$ at 295 ± 2 K for X = OH, CO_2H, and CN.[4] The thermally stable $[(H_2O)_5CrCHCl_2]^{2+}$ and $[(H_2O)_5CrCH_2OMe]^{2+}$ ions also undergo homolysis when irradiated with light from an unfiltered xenon lamp [equation (1)]. In the presence of H_2O_2, the rate-determining step is reaction (2), and at 298.2 K the rate constant of this reaction is $(7.06 \pm 0.04) \times 10^4\ dm^3\ mol^{-1}\ s^{-1}$:

$$CrR^{2+} \xrightarrow{h\nu} Cr^{2+} + R^{\cdot} \tag{1}$$

$$Cr^{2+} + H_2O_2 \rightarrow CrOH^{2+} + {\cdot}OH \tag{2}$$

The OH radicals react rapidly with organic and inorganic solutes to yield radicals which react further with Cr^{2+} to give intensely colored Cr^{3+} products.[5] When $[Cr(H_2O)_6]^{2+}$ is reacted with H_2O_2 in an aqueous solution saturated with cyclopentane, the $[(H_2O)_5Cr\text{-}c\text{-}C_5H_9]^{2+}$ ion is formed. Acidolysis of this ion to give $[(H_2O)_5CrOH]^{2+}$ and $c\text{-}C_5H_{10}$ is characterized at 298.2 K by $10^4k = 4.86 \pm 0.12\ s^{-1}$, $\Delta H^{\ddagger} = 73.5 \pm 2.4\ kJ\ mol^{-1}$, and $\Delta S^{\ddagger} = -61.5 \pm 7.9\ J\ K^{-1}\ mol^{-1}$. Homolysis gives Cr^{2+} and ${\cdot}C_5H_9$ with $10^4k = 1.07 \pm 0.16\ s^{-1}$, $\Delta H^{\ddagger} = 126 \pm 2.9\ kJ\ mol^{-1}$, and $\Delta S^{\ddagger} = 102.5 \pm 9.3\ J\ K^{-1}\ mol^{-1}$ at 298.2 K. The $[(H_2O)_5Cr\text{-}c\text{-}C_5H_9]^{2+}$ ion is also formed pulse radiolytically by the reaction between Cr^{2+} and the cyclopentyl radical ($10^7k = 8 \pm 1\ s^{-1}$), and at 298.2 K it reacts electrophilically in an S_E2 mechanism with Hg^{2+} ($k = 1.08 \pm 0.2\ dm^3\ mol^{-1}\ s^{-1}$), $Br_2(10^{-4}k = 1.30 \pm 0.19\ dm^3\ mol^{-1}\ s^{-1})$, and $I_2(k = 8.2\ dm^3\ mol^{-1}\ s^{-1})$.[6]

Acetate ion has been found to assist the cleavage of the Cr–C bond in $[(H_2O)_5CrCH_2OH]^{2+}$:[7]

$$[(H_2O)_5CrCH_2OH]^{2+} + CH_3CO^- \rightleftharpoons [(CH_3CO_2)(H_2O)_4CrCH_2OH]^+ \tag{3}$$

$$[(CH_3CO_2)(H_2O)_4CrCH_2OH]^+ \rightarrow [(H_2O)_5CrO_2CCH_3]^{2+} + CH_3OH \tag{4}$$

The mechanism is very similar to that observed for $[Cr(H_2O)_5I]^{2+}$ reacting with HCl, in which the *trans*-labelizing iodide ion gives rise to $[(H_2O)_4CrClI]^+$ which then rapidly loses I^- ion.[10] It is to be expected that coordinating anions (X^-) will always assist loss of a good *trans*-labilizing group R in complexes of the type $[Cr(H_2O)_5R]^{2+}$, via the formation of *trans*-$[X(H_2O)_4CrR]^+$.

The electrophilic reactions of $[(H_2O)_5CrR]^{2+}$ have been investigated in D_2O/H_2O mixtures. The rate decreases with increasing $[D_2O]$, especially when H^+ is the electrophile.[9]

Adenosine-5'-triphosphate (ATP) forms bidentate Δ- and Λ- complexes of the type $[Cr(H_2O)_4(ATP)]$ [structures (1) and (2)]. Hydrolysis gives predominantly free ATP with very little of the diphosphate, ADP, being formed. The rate constant is $5 \times 10^{-4}\,s^{-1}$ at 310 K and pH 11. The terdentate ATP complex $[Cr(H_2O)_3(ATP)]$ is also formed, and interestingly this hydrolyzes to give mostly ADP and very little free ATP ($k \sim 5 \times 10^4\,s^{-1}$ at pH 11 and 310 K). The rate constant is $\sim$5000 times as large as the value observed for the alkaline hydrolysis of ATP in the absence of metal ions.[11]

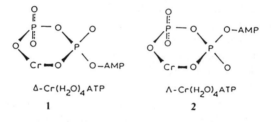

$$\Delta\text{-Cr}(H_2O)_4\text{ATP} \qquad \Lambda\text{-Cr}(H_2O)_4\text{ATP}$$
$$\mathbf{1} \qquad\qquad\qquad \mathbf{2}$$

Several chromium(III) complexes have been made with amino alcohols of the type $[Cr(LL_3)]$ $[LL^- = NH_2CH_2CHRO^-$, where $R = H$, Me, or $NH_2CH(Me)CH_2O^-]$. Acidolysis proceeds in three stages, as in equations (5-7):

$$[Cr(LL)_3] + 3H^+ \rightleftharpoons [Cr(LLH)_3]^{3+} \tag{5}$$

$$[Cr(LLH)_3]^{3+} + 3H_2O \rightarrow [Cr(LLH)_3(OH_2)_3]^{3+} \tag{6}$$

$$[Cr(LLH)(OH_2)_3]^{3+} + 3H_2O \rightarrow [Cr(H_2O)_6]^{3+} + 3LLH \tag{7}$$

Reaction (5) involves rapid protonation of all three alkoxide groups. Reaction (6) takes $\sim$10 min at 293 K and involves the formation of monodentate

Table 6.1. *Rate Data at 298.2 K ($I = 0.1\ mol\ dm^{-3}$) for the Three Steps Involved in Reaction (7)[a]*

	$10^6 k\ (s^{-1})$	$\Delta H^{\ddagger}(kcal\ mol^{-1})$	$\Delta S^{\ddagger}(cal\ K^{-1}\ mol^{-1})$
First step	227	18.0	−15
Second step	16.8	21.3	−9
Third step	0.64	21.2	−16

[a] LLH = ethanolamines; data from Reference 12.

N-bonded tris(amino–alcohol) complexes. The final stage is slowest and proceeds in three steps. The rate data for the successive loss of the three ethanolamine molecules are summarized in Table 6.1.

6.2.1.2. Ammine and Amine Complexes

It is well known from the work of H. Taube and co-workers that triflate ion, $CF_3SO_3^-$, forms a very weak complex with chromium(III), of the type $[(H_2O)_5CrOSO_2CF_3]^{2+}$, which is only marginally more stable than the corresponding perchlorato complex. It is not surprising, therefore, that complexes of the type $[M(NH_3)_5OSO_2CF_3]^{n+}$ have been found to be useful precursors in the preparation of $[M(NH_3)_5L]^{n+}$ and $[(H_3N)_5M(L\text{-}L)M(NH_3)_5]^{m+}$ ions. At 298.2 K, rate constants for the loss of the triflate ion in 0.1 mol dm^{-3} CF_3SO_3H are $10^4k = 268$, 187, 1.61, 124, 930, 16, and 21 for $M = Co^{3+}$, Rh^{3+}, Ir^{3+}, Cr^{3+}, Ru^{3+}, Os^{3+}, and Pt^{4+}. The Cr^{3+} complex is slightly less labile than the Rh^{3+} complex, which in turn is much more labile than the Ir^{3+}, Os^{3+}, and Pt^{4+} complexes.[13]

Acid hydrolysis of $[Cr(NH_3)_5NCO]^{2+}$ in perchloric acid (0.04–0.60 mol dm^{-3} at 313–338 K; $I = 0.6$ mol dm^{-3}, $NaClO_4$) gives largely $[Cr(NH_3)_6]^{3+}$ and CO_2 plus other chromium(III) products; at 328.2 K, $k_{obs} = a[H^+]^2/(1 + b[H^+])$ with $a = 0.36$ dm^3 mol^{-1} s^{-2} and $10^3b = 6.9$ dm^3 mol^{-1} s^{-1}. The maximum kinetic yield of $[Cr(NH_3)_6]^{3+}$ was only 35%, but a synthetic route was developed which gave a 60% yield.[14]

Acid hydrolysis of $[Cr(NH_3)_5F]^{2+}$ ion has been reinvestigated to establish the stereochemistry of the reaction involving loss of NH_3. The reactions were followed by directly monitoring both the released NH_3 and the HF. The ratio of the *cis-trans*-fluoroproducts $[Cr(NH_3)_4(OH_2)F]^{2+}$ was found to be 3:1 and independent of the acidity used. Taking into account the statistical factor of 4, the calculated rate constants for loss of a single NH_3 molecule are $10^6k_{trans} = 1.7$ s^{-1} and $10^6k_{cis} = 1.3$ s^{-1}, and there is no *cis*-effect of the F^- ion. The rate constants for F^- loss were found to be much smaller than previously reported; in a ClO_4^- medium at 323.2 K, $10^7k/s^{-1} = 4.4 + 81[H^+]$. Loss of NH_3 is slower in the presence of Cl^- ion by a factor of ~2.[15]

High-pressure studies of the aquation of $[M(NH_3)_5Br]^{2+}$ (M = Cr, Co) in the presence and absence of polyelectrolytes (sodium polyethylene sulfonate or polystyrenesulfonate) indicates a dissociative mechanism for the cobalt(III) complex, and an associative mechanism for the chromium(III) complex. The chromium(III) data in various polyelectrolyte media are summarized in Table 6.2.[16] The Hg^{2+}-induced aquation of $[Cr(NH_3)_5Br]^{2+}$ has also been studied in the presence of polyelectrolytes and at high pressures.[17]

Table 6.2. *Rate Data at 298.2 K (I = 0.1 mol dm^{-3}) for the Acid Hydrolysis of* [Cr(NH$_3$)$_5$Br]$^{2+}$

Polyelectrolyte[a]	$\Delta H^{\ddagger}$ (kcal mol^{-1})	$\Delta S^{\ddagger}$ (cal K^{-1} mol^{-1})	$\Delta V^{\ddagger}$ (cm^3 mol^{-1})
None	20.7 ± 0.4	−7 ± 2	−9.3 ± 2.0
Na PES (0.001)	20.8	−6	−7
Na PES (0.005)	21.3	−4	−4
(C$_4$H$_9$)$_4$N PES (0.001)	20.8	−6	−4
Na PPS (0.005)	23.0	−1	

[a] PES = polyethylene sulfonate; PPS = polystyrenesulfonate (polyelectrolyte equivalents in parentheses).

Initial-state transition-state dissection of the rates of the Hg^{2+} induced aquations of *trans*-[Rh(en)$_2$Cl$_2$]$^+$, [Cr(NH$_3$)$_5$Cl]$^{2+}$, and *cis*-[Cr(NH$_3$)$_4$(OH$_2$)Cl]$^{2+}$ has been reported in aqueous MeOH, EtOH, and MeCN.[18] For [M(NH$_3$)$_5$Cl]$^{2+}$ (M = Co, Cr) the solvation energies for the chromium(III) complexes are smaller than for the corresponding cobalt(III) ions, and the rates of aquation of the chromium(III) complexes are less sensitive to solvent variation.

A high-pressure electrical-conductivity apparatus has been used to study the aquation of [Cr(en)$_2$Cl$_2$]$^+$ at pressures up to 2 kbar (T = 288–303 K). At 298.2 K, $\Delta V^{\ddagger}$ = +1.82 cm^3 mol^{-1}, whereas $\Delta S^{\ddagger}$ = −9.019 cal K^{-1} mol^{-1}, and an I_d mechanism is favored.[19]

Reaction of *cis*-[Cr(en)$_2$FCl]I with dry liquid ammonia gives >70% *cis*-[Cr(en)$_2$F(NH$_3$)]ICl, and the rest *trans*-[Cr(en$_2$)F(NH$_3$)]ICl as shown by HPLC experiments. The *cis/trans* ratio is ~4:1 indicating that a trigonal bipyramidal intermediate is formed, since one would expect a 2:1 ratio if a *trans* attack in an S_N2 mechanism were involved.[20]

Hydrolysis of *cis*- and *trans*-[Cr(en)$_2$L$_2$]$^{3+}$ ions (L = NH$_3$, MeNH$_2$) has been studied in HClO$_4$, HCl, and HBr. The products are [Cr(en)$_2$(OH$_2$)$_2$]$^{3+}$ in all cases, and below pH 6 the reactions are acid independent and first order. The rate of loss of MeNH$_2$ is similar to the loss of en from [Cr(en)$_3$]$^{3+}$ ion.[21]

The *cis*-β-(*RR*, *SS*)[Cr(ox)(L)]$^+$ ion (L = 1,4,8,11-tetraazaundecane, 2,3,2-tet) has been prepared and resolved, together with *cis*-α-[CrCl$_2$(L)]$^+$ and *trans*-(*RS*)[CrX$_2$(L)]$^+$ (X = Cl, Br). Rate data for the acid and base hydrolysis of the *cis*-dichloro complex, and for the base hydrolysis of the *trans*-dichloro and dibromo complexes, are compared with data for the dichloro and chlorohydroxo complexes with L = N,N'-1,2-ethanediylbis(1,3-propane-diame) (trivial name 3,2,3-tet),[23] and with data for the acid hydrolysis of *trans*-[Cr(L)X$_2$] [X = Br; L = *RR,SS*-3,2,3-tet, *R,S*-2,3,2-tet and 5,7,7,12,14,14-hexamethyl-1,4,8,11-tetra-

azacyclotetradecane (trivial name teta)[24]; X = Cl; L = teta[25]] in Table
6.3. With the exception of the teta complexes for chromium(III) the rate
of hydrolysis of Br^- is 10–17 times as fast as Cl^-, whereas for analogous
cobalt(III) complexes the rate ratio is in the range 1.4–15 (usually much
smaller than 15). A dissociative mechanism is favored for both
chromium(III) and cobalt(III) complexes of this type, with stereochemical
changes more common for the cobalt(III) species.[24] Base hydrolysis of
the chromium(III) complexes is much slower than for cobalt(III) complexes
and competition with background hydrolysis is often observed. If the
conjugate-base mechanism applies to chromium(III) complexes, as seems
likely, then either the NH protons are less acidic or the conjugate base is
less labile. The former explanation applies to the 2,3,2-tet complexes.[23]

Reaction of cis-$[Cr(L)Cl_2]^+$ (L = 2,3,2-tet) with thiocyanate ion surprisingly gave trans-$(RS)[Cr(L)(NCS)_2]^+$ as established by a crystal structure.
At 326.2 K, aquation of the first thiocyanate ion to give trans-$[Cr(L)(NCS)$-
$(H_2O)]^{2+}$ is characterized by a first-order rate constant, $10^8 k = 6.6 \, s^{-1}$. This
very low rate is much less than that of trans-$[Cr(en)_2(NCS)_2]^+$ for which
$10^6 k = 6.6 \, s^{-1}$ at 326.2 K. Solvation effects may be important in directing
the stereochemical change in the initial reaction.[26]

Following the synthesis of the new edta complex, $Na_3[Cr(L)$-
$(CN)_2OH_2]\cdot 4H_2O$ (H_4L = edta) in which edta is terdentate with three carboxylate groups uncoordinated, the kinetics of the acid-catalyzed aquation
of the first cyanide group has been measured. The loss of the second cyanide
group is rapid, and acid catalysis involves protonation of one of the uncoordinated carboxylate groups:

$$[Cr(HL)(CN)_2OH_2]^{2-} \underset{}{\overset{K_1}{\rightleftharpoons}} [Cr(L)(CN)_2OH_2]^{3-} + H^+ \qquad (8)$$

$$[Cr(HL)(CN)_2OH_2]^{2+} \xrightarrow{k_1} [Cr(HL)(CN)OH_2]^- + CN^- \qquad (9)$$

$$[Cr(HL)(CN)OH_2]^- \xrightarrow{\text{rapid}} [Cr(HL)OH_2] + CN^- \qquad (10)$$

At 298.2 K the rapid protonation reaction (8) is characterized by
$10^6 K_1/dm^3 \, mol^{-1} = 3.69$. Reaction (9), which involves chelation as well as
CN^- loss, is rate determining, and at 298.2 K, $10^2 k_1/s^{-1} = 4.65$ ($\Delta H^{\ddagger} =$
18.88 kcal mol^{-1}, $\Delta S^{\ddagger} = -11.39$ cal $mol^{-1} \, K^{-1}$). Although the usual protonation of the cyano groups was not observed, it is postulated that aquation
is assisted by internal protonation of a cyano group by an uncoordinated
but protonated carboxylate group (a sort of anchimeric assistance).
Photoaquation was also investigated.[21]

6.2.2. Multidentate Leaving Groups

Rates and activation parameters have been measured for reaction (11)
in the pH range 2–3 (L = histamine). Chelate-ring opening of the bidentate

Table 6.3. Rate Data at 298.2 K for the Acid and Base Hydrolysis of $[CrX_2L]^+$ Ions

Geometry	L^a	X	Acid hydrolysis			Base hydrolysis			References
			$10^6 k_{aq}$ (s⁻¹)	E_a (kJ mol⁻¹)	$\Delta S^{\ddagger}$ (JK⁻¹ mol⁻¹)	k_{OH} (dm³ mol⁻¹ s⁻¹)	E_a (kJ mol⁻¹)	$\Delta S^{\ddagger}$ (JK⁻¹ mol⁻¹)	
cis-α (?)	2,3,2-tet	Cl	412	74.2	−69	0.52			22
trans-(RS)		Cl	3.21	107	0	0.28	97.9	+65	22
		Br	42.7	100	+1	91.8	114	+167	24
trans-(RRSS)	3,2,3-tet	Cl	1.06	94.7	−50	0.58^b	87.7	+36	23
		Br	10.2	94	−35				24
cis-β (RRSS)		Cl	102	88.7	−32	11.2^c			23
trans-(RSSR)	teta	Cl	12.6	92.6	−37				25
		Br	1700	74	−58				24

^{a} For ligand abbreviations, see text.
^{b} For trans-[Cr(3,2,3-tet)Cl(OH)]⁺, $10^3 k_{OH} = 8.81$ dm³ mol⁻¹ s⁻¹, $E_a = 97.3$ kJ mol⁻¹, and $\Delta S^{\ddagger} = +33$ JK⁻¹ mol⁻¹.
^{c} For cis-[Cr(3,2,3-tet)Cl(OH)]⁺, $10^2 k_{OH} = 4.0$ dm³ mol⁻¹ s⁻¹.

Table 6.4. Second-Order Rate Constants and Activation Parameters at 298.2 K for the Metal-Ion (M^{n+})-Assisted Loss of Oxalate from Oxalato Chromium(III) Complexes[a]

[Cr(ox)₃]³⁻ M^{n+}	Fe^{3+}	Cu^{2+}	Ni^{2+}	Zn^{2+}	Co^{2+}	Mn^{2+}	H^+
$10^4 k_M$ (dm³ mol⁻¹ s⁻¹)	567	10.3	0.35	0.18	0.126	0.022	0.162
$\Delta H^{\ddagger}$ (kJ mol⁻¹)	95.4	90.3	89.9	90.3	90.3	90.3	90.3
$\Delta S^{\ddagger}$ (JK⁻¹ mol⁻¹)	50.2	−0.4	−30.1	−34.3	−37.2	−51.8	−35.1

cis-[Cr(ox)₂(OH₂)₂]⁻ M	Fe^{3+}	H^+	[Cr(ox)(OH₂)₄]⁺ Fe^{3+}
$10^5 k_M$ (dm³ mol⁻¹ s⁻¹)	103	0.24	0.0268
$\Delta H^{\ddagger}$ (kJ mol⁻¹)	101.2	99.9	113.6
$\Delta S^{\ddagger}$ (J K⁻¹ mol⁻¹)	37.2	−18.8	9.2

[a] Data from Reference 29.

Table 6.5. *Second-Order Rate Constants at 303 K and Activation Parameters at 298.2 K for the Metal Ion (M^{n+}) Catalyzed Loss of Malonate Ion (mal) from $[Cr(mal)_3]^{3-}$*

M^{n+}	H^+	Mn^{2+}	Co^{2+}	Ni^{2+}	Cu^{2+}	Zn^{2+}
$10^4 k_M$ (dm^3 mol^{-1} s^{-1})	10.6	1.7	4.7	5.4	6.7	3.5
$\Delta H^\ddagger$ (kJ mol^{-1})	109.2	118.9	67.2	73.5	67.6	80.2
$\Delta S^\ddagger$ (J K^{-1} mol^{-1})	+55.0	+72.7	−89.5	−67.6	−86.1	−50.0

histamine is involved in the rate-determining step, and in the pH range studied the rate is pH independent.[28]

$$[Cr(ox)_2(L)]^- + 2H_3O^+ \rightarrow cis\text{-}[Cr(ox)_2(OH_2)_2]^- + H_2L^{2+} \quad (11)$$

The metal-ion-catalyzed dissociation of oxalato complexes of the type $[Cr(ox)(OH_2)_4]^+$, $[Cr(ox)_2(OH_2)_2]^-$, and $[Cr(ox)_3]^{3-}$ have been investigated.[29] The pseudo-first-order rate constant (k_{obs}) is given by $k_{obs} = k + k_M[M^{n+}]$, and k_M varies for different metal ions (M^{n+}) in the order $Fe^{3+} \gg Cu^{2+} > Ni^{2+} > Zn^{2+} > Co^{2+} > Mn^{2+}$. This order reflects the ease of binding of M^{n+} to a chromium(III)-coordinated oxalato group. Protons also assist oxalate dissociation, with k_H between the values of k_M obtained for Co^{2+} and Mn^{2+}. The rate data are summarized in Table 6.4. A very similar study of the metal-ion-catalyzed aquation of $[Cr(mal)_3]^{3-}$ ion (mal = malonate ion) has appeared.[30] The data are presented in Table 6.5.

The acid hydrolysis of $[Cr(acac)(OH_2)_4]^{2+}$ has been studied in 3.0–4.5 mol dm^{-3} perchloric acid solutions. The activation parameters are $\Delta H^\ddagger = 123.7$ kJ mol^{-1} and $\Delta S^\ddagger = +13.4$ J K^{-1} mol^{-1}, and an associative process is favored.

6.2.2.1. Tris(Chelates)

In addition to studies of $[Cr(ox)_3]^{3-}$ and $[Cr(mal)_3]^{3-}$ (previous section), the acid-catalyzed loss of LL from $[Cr(LL)_3]^{3-}$ [LL = $(O_2CCR_2CO_2)_2$, where R_2 = Me,H; Et,H; Bz,H; Me$_2$; H$_2$)] has been investigated. In general $k_{obs} = k_0 + k_1[H^+]$, with the acid-independent term only important for the cases where R_2 = Me,H, and Me$_2$. The data are summarized in Table 6.6.[32]

The acid hydrolysis of $[Cr(LL)_3]$ [LL = NH$_2$CH$_2$CH$_2$OH, HOCH$_2$CH(NH$_2$)CH$_3$, and NH$_2$CH$_2$CH(OH)CH$_3$] was discussed previously (Reference 12). A study of the tris(ethanolamine) complex in aqueous/acetone mixed solvents has also appeared.[33] The acid hydrolysis of the low-spin tris(chelates) of chromium(II) with 2,2′-bipyridine and

Table 6.6. Rate Data at 298.2 K for the Acid-Catalyzed Aquation of Substituted tris(Malonato)Chromium(III) Complexes[a]

Complex[b]	$10^4 k_0$ (s^{-1})	$\Delta H^\ddagger$ (kJ mol^{-1})	$\Delta S^\ddagger$ (JK^{-1} mol^{-1})	$10^4 k_1$ (dm^3 mol^{-1} s^{-1})	$\Delta H_1^\ddagger$ (kJ mol^{-1})	$\Delta S_1^\ddagger$ (JK^{-1} mol^{-1})
[Cr(Etmal)$_3$]$^{3-}$				15.8	102	+43
[Cr(Bzmal)$_3$]$^{3-}$				32	94	+22
[Cr(Me$_2$mal)$_3$]$^{3-}$		77.0	−85	300	52	−120
[Cr(Memal)$_3$]$^{3-}$	1.0	242.7	+473	5.88	82.8	−30.5
[Cr(mal)]$^{3-}$	0.967				109.6	+58

[a] Data from Reference 31.
[b] Etmal = [$O_2CCH(C_2H_5)CO_2$]$^{2-}$; Bzmal = [$O_2CCH(CH_2Ph)CO_2$]$^{2-}$; Me$_2$mal = [$O_2CC(Me_2)CO_2$]$^{2-}$; Memal = [$O_2CCH(Me)CO_2$]$^{2-}$.

Table 6.7. *Rate Data at 298.2 K for the Loss of the First 2,2'-Bipyridine (bipy) or 1,10-Phenanthroline (phen) Molecule from Low-Spin $[Cr(LL)_3]^{2+}$ Ions*[a]

LL	k (s^{-1})	$\Delta H^{\ddagger}$ (kJ mol^{-1})	$\Delta S^{\ddagger}$ (JK^{-1} mol^{-1})
bipy	0.390	84.4	30.6
phen	0.0643	89.0	30.3

[a] Data from Reference 34.

1,10-phenanthroline has been reexamined by stopped-flow spectrophotometry. The data for the loss of one bipy or phen molecule are given in Table 6.7.[34] Data for the thermal aquation of $[Cr(phen)_3]^{3+}$ are considered together with the photochemical data in Section 6.4 (Reference 69).

6.2.3. Bridged Dichromium(III) Complexes

The kinetics of the acid hydrolysis of the (μ-oxo)bis(pentaaquochromium(III)) ion $[(H_2O)_5Cr-O-Cr(OH_2)_5]^{4+}$ has been studied. $[Cr(H_2O)_6]^{3+}$ is the sole chromium(III) product, with $k_{obs} = k_0 + k_1[H^+]$. At 298.2 K, $10^5 k_0 = 5$ s^{-1}, $\Delta H^{\ddagger} = 22$ kcal mol^{-1}, and $\Delta S^{\ddagger} = 5$ cal K^{-1} mol^{-1}, and $10^3 k_1 = 1.61$ dm^3 mol^{-1} s^{-1}, $\Delta H^{\ddagger} = 12.9$ kcal mol^{-1}, and $\Delta S^{\ddagger} = -28$ cal K^{-1} mol^{-1}. The k_0 pathway is believed to involve associative attack by the ion-coming H_2O molecule during Cr–O bond cleavage, and the k_1 pathway involves prior protonation of the bridging μ-oxo group.[35] The value of k_0 is 10^2 larger than the rate of water exchange with $[Cr(H_2O)_6]^{3+}$ ion, and this is due to $\Delta H^{\ddagger}$ being 4 kcal mol^{-1} smaller. The values of $\Delta S^{\ddagger}$ are very similar for k_0 and for water exchange with $[Cr(H_2O)_6]^{3+}$. The protonation of the bridging oxo group is not very easy, presumably due to π bonding in the bridge, and the concentration of $[(H_2O)_5Cr(OH)Cr(OH_2)_5]^{5+}$ present is very low.

6.3. Formation of Chromium(III) Complexes

6.3.1. Reactions of $[Cr(H_2O)_6]^{3+}$

Several studies of the reactions of $[Cr(H_2O)_6]^{3+}$ and $[Cr(H_2O)_5OH]^{2+}$ with amino acids and related ligands have appeared.[36-40] The reaction of $[Cr(H_2O)_5OH]^{2+}$ with *dl*-alanine has been studied between pH 4.5 and 5.4

and the usual ion-pair mechanism established. The rate is higher than the rate of water exchange with $[Cr(H_2O)_5OH]^{2+}$, and the rate of other substitution reactions involving $[Cr(H_2O)_5OH]^{2+}$. From the activation parameters it is deduced that both bond making and bond breaking are involved in the transition state.[36] The reaction of quinolinic acid (H_2L) with $[Cr(H_2O)_6]^{3+}$ has been studied between pH 3.2 and 4.3 in aqueous ethanol, and the rate of anation found to be of the form, rate $= kK[Cr^{3+}][HL^-]/(1 + K[HL^-])$, as expected for an ion-pair mechanism.[37] An I_a mechanism is claimed for the reaction of $[Cr(H_2O)_6]^{3+}$ with serine in aqueous acidic solution[38] and for the reaction with 2-picolinic acid N-oxide in 30% (v/v) aqueous ethanol.[40] Ion pairing with the zwitterion of serine is involved, and the rate of the reaction is not very sensitive to ionic strength (0.2–1.0 mol dm^{-3}). Activation parameters were determined from studies over a narrow temperature range (312.7–329.2 K).[38] For the formation of the 1:1 complex of $[Cr(H_2O)_6]^{3+}$ with furfuryliminodiacetic acid, the equilibrium constant is $10^{3.32}$, and the Arrhenius parameters for complex formation are $A = 3 \times 10^{-10}$ s^{-1} and $E = 20.8$ kcal mol^{-1}.[39] Two studies of the reactions of $[Cr(H_2O)_6]^{3+}$ with the colored ligands, violuric acid[41] and semimethylthymol blue[42], have appeared. Reaction of $[Cr(H_2O)_6]^{3+}$ with $[Mo(CN)_8]^{4-}$ gives $[CrMo(CN)_6(H_2O)(OH)]$ via $[Cr(H_2O)_5CN]^{2+}$ as an intermediate. The reaction is second order overall, first order in each of the two reactants.[43]

The interesting complex ion $[CrF_5OH_2]^{2-}$ is reported to be kinetically inert. It is formed by reduction of CrO_3 by formaldehyde in 40% aqueous HF.[44]

Formation of metastable $[Cr(H_2O)_5L]^{n+}$ ions is often achieved by inner-sphere redox reactions, usually by the reaction of $[Cr(H_2O)_6]^{2+}$ with inert-metal complexes such as $[Co(NH_3)_5L]^{n+}$. Recent examples include cases where L = tetrazoles,[45] $^-OSO_2NH_2$,[46] 3- or 4-$CNC_6H_4CO_2^-$,[47] and $[OCC(CH_3)C(CN)C(CH_3)CO]^-$.[48]

The early stages of the oligomerization of chromium(III), following the reaction between $[Cr(H_2O)_6]^{3+}$ and OH^-, have been investigated. The dimer, $[Cr_2(OH)_2(aq)]^{4+}$, trimer, $[Cr_3(OH)_4(aq)]^{5+}$, tetramer $[Cr_4(OH)_6(aq)]^{6+}$, as well as higher polymers (probably a pentamer and a hexamer), were separated by Sephadex SP C–25 chromatography. In acidic solution the trimer is the most robust, and this forms rapidly from the tetramer upon acidification. The structures of these oligomers are discussed together with the mechanism of their formation and dissociation.[49] The formation of other dimers is discussed in the next section. In strong perchloric acid solution the aqua dimer gives $[(H_2O)_4Cr(OH)_2Cr(OH_2)_3OClO_3]^{3+}$, analogous to the well-known perchlorato-complex $[(H_2O)_5CrOClO_3]^{2+}$.[50]

6.3.2. Formation of Mixed-Ligand Complexes

6.3.2.1. Ammine and Amine Complexes

Anation of $[Cr(NH_3)_5OH_2]^{3+}$ by $H_3PO_2/H_2PO_2^-$ has been studied in aqueous solution 313.2-333.2 K and $I = 1.0$ mol dm^{-3} (LiClO$_4$). The usual interchange mechanism applies with ion-pair constants of 0.13 and 0.33 dm^3 mol^{-1} for H_3PO_2 and $H_2PO_2^-$, respectively. The rate constants and activation parameters for the interchange process involving both outer-sphere complexes were found to be identical: $10^4 k/s^{-1} = 1.39$ at 313.2 K, and $\Delta H^{\ddagger} = 28.7$ kcal mol^{-1} and $\Delta S^{\ddagger} = 14.8$ cal K^{-1} mol^{-1}. The reaction is believed to have a borderline I_a-I_d mechanism.[51]

A pink complex, previously thought to be cis-$[Cr(bipy)_2(OH_2)_2]^{3+}$ has been shown to be cis-$[Cr(bipy)_2(OH_2)Cl](ClO_4)_2 \cdot 2H_2O$. It forms by slow oxidation of CrCl$_2$ in the presence of bipy.[52] Synthesis of cis- and $trans$-$[Cr(en)_2L_2]^{3+}$ (L = MeNH$_2$[53,54] and NH$_3$[54]), cis-$[Cr(LL)_2(ox)]^+$ (LL = en, 1,2-pn, and 1,3-pn),[55] and $[Cr(LL)X_4]^-$ (LL = diphosphine or diarsine; X = Cl, Br)[56] have been reported. Following the synthesis of the μ-dihydroxy-bridged dimer, $[(LL)_2Cr(OH)_2Cr(LL)_2]^{4+}$ (LL = 1,3-pn), the kinetics of the equilibrium reaction (12) was investigated. At 298.2 K,

$$[(LL)_2Cr(OH)_2Cr(LL)_2]^{4+} + H_2O \underset{k_{-1}}{\overset{k_1}{\rightleftharpoons}}$$

$$[(LL)_2(H_2O)Cr(OH)Cr(OH)(LL)_2]^{4+} \quad (12)$$

$I = 0.1$ mol dm^{-3} (NaClO$_4$/HClO$_4$) and $10^4 k_{obs} = 6$ s$^{-1} = k_1 + k_{-1}$. This rate is very similar to that obtained for related systems.[57] Formation of other chromium(III) dimers have been investigated including $[(acac)_2Cr(OMe)_2Cr(acac)_2]$,[58] $[(H_3N)_4Cr(OH)_2Cr(NH_3)_4]^{4+}$ ion[59], and an antiferromagnetically coupled μ-oxo-bridged porphyrin dimer of the type $[LCr(O)FeL']$ (L = $\alpha,\beta,\gamma,\delta$-tetraphenylporphyrinate, L = L' or tetra-p-tolylphenylporphyrinate or tetra-p-fluorophenylporphyrinate).[60] A mixed Co$_2$(III)Cr(III) trimer has been made by reacting $[Cr(Hedta)OH_2]$ with $[(H_3N)_3Co(OH)_3Co(NH_3)_3]^{3+}$ ion. The uncoordinated carboxylate group of the chromium(III)-coordinated edta forms a bridge between the two cobalt(III) centers in a μ-dihydroxy-μ-carboxylato-bridged dicobalt(III) structure.[61]

Formation of $[Cr(edta)]^-$ from the reaction of edta^{4-} with the cis-$[Cr(ox)_2(OH_2)_2]^-$ ion has been studied in alkaline solutions. Both the uncatalyzed and HCO$_3^-$-catalyzed reactions were investigated. The initial step involves deprotonation of a coordinated water molecule, or its reaction with HCO$_3^-$ ion, followed by oxalate displacement by Hedta^{3-}, either directly or via oxalate chelate-ring opening.[62]

[Cr(Hedta)(OH$_2$)] is the main product of the reaction between [Ca(edta)]$^{2-}$ and [Cr(H$_2$O)$_6$]$^{3+}$. The reaction was studied at pH 3.5 and 297.2–318.2 K in the presence of excess [Ca(edta)]$^{2-}$. Under these conditions, below pH 4.1 a dissociative, first-order process was observed, with some evidence for an associative pathway above pH 4.1.[63]

As discussed in an earlier section, the trans-(R,S)[Cr(2,3,2-tet)-(NCS)$_2$]$^+$ ion is the surprising product of the reaction between cis-[Cr(2,3,2-tet)Cl$_2$]$^+$ and thiocyanate ion.[26]

Formation of several chromium(III) complexes with tetra-azamacrocycles have been reported.[64-66] The trans-[Cr(L)(OH$_2$)$_2$]$^{3+}$ ion (L = 5, 57,12,14-tetramethyl-1,4,8,11-tetra-azacyclotetradeca-4,6,11,13-tetraene) is formed via the reaction of [Cr(H$_2$O)$_6$]$^{2+}$ and L·2HClO$_4$. The trans-[Cr(L)-(SH)(OH$_2$)]$^{2+}$ ion was also prepared, and the kinetics of the reactions of this and trans-[Cr(L)(OH$_2$)$_2$]$^{3+}$ with the unidentate ligands NCS$^-$, N$_3^-$, Cl$^-$, pyridine, and imidazole investigated. The coordinated water molecules are less acidic than those in [Cr(H$_2$O)$_6$]$^{3+}$, and more labile. Interestingly, the trans-[Cr(L)(SH)(OH$_2$)]$^{2+}$ ion reacts more slowly than trans-[Cr(L)(OH)-(OH)$_2$]$^{2+}$ with the azide ion.[64]

A high-yield synthetic route to the trans-[Cr(cyclam)(CN)$_2$]$^+$ ion has appeared from the reaction of cis- or trans-[Cr(cyclam)Cl$_2$]$^+$ with the CN$^-$ ion in DMSO solution.[65] Conversion to trans-[Cr(cyclam)(NCS)$_2$]$^+$ was possible. The photochemistry of trans-[Cr(cyclam)(CN)$_2$]$^+$ is considered later.

There are very few carbonato complexes of chromium(III) known, and so the isolation of cis-[Cr(L)(CO$_3$)]$^+$ (L = rac-5,5,7,-12,12,14-hexamethyl-1,4,8,11-tetra-azacyclotetradecane) is especially interesting.[66]

Table 6.8. Rate Data from Variable-Pressure and Variable-Temperature Studies of the First-Stage Anation of the trans-[M(L)(OH$_2$)$_2$]$^{3-}$ Ion (L = meso-Tetrakis(p-Sulfonatophenyl)Porphyrinate Ion) by NCS$^-$ Ion[a]

M^{3+}	T (K)	k (dm^3 mol^{-1} s^{-1})	$\Delta H^{\ddagger}$ (kcal mol^{-1})	$\Delta S^{\ddagger}$ (cal K^{-1} mol^{-1})	$\Delta V^{\ddagger}$ (cm^3 mol^{-1})
Co	293.2	103	18.4	+14.4	+15.4
Rh	288.2	12.1 × 10^{-3}	16.5	−10.3	+8.8
Cr	288.2	9.47 × 10^{-4} [b]	16.8	−12.8	+7.4

[a] Data from Reference 67.
[b] The dissociation rate constant is 6.38 × 10^{-4} s^{-1}, $\Delta H^{\ddagger}$ = 15.7 kcal mol^{-1}, $\Delta S^{\ddagger}$ = −18.5 cal K^{-1} mol^{-1}, and $\Delta V^{\ddagger}$ = +8.2 cm^3 mol^{-1}

6.3.2.2. *Porphyrins*

Anation of *trans*-$[M(L)(OH_2)_2]^{3-}$ ions [L = *meso*-tetrakis(*p*-sulfonato-phenyl)porphyrinate ion; M = Co, Rh, Cr] by NCS^- has been studied at high pressures. The volumes of activation and other activation parameters are compared in Table 6.8. Whereas the values of $\Delta S^{\ddagger}$ would suggest a change in mechanism from dissociative for Co^{3+} to associative for Rh^{3+} and Cr^{3+}, the values of $\Delta V^{\ddagger}$ show that all three metals react dissociatively. The cobalt system is so labile that high-pressure stopped flow was needed to measure $\Delta V^{\ddagger}$, and a D mechanism was found to operate in this case. The Rh^{3+} and Cr^{3+} complexes anate with an I_d mechanism.[67]

6.3.2.3. *Oxalato Complexes*

Anation of *cis*-$[Cr(ox)_2(OH_2)_2]^-$ by oxalate has been studied in great detail at pH 3-5, including the catalytic effect of nitrate ion. Good evidence is presented for an initial isomerization reaction involving one oxalate ion changing its mode of coordination from a five-membered chelate ring (involving one oxygen atom from each carboxylate group) to a four-membered "end-on" chelate in which chelation to only one of the two oxalate CO_2^- groups is involved. The latter species is more reactive than the former, and substitution proceeds in competition with the reverse isomerization process. Nitrate catalysis occurs by an assisted dissociation of the water molecule of *trans*-$[Cr(ox)(oxH)(OH_2)]^{2-}$ to give the (bisoxalto-O,O)bis(oxalato-O,O')chromate(III) intermediate.[68]

6.4. *Chromium(III) Photochemistry*

The thermal, photochemical, and photophysical behavior of $[Cr(phen)_3]^{3+}$ is very similar to that of $[Cr(bipy)_3]^{3+}$. Following photoexcitation into the spin-allowed $^4T_2$ state, very efficient intersystem crossing into $^2T_1/^2E$ states occurs. These doublet states depopulate either by nonradiative decay, including ground-state quenching, or interact with H_2O/OH^- to give $[Cr(phen)OH_2]^{3+}/[Cr(phen)_3OH]^{2+}$. The latter are the precursor intermediates of aquation, either thermal or photochemical. Thermal aquation was only observed above pH 5, and in the pH range 6.6–10.5 equation (13) holds;

$$k = [OH^-]/(a + b[OH^-]) \tag{13}$$

at 304.3 K, $a = 2.8$ s and $b = 5.4 \times 10^5$ dm^3 mol^{-1} s. At pH 11.1–12.2, the observed rate constant for thermal aquation varies according to equation

(14), and with [OH]$^-$ in the range 0.1–1.0 mol dm^{-3} equation (15) holds at 304.3 K:

$$10^6 k/s^{-1} = 320[OH^-] + 3.6 \qquad (14)$$

$$10^6 k/s^{-1} = 690[OH^-]^2 + 3 \qquad (15)$$

The following mechanism accounts for these rate laws:

$$[Cr(phen)_3]^{3+} + H_2O \rightleftharpoons [Cr(phen)_3OH_2]^{3+} \qquad (16)$$

$$[Cr(phen)_3OH_2]^{3+} \rightleftharpoons [Cr(phen)_3OH]^{2+} + H^+ \qquad (17)$$

$$[Cr(phen)_3OH]^{2+} + OH \rightarrow [Cr(phen)_2(OH)_2]^+ + phen \qquad (18)$$

Ion pairing is invoked to account for the dependence on [OH$^-$]^2 at very high [OH$^-$], as in equation (19):

$$[Cr(phen)_3]^{3+} + OH^- \rightleftharpoons \{[Cr(phen)_3]^{3+}, OH^-\} \qquad (19)$$

$$\{[Cr(phen)_3]^{3+}, OH^-\} + OH^- \rightleftharpoons \{[Cr(phen)_3OH]^{2+}, OH^-\}$$

$$\text{or } \{[Cr(phen)_2(phen-)(OH)_2]\}^+ \qquad (20)$$

Activation parameters associated with k were determined at pH 10, 12, and high [OH$^-$], but no dissection into values for the elementary steps was attempted.[69] The complete thermal, photophysical, and photochemical pathways are summarized in Scheme 1. Very similar results have been obtained for [Cr(bipy)$_3$]$^{3+}$ using laser flash photolysis with conductivity and visible spectral detection. However, the pK_a of the aquointermediate was found to be less than 2, and it is suggested that this may indicate that this intermediate has a Gillard-type covalent hydrate structure rather than a seven-coordinate chromium(III) structure. The preference, however, is for a seven-coordinate intermediate.[70] Energy transfer from the ^2E_g excited state of [Cr(bipy)$_3$]$^{3+}$ to a series of cobalt(III) complexes has also been studied.[71] The wavelength dependence of the phosphorescence yields of [Cr(en)$_3$]$^{3+}$ is also available.[72]

Photoaquation of *cis*-[Cr(NH$_3$)$_4$(CN)$_2$]$^+$ by ligand field excitation involves predominantly loss of NH$_3$, and only ~10% loss of CN$^-$. The wavelength dependence of the quantum yields were determined and found to be ~0.24–0.30 mol einstein^{-1} for loss of NH$_3$, and 0.010–0.022 mol einstein^{-1} for loss of CN$^-$. The product of NH$_3$ release [Cr(NH$_3$)$_3$(H$_2$O)(CN)$_2$]$^+$ was found to be a mixture of the 1,2-CN·3H$_2$O and 1,2-CN·6H$_2$O isomers, the ratio varying from ~2:1 to 1:1 in moving the wavelength from the first to the second ligand field band.[73] To avoid amine loss, the *trans*-[Cr(cyclam)(CN)$_2$]$^+$ ion has been studied photochemically. No photoaquation was observed in this case after prolonged irradiation, and

Scheme 1

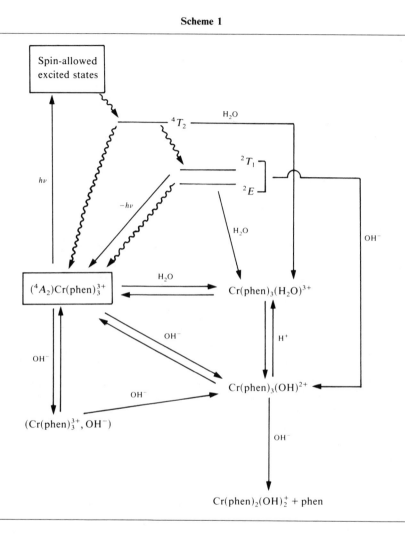

the quantum yields for loss of cyclam and CN^- ion are less than 10^{-5}. A strong *intramolecular* deuterium isotope effect was observed for the lifetime of the ^2E_g state, and three possible reasons for this unusual behavior are discussed.[74]

The binuclear complex $[(NC)_5Co(CN)Cr(NH_3)_5]$ has been prepared from the reaction between $[Cr(NH_3)_5OH_2]^{3+}$ and $[Co(CN)_6]^{3-}$ ions. Irradiation of the $Co-C_6$ chromophore at 313 nm in 0.02 mol dm^{-3} HClO$_4$ leads to bridge cleavage and $[Co(CN)_5OH_2]^{2-}$ formation with a quantum yield of 0.08. This is four times less than the quantum yield for photoaquation of $[Co(CN)_6]^{3-}$, and is indicative of intramolecular $Co-C_6 \rightsquigarrow Cr-N_6$ energy transfer. Further evidence for this comes from studies of $Cr-N_6$ sensitization.[75]

Photoaquation of CN^- from $[Cr(edta)(CN)_2(OH_2)]^{3-}$ has been studied at pH 7.6, 293.2 K, and 514 nm. The pH dependence of the quantum yield ($\phi \sim 0.018$) shows that the protonated form $[Cr(Hedta)(CN)_2(OH_2)]^{2+}$ is not significantly more reactive than the unprotonated form.[27]

6.5. *Isomerization and Racemization Reactions*

Complexes of the type $[Cr(NH_3)L]^{3+}$ containing the O-bonded ligands $OCHNH_2$, $OC(NH_2)_2$, and $OC(NHMe)_2$ have been synthesized from the $[Cr(NH_3)_5OSO_2CF_3]^{2+}$ ion which was discussed previously (Reference 13). Base-catalyzed linkage isomerism of the $[Cr(NH_3)_5L]^{3+}$ ions occurs to the deprotonated N-bonded isomers, and without competitive hydrolysis. A

$$[Cr(H_3N)_5CrOCHNH_2]^{3+} + OH^-$$

$$[(H_3N)_5CrNHCHO]^{2+} + H_2O \tag{21}$$

comparison of the pK_a's of the coordinated ligands, and the rates of these base-catalyzed isomerizations, are given in Table 6.9 for complexes of chromium(III), cobalt(III), and rhodium(III). A mechanism is proposed (Scheme 2) in which deprotonation of a coordinated ammonia molecule, rather than the ligand L is responsible for O-N isomerization. There was no evidence found for the formation of $[Cr(NH_3)_5OH]^{2+}$ ion.[76]

Scheme 2

$$[(H_3N)_5CrOCRNH_2]^{3+} + OH^- \rightleftharpoons [(H_3N)_5(H_2N)CrOCRNH_2]^{2+}$$

$$\Big\Updownarrow OH^-, K_a \qquad\qquad k_b \Big\Updownarrow k_f$$

$$[(H_3N)_5CrOCRNH]^{2+} \qquad\qquad [(H_3N)_4(H_2N)CrNH_2RO]^{2+}$$

$$\Big\downarrow rapid$$

$$[(H_3N)_5CrNHCRO]^{2+}$$

At 295.2 K, in aqueous solution the isomerization reaction (22) of the

$$lel_2\Delta(\lambda\lambda), \Lambda(\delta\lambda) \rightleftharpoons lel\ ab\ (\Delta(\lambda\delta), \Lambda(\delta\lambda)) \tag{22}$$

Table 6.9. Rate Constants and Kinetically Determined pK Values at 298.2 K Associated with Isomerization Reactions of the Type Shown in Equation (21)[a]

Ligand	M^{3+}	k (dm^3 mol^{-1} s^{-1})	pK_a
OCHNH$_2$	Cr	37.0	11.72
OC(NH$_2$)$_2$	Cr	0.324	13.50
	Rh	0.112	13.67
	Co	0.11	13.19
OC(NHMe)$_2$	Cr	0.0375	14.0
OC(NH$_2$)(NMe$_2$)	Co	0.05	13.48

[a] Data from Reference 76.

complex [Cr(acac)L$_2$]$^{2+}$ (L = 2,2'-bipyridine-NN'-dioxide) is characterized by a first-order rate constant $10^3 k \doteq 3.31$ s^{-1}. The racemization reaction, $\Delta \rightleftharpoons \Lambda$, is a factor of 3 slower ($10^4 k = 1.08$ s^{-1}).[77]

(−)$_{589}$-*trans*-(RR)-[Cr(3,2,3-tet)Cl$_2$]$^+$ ion racemizes in basic solution about 10 times faster than the rate of loss of the first Cl$^-$ ligand[23]; at 298.2 K, $k = 5.93$ dm^3 mol^{-1} s^{-1}, $E_a = 88.7$ kJ mol^{-1}, and $\Delta S^{\ddagger} = +59$ J K^{-1} mol^{-1}.

6.6. Base Hydrolysis of Chromium(III) Complexes

Loss of the first NCS$^-$ ion from [Cr(NCS)$_6$]$^{3-}$ in the presence of base is independent of the concentrations of OH$^-$ and NCS$^-$. From the rate and activation parameters an I_d mechanism is proposed.[78]

The base hydrolysis of [Cr(dmso)$_6$]$^{3+}$ has also been studied in aqueous NaOH. ^{18}O exchange between the coordinated DMSO and ^{18}OH$^-$ was observed, and the slow step found to be loss of the first DMSO molecule.[79]

Base hydrolysis of [M(NH$_3$)$_5$NCS]$^{2+}$ ions (M = Co, Cr, Rh) has been studied in aqueous organic solvents (EtOH, Me$_2$CO). $k_{obs} = kK[OH]^-/(1 + K[OH^-])$, but with K ascribed to an ion-pair constant and not conjugate-base formation. The activation parameters were determined and the mechanism discussed.[80] Data for the base hydrolysis of [CrX$_2$L]$^+$ ions (L = 2,3,2- or 3,2,3-tet) were considered previously (Table 6.3), as were results for the base-catalyzed oligomerization of [Cr(H$_2$O)$_6$]$^{3+}$ (Reference 49) and the photolysis of [Cr(phen)$_3$]$^{3+}$ under basic conditions (Reference 69).

6.7. Solids

Studies of the thermal deamination of $[Cr(aa)_3]X_3 \cdot nH_2O$, $[Cr(aa)_2(bb)]X_3 \cdot nH_2O$, and $[Cr(aa)(bb)(cc)]X_3 \cdot nH_2O$ have been made where aa, bb, and cc are the bidentate amines en, 1,2-pn, and 1,3-pn (or tn), $X^- = Cl^-$ or NCS^-, and $n = 0-3$. The lowest bp diamine leaves first, and when $X^- = Cl^-$, the products are *cis*-dichlorobis(diamine)chromium(III) complexes, whereas when X = NCS^- the *trans*-bis(diamine)bis(isothiocyanato)chromium(III) compounds are obtained. The *cis*-$[CrCl_2(tn)_2]Cl$ which is obtained from $[Cr(tn)_3]Cl_3 \cdot H_2O$, $[Cr(en)(tn)_2]Cl_3 \cdot 2H_2O$, and $[Cr(pn)(tn)_2]Cl_3 \cdot H_2O$ readily isomerizes to the transform on further heating.[81]

Several studies of dehydration–anation reactions have appeared including studies of $[M(NH_3)_5OH_2][M(CN)_6]$ (M = Co,Cr),[82] K(*trans*-$[CrF(OH_2)(LL)_2])[Cr(CN)_6] \cdot H_2O$ and K(*trans*-$[CrF(OH_2)(LL)_2])$-$[CrNO(CN)_5] \cdot H_2O$ (LL = en,tn),[83] *trans*-$[Cr(tn)_2F(H_2O)][M(CN)_4]$ M = Ni,[84] Pd, Pt[85]), K(*trans*-$[Cr(LL)_2F(H_2O)])[Co(CN)_6]$ (LL = en,tn),[86] and $[Cr(H_2O)_4X_2] \cdot 2H_2O$ (X = Cl,Br).[87] For the cyano complexes activation parameters were determined.

Thermal degradation of $M_3[Cr(NCS)_6]$ (M = alkali metal or NH_4^+) gives $[Cr(NCS)_3]$ endothermically, and then $[Cr_2O_3]$ exothermically.[88]

The mechanisms involved in the racemization of optically active coordination compounds (mainly tris(chelates) such as $[Cr(ox)_3]^{3+}$) have been discussed.[89] Kinetic data for the thermal degradation of $[M(LL)_3] \cdot nH_2O$ (M = Al, Cr, Fe; LL = lapachol) to the metal oxides have been reported.[90]

6.8. Other Chromium Oxidation States

6.8.1. Chromium(I)

The chromium(I) nitrosyls, *trans*-$[Cr(LL)_2(NO)(OH)_2]$ and 1,2,3,6-$[Cr(NO)(OH_2)_3(LL)]$ are the products of the reaction of the chromium(V) complex $Na[CrO(LL)_2]$ (LL = $[OC(Et)_2CO_2]^{2-}$) with $[NH_3OH]^+$ at pH 3.6–4.7 in the presence of excess LL. The detailed mechanism was investigated.[91]

6.8.2. Chromium(II)

The kinetics of the dissociation of $[Cr(LL)_3]^{2+}$ (LL = bipy or phen) were considered previously.[33]

6.8.3. Chromium(V) and Chromium(VI)

ESR and visible spectroscopic evidence has been found for chromium(V) intermediates during the oxidation of [Cr(L)(OH)] (L = hydroxyethylenediaminetriacetate or edta) and [Cr(LL)$_2$(OH)$_2$] (LL = ox, mal) with H$_2$O$_2$.[92] Oxidation of [Cr(Hedta)(OH$_2$)] with IO$_4^-$ to CrO$_4^{2-}$ has also been studied.[93] The photochemical or thermal formation of chromium(V) complexes with crown ethers is reported from reactions of Cr$_2$O$_7^{2-}$ in nonaqueous solvents in the presence of crowns.[94]

Studies of oxygen exchange between [Cr(NH$_3$)$_5$OCrO$_3$]$^+$ and H$_2$O at pH 6.7 shows that Cr–O bond fission is involved in the loss of the chromate ion from the inner sphere of the cobalt(III).[95]

The formation of the chromium(VI) violet diperoxo complex, [HOCr(O)(O$_2$)$_2$] parallels that found for the blue [H$_2$OCr(O)(O$_2$)$_2$] in being a third-order process. Activation parameters for the rate-determining step are $\Delta H^{\ddagger} = 6.6$ k cal mol^{-1} and $\Delta S^{\ddagger} = -14$ cal K^{-1} mol^{-1}.

Chapter 7

Substitution Reactions of Inert-Metal Complexes— Coordination Numbers 6 and Above: Cobalt

Activity in this area remains high, with a considerable volume of interesting work appearing. Review articles of interest include a general discussion of the chemistry of cobalt(III) complexes covering the 1981 literature,[1] and an excellent review by Tobe[2] on the base hydrolysis of metal complexes [predominantly Co(III) and Cr(III) complexes] which will be of great value in defining future research activities. Dixon and Sargeson[3] have also reviewed many aspects of the reactions of coordinated ligands on cobalt(III).

7.1. Aquation

The pressure dependence of the spontaneous aquation of a series of pentaammine complexes $[Co(NH_3)_5L]^{3+}$, where L = the neutral ligands, methanol, ethanol, 2-propanol, formamide, N-methylformamide, N,N-dimethylformamide urea, N-methylurea, N,N'-dimethylurea, and dimethylsulfoxide, have been determined.[4] The activation volume $(\Delta V^{\ddagger})$ lies in the limited range 0.3–3.8 $cm^3 mol^{-1}$ across the series and is essentially independent of the size of the leaving group. The average $\Delta V^{\ddagger}$ of

Table 7.1. Values of k_{aq} and Steric Course for trans-$[Co(en)_2X_2]^+$ Ions in 0.1 M CF_3SO_3H at 25°C [5]

X	$k_{aq}(s^{-1})$	Products
Cl	$(4.2 \pm 0.1) \times 10^{-5}$	$74 \pm 1.5\%$ trans[a]
Br	$(4.13 \pm 0.07) \times 10^{-4}$	$84.5 \pm 1.5\%$ trans[b]

[a] $79 \pm 1\%$ cis at equilibrium.
[b] $74 \pm 1.5\%$ cis at equilibrium.

~2 cm^3 mol^{-1} is an estimate of the solvent-independent intrinsic component of $\Delta V^{\ddagger}$ and is interpreted as being indicative of a dissociative interchange (I_d) mechanism.

The rates and steric course of aquation of trans-$[Co(en)_2X_2]^+$ ions in 0.1 M CF_3SO_3H at 25°C have been investigated[5] (see Table 7.1). The aquation rate for the bromo complex and the steric-course data differ significantly from earlier values. The rates of isomerization of trans- and cis-$[Co(en)_2(OH_2)Br]^{2+}$ have also been determined in 0.01–1.0 M $HClO_4$ and 0.1 M CF_3SO_3H at 25°C, the rate constant $k_i = 2.32 \times 10^{-4} s^{-1}$ is independent of the anion and ionic strength.

House and Nor[6] have studied the aquation (first stage) of trans-RSSR-$[CoBr_2(cyclam)]^+$ and a variety of chromium(III) complexes (trans-RR,SS-$[CrBr_2(3,2,3-tet)]^+$, trans-R,S-$[CrBr_2(2,3,2-tet)]$, and trans-RSSR-$[CrBr_2(tet\ a)]^+$[tet a = C-meso-Me$_6$cyclam = 5,7,7,12,14,14-hexamethyl-1,4,8,11-tetraazacyclotetradecane)] (see Table 7.2). Dissociative mechanisms are proposed for both the Cr(III) and Co(III) systems. The bromo/chloro aquation rate ratio is about 4 for Co(III) and about 15 for Cr(III).

Poon and co-workers[7] have studied the aquation of some $[ML(A)X]^{n+}$ complexes, where M = Co(III), Ru(III), or Ru(II) and L = two bidentate

Table 7.2. Aquation of Some trans-$[MBr_2(N_4)]^+$ Complexes (M = Co,Cr) at I = 1.0 M (HNO_3) [6]

Complex	$10^5 k_{aq}$ (25°) (s^{-1})	E_a (kJ/mol^{-1})	$\Delta S^{\ddagger}_{298}$ (J K^{-1} mol^{-1})
trans-RR,SS-$[CrBr_2(3,2,3-tet)]^+$	1.02	94 ± 4	-35 ± 8
trans-R,S-$[CrBr_2(2,3,2-tet)]^+$	4.27	100 ± 2	$+1 \pm 4$
trans-RSSR-$[CoBr_2(cyclam)]^+$	1.53	115 ± 2	$+41 \pm 4$
trans-RSSR-$[CrBr_2(tet\ a)]^+$	1.70	74 ± 2	-58 ± 4

diamines, or one quadridentate linear or macrocyclic tetramine, with $(A)X =$ $(Cl)Cl$, $(Cl)Br$, $Br(Cl)$, or $(Br)Br$. These reactions may be represented by equation (1):

$$trans\text{-}[ML(A)X]^+ + Y^{n-} \rightarrow trans\text{-}[ML(A)Y]^{(2-n)+} + X^- \qquad (1)$$

where $Y = H_2O$ and various anions. The main conclusion is that the d^6 Ru(II) complexes behave more like the d^5 Ru(III) rather than the d^6 Co(III) and Rh(III) systems.

The activation volumes $(\Delta V^{\ddagger})$ and the reaction volumes (ΔV) have been determined[8] for five aquation reactions on Co(III) (see Table 7.3). The magnitude of $\Delta V^{\ddagger}$ is considered as resulting from the counterbalance of the volume increase attributable to the dissociation of the leaving group, and the volume decrease due to the partial incorporation of one water molecule in the activated complex.

At elevated temperatures ($\geq 70°C$), the complex $H[Co(dmgH)_2(NCS)_2]$ aquates in aqueous acid releasing one thiocyanate.[9] This step is not acid dependent, and $k_{aq} = 1.2 \times 10^{-8} \, s^{-1}$ ($\Delta H^{\ddagger} = 132 \, kJ \, mol^{-1}$; $\Delta S^{\ddagger} = 59 \, J \, K^{-1} \, mol^{-1}$). This first step is followed by a slower second step in which the second thiocyanate is released and the $Co(dmgH)_2$ moiety cleaved. The second reaction is catalyzed by Hg(II) giving $[Co(dmgH)_2(OH_2)_2]^+$ and thiocyanate without competing side reactions. For loss of the first thiocyanate ligand, the k_{Hg} value is $2.85 \, M^{-1} s^{-1}$ at 25°C ($\Delta H^{\ddagger} = 89 \, kJ \, mol^{-1}$; $\Delta S^{\ddagger} = 60 \, J \, K^{-1} \, mol^{-1}$) and for the second, $k_{Hg} = 1.48 \, M^{-1} s^{-1}$ ($\Delta H^{\ddagger} = 63 \, kJ \, mol^{-1}$; $\Delta S^{\ddagger} = -30 \, J \, K^{-1} \, mol^{-1}$). It is considered that Co(III)–NCS bond fission occurs by direct electrophilic ($S_E 2$) attack of Hg(II) on the leaving thiocyanate ligand without the involvement of a long-lived Co(III)–NCS–Hg(II) binuclear intermediate.

The complexes $trans\text{-}[Co(en)_2Cl(S_2O_3)]$, $trans\text{-}[Co(en)_2(C_2O_4)\text{-}(S_2O_3)]^-$, and $trans\text{-}[Co(en)_2(NCO)(S_2O_3)]$ have been characterized.[10] The

Table 7.3. Activation Volumes ($\Delta V^{\ddagger}$) and Reaction Volumes (ΔV) for Several Aquation Reactions of Cationic Cobalt(III) Complexes[8]

Complex	$\Delta V^{\ddagger}$ $(cm^3 \, mol^{-1})^a$	ΔV $(cm^3 \, mol^{-1})^a$
$trans\text{-}[Co(en)_2(NO_2)Cl]^+$	0.1 (15)	−10.4 (15)
$cis\text{-}[Co(en)_2(NO_2)Cl]^+$	0.9 (30)	−9.3 (25)
$cis\text{-}[Co(bipy)_2(NO_2)Cl]^+$	2.9 (30)	−9.1 (25)
$cis\text{-}[Co(bipy)_2(NO_2)Br]^+$	11.3 (30)	−6.7 (25)
$cis\text{-}[Co(phen)_2(NO_2)Br]^+$	3.3 (35)	−12.5 (30)

^{a} Figure in parentheses is the temperature in °C.

chloro complex aquates rapidly at pH 5–6. The oxalato and cyanato complexes undergo acidolyses in dilute $HClO_4$ to give the common intermediate *trans*-$[Co(en)_2(OH_2)(S_2O_3)]^+$, which subsequently equilibrates to a mixture of *cis* and *trans* isomers.

7.2. Catalyzed Aquation

A number of studies have appeared dealing with acid-catalyzed aquation, and metal-ion [Hg(II) and Al(III)]-catalyzed aquations. In acidic solution, a chelate ligand with a basic site may undergo protonation leading to ring opening. The aquation of $[Co(BH)_2(gly)]^{2+}$ (BH = biguanide) in acidic solution has been studied in detail.[11] The presence of the unsymmetrical glycinato ligand makes the positions of the two biguanides nonequivalent. The acid-catalyzed dissociation of one or other of them, gives two different isomeric (α- and β-) diaquo products. Up to ~50°C, the dissociation of only one biguanide takes place with the formation of the α-diaquo product. The α-diaquo form isomerizes to the β-form at higher temperature, and direct formation of the β-form does not occur. Aquation of the bis(biguanide) complex follows a dissociative pathway up to ~50°C. Isomerization of the α-product takes place intramolecularly by two pathways, one dependent and the other independent of the acid concentration.

The preparation of *cis*-$[Co(en)_2(\beta\text{-alaO})]I_2$ has been described and the kinetics of ring opening in acidic solution studied.[12] Further studies[13] have indicated that the kinetics may be in error, due to spectral changes associated with the oxidation of the iodide counterion occurring in the nitric acid solutions employed for the kinetic measurements.

The bismalonato complex $[Co(en)(mal)_2]^-$ aquates[14] in acidic solution initially to give $[Co(mal)(H_2O)_2(en)]^+$, then subsequently $[Co(H_2O)_4(en)]^{3+}$. The intermediate diaquo complex has been characterized in solution. The first step only occurs by an acid-catalyzed pathway, while the second step occurs by both an acid-independent and an acid-dependent pathway (i.e., $k_{obs} = k_0 + k_H[H^+]$).

The presence of multivalent ions such as SO_4^{2-}, ox^{2-}, and mal^{2-} has been found[15] to accelerate the Al(III) assisted aquation of $[CoF(NH_3)_5]^{2+}$ giving the anionopentaamminecobalt(III) ion in addition to $[Co(NH_3)_5(OH_2)]^{3+}$. The maximum product percentage of $[Co(SO_4)(NH_3)_5]^+$ (33%) occurred at $[SO_4^{2-}] = 0.1$ mol dm^{-3}, but higher sulfate concentrations were required for the reaction rate to become independent of the sulfate concentration. This behavior is attributed to the presence of two different intermediates $[CoF(NH_3)_5]^{2+} \cdot SO_4^{2-} \cdot Al^{3+}$ and

$SO_4^{2-} \cdot [CoF(NH_3)_5]^{2+} \cdot SO_4^{2-} \cdot Al^{3+}$. In the presence of oxalate and malonate very high yields ($>90\%$) of $[CoA(NH_3)_5]^+$ (A^{2-} = oxalate or malonate) occurred when $[A^{2-}]_{tot}/[Al^{3+}]_{tot} > 2$ and the reaction rates showed a maximum at approximately $[A^{2-}]_{tot}/[Al^{3+}]_{tot} = 2$. The strong affinity of the dicarboxylate ions for Al(III) and the availability of a free carboxylate group to coordinate to cobalt(III) are suggested as possible reasons for these specific effects.

Attempts to detect the possible five-coordinate intermediate $[Co(NH_3)_5]^{3+}$ involve trapping it with nucleophiles. If $[Co(NH_3)_5]^{3+}$ exists for a sufficient period to qualify as an intermediate it can meet and react with nucleophiles not originally present in the solvation shell and the mechanism of the ligand exchange is D. If this is not the case, an I_d mechanism applies. Reynolds and co-workers[16] have studied the product ratios for competition reactions accompanying the Hg(II)-assisted removal of Cl^- from $[Co(NH_3)_5Cl]^{2+}$ in aqueous solutions of $NaNO_3$, $NaHSO_4$, H_3PO_4, and CH_3CO_2H and in mixtures of water and one of the nonaqueous solvents DMSO, DMF, or CH_3CN at 0, 25 and 55°C. The temperature dependence of the product ratios was very small or undetectable. An I_d interchange of leaving and incoming ligands appears to adequately account for the experimental data. In mixtures of water and nonaqueous solvent the product ratios were found to be equal for equivalent bulk mole fractions of nonaqueous solvent. The three nonaqueous solvents are thus equally available in the solvation shell of the encounter complex preceding the formation of products, and preferential solvation by one solvent component does not occur.

The nitrosation of $[Co(NH_3)_5N_3]^{2+}$ in the presence of added anions $X = Cl^-$, NO_3^-, and ClO_4^- has been studied in detail.[17] The kinetic data are interpreted in terms of both NO^+ and XNO reacting with $[Co(NH_3)_5N_3]^{2+}$ with rate constants of 3.3×10^9 (NO^+), 1.9×10^8 (ClNO), and 3.5×10^7 (BrNO) $M^{-1} s^{-1}$. The X^- species (Cl^-, NO_3^-) are included in the products of the reaction as $[Co(NH_3)_5X]^{2+}$ but this does not occur via the XNO species. The lifetime of the intermediate $X^- \cdot [Co(NH_3)_5N_3(ONX)]^{2+}$ is considered.

The Hg(II)-catalyzed aquation of a number of cis-$[CoCl(en)_2(py-X)]^{2+}$ complexes [X = H, 4-Me, 4-NH_2, and 4-$N(Me)_2$, py = pyridine] has been investigated[18] in 1.0 M trifluoroacetic acid or perchloric acid. Ligand binding via the pyridine nitrogen is believed to occur in all the complexes. The order of increasing reactivity at 25°C in 1.0 M trifluoroacetic acid is H < 4-Me < 4-NH_2 < 4-NMe_2. This order parallels the base strength of the pyridine ($pK_a = 5.23, 6.02, 9.13, 9.61$ for X = H, 4-Me, 4-NH_2, and 4-NMe_2, respectively), and a reasonably linear plot of pK_a vs. ln k_{Hg} is obtained with a slope of 0.5.

The complete decarboxylation of the bis(carbonato) complexes $K[Co(en)(CO_3)_2]\cdot H_2O$ and $Na[Co(en)(CO_3)_2]\cdot H_2O$ in acidic solution has been studied.[19] Only the second decarboxylation step was studied under the experimental conditions employed. For loss of the second CO_3^{2-} ligand, $k_{obs} = k_0 + k_H[H^+]$ and at 5°C, $k_0 = 5.7 \times 10^{-4} s^{-1}$ and $k_H = 4.37 \times 10^{-2} M^{-1} s^{-1}$ at $I = 1.0\ M$ (NaClO$_4$). For the acid-catalyzed pathway, $\Delta H^{\ddagger} = 12 \pm 2\ kcal\ mol^{-1}$ and $\Delta S^{\ddagger} = -21\ cal\ deg^{-1}\ mol^{-1}$. The kinetic parameters apply to the ring-opening step as loss of the monodentate HCO_3^- ligand is rapid. An induction period noted during the decarboxylation is attributed to the formation of the ring-opened monodentate carbonato–aquo species $[Co(en)(CO_3)(HCO_3)(H_2O)]$ or $[Co(en)(HCO_3)(H_2O)_3]^{2+}$, but attempts to isolate a stable bidentate $[Co(en)(CO_3)(H_2O)_2]^+$ species were unsuccessful.

Acid-catalyzed aquation of $[Co(NH_3)_5OCO_2]^+$, *cis*-β-$[Co(edda)CO_3]^-$, and $[Co(nta)CO_3]^{2-}$ has been reinvestigated using rapid scan spectrophotometry.[20] Direct spectral evidence for the participation of protonated and ring-opened carbonato species was obtained. The spectral observations are consistent with previously suggested mechanisms for the decarboxylation of monodentate and bidentate carbonato complexes.

Rate constants have been obtained for the aquation of a series of pentaammine(carboxylato)cobalt(III) complexes $[Co(NH_3)_5L]^{n+}$ in LiClO$_4$–HClO$_4$ media.[21] In many cases the values of k_{obs} are not proportional to $[H^+]$. The data have been used to calculate association constants of protonated forms of the complexes and the corresponding rate constants for aquation. Association constants (K_{ML}) of ion pairs formed by a series of divalent metal ions and also lanthanum(III) ions with $[Co(NH_3)_5OOCCO_2]^{2+}$ have been determined from spectrophotometric UV absorbance data at 50 and 60°C.[22] Rate constants for aquation of the complex in the presence of the metal ions were also obtained and rate constants estimated for the metal ion–oxalato complex ion pairs (k_{aq}). There is an approximately linear relationship between K_{ML} and k_{aq} for metal ions of the first transition series.

7.3. Base Hydrolysis

The base hydrolysis of cobalt(III) complexes has been reviewed by Tobe.[2] Anion competition data have been reported for the base hydrolysis (pH 8.5–9) of the $[Co(NH_3)_5O_3SCF_3]^{2+}$ cation in 1 M NaX (X$^-$ = F$^-$, Cl$^-$, Br$^-$, I$^-$, and NO$_3^-$).[23] At 25°C, $[Co(NH_3)_5X]^{2+}$ is formed in conjunction with $[Co(NH_3)_5OH]^{2+}$ (X = F$^-$, <0.2%; Cl$^-$, 7.5%; Br$^-$, 7.1%; I$^-$, 7.5%; NO$_3^-$, 11.9%). Poorly solvated anions such as NO$_3^-$ appear to be the best competitors. Ions such as F^- and OH^-, which are strongly solvated, do not compete effectively for the five-coordinate intermediate produced in the S_N1 CB (CB = conjugate base) process. Competition by the neutral competitor NH_3 (at the 1 M level) is shown to be negligible. Acetate ion (1 M) competition was also studied for the base hydrolysis of $[Co(NH_3)_5L]^{n+}$ using seven different leaving groups. The results obtained remove a previous anomaly in the literature and show that the extent of anion competition is essentially independent of the leaving group L. The relative nucleophilicity of Cl^- (and NO_3^-) towards the intermediates $[Co(NH_3)_5]^{3+}$ and $[Co(NH_3)_4(NH_2)]^{2+}$, believed to be involved in the aquation and base hydrolysis of $[Co(NH_3)_5X]^{n+}$, is thought to reside primarily in the formal charge of the intermediates. Anion capture occurs at somewhat less than a diffusion-controlled rate from an ion atmosphere largely determined by the degree of ion association with the intermediate.

Nitrite ion competition has been studied[24] for base hydrolysis of a series of $[Co(NH_3)_5X]^{n+}$ ions [X = $CF_3SO_3^-$, $CH_3SO_3^-$, NO_3^-, $(CH_3)_2SO$, $OP(OCH_3)_3$, $OC(NH_2)_2$, and $OC(NH_2)N(CH_3)_2$] in 1.0 M $NaNO_2$. Contrary to previous reports, both O- and N-bonded $[Co(NH_3)_5NO_2]^{2+}$ are produced. A constant O-isomer/N-isomer ratio of 2/1 occurs for both anionic and neutral leaving groups X. The total NO_2^- capture shows a slight dependence on the overall charge of the complex (7.0 ± 0.5% for 2+ cations; 8.3 ± 0.5% for 3+ cations) but is less dependent on the nature of X. The results are consistent with a common reactive intermediate of reduced coordination number, $[Co(NH_3)_5NH_2]^{2+}$ in the base-hydrolysis process, if it is assumed that the intermediate captures the environment of its precursor.

Base hydrolysis of $[Co(NH_3)X]^{n+}$ in 1.0 M $Na_2S_2O_3$ leads to both S- and O-bonded $[Co(NH_3)_5(S_2O_3)]^+$ for seven different leaving groups.[25] Isomerization of the O-bonded isomer to the S-bonded isomer has $k_{obs} = k_0 + k_{OH}[OH^-]$, where $k_0 = 4.3 \times 10^{-4} s^{-1}$ and $k_{OH} = 0.185\ M^{-1}s^{-1}$ at 25°C and $I = 3\ M$. The O isomer isomerizes intramolecularly to the S isomer by both spontaneous (k_0) and base-catalyzed (k_{OH}) pathways. Competitive hydrolysis (43 ± 5%) also occurs in the k_{OH} route. The extent of total $S_2O_3^{2-}$ capture and the S-bonded/O-bonded isomer ratios have been studied in detail for a variety of leaving groups. For the 2+ cations, a constant percent capture is observed and a slightly higher percent for the 3+ cations. In all cases a constant S-isomer/O-isomer ratio is observed (70 ± 3% S bonded and 30 ± 3% O bonded). This work complements the above study,[24] where a fixed O/N ratio for NO_2^- capture was observed and provides further support for a reactive five-coordinate intermediate in base-hydrolysis reactions.

Two isomeric forms (red and orange) of $[CoCl(AA)(dien)]^{2+}$ (AA = phen,bipy) have been characterized.[26] ^{13}C NMR data indicate that the red

Table 7.4. Values of k_{Hg} and Activation Parameters[26] for Mercury(II)Catalyzed Aquation of unsym-fac-cis and mer Isomers of [CoCl(AA)(dien)]²⁺ (AA = phen or bipy) in HNO₃ (I = 1.0 M)

Isomer	k_{Hg} (25°C) $(M^{-1} s^{-1})$	E_a (kJ mol⁻¹)	$\Delta S^{\ddagger}$ (J K⁻¹ mol⁻¹)
unsym-fac-cis(phen)	1.65×10^{-2}	66	−66
unsym-fac-cis(bipy)	1.62×10^{-2}	73	−44
mer(phen)	1.74×10^{-4}	93	−15
mer(bipy)	6.87×10^{-4}	99	+16

isomer has the *unsym-fac-cis* configuration (**1**) and the red isomer is *mer* (**2**). Base hydrolysis was studied by pH stat for the *unsym-fac-cis* isomer. At 25°C and $I = 0.1\ M$ NaCl, $k_{OH} = 1920\ M^{-1} s^{-1}$, $E_a = 92\ kJ\ mol^{-1}$, and $\Delta S^{\ddagger} = +120\ J\ K^{-1}\ mol^{-1}$ (phen derivative), and $k_{OH} = 1740\ M^{-1} s^{-1}$, $E_a = 90\ kJ\ mol^{-1}$, and $\Delta S^{\ddagger} = +111\ J\ K^{-1}\ mol^{-1}$ (bipy derivative). Data for the Hg(II)-catalyzed aquation are summarized in Table 7.4.

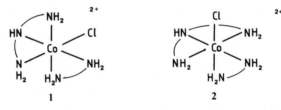

　　　　　　　　1　　　　　　　　　　　　　　　　**2**

Cobalt(III) with the pentadentate pyridyl containing ligands 1,9-bis(2-pyridyl)2,5,8-triazanonane [picdien = (**3**)] and 1,11-bis-(2-pyridyl)2,6,10-triazaundecane [picditn = (**4**)] gives acidopentamine complexes of the type [Co(picdien)X]⁺ and [Co(picditn)X]²⁺ (X = Cl,Br,NO₂), which are of special interest due to their extremely rapid base hydrolysis.[27] A number

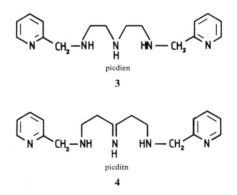

picdien

3

picditn

4

of crystal structures have now been published, including an isomer of $[Co(picdien)Br](ClO_4)_2$,[28] and $[Co(picdien)(NO_2)](ClO_4)_2$,[28] where the picdien ligand has the α,β-configuration in both cations. The structure of a red form of the $[Co(picdien)NO_2]^+$ has also been published,[29] as has that of a further isomer of $[Co(picdien)Br](ClO_4)_2$,[30] where the picdien ligand has the α,β-configuration and the bromine lies *trans* to an "angular" secondary nitrogen donor.

The synthesis of 2-(methylamino)-2',2''-diaminotriethylamine [Me(tren) = (5)] and a number of cobalt(III) complexes, including

$$\begin{array}{c} CH_2CH_2NHMe \\ \diagup \\ N-CH_2CH_2NH_2 \\ \diagdown \\ CH_2CH_2NH_2 \end{array}$$

Me(tren)

5

$[Co(Me(tren))(NO_2)_2]Cl$, $[Co(Me(tren))Cl_2]Cl$, *t*-, *anti-p*-, *syn-p*-, and *s*-$[Co(Me(tren))(NH_3)Cl]^{2+}$ salts, and *anti-p*-, *syn-p*-, and *s*-$[Co(Me(tren))(NH_3)X]ClO_4$ (X = Br,N$_3$) complexes, have been described.[31] The rates of aquation (k_{aq}), mercury(II)–induced aquation (k_{Hg}), and base hydrolysis (k_{OH}) were also determined. The aim of the kinetic work was to further investigate the possibility of discrete intermediates of reduced coordination number in the substitution chemistry of cobalt(III) complexes. The general consensus now appears to be that such species, if they exist at all, have lifetimes of $\sim 10^{-11}$ s at room temperature and exist at the most for a few collisions within the solvent cage. The present results appear to reduce this lifetime even further. The tren ligand prefers the symmetrical trigonal bipyramidal geometry in its complexes (**6**), and this stereochemistry has been observed for several Cu(II), Ni(II), Zn(II), and Co(II) complexes.[32]

$$\begin{array}{c} N \\ | \\ H_2N-M \diagdown NH_2 \\ \diagdown NH_2 \\ | \\ X \end{array}$$

6

Studies of the substitution chemistry[33,34] of the *p*- and *t*-$[Co(tren)(NH_3)X]^{n+}$ ions suggest that there is a driving force towards this geometry during replacement of the X ligand. (It is noteworthy that a low-spin d^6 ion in a trigonal bipyramidal field would be paramagnetic to the extent of two unpaired electrons, while the square pyramidal stereochemistry would lead to diamagnetism.) Using the Me(tren) ligand in the three closely related *syn-p-*, *anti-p-*, and *s*-[Co(Me(tren)(NH$_3$)X]$^{n+}$ isomers (**7**)–(**9**), respectively,

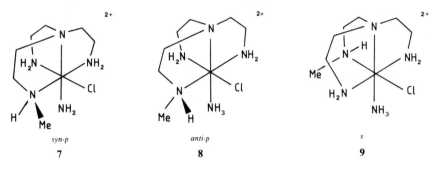

it is possible to distinguish the different octahedral faces of a trigonal bipyramidal intermediate. As only ~30° bond rotations about the metal are required to give a common intermediate presumably any significant lifetime should lead to a common set of products. The Hg(II)-, Ag(I)-, NO$^+$-, and OH$^-$-induced reactions on the pure isomers (X = Cl,Br,NH$_3$,N$_3$) give largely retention thus eliminating a common intermediate.

Isomerization and base hydrolysis of *cis* and *trans* isomers of [Co(en)$_2$(SO$_2$O$_3$)$_2$]$^-$ has been studied.[35] The *cis* isomer isomerizes to the *trans* isomer in neutral solution with $k = 2.30 \times 10^{-5}$ s^{-1} at 50°C. The *trans* isomer isomerizes to the *cis* only in basic solution and the pH dependence of the isomerization has been investigated. For hydrolysis of the *trans* isomer $k_{obs} = k_0 + k_{OH}[OH^-]$, but only a base-dependent pathway is observed with the *cis* derivative. Base hydrolysis of [Co(en)$_2$(S$_2$O$_3$)]$^+$, in which the thiosulfate ligand bonds via oxygen and sulfur, has also been investigated[36] using NaOH concentrations up to 2.0 *M* and temperatures of 35–65°C. Dechelation is the slow step followed by rapid loss of the monodentate ligand. Activation parameters for base hydrolysis are $\Delta H^\ddagger = 77.8$ kJ mol^{-1} and $\Delta S^\ddagger = -75 \pm 20$ J K^{-1} mol^{-1}.

Base hydrolysis of [Co(NH$_3$)$_5$(NCS)]$^{2+}$ in water–ethanol and water–acetone corresponds to the equation $1/k_{obs} = 1/k + 1/kK[OH^-]$, where K is suggested to relate to pre-equilibrium ion-pair formation with hydroxide ion, and k is the rate constant for breakdown of the ion pair to give products.[37] Similar kinetic behavior is observed with the Cr(III) and Rh(III) derivatives.

Ring opening of carbonato complexes in basic solution has attracted considerable attention. The kinetics of base hydrolysis of *cis*-[Co(cyclam)CO$_3$]$^+$ has been studied using [OH$^-$] in the range 0.29–0.95 *M* at $I = 1.0$ *M* (LiClO$_4$).[38] Plots of k_{obs} vs. [OH$^-$] are linear with $k_{OH} = 1.34$ *M*$^{-1}$ s^{-1} at 31.2°C. The slow step of the reaction is believed to be as in equation (2) with base hydrolysis of *cis*-[Co(cyclam)(OH)(CO$_3$] being

$$cis\text{-}[Co(cyclam)CO_3]^+ + OH^- \xrightarrow{k_{OH}} cis\text{-}[Co(cyclam)(OH)CO_3] \quad (2)$$

rapid to give *cis*-[Co(cyclam)(OH)$_2$]$^+$ and CO$_3^{2-}$. The unusually fast base hydrolysis of *cis*-[Co(cyclam)(OH)CO$_3$] is in sharp contrast with the hydrolytic behavior of the corresponding derivatives of en and tren, and has been attributed to a "kinetic nephelauxetic effect."

The kinetics of carbonato ring opening of *cis*-[Co(cyclen)CO$_3$]$^+$ (cyclen = 1,4,7,10-tetraazacyclododecane) in basic solution has also been studied,[39] using sodium hydroxide solutions in the range 0.04–0.35 M at $I = 0.5\ M$ (NaClO$_4$). In this case the reaction shows a more complex dependence on the hydroxide ion concentration with both first- and second-order terms in the hydroxide ion concentration occurring. The hydroxide ion dependence can be rationalized in terms of the reactions shown in Scheme 1. Comparable kinetic behavior may be expected for ring opening

Scheme 1

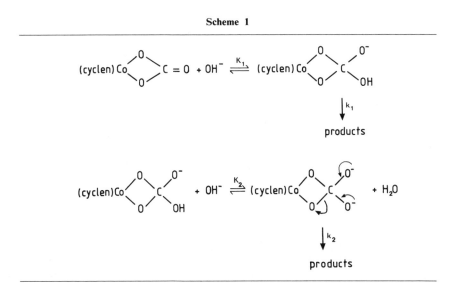

of oxalato and malonato complexes in basic solution, where a mechanism involving a tetrahedral intermediate and carbon–oxygen bond cleavage is also expected to occur. Early work by Andrade and Taube[40] and Harris and co-workers[41] has shown that base hydrolysis of *cis*-[Co(en)$_2$C$_2$O$_4$]$^+$ involves rate-determining ring opening occurring with C–O bond cleavage. Base hydrolysis of [Co(NH$_3$)$_5$OOCCF$_3$]$^{2+}$ with carbon–oxygen bond cleavage shows a second-order dependence on $[OH^-]$.[42] A single hydroxide ion does not provide sufficient driving force to expel the very poor leaving group $[Co(NH_3)_5O]^+$, but this process becomes easier when deprotonation of the hydroxyl group occurs in the presence of additional base, see Scheme 2).

Scheme 2

$$(NH_3)_5Co-O-\underset{\underset{O}{\|}}{C}-CF_3 \overset{2+}{} + OH^- \underset{}{\overset{K_1}{\rightleftharpoons}} (NH_3)_5Co-O-\underset{\underset{O^-}{|}}{\overset{\overset{OH}{|}}{C}}-CF_3 \,^+$$

$$(NH_3)_5Co-O-\underset{\underset{O^-}{|}}{\overset{\overset{OH}{|}}{C}}-CF_3 \,^+ + OH^- \overset{K_2}{\rightleftharpoons} (NH_3)_5Co-O-\underset{\underset{O^-}{|}}{\overset{\overset{O^-}{|}}{C}}-CF_3 + H_2O$$

$$\downarrow$$

$$[(NH_3)_5CoOH]^{2+} \overset{H^+}{\leftarrow} [(NH_3)_5CoO]^+ + CF_3CO_2^-$$

The differing kinetic behavior observed with $[Co(cyclam)CO_3]^+$ and $[Co(cyclen)CO_3]^+$ could be due to differing leaving-group abilities or to differing mechanisms. For example, ring opening of cis-$[Co(cyclam)CO_3]^+$ could occur by an S_N1 CB mechanism leading to Co–O cleavage. Additional experimental work is clearly required.

The reactions of $[Co(tren)CO_3]^+$ in basic solution have been studied in detail.[43] For opening of the carbonato ring, $k_{obs} = k_0 + k_{OH}[OH^-]$. At 30°C, $k_0 = 7.0 \times 10^{-4} \, s^{-1}$ and $k_{OH} = 4.7 \times 10^{-3} \, M^{-1} s^{-1}$ at $I = 1.0 \, M$ (KNO$_3$). The activation parameters for the base hydrolysis pathway are $\Delta H^{\ddagger} = 34.1$ kcal mol^{-1} and $\Delta S^{\ddagger}_{298} = +43.5$ cal K^{-1} mol^{-1}. The monodentate carbonato complex $[Co(tren)(OH)OCO_2]$ can be characterized in solution and its base hydrolysis studied in isolation. Only a base-hydrolysis pathway was observed with $k_{OH} = 2.8 \times 10^{-4} \, M^{-1} s^{-1}$ at 50.2°C and $I = 1.0 \, M$ (NaClO$_4$). The requisite activation parameters are $\Delta H^{\ddagger} = 27.8$ kcal mol^{-1} and $\Delta S^{\ddagger}_{298} = 10.9$ cal K^{-1} mol^{-1}.

Base hydrolysis of $[Co(en)_2(acac)]^{2+}$ involves ring opening as the rate-determining step with $k_{obs} = k_{OH}[OH]^2$.[44] Activation parameters are $\Delta H^{\ddagger} = 70.6$ kJ mol^{-1} and $\Delta S^{\ddagger} = -119$ J K^{-1} mol^{-1}. Previous work has established that there is no evidence for formation of a monodentate acetyl-acetonato complex in basic solution.

The cobalt(III) complex [Co([15]aneN$_5$)Cl](ClO$_4$)$_2$ has been prepared and base hydrolysis and aquation studied.[45] The ligand [15]aneN$_5$ (1,4,7,10,13-pentaazacyclopentadecane has the structure (**10**) and gives the complex (**11**) which contains two chiral nitrogen centers. As a result two

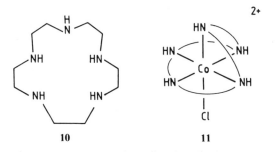

10   11

N-meso diastereoisomers and one *N-racemic* diastereoisomer can occur. Carbon-13 NMR measurements suggest that the complex is a mixture of the possible stereoisomers. Base hydrolysis of [Co([15]aneN$_5$)Cl]$^{2+}$ is quite rapid with $k_{OH} = 2.45 \times 10^4 \ M^{-1} s^{-1}$ at 30°C and $I = 0.1 \ M$, a result which can be rationalized in terms of the *mer*-N$_3$ donor set.[46] Acid aquation with $k_{aq} = 1.66 \times 10^{-4} s^{-1}$ at 50°C is, as expected, quite slow.

A variety of branched cyclononane macrocycles with pentaamine and tetraamine–thioether donor sets have been prepared.[47] These include datn[1,4-bis(2-aminoethyl)-1,4,7-triazacyclononane = (**12**)], dats[4,7-bis(2-aminoethyl)-1-thia-4,7-diazacyclononane = (**13**)], and the new macrocyclic ligand tasn[1-thia-4,7-diazacyclononane = (**14**)]. Synthetic routes to the

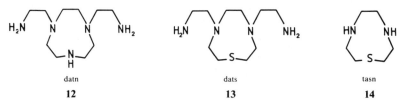

datn   dats   tasn
12   13   14

resolved unsymmetrical (μ) isomers of [Co(dats)Cl](ClO$_4$)$_2$ and [Co(datn)Cl](ClO$_4$)$_2$ have been developed and *cis*-[Co(tasn)$_2$]-(ClO$_4$)$_3$ prepared. Base hydrolysis of both chloropentamine complexes occurs with two consecutive steps. The first reaction in each case is chloride hydrolysis with k_{OH}(dats) $= 3.6 \times 10^4 \ M^{-1} s^{-1}$ and k_{OH}(datn) $= 2.85 \times 10^3 \ M^{-1} s^{-1}$ at 25°C and $I = 0.1 \ M$. The second reaction for the dats complex is base-catalyzed dissociation of the thioether to the *cis*-[Co(dats)(OH)$_2$]$^+$ complex, characterized by ^{13}C NMR measurements, and occurring ~20 times slower than chloride hydrolysis. The datn complex undergoes base-catalyzed terminal-ring opening to the cis-$[Co(datn)(OH)_2]^+$ complex at a rate ~30 times slower than chloride hydrolysis.

The cis-α and cis-β_2 isomers of the $[Co(trien)(aniline)Cl]^{2+}$ cations (**15**) and (**16**), respectively, have been characterized and their aquation and base-hydrolysis rates studied.[48] In the pH range 1–3, $k_{obs} = k_{aq} + k_{OH}[OH]$.

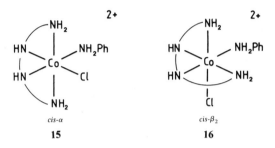

The k_{aq} term is assigned to aquation and k_{OH} to base hydrolysis. Kinetic parameters are summarized in Table 7.5. Both isomers are extremely reactive in base hydrolysis with the rate constants k_{OH} at 50°C being 3.3×10^5 (cis-α) and 1.38×10^7 (cis-β_2) $M^{-1} s^{-1}$. The base hydrolysis of cis-β_2-$[Co(trien)amine)Cl]^{2+}$ complexes is believed to involve the conjugate base formed by deprotonation of the sec-NH group of the trien ligand $trans$ to the monodentate amine ligand,[49] and normally the cis-β_2 isomer reacts ~10^3 times faster than the cis-α isomer. In the aniline derivatives the cis-β_2 isomer is only ~40 times more reactive than the cis-α isomer; this small difference may arise because the conjugate base in these complexes is formed mainly by deprotonation of coordinated aniline. The pK_a of coordinated aniline in cis-$[Co(en)_2(NH_2Ph)Cl]^{2+}$ is known to be ~10[50] with $k_{OH} = 2.6 \times 10^4 M^{-1} s^{-1}$ at 25°C.[51]

Table 7.5. Values[48] of k_{aq}, k_{OH}, and Activation Parameters for the Hydrolysis of cis-α- and cis-β_2-[Co(trien)(aniline)Cl]$^{2+}$

Isomer[a]	k_{aq} (50°C) (s^{-1})	k_{OH} (50°C) $(M^{-1} s^{-1})$
cis-α	6×10^{-6}	3.3×10^5
cis-β_2	3.37×10^{-5}	1.38×10^7

[a] For the cis-α isomer, $\Delta H^{\ddagger} = 135$ kJ mol^{-1} and $\Delta S^{\ddagger} = 78 \pm 6$ J K^{-1} mol^{-1} (aquation); $\Delta H^{\ddagger} = 153$ kJ mol^{-1} and $\Delta S^{\ddagger} = 272 \pm 22$ J K^{-1} mol^{-1} (base hydrolysis) ($I = 0.1 M$). For the cis-β_2 isomer, $\Delta H^{\ddagger} = 150$ kJ mol^{-1} and $\Delta S^{\ddagger} = 126 \pm 38$ J K^{-1} mol^{-1} (aquation); $\Delta H^{\ddagger} = 51 \pm 14$ kJ mol^{-1} and $\Delta S^{\ddagger} = 54 \pm 47$ J K^{-1} mol^{-1} (base hydrolysis) ($I = 0.3 M$).

Rapid scan spectrophotometry has been used to study the base hydrolysis of $\alpha\beta S$-(salicylato)(tetraethylenepentamine) cobalt(III)[52] (**17**). The instantaneous color change observed on addition of base to the complex has been attributed[53] to formation of the phenoxide species and this point has been confirmed. Subsequent aquation and base hydrolysis of the phenoxide species then occurs with $k_{aq} = 0.116$ s^{-1} and $k_{OH} = 3.32$ M^{-1} s^{-1} at 25°C.

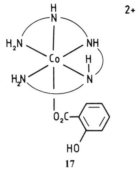

17

Stereochemical change in base hydrolysis is still to a large extent an unsolved problem. Early studies assumed that the five-coordinate intermediate formed in the S_N1 CB process would have a fixed geometry and that stereospecific entry of the incoming ligand was directed by the amido group. Recent work (discussed above[31]) with isomers of [CoL(NH$_3$)X]$^{n+}$ [X = Cl,Br,N$_3$ or NH$_3$; L = Me(trien)] indicates that the lifetime of the five-coordinate intermediate is such that it does not relax towards a common trigonal bipyramidal structure, although this process would require only a small change in bond angles. Immediate capture of a solvent molecule occurs, before bond angles can equilibrate. Balt and co-workers[54] have argued that if indeed the process of capture of the entering group by the five-coordinate species competes in rate with angle equilibration and solvent-structure rearrangement, relative rates, and consequently product stereochemistry, should be affected by the nature of the solvent.

Studies of the *cis trans* product ratio for the base-catalyzed solvolysis of *trans*-[Co(NH$_3$)$_4$(^{15}NH$_3$)Cl]$^{2+}$ in various solvents have been carried out by Balt *et al.*[54] The solvents used were H$_2$O–MeOH and H$_2$O–DMSO mixtures and anhydrous MeOH and MeNH$_2$. Studies were also made with *trans*-[Co(NH$_3$)$_4$(^{15}NH$_3$)(DMSO)]$^{3+}$ in order to vary the leaving group. The results indicate that all *trans* systems give $(44 \pm 1)\%$ rearrangement, although small but significant differences are shown by *trans*-[Co(NH$_3$)$_4$(^{15}NH$_3$)Cl]$^{2+}$ in MeNH$_2$ (50% rearrangement) and by the DMSO complex (48% rearrangement in H$_2$O and 46% in MeOH). The five-coordinate intermediate in this case appears to have a significant lifetime which allows it to equilibrate to a common stereochemistry. It has also been previously observed[55,56] that there is a close agreement in product stereochemistry for the base-catalyzed reactions of $[Co(en)_2(X)Y]^{n+}$ (X = Cl, Br, or DMSO; Y = Cl, NCS, NO_2, N_3, or NH_3) complexes in liquid ammonia and in aqueous solution.

The rate of substitution of the axial water ligand by ammonia in the complexes $[RCo(dmgH)_2(OH_2)]$ (18) with R = Me, Et, and CH_2Cl has been

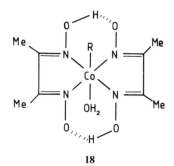

18

studied[57] as a function of the hydroxide ion concentration. Two counter-balancing effects occur in basic solution, a decrease in reactivity due to the formation of the inert hydroxo complex and an enhancement of the reactivity due to formation of the conjugate base. Conjugate-base formation involves deprotonation of the equatorial ligand system. The hydrolysis of the two dimethylglyoxime complexes of cobalt(III), $[Co(dmgH)_2Cl(py)]$ and $[Co(dmgH)_2Cl(NH_3)]$, in basic solution has also been studied in detail[58] [see equation (3)]:

$$[Co(dmgH)_2Cl(L)] + OH^- \rightarrow [Co(dmgH)_2(OH)L] + Cl^- \qquad (3)$$

$$(L = NH_3 \text{ or py})$$

The oxime proton is sufficiently acidic that it can be titrated to give the conjugate base, as in equation (4), and it is possible to consider the relative reactivities of the neutral (19) and deprotonated (20) complexes in hydrolysis.

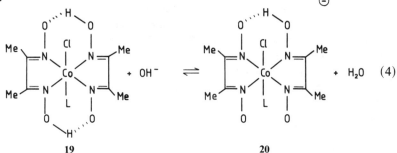

$$+ OH^- \rightleftharpoons \qquad\qquad + H_2O \qquad (4)$$

19   **20**

The ammoniation of the $trans$-$[Co(NH_3)_4Cl_2]^+$ cation to $[Co(NH_3)_5Cl]^{2+}$, studied in liquid ammonia between -40 and $-60°C$ involves a conjugate-base mechanism and gives over 98% stereochemical retention as shown by $^{15}NH_3$ labeling.[59] Activation parameters for the overall reaction ($K_{CB}k$) are $\Delta H^{\ddagger} = 59$ kJ mol^{-1} and $\Delta S^{\ddagger} = -55$ J K^{-1} mol^{-1}. At NH$_4$ClO$_4$ concentrations below 0.01 mol kg^{-1} deprotonation is rate determining and general base catalysis is observed with NH$_3$ and NH$_2^-$ as proton abstracting bases. ^{1}H NMR measurements in ND$_3$ during ammoniation show the loss of one proton in the formation of the pentammine.

7.4. Solvolysis

The kinetics of solvolysis of $trans$-$[Co(4\text{-}Mepy)_4Cl_2]^+$ has been studied using water and water propan-2-ol as solvent.[60] Correlation of the free energies of activation in the mixture, with that in water, using Gibbs free energies of transfer of individual ionic species, suggests that the effect of changes in solvent structure on the complex ion in the transition state dominates over that on the complex ion in the initial state. Similar studies have also been carried out with $trans$-$[Co(py)_4Cl_2]^+$ using mixtures of water propan-2-ol as solvent.[61] A linear plot of log k $vs.$ the Grunwald–Winstein Y factor suggests that solvolysis is D in character with considerable extension of the Co–Cl bond in the transition state.

7.5. Anation

The kinetics of anation of $trans$-$[Co(N_4)(OH)(OH_2)]^{2+}$ complexes by acetate ions have been studied[62] at 25°C and $I = 0.1$ M (LiClO$_4$) over the pH range 3.6–5.2; N$_4$ represents the series of macrocyclic ligands cyclam, C-$meso$-Me$_6$[14]aneN$_4$, C-rac-Me$_6$[14]aneN$_4$, Me$_6$[14]4,11-dieneN$_4$ (**21**), and Me$_4$[14]tetraeneN$_4$ (**22**). The reaction is considered to proceed by a D

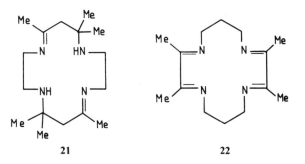

21   22

mechanism. The rate constants estimated for the loss of coordinated water are summarized in Table 7.6. Steric acceleration by the ring methyl substituents is observed. The anation of the cyclam complex by Cl^- and Br^- in acetate buffer solution was also studied. The relative effectiveness of entering groups to the pentacoordinate intermediate follow the order $H_2O < OAc^- < Cl^- < Br^-$.

The kinetics of water substitution of some alkylaquocobaloximes $[RCo(dmgH)_2(H_2O)]$ (R = Me, CH_2Cl, CF_3CH_2) has been investigated[63] in aqueous solution. Nucleophiles included NH_3, imidazole, morpholine, and SCN^-. Evidence for pre-equilibrium association of the alkylaquocobaloxime with some ligands (NH_3, imidazole, morpholine) was obtained using high nucleophile concentrations. Normally, strictly second-order kinetics (first order in the complex and first order in the nucleophile) is observed for such reactions.

The kinetics of anation of cis-$[Co(tmd)_2(OH_2)_2]$ by $H_2C_2O_4$ and $HC_2O_4^-$ have been studied[64] at 35, 40, and 45°C at $I = 1.0$ M ($NaClO_4$). Formation of $[Co(tmd)_2C_2O_4]^+$ corresponds to the rate expression (5):

$$k_{obs} = \frac{k_1 K_1[H_2C_2O_4] + k_2 K_2[HC_2O_4^-]}{1 + K_1[H_2C_2O_4] + K_2[HC_2O_4^-]} \quad (5)$$

The anation rate constants k_1 and k_2 for the cis-$[Co(tmd)_2(OH_2)_2]$·$H_2C_2O_4$ and cis-$[Co(tmd)_2(OH)_2)_2]$·HC_2O_4 ion pairs, respectively, are 10.7×10^{-4} and 11.0×10^{-4} s^{-1} at 40°C. The ion-pair constants with $H_2C_2O_4$ ($K_1 = 1.7 \pm 0.3$ M^{-1}) and $HC_2O_4^-$ ($K_2 = 11.5 \pm 0.7$ M^{-1}) are essentially independent of temperature. An I_d mechanism is suggested, and the reaction is shown to be catalyzed by NO_3^-.

The reaction of cis-$[Co(en)_2(H_2O)_2]^{3+}$ and $S_2O_3^{2-}$ at pH 6.3 gives $trans$-$[Co(en)_2(S_2O_3)_2]^-$ as the final product.[65] Only the cis-diaquo species is reactive and no reaction is observed with the hydroaquo and dihydroxy species. The reaction is believed to proceed via the intermediate cis-

Table 7.6. Rate Constants for the Loss of Coordinated Water from Some Macrocyclic $trans[Co(N_4)(OH)(OH_2)]^{2+}$ Complexes at 25°C and $I = 0.1$ M $(LiClO_4)$[62]

N_4	k (s^{-1})
cyclam	$(1.4 \pm 0.2) \times 10^{-2}$
C-$meso$-$Me_6[14]aneN_4$	$(1.0 \pm 0.1) \times 10^{-1}$
C-rac-$Me_6[14]aneN_4$	$(4.0 \pm 0.1) \times 10^{-1}$
$Me_6[14]4,11$-dieneN$_4$	$(8.9 \pm 1.6) \times 10^{-2}$

$[Co(en)_2(S_2O_3)OH_2]^+$. Activation parameters for formation of the intermediate ($\Delta H^{\ddagger} = 109 \text{ kJ mol}^{-1}$; $\Delta S^{\ddagger} = 104 \text{ J K}^{-1} \text{mol}^{-1}$) are similar to those previously observed for replacement of an aquo ligand in cis-$[Co(en)_2(OH_2)_2]^{3+}$. Activation parameters for the second step ($\Delta H^{\ddagger} = 53 \text{ kJ mol}^{-1}$; $\Delta S^{\ddagger} = -117 \text{ J K}^{-1} \text{mol}^{-1}$) indicate high trans labilization by the first $S_2O_3^{2-}$ group.

Substitution of coordinated water in trans-$[Co(CN)_4(SO_3)(OH_2)]^{3-}$ by NH_3, py, N_3^-, CNS^-, and SO_3^{2-} has been investigated in detail and the aquation of trans-$[Co(CN)_4(SO_3)NH_3]^{3-}$ and trans-$[Co(CN)_4(SO_3)py]^{3-}$ also studied.[66] Substitution of coordinated water in trans-$[Co(CN)_4(SO_3)OH_2)]^{3-}$ by 4-Mepy, 4-OAcpy, I^-, NO_2^-, $MeNH_2$, HSO_3^-, and $S_2O_3^{2-}$ has been the subject of a further investigation.[67]

A reinvestigation of the kinetics of the reaction of $[Co(CN)_5OH_2]^{2-}$ with N_3^- has been reported.[68] The earliest kinetic studies[69] of the reaction of $[Co(CN)_5OH_2]^{2-}$ with N_3^- or SCN^- gave values of k_{obs} which varied with $[X^-]$ according to equation (6):

$$k_{obs} = k_1[X^-]/(k_2/k_3 + [X^-]) \qquad (6)$$

However, measurements by Burnett and Gilfillan,[70] although giving good agreement with equation (6) for SCN^- (with $k_1 = 2.3 \times 10^{-3} \text{ s}^{-1}$ and $k_2/k_3 = 3.7 \, M$), gave a first-order dependence of k_{obs} upon $[N_3^-]$ with $k = k_{obs}/[N_3^-] = 6.6 \times 10^{-4} \, M^{-1} \text{s}^{-1}$ at 40°C. The new results[68] support this view.

Anations of cis- and trans-$[Co(en)_2(NO_2)OH_2]^{2+}$ with NO_2^- and CNS^- are stereoretentive.[71] Use of this fact has been made, to prepare cis- and trans-$[Co(en)_2Co(NO_2)(C_2O_4)]$ containing the monodentate oxalato ligand.[72] Anation of cis-$[Co(NH_3)_4(OH_2)_2]^{3+}$ by NO_2^- in $HNO_2-NO_2^-$ buffers has been studied kinetically.[73] A two-step reaction was observed, with a second-order rate law applying to each step. No evidence was obtained for any nitrito intermediates in the reaction.

Rate constants for the reaction of aquocobalamin with thiosulfate have been determined as a function of ionic strength and solvent composition using dioxane–water mixtures.[74] The solvent effects are small and are of the same order as the ionic-strength effects. A linear free-energy relationship between the rate constants and the dissociation constants of the thiosulfatocobalamin in dioxane–water mixtures was observed. The water substitution reactions of the complex $[Co(III)(salen)(H_2O)_2]^+$ [salen = N,N'-ethylene-bis(salicylideneiminate)] have been studied in aqueous solution.[75] The nucleophiles employed were pyridine, thiourea, morpholine, aniline, imidazole, NCO^-, HSO_3^-, and $S_2O_3^{2-}$. The kinetics of monosubstitution were found to be very rapid, with the second-order rate constants being nearly independent of the nature of the incoming ligand.

7.6. Solvent Exchange, Racemization, Isomerization, and Ligand Exchange

Although reactions of coordination compounds in the solid state have been known since the time of Werner, detailed studies of the kinetics and mechanism of such reactions remain limited. A recent review[76] discusses mechanisms for the racemization of optically active coordination compounds in the solid state and should help to revive interest in this area.

When Λ-α-[Co(R-picpn)Cl$_2$]ClO$_4$ [R-picpn = (3R)-3-methyl-1,6-di(2-pyridyl)-2,5-diazahexane] is reacted with sodium oxalate in aqueous solution, the complex Λ-α-[Co(R-picpn)ox]ClO$_4$ is obtained.[77] This reaction represents a total inversion of absolute configuration with respect to the metal center. When isolated as the iodide salt, the cationic product [Co(R-picpn)(ox)]I,1.5H$_2$O is obtained as a mixture of Λ-α and Λ-β diastereoisomers with a minor quantity of the Λ-α form. Isomeric ratios appear to be influenced by differences in reaction conditions, including exposure to light.

A variety of studies relating to linkage isomerization have appeared. Linkage isomerization reactions of cis- and trans-[Co(en)$_2$(ONO)$_2$]$^+$ have been studied as a function of [OH$^-$], temperature, and pressure.[78] The volumes of activation for the base-catalyzed isomerization are $+19.7 \pm 1.1$ and $+13.6 \pm 1.2$ cm^3 mol^{-1} at 25°C for the cis and trans species, respectively. The results are interpreted in terms of a conjugate-base mechanism. The corresponding nitro complexes are shown to undergo slow cis to trans isomerization ($t_{1/2} \sim 20$ h at 86°C) in neutral solution. The spontaneous nitrito-to-nitro linkage isomerization of a variety of octahedral (nitrito)-(amine)cobalt(III) complexes has been studied to investigate the role of the "inert" amine ligand on the isomerization rate.[79] The following general conclusions were reached: (1) isomerization is retarded relative to that of [Co(NH$_3$)$_5$ONO)]$^{2+}$ for complexes with chelate rings trans to each other, but is accelerated for those with other arrangements; (2) chelate-ring size has little effect on the rate constant; (3) isomerization of cis-[Co(en)$_2$(X)-(ONO)]$^{n+}$ is relatively insensitive to the nature of X, but that of trans-[Co(en)$_2$(X)(ONO)]$^{n+}$ is very dependent on the nature of the ligand trans to the ONO group, and (4) the net charge carried by the complex has little effect on the isomerization rate.

The nitrito–nitro linkage isomerization of [Co(NH$_3$)$_5$(ONO)]$^{2+}$ in liquid ammonia has been shown to proceed entirely by an intramolecular conjugate-base mechanism.[80] The use of liquid ammonia as solvent allowed the separate determination of the pre-equilibrium constant K_{CB} ($\Delta H = 27$ kJ mol^{-1}; $\Delta S = -14$ J K^{-1} mol^{-1}) and the rate constant k_{CB} for the rate-determining step ($\Delta H^{\ddagger} = 78$ kJ mol^{-1}; $\Delta S^{\ddagger} = 29$ J K^{-1} mol^{-1}).

Linkage isomerization of *cis*-[Co(en)$_2$(ONO)$_2$]$^+$ in aqueous acid involves two reactions: a fast step ($k = 3.6 \times 10^{-3}$ s^{-1} at 35°C, $\Delta H^\ddagger = 15.5 \pm 0.5$ kcal mol^{-1}, and $\Delta S^\ddagger = -19.1 \pm 1.7$ cal K^{-1} mol^{-1}) to give *cis*-[Co(en)$_2$NO$_2$)(ONO)]$^+$ followed by a slower step to give *cis*-[Co(en)$_2$(NO$_2$)$_2$]$^+$ ($k = 7.1 \times 10^{-4}$ s^{-1} at 35°C, $\Delta H^\ddagger = 19.5 \pm 0.2$ kcal mol^{-1}; $\Delta S^\ddagger = -9.4 \pm 0.6$ cal K^{-1} mol^{-1}).[81] The *trans*-[Co(en)$_2$(ONO)]$^+$ species undergoes a slower isomerization ($k = 8.4 \times 10^{-4}$ s^{-1} at 35°C) to give *trans*-[Co(en)$_2$(NO$_2$)(ONO)]$^+$ followed by a rapid reaction to give *trans*-[Co(en)$_2$(NO$_2$)$_2$]$^+$. Volumes of activation for these reactions in acidic solution lie between -3.4 and -6.9 cm^3 mol^{-1} and are interpreted in terms of an intramolecular isomerization process. Base-catalyzed isomerization of [Co(NH$_3$)$_5$ONO]$^{2+}$ has $\Delta V^\ddagger = 27 \pm 1.4$ cm^3 mol^{-1} assigned to a conjugate-base mechanism.

The spontaneous rate of nitrito-to-nitro isomerism of [(NH$_3$)$_5$Co(ONO)]$^{2+}$ has been shown[82] to span two orders of magnitude for the 16 solvents studied. The volume of activation is small and negative, -3.5 to -7 cm^3 mol^{-1} for water, DMSO, *N*-methylformamide, and sulfolane. The observation of negligible competitive solvolysis, and of an isokinetic plot of $\Delta H^\ddagger$ *vs.* $\Delta S^\ddagger$ also indicates that isomerization is intramolecular in all solvents studied. The solvent dependence is interpreted in terms of a dual parameter equation involving terms for the Lewis basicities (D_N) and the Lewis acidities and polarity (E_T) of the solvents. The base-catalyzed nitrito-to-nitro isomerization is shown to be subject to weak catalysis by F$^-$, OAc$^-$, and Et$_3$N in DMSO. The metal ions Hg(II), Ag(I), and Cd(II) also catalyze the isomerization in water ($k_{Hg} = 1.16 \times 10^{-2}$ M^{-1} s^{-1}, $k_{Ag} = 2.85 \times 10^{-4}$ M^{-1} s^{-1}, and $k_{Cd} = 4.4 \times 10^{-5}$ M^{-1} s^{-1} at $I = 0.1$ M and 25°C). Competition by nucleophiles such as OAc$^-$, NO$_3^-$, and H$_2$O in the Hg(II)-catalyzed isomerization is not observed, suggesting that the reaction occurs by an intramolecular process.

Calorimetric measurements of the enthalpies of reaction of solid *cis*- and *trans*-[Co(en)$_2$X$_2$]X (X = Cl, Br) with basic sodium sulfide solution have allowed the heats of isomerization of *cis*-[Co(en)$_2$Cl$_2$]Cl(S) → *trans*-[Co(en)$_2$Cl$_2$]Cl(S) and *cis*-[Co(en)$_2$Br$_2$](S) → *trans*-[Co(en)$_2$Br$_2$]Br(S) to be determined ($\Delta H = -11.4$ kJ mol^{-1} and -6.4 kJ mol^{-1}, respectively).[83] Assuming ΔS for *cis* → *trans* is -11.6 J K^{-1} mol^{-1} at 25°C, the respective values of ΔG are -8.0 kJ mol^{-1} and -3.0 kJ mol^{-1} for X = Cl and Br. Isomerization in the solid state was verified experimentally.

The kinetics of the linkage isomerization for the N(1)-bonded (5-methyltetrazolato) pentamminecobalt(III) ion (**23**) to the N(2) isomer (**24**) has been studied as a function of temperature and pH.[84] Rate constants for isomerization of the protonated (3+) complex and the deprotonated (2+) complex differ appreciably, and a base-catalyzed pathway for the (2+)

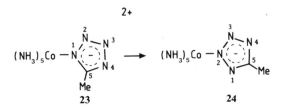

23   24

complex was observed. At 25°C and $I = 1.0\ M$, k for the (3+) ion is $(3.1 \pm 2.5) \times 10^{-4}\ s^{-1}$ ($\Delta H^{\ddagger} = 15 \pm 4\ kcal\ mol^{-1}$; $\Delta S^{\ddagger} = -24 \pm 13\ cal\ K^{-1}\ mol^{-1}$), while k for the 2+ ion is $(1.9 \pm 0.5) \times 10^{-6}\ s^{-1}$ ($\Delta H^{\ddagger} = 26.0 \pm 1.8\ kcal\ mol^{-1}$; $\Delta S^{\ddagger} = 2.5 \pm 5.6\ kcal\ K^{-1}\ mol^{-1}$). In basic solution, the isomerization of the 2+ ion conforms to a rate equation $k_{obs} = k + k_{OH}[OH^-]$ with $k_{OH} \doteq 3.1 \times 10^{-3}\ M^{-1}\ s^{-1}$ at 41.2°C.

Isomers of (4-methylimidazole)pentaamminecobalt(III) have been isolated in which the methyl group of the imidazole ligand is directed away from the five NH_3 ligands [R = remote = (**25**)] and near the NH_3 ligands [A = adjacent = (**26**)].[85] In tris or pyridine buffers two pathways are

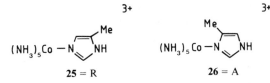

25 = R   26 = A

observed for the isomerization of A to R. One pathway, $k_0 = 5.8 \times 10^{-8}\ s^{-1}$ is assigned to the isomerization of the parent 4-methylimidazole species A, while a second pathway, first order in $[OH^-]$, is attributed to the deprotonated imidazolato species (A–H) with $k_{OH} = 3.2\ M^{-1}\ s^{-1}$ ($\Delta H^{\ddagger} = 32.4\ kcal\ mol^{-1}$; $\Delta S^{\ddagger} = 16.8 \pm 11.7\ cal\ K^{-1}\ mol^{-1}$). Comparisons are made with the linkage isomerization reactions of $(NH_3)_5Co(III)$ coordinated to ONO^- and $N(1)$ of the 5-methyltetrazolato ligand. Isomerization of the 5-methyltetrazole ligand is faster by a factor of 10^5 relative to that for 4-methylimidazole, but isomerization of the deprotonated 4-methylimidazolato species is faster by a factor of 4×10^2 compared with the analogous 5-methyltetrazolato system.

The reaction of $[Co_2(Et_2dtc)_5]BF_4$ ($Et_2dtc = N,N$-diethyldithiocarbamate) with a variety of substituted dithiooxamides (DTO) has been studied kinetically.[86] The reactions have the stoichiometry $[Co_2(Et_2dtc)_5]BF_4 + DTO \rightarrow [Co(Et_2dtc)_2DTO]BF_4 + [Co(Et_2dtc)_3]$, and the mixed-ligand complexes $[Co(Et_2dtc)_2DTO]BF_4$ have been isolated and characterized.[87] The kinetics correspond to the expression $k_{obs} = k_0 + k[DTO]$ with $k \gg k_0$. A mechanism involving a one-ended reversible dissociation of the $[Co_2(Et_2dtc)_5]^+$ unit preceding the rate-determining step is suggested.

The kinetics of the reaction of $[Co_2(Et_2dtc)_5]BF_4$ with some N-substituted ethylenediamines using CH_2Cl_2 as solvent have also been studied.[88] As with the dithio-oxamides these reactions appear to involve a pre-equilibrium dissociation (Scheme 3) followed by rate-determining attack by the ligand or solvent on the intermediate.

Scheme 3

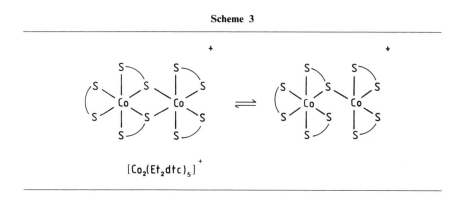

$[Co_2(Et_2dtc)_5]^+$

One of the most convincing demonstrations of a D-type mechanism involves the measurement of nucleophile competition during substitution proceeding at the limiting dissociative rate.[89] It has recently been observed[90] that the reaction of the nucleophiles (N_3^-, NCS^-, and DMSO) with $[Co(CN)_5(OH_2)]^{2-}$, $[Co(CN)_5Cl]^{3-}$, and $[Co(CN)_5(N_3H)]^{2-}$ occurs at different rates, suggesting that substitution takes place via an I_d mechanism and not a D mechanism involving the common intermediate $[Co(CN)_5]^{2-}$. The exchange reaction between *trans*-$[Co(en)_2(SO_3)CN]$ and K ^{14}CN has been studied using ethylene glycol as solvent over the temperature range 40–60°C.[91] The activation parameters are $\Delta H^\ddagger = 18.3$ kcal mol^{-1} and $\Delta S^\ddagger = -23.9$ cal K^{-1} mol^{-1}. The exchange rate is independent of the KCN concentration indicating a dissociative mechanism. Previous measurements[92] using water as a solvent indicated a first-order dependence on the KCN concentration.

7.7. Decarboxylation of Carbonato Complexes in Acidic Solution

The reactions of carbonato complexes in basic solution are considered in Section 7.3, and the present section deals with reactions in acidic media.

The acid-catalyzed decarboxylation of carbonato complexes of the type *cis*-[CoN$_4$CO$_3$]$^+$ (N$_4$ = a series of four nitrogen donors of the uni-, bi-, or quadridentate type) proceeds in two steps. The first step involves ring opening to give the bicarbonato complex *cis*-[CoN$_4$(OCO$_2$H)OH$_2$]$^{2+}$ which undergoes rapid decarboxylation (loss of CO$_2$) in the second step to give *cis*-[CoN$_4$(OH$_2$)$_2$]$^{3+}$. The first step is acid catalyzed and results in a linear dependence of k_{obs} on [H$^+$]. In some cases a small intercept is observed in such plots assigned to the water-promoted ring-opening process. Normally, the first step is rate determining, and the second step is comparatively fast with k of the order of 1 s^{-1} at 25°C, a value typical for the decarboxylation of bicarbonato complexes.

Solvent deuterium isotope effect studies[93,94] indicate that ring opening involves rapid pre-equilibrium protonation of the carbonato ligand followed by rate-determining ring opening (Scheme 4). As K_1 is small, k_{obs} = $k_1 K_1$[H$^+$] and in general is significantly smaller than k_2. However, in a number of systems[95] it is observed that plots of k_{obs} *vs.* [H$^+$] reach a limiting value at high [H$^+$] which could be due to the second step (i.e., loss of CO$_2$) becoming rate determining. Rapid scan spectrophotometry has now been used[96] to detect the ring-opened bicarbonato intermediate *cis*-[Co(en)$_2$(OCO$_2$H)OH$_2$]$^{2+}$ produced during the acid-catalyzed decarboxylation of *cis*-[Co(en)$_2$CO$_3$]$^+$ in aqueous solution. Direct spectral evidence for a change in rate-determining step is obtained.

Scheme 4

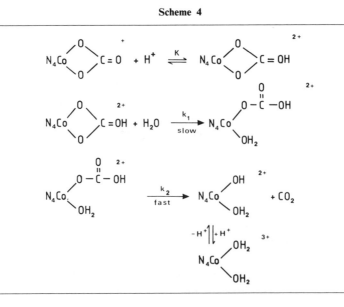

The μ-carbonato complex $[(NH_3)_5CoCO_3Co(NH_3)_5](SO_4)_2\cdot 4H_2O$ has been characterized and its acid-catalyzed decomposition over the pH range 1–7 studied by flow techniques at $I = 0.5\ M$ and 25°C.[97] The rate-determining step is reaction (7).

$$[(NH_3)_5CoOCO_2H]^{2+} \xrightarrow{k_1} [Co(NH_3)_5OH]^{2+} + CO_2 \tag{7}$$

At pH < 5 values of k_{obs} are independent of pH. Above pH 5 values of k_{obs} fall off rapidly becoming negligible beyond pH 7. The data fit the equation $k_{obs} = k_1[H^+]/([H^+] + K_1)$ with $k_1 = 1.09\ s^{-1}$ and $K_1 = 4.4 \times 10^{-7}\ M$ (K_1 is the acid dissociation of $[(NH_3)_5CoOCO_2H]^{2+}$). These values are very similar to those previously determined for the acid-catalyzed decarboxylation of $[Co(NH_3)_5CO_3]^+$.

The kinetics of the complete decomposition of *cis*-$[Co(py)_2(CO_3)_2]^-$ in aqueous $HClO_4$ have been studied.[98] Decomposition occurs in two successive decarboxylation steps and a final reduction step which gives Co(II) as the only cobalt-containing product.

7.8. Formation

The formation kinetics of $[Co(NH_3)_5SO_3]^+$ and *trans*-$[Co(NH_3)_4(SO_3)_2]^-$ have been studied at pH > 9 as a function of temperature, pressure, $[SO_3^{2-}]$, and pH in ammonia buffer solutions.[99] The data obtained can be summarized by equation (8). The k' and k'' terms

$$k_{obs} = k'[H^+] + k''[SO_3^{2-}] \tag{8}$$

represent contributions from the rate-determining hydrolysis and formation reactions of $[Co(NH_3)_5SO_3]^+$, respectively. The $[Co(NH_3)_5SO_3]^+$ species undergoes subsequent rapid substitution to produce *trans*-$[Co(NH_3)_4(SO_3)_2]^-$ for which k_{obs} is independent of $[SO_3^{2-}]$. It has previously been suggested[100,101] that $[Co(NH_3)_5OH]^{2+}$ reacts with SO_2 in aqueous solution to produce an O-bonded sulfito complex $[Co(NH_3)_5OSO_2]^+$, which on acidification releases SO_2 and gives the corresponding aquo complex. The kinetics of both processes indicate that Co–O bond cleavage did not occur. Recent ^{17}O-exchange experiments using ^{17}O NMR provide evidence that the earlier suggestions are correct.[102]

The formation reactions of $[Co(NH_3)_5OCO_2]^+$ or $[Co(NH_3)_5OSO_2]^+$ proceed by simple CO_2 or SO_2 addition to $[Co(NH_3)_5OH]^{2+}$ in which nucleophilic attack by coordinated hydroxide occurs. The Co–O bond remains intact.[103] Retention of the Co–O bond in the formation of $[Co(NH_3)_5ONO]^{2+}$ has also been confirmed[104] by a recent NMR study. A comment on the formation and acid-catalyzed aquation reactions of car-bonato-, O-bonded sulfito-, and nitritopentamminecobalt(III) ions has appeared.[105] The acid-catalyzed aquation of $[Co(NH_3)_5ONO]^{2+}$ is very rapid and Co-O bond fission does not appear to be involved.

A variety of transition-metal hydroxo complexes, $[M(NH_3)_5OH]^{2+}$ [M = Co(III), Rh(III), and Ir(III)], react with CO_2 to give the corresponding monodentate carbonato complexes $[M(NH_3)_5OCO_2]^+$. The formation and decarboxylation kinetics of these complexes have now been studied as a function of pressure up to 1000 bar.[106] The volumes of activation for CO_2 uptake are -10.1 ± 0.6 (Co(III)), -4.7 ± 0.8 (Rh(III)), and -4.0 ± 1.0 (Ir(III)) cm³ mol⁻¹, whereas the corresponding values for decarboxylation are $+6.8 \pm 0.3$, $+5.2 \pm 0.3$, and $+2.5 \pm 0.4$ cm³ mol⁻¹, respectively. Com-bined with partial molar volume measurements, these values enable the construction of overall reaction volume profiles. Bond formation during CO_2 uptake and bond breakage during decarboxylation are approximately 50% completed in the transition state of these processes.

Complex formation between iron(III) and some (salicylato)pen-taminecobalt(III) ions [$N_5 = (NH_3)_5$, $en_2(NH_3)$, or tetren) has been studied kinetically.[107] Analogous measurements with Al(III) have also been reported.[108]

7.9. Photochemistry

Photochemistry of cis-$[Co(en)_2(NO_2)_2]^+$ in neat acetonitrile differs from that in water.[109] In aqueous solution excitation of the complex ion in the ligand-to-metal charge transfer band gives the cobalt(II) ion and the nitrito linkage isomer. In acetonitrile, an intermediate cobalt(II) complex coordi-nated to NO_2 is produced. As previous studies have shown, the solvent employed has a marked effect on the reactivity pattern of excited states. The viscosity dependence of the photoaquation of $[Co(CN)_6]^{3-}$ and a variety of Cr(III) complexes has been studied using glycerol–water mixtures.[110] Cage effects are not important in the photosubstitution of CN^- in the systems studied.

The photolysis of $[Co(Me_4[14]tetraeneN_4)(OH_2)_2]^{3+}$ (where $Me_4[14]tetraeneN_4$ is the macrocycle (27)) has been studied,[111] employing

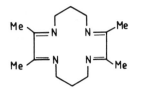

27

an aqueous acidic medium and a variety of incident wavelengths. Various physical techniques (UV, IR, visible, NMR) show that reduction of the metal center occurs, but the macrocyclic ligand remains intact. The quantum yield of the cobalt(II) complex is small, but can be increased by the addition of alcohols. Mechanisms for the photoredox reaction are considered. The vitamin B_{12}-derived cobalt(III) complex "pyrocobester" (**28**) acts as a photosensitizer in its light-induced oxygenation,[112] which involves singlet

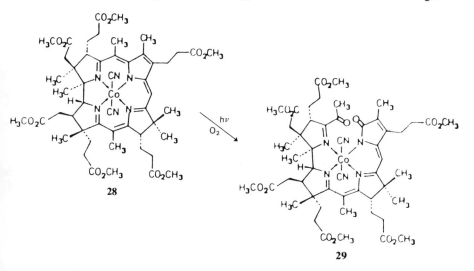

oxygen ($^1O_2$), giving 5,6-dioxosecopyrocobester (**29**). Previous mechanistic proposals regarding the photochemistry of tris(β-diketonate)cobalt(III) complexes may require modification in the light of these results.

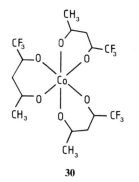

30

The complex [Co(acac)$_2$(N$_3$)NH$_3$] decomposes on photolysis to give [Co(acac)$_2$] and azide radicals.[114] The threshold energy for this process is $\sim 20 \times 10^3$ cm^{-1} and a triplet charge transfer photoactive state is implicated. The action of various sensitizers is considered in detail.

7.10. Reactions of Coordinated Ligands

Interesting work in this area continues, and a number of useful synthetic applications of cobalt(III) complexes have been described.

7.10.1. Nitrile Hydrolysis

Coordination compounds with organonitrile ligands are of considerable interest due to the increased susceptibility of the coordinated nitrile to nucleophilic attack by reagents such as hydroxide ion. Rate accelerations of 10^6 in base hydrolysis (nitrile → amide) have been observed. Two new aromatic organonitrile complexes of $[Co(NH_3)_5]^{3+}$ with 2- and 4-nitrobenzonitrile have been prepared, and their base hydrolysis to coordinated carboxamides studied.[115] The second-order rate constants for base hydrolysis are 180 ± 4 and 510 ± 90 $M^{-1}s^{-1}$ at 25°C and $I = 1.0$ M for the ortho and para isomers, respectively. Rate constants for the hydrolysis of 11 coordinated aromatic nitriles follow a Hammett-type correlation with log $k_{OH} = 1.93\sigma + 1.30$ at 25°C and $I = 1.0$ M. Carbon-13 NMR studies of the free and coordinated nitriles indicate similar chemical shifts, so that the variation in the slope of the Hammett plots for free- and coordinated-nitrile hydrolysis is a transition-state effect rather than a ground-state phenomenon.

7.10.2. Amino Acid Synthesis

Intramolecular imine formation between a coordinated aminate ion and a 2-oxo acid has been utilized to synthesize two racemic amino acids: 2-cyclopropylglycine and proline.[116] Thus anation of $[Co(NH_3)_5OH_2]^{3+}$ by $Br(CH_2)_3COCO_2H$ at pH 5 gives two major products: (**31**) and (**32**). Both are converted to tetraamine-iminocarboxylato chelates by attack of an adjacent deprotonated ammonia. The cyclopropylimine complex can, for example, be reduced by alkaline BH_4^- to give the (RS)-2-cyclopropylglycine complex.

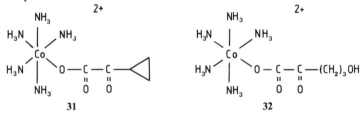

31   32

The cobalt(III)-promoted synthesis of β-carboxyaspartic acid has also been reported[117] by the intramolecular addition of coordinated amide ion to the olefinic center in the (3,3-bis(ethoxycarbonyl)-2-propenoato)pentaamminecobalt(III) ion, $[(H_3N)_5CoOOCCH=C(CO_2Et)_2]^{2+}$. In aqueous solution, an intramolecular addition of the *cis*-aminate ion at the olefinic center gives the N,O-chelated diester of β-carboxyaspartic acid.

7.10.3. Peptide Synthesis

Kinetically inert cobalt(III) complexes have been employed as N-protecting and N-activating groups for peptide synthesis.[118] Pentaamminecobalt(III) has now been shown to be a useful C-terminal protecting group for sequential peptide synthesis.[119] The reaction of complexes of

$$\left[(NH_3)_5Co-O_2C-\overset{\overset{\displaystyle R}{\displaystyle |}}{C}H-NH_3\right](BF_4)_3$$

$$\textbf{33} = [Co(AA)_1]$$

the type (**33**) with BOC-amino-acid active esters or BOC symmetric anhydrides gives $[Co(AA)_1(AA)_2BOC]$. The BOC group is removed with 95% CF_3CO_2H to give $[Co(AA)_1(AA)_2]$ which can be used for sequential peptide synthesis. Alternatively, the $(NH_3)_5Co$ group is selectively removed by rapid reduction with $NaBH_4$ or NaHS to give the N-protected peptide fragment $[(AA)_1(AA)_2BOC]$ under very mild conditions. The general procedures are discussed and the preparation of a number of pentapeptides and hexapeptides by the route described.

7.10.4. Imine Formation

Some of the synthetic aspects of intramolecular imine formation are discussed in Section 7.10.2. The complex *cis*-$[Co(en)_2\{NH_2CH_2CH-(OCH_3)_2\}Cl]^{2+}$ undergoes hydrolysis in dilute HCl solution to give the aminoacetaldehyde complex (**34**) which was isolated in an equilibrium mixture with its hydrated adduct, *cis*-$[Co(en)_2\{NH_2CH_2CH(OH)_2\}Cl]^{2+}$.[120] In aqueous solution intramolecular imine formation occurs to give (**35**). Reaction takes place with an amino group *cis* to Cl$^-$ so that the tridentate imine ligand adopts a *fac* stereochemistry.

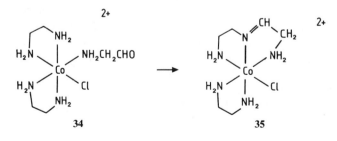

7.10.5. Base-Catalyzed Exchange Reactions

The activation of α-hydrogens of amino acids and peptides by coordination to metal ions has been the subject of many investigations and the topic has been reviewed.[121] The rate of the α-hydrogen exchange reaction of glycinato ligands in *mer*-[Co(gly)$_3$] has recently been found to be different for the three stereochemically different glycinato ligands.[122] This rate difference is attributed to the fact that the *trans* influence of the carboxyl oxygen is more effective than that of the amino nitrogen in labilizing the trans α-hydrogen.

In the reaction of [Co(glygly)(en)NO$_2$] with acetaldehyde in aqueous solution at pH 11, four condensation products have been obtained.[123] These products have been characterized as [Co(threogly)(en)NO$_2$], [Co(allothreogly)(en)(NO$_2$)], [Co(glygly)NO$_2$(CH$_3$CH=en)] (CH$_3$CH= en is *N*-ethylideneethylenediamine), and [Co(CH$_3$CH=glygly)(en)NO$_2$] (CH$_3$CH=glyglyH$_2$ is *N*-ethylideneglycylglycine).

The kinetics of iodination of malonate and pyruvate in the complexes [Co(NH$_3$)$_5$O$_2$CCH$_2$CO$_2$H]$^{2+}$, [Co(NH$_3$)$_5$O$_2$CCOCH$_3$]$^{2+}$, and [Co(en)$_2$(mal)]$^+$ has been studied at 35°C and $I = 0.3$ M.[124] The reaction is catalyzed by H$_2$O, OH$^-$, and by the buffer anions employed, and is independent of [I$_2$] but first order in the complex. Bidentate malonate is considerably more active towards electrophilic substitution than the monodentate ligand.

7.10.6. Phosphato Complexes

The hydrolytic reactions of coordinated polyphosphates have been studied by a number of groups due to the important role which these and other phosphates play in biological systems. The Cr(III) and Co(III) complexes of ATP (adenosine 5'-triphosphate) have been shown to be very good analogues for MgATP. The hydrolysis and decomposition of these complexes has now been studied in some detail.[125] Decomposition of [Cr(H$_2$O)$_4$ATP] proceeds predominantly with release of ATP producing lesser amounts of ADP, while the tridentate complex [Cr(H$_2$O)$_3$ATP] produces exclusively ADP. The complex Co(NH$_3$)$_4$ATP decomposes more slowly to give amounts of ATP and ADP which are lower than those produced with the analogous [Cr(H$_2$O)$_4$ATP]. However, the breakdown of the tridentate Co(NH$_3$)$_3$ATP is rapid, producing high levels of free ATP and lesser amounts of ADP. Rate constants for hydrolysis are 100–500 times greater than those for uncomplexed ATP.

The hydrolysis of β,γ-[Co(NH$_3$)$_4$H$_2$P$_3$O$_{10}$] (**36**) is greatly accelerated in the presence of *cis*-[Co(cyclen)(OH$_2$)$_2$]$^{3+}$ and the intermediate (**37**) has been suggested as the active complex.[126] Phosphorus-31 NMR studies have now provided evidence in support of such an intermediate.[127]

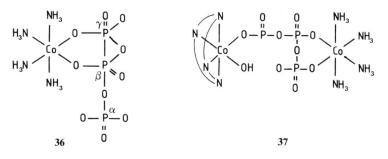

36   37

The preparation of cis-[Co(cyclen)PO$_4$] has been described and its reactivity in acidic and basic solution studied.[128] This complex undergoes rapid ring opening in acidic or basic solution to give the monodentate phosphato species. Loss of monodentate phosphate in acidic solution follows a rate expression of the form $k_{obs} = k_0 + k_H[H^+]$, where $k_0 = 5 \times 10^{-4}$ s^{-1} and $k_H = 4.05 \times 10^{-2}$ M^{-1} s^{-1} at 25° ($I = 0.49$ M). The acid-catalyzed reaction displays a solvent deuterium isotope effect k_{D_2O}/k_{H_2O} of ~1.4 consistent with a mechanism involving a rapid pre-equilibrium protonation followed by rate-determining loss of monodentate phosphate. Loss of monodentate phosphate from the complex [Co(cyclen)OH(OPO$_3$)]$^-$ also takes place in basic solution, and the reaction shows a first-order dependence on the hydroxide ion concentration with $k_{OH} = 2.7 \times 10^{-2}$ M^{-1} s^{-1} at 25°C and $I = 0.49$ M. Similar studies with [Co(en)$_2$PO$_4$] are also described in the same work. This work provides supporting evidence for the participation of the intermediate (37) in the [Co(cyclen)(OH$_2$)$_2$]$^{3+}$ promoted by $\beta\gamma$-[Co(NH$_3$)$_4$H$_2$P$_3$O$_{10}$]. The monodentate intermediate will be sufficiently long lived in solutions of pH 8–11, to participate in the reaction.

Sargeson and co-workers[129] initially reported that 4-nitrophenyl phosphate in (38) underwent base hydrolysis some 10^9-fold faster than the uncoordinated ester. Subsequent X-ray work[130] has now shown that (38) and the analogous ethylenediamine derivative do not contain chelated

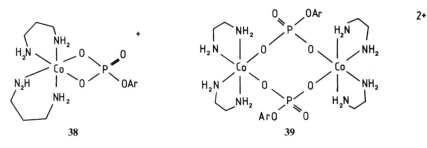

38   39

phosphate esters but are dimeric species (**39**) with a surprisingly stable eight-membered chelate ring. The reactivity of such bridged complexes in basic solution has been discussed in a review.[3] Ester hydrolysis occurs primarily via intramolecular attack by coordinated hydroxide in the ring-opened complex (**40**).

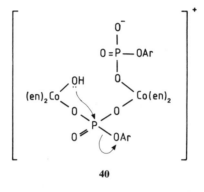

40

7.10.7. B_{12} and B_{12} Models

Methyl transfer from a number of *trans*-dimethylcobalt(III) complexes of synthetic macrocycles to Pb(II) and $(CH_3)_3Pb^+$ in acetonitrile has been studied.[131] A monomethylcobalt(III) complex is always a product, but different lead ion electrophiles and reaction stoichiometries give different lead products. The stoichiometries and kinetics of reactions of methyl-cobalamin (CH_3–B_{12}) with $[AuX_4]^-$ (X = Cl, Br) in acidic solution has also been examined.[132] Under anaerobic conditions, the reactions occur with a 2:1 stoichiometry ($[AuX_4]^-$: CH_3–B_{12}), producing aquocobalamin with an oxidized corrin ring, CH_3X, and metallic gold. The kinetic data support a mechanism which involves an equilibrium prior to the electron transfer between CH_3–B_{12} and $[AuX_4]^-$. Effects of pH and ionic strength on the kinetics are also examined and the detailed mechanism for the Co–C bond cleavage discussed.

The factors influencing cobalt–alkyl bond dissociation energies are of considerable interest in view of coenzyme B_{12}-dependent enzymatic proces-ses which are triggered by such bond dissociation. Steric influences on cobalt–alkyl bond dissociation energies have now been considered.[133] For reactions of the type in equation (9), where L is an axial tertiary phosphine

$$[LCo(dmgH)_2CH(CH_3)C_6H_5] \rightarrow [LCo(II)(dmgH)_2]$$

$$+ C_6H_5CH{=}CH_2 + \tfrac{1}{2}H_2 \qquad (9)$$

Table 7.7. *Decomposition of [LCo(dmgH)₂CH(CH₃)-*
C₆H₅] in Acetone at 25°C [133]

L	$\Delta H^{\ddagger}$ (kcal mol^{-1})	D_{Co-R} (kcal mol^{-1})
$P(CH_3)_2C_6H_5$	25.9	24
$P(n\text{-}C_4H_9)_3$	22.8	21
$P(CH_2CH_2CN)_3$	22.1	20
$P(C_2H_5)(C_6H_5)_2$	21.3	19
$P(C_6H_5)_3$	19.3	17

ligand, values of the bond dissociation energies (D_{Co-R}) have been estimated from the activation enthalpies (see Table 7.7). Homolysis of [RCo(7-Mesalen)(en)]$^+$ complexes (R = Me, Et, Bu^n, Pr^i) occurs quite readily in aqueous solution and the various hydrocarbon products have been characterized.[134]

7.10.8. Ligand Oxidation

The kinetics of oxidation of [Co(CysOS)(en)₂]$^+$ (**41**) by peroxodisulfate has been investigated and activation paramcters obtained.[135] Gibbs

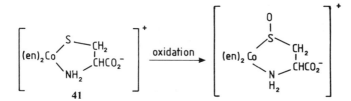

free energies of transfer for [Co(CysOs)(en)₂]$^+$ and $S_2O_8^{2-}$ have been estimated and the solvent effects on the oxidation (mixtures up to 40% acetone, *t*-butanol, isopropanol, or acetonitrile) have been divided into their initial-state and transition-state components.

Chapter 8

Substitution Reactions of Inert-Metal Complexes— Coordination Numbers 6 and Above: Other Inert Centers

As in previous volumes of this series, this chapter deals at some length with low-spin iron(II) complexes and with complexes of ruthenium(II) and (III) and rhodium(III); more briefly with complexes of iridium(III), osmium, and platinum(IV); and touches on a few other centers of marginal interest in the present context.

8.1 Groups VI and VII

In the period covered by this volume there have been very few papers published which concern inert octahedral complexes of metals in these two groups.

8.1.1. Molybdenum

The rate law has been established for the ready isomerization of the molybdenum(V) dimer (1) from the *trans*-bis-oxo- to the usually encountered *cis*-bis-oxo- form (**2**); the ligand LLL is 1,4,7–triazacyclononane, (**3**). There is marked acid catalysis, involving a protonated form of the trans complex in the rate-determining step.[1]

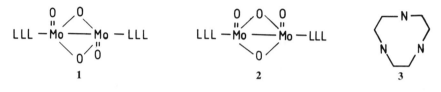

1   2   3

8.1.2. Technetium

There have been several semiquantitative investigations of redox reactions of complexes of this element, but little on kinetics and mechanisms of substitution. However, there has been a detailed examination of ligand-exchange kinetics for several oxotechnetate(V) anions. Reaction of $[TcOCl_4]^-$ with toluene-2,3-dithiolate (tdt) in methanol gives $[TcO(tdt)_2]^-$ in a reaction showing clean one-step kinetics. Similarly, there is no evidence for any intermediates of significant lifetime in the reaction of $[TcO(LL)_2]^-$, where $LL = OCH_2CH_2O^{2-}$, with ethane-1,2-dithiolate, but the kinetics do indicate transient intermediates for the complexes with $LL = 1,2\text{-}OC_6H_4O^{2-}$ or $SCH_2CH_2O^{2-}$. In the latter case spectroscopic evidence suggests that the intermediate may be a dinuclear species.[2] These substitutions are relatively slow as, predictably, is substitution at the d^6 center technetium(I) in hexaisocyanide complexes, which can be generated electrochemically.[3] Replacement of bromide in $[Tc(dppe)_2Br_2]^+$ by chloride is extremely slow; the bromo complex has to be refluxed with lithium chloride in ethanol for several hours.[4] The appearance of the first thorough treatment of technetium by Gmelin (the first coverage appeared in 1941, consisting of 10 pages entitled "Masurium") permits the locating of the limited published kinetic data and of much descriptive chemistry of ligand-exchange reactions.[5]

8.2. Iron

8.2.1. Pentacyanoferrates(II)

There has been a marked resurgence of interest in this area. Recently published kinetic parameters for ligand dissociation from pentacyanoferrates(II), $[Fe(CN)_5L]^{n-}$, are shown in Table 8.1,[6-19] which also contains earlier kinetic data for a few key complexes for comparison. It may be presumed that these dissociations take place by the limiting D mechanism, though the characteristic pattern of k_{obs} as a function of nature and concentration of the incoming ligand is explicitly reported in only a few cases.[10,16,18]

Most explanatory effort is applied to the dependence of reactivity on the nature of the leaving ligand. Electrostatic (charge) effects are small, though the position of charged ligands on a plot of rate constants *vs.* charge transfer frequencies shows that these charge effects are real enough.[7] Solvation effects have been shown to be important, for example, for piperidine as the leaving group.[12,13] Clearly, the strength of the iron–ligand bonding is paramount in determining dissociation rates, but this is compounded by σ- and π-contributions. The latter are generally assessed from frequencies of metal–ligand charge transfer bands[8,11,18,20,21] and the former from ligand pK_a values,[11] while the combined effect can be gauged from stability constants.[11] Other comparisons and correlations have included redox potentials,[11,18] ^{1}H NMR shifts,[20] and temperatures of ligand loss in thermogravimetric analysis.[21]

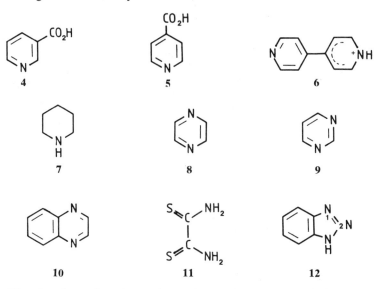

Kinetic data for formation of pentacyanoferrate(II) complexes $[Fe(CN)_5L]^{n-}$ from $[Fe(CN)_5(OH_2)]^{3-}$ are shown in Table 8.2. The Eigen-Wilkins mechanism applies here; overall formation rate constants k_f are determined by outer-sphere association constants K_{os}, that is, mainly by electrostatics. However, there is an unexpected inconsistency if one compares $N_2H_4/N_2H_5^+$ with en/enH$^+$, though this may well arise in part from solvation effects (en much less hydrophilic and thus less heavily solvated than N_2H_4) (cf. previous paragraph).[9] There seems to be some difficulty

Table 8.1. Kinetic Parameters for Substitution at Pentacyanoferrate(II) Complexes $[Fe(CN)_5L]^{n-}$, in Aqueous Solution at 298.2 K

L	$10^3 k$ (s^{-1})	$\Delta H^\ddagger$ (kJ mol^{-1})	$\Delta S^\ddagger$ (J K mol^{-1})	$\Delta V^\ddagger$ (cm^3 mol^{-1})	Conditions	Reference
Ammonia	17.5	94	33			6
Nicotinate	0.48	116	82		$I = 1.0\,M$ (NaCl)	7
Isonicotinate	0.29	114	70			7
Methyl isonicotinate	0.67				$I = 1.0\,M$ (NaCl)	8
Ethyl isonicotinate	0.57					8
Nicotinic acid (4)	1.42	96	21			7
Isonicotinic acid (5)	0.71	97	17			7
Nicotinamide	0.90	106	53			7
Pyridylpyridinium (6)	1.78	110	76			7
Hydrazine	1.6	107	57		$I = 1.0\,M$ (NaClO$_4$)	9
Hydrazinium	50	108	94			9
1,2-Dimethylethane-1,2-diamine	4.4	94	27		$I = 1.0\,M$ (NaCl)	10
Amino acids^a	2.7^b–26^c	92–99	28–51		$I = 1.0\,M$ (LiClO$_4$)	11
Piperidine (7)	8.8	94	25		$I \sim 0$	12
	6.2	88	5		$I = 1.0\,M$ (NaCl)	13

					Conditions	Ref.
Pyridine	1.1	104	46		$I = 1.0\ M$ (LiClO$_4$); pH 7	6
3-Acetylpyridine	0.86	106	59		$I = 1.0\ M$ (LiClO$_4$)	14
4-Acetylpyridine	1.61	113	75			14
3-Cyanopyridine	2.2	93	16	20.6	$I \sim 0$	15
4-Cyanopyridine	1.0	105	50	20.6		15
2-Methylpyrazine	0.77	114	44	19.4		15
Pyrazine (8)	0.42	110	59		$I = 1.0\ M$ (LiClO$_4$); pH 7	6
Pyrazine-N-oxide	9.0	111	70			7
Pyrimidine (9)	1.3	91	8		$I = 0.1\ M$ (LiClO$_4$); pH 7–8	16
Quinoxaline (10)	620	82	17			16
Dimethyl sulfoxide	0.075	110	46			17
Thiourea	12.60	88	21		$I = 0.1\ M$ (LiClO$_4$)	18
Thioacetamide	27.1	106	64			18
Dithiooxamide (11)[d]	46	108	70			18
N-1-benzotriazole (12)	3.7	93	19		$I = 0.1\ M$ (KCl)[e]	19
N-2-benzotriazole (12)	9.0	89	11			19
Carbon monoxide	10^{-5}[f]					19

[a] Rate constants for 12 amino acids.
[b] Glycine.
[c] Leucine.
[d] Rubeanic acid.
[e] Acetate buffer ($10^{-2}\ M$).
[f] Estimated from k_f (cf. Table 8.2).

Table 8.2. *Kinetic Parameters for Complex Formation between* $[Fe(CN)_5(OH_2)]^{3-}$ *and Ligands L, in Aqueous Solution at 298.2 K*

L	k_f $(M^{-1} s^{-1})$	$\Delta H^{\ddagger}$ (kJ mol^{-1})	$\Delta S^{\ddagger}$ (J K mol^{-1})	References
Ethane-1,2-diamine	230			9
Ethane-1,2-diamineH$^+$	410			9
Hydrazine	400			9
Hydrazinium$^+$	250			9
aa^{2-} ^{a}	9			11
aa$^-$ ^{a}	18–30	59–74	−14 to +21	11
aa^a	240–370			11
3-Acetylpyridine	473^b	64	79	15
4-Acetylpyridine	471^b	67	33	15
Quinoxaline (10)	425	64	25	16
Thiourea	194	60	1	18
Thioacetamide	271	65	21	18
Dithiooxamide (11)	460	65	22	18
Benzotriazole (12)	330	60	6	19, 22
Carbon monoxide	310	63	15	23

^{a} aa = amino acids in dianionic (glu^{2-}, tyr^{2-}), monoanionic, and zwitterionic forms.
^{b} At 297.2 K.

in establishing reliable kinetic data for the reaction of $[Fe(CN)_5(OH_2)]^{3-}$ with thiourea (cf. Table 8.2 with pp. 129–130 of Volume 1 and p. 191 of Volume 2).

The formation of a binuclear complex precedes electron transfer in the reaction of $[Fe(CN)_5(OH_2)]^{3-}$ with $[Co(NH_3)_5(LL)]^{3+}$, where LL = 3- or 4-cyanopyridine cyano-N or pyridine-N bonded (Scheme 1).[24] The range of K_{os} values is 477–872 (molar scale) and the range of k_i values is 13–42 s^{-1}. The high values of K_{os} and k_f ($\sim 10^4 M^{-1} s^{-1}$; cf. Table 8.2) are the consequence of the unusually high charge product here, $|z_+ z_-| = 9$! Outer-sphere preassociation and complex formation are also important preliminaries in the redox reaction between the amino acid anions $[Fe(CN)_5(aa)]^{n-}$ and horseheart ferricytochrome c.[25] Thermal decomposition of

Scheme 1

$$[Fe(CN)_5(OH_2)]^{3-} + [Co(NH_3)_5(LL)]^{3+} \xrightleftharpoons{K_{os}} [Fe(CN)_5(OH_2)]^{3-}, [Co(NH_3)_5(LL)]^{3+}$$

$$\xrightarrow{k_i} [(NC)_5Fe(\mu\text{-}LL)Co(NH_3)_5] \rightarrow \text{redox products}$$

$[Fe(CN)_5(OH_2)]^{3-}$ can be studied in aqueous solution if air is rigorously excluded [equation (1)]:

$$6[Fe(CN)_5(OH_2)]^{3-} \rightarrow 5[Fe(CN)_6]^{4-} + Fe^{2+} + 6H_2O \qquad (1)$$

The mechanism involves successive loss of the five cyanides from one aquopentacyanoferrate(II) anion, with each released cyanide being scavenged by another $[Fe(CN)_5(OH_2)]^{3-}$ to give $[Fe(CN)_6]^{4-}$.[26]

Irradiation of pentacyanoferrate(II) complexes $[Fe(CN)_5L]^{n-}$ in aqueous solution in the presence of a potential ligand L' normally gives $[Fe(CN)_5L']^{n-}$; cyanide loss to give $[Fe(CN)_4(OH_2)L]^{(n-1)-}$ is a minor or negligible path. But irradiation of $[Fe(CN)_5(en)]^{3-}$ does result in cyanide loss, to give the chelate $[Fe(CN)_4(en)]^{2-}$ with a high quantum yield.[27] Irradiation of the special case of $[Fe(CN)_5L]^{n-}$ where $L = CN^-$, perforce results in cyanide loss. Such loss is reversible in neutral or in alkaline solution; forward and reverse rate constants, quantum yields, and a mechanistic scheme have been reported.[28,29] Kinetics of photoaquation (at 365 nm) in acid solution have also been described.[29] For binuclear complexes $[(NC)_5Fe(\mu\text{-}LL)ML_5]^{n-}$, where $LL = $ pyrazine (**8**) or 4,4'-bipyridyl (**13**) and $ML_5 = Co(CN)_5^{2-}$ $(n = 5)$ or $Rh(NH_3)_5^{3+}$ $(n = 0)$, quantum yields range between 0.1 and 0.6 for iron to μ-LL bond breaking, but are negligible for μ-LL to cobalt or rhodium bond breaking.[30] Comparison of photochemical and thermal reactions of $[Fe(CN)_5(NO_2)]^{4-}$ shows the same major reaction product, $[Fe(CN)_5(OH_2)]^{3-}$, but the mechanisms are thought to differ, with some redox character to the photoaquation mechanism.[31]

Such ligands as the cyanopyridines can coordinate at two different positions to the iron of the $[Fe(CN)_5]^{3-}$ moiety. The kinetics of isomerization (**14**) → (**15**) (and the analogues 3-cyanopyridine reaction) were studied several years ago.[32] In alkaline solution, two linkage isomers are possible with the anions of methionine (**16**) or lysine (**17**), and three with that of histidine (**18**). Kinetics of isomerization processes involving these complexes have been reported and analyzed.[10] Benzotriazole [bt, (**12**)] can also give isomers, bonding to the iron through N-1(3) or N-2, as shown by NMR

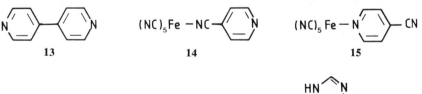

13   14   15

$MeSCH_2CH_2CH(NH_2)CO_2^-$ $H_2N(CH_2)_4CH(NH_2)CO_2^-$

16   17   18

spectroscopy.[22] Rate constants for isomerization at 298.2 K are given in equation (2):

$$[Fe(CN)_5(N\text{-}2\text{-}bt)]^{3-} \underset{0.65\,s^{-1}}{\overset{1.55\,s^{-1}}{\rightleftharpoons}} [Fe(CN)_5(N\text{-}1\text{-}bt)]^{3-} \qquad (2)$$

These are much larger than dissociation rate constants (Table 8.1), so isomerization is deduced to be intramolecular.[19,22]

Leaving-group solvation has already been invoked above as one factor affecting dissociation rates of these pentacyanoferrate(II) complexes. It can be further probed by monitoring rate constants and activation parameters in binary aqueous solvent mixtures, where, in general, leaving of a predominantly hydrophobic ligand is facilitated by increasing the organic content of the solvent.[10,12,13] Such trends are reflected in negative values for Grunwald–Winstein m values, for example, that of -0.3 for 1,2-dimethyl-1,2-diaminoethane as the leaving group.[10] The trend of $\delta_m \Delta G^{\ddagger}$ with G^E values for the solvent mixtures for $[Fe(CN)_5(Me_2en)]^{3-}$ substitution is also consistent with the transition state being more hydrophobic than the initial state.[10] Correlations between kinetic parameters and excess functions for solvent mixtures have also been established (enthalpies and Gibbs functions) for piperidine as the leaving group.[13] All these treatments deal with kinetic parameters that represent *differences* between initial states and transition states. Solvent effects can be understood more fully if they can be separated into initial-state and transition-state components. This was attempted for $[Fe(CN)_5(Me_2en)]^{3-}$. The treatment here was greatly hampered by the lack of thermodynamic data, but the authors were able to deduce, albeit somewhat skeptically, that transition-state solvation was more important than initial-state solvation in determining reactivity.[10] By conducting a series of solubility and calorimetry measurements on appropriate salts of $[Fe(CN)_5(4CNpy)]^{3-}$, as well as kinetic measurements on the substitution of 4-cyanopyridine, in methanol–water mixtures, a complete initial-state/transition-state analysis for this system was possible. The striking result is that the very small effects of solvent composition on rate constants hide large but compensating effects on the initial and transition states, both being markedly destabilized as the methanol content of the medium increases.[15]

A complementary approach to solvent effects is to use rate-constant or activation-parameter trends as probes for solvent structural changes. In recent months this has been attempted for the $[Fe(CN)_5(Me_2en)]^{3-}$ and $[Fe(CN)_5(pip)]^{3-}$ complexes in a range of binary aqueous solvent mixtures.[10,12,13] As is usual, $\Delta H^{\ddagger}$ and $\Delta S^{\ddagger}$ are better indicators than rate constants, but due to their relatively low sensitivity here it is difficult to establish extrema which are significant in relation to the uncertainties in these derived kinetic parameters. Indeed, in relation to the comments at the end of the previous paragraph, it is not surprising that initial-state properties, for instance, solution enthalpies, are better indicators of solvent structural changes than kinetic parameters for this particular class of complexes [contrast nickel(II) complex formation reactions].[15] The small effects of added sodium chloride or of added tetramethylammonium bromide on the dissociation rate constant for $[Fe(CN)_5(ethyl\ nicotinate)]^{3-}$ in aqueous solution have been explained in terms of solvent structural effects.[8] A more interesting aspect of this work is the effect of the $[Fe(CN)_5]^{3-}$ moiety on the rates of hydrolysis of this ester and of methyl nicotinate.[8] These ester hydrolyses are coordinated ligand reactions, as is the reaction of nitroprusside, $[Fe(CN)_5(NO)]^{2-}$, with morpholine. The rate law for this reaction is first order in the complex and second order in morpholine. A 1:1 adduct, in which the morpholine is attached to the nitrosyl-N, is the key transient intermediate, en route to the products $[Fe(CN)_5(morpholine)]^{3-}$ and N-nitrosomorpholine.[33] Reaction (3) is a ligand oxidation, effected in alkaline solution by the action of hexacyanoferrate(III) or simply air, but as far as the iron is concerned the reaction is formally ligand replacement, for the metal is in oxidation state 2+ before and after.[34] It is interesting that it seems to be impossible to get more than one molecule of this simplest diimine ligand onto iron(II), or indeed onto ruthenium(II) (e.g., in $[Ru(en)_2(HN:CHCH:NH)]^{2+}$), when the cation $[Fe(MeN:CHCH:NMe)_3]^{2+}$ is perfectly stable.

$$[Fe(CN)_4(en)]^{2-} \rightarrow [Fe(CN)_4(HN:CHCH:NH)]^{2-} \qquad (3)$$

Although the edition of Coordination Chemistry Reviews dedicated to the memory of W. K. Wilmarth in fact contains no new material on $[Fe(CN)_5L]^{3-}$ complexes, it is relevant to mention it here both since it includes a list of Wilmarth's publications in the $[Fe(CN)_5L]^{3-}$ kinetics field and since it deals with several cobalt(III) and platinum(IV) systems that are very close analogues.[35] It has always seemed surprising that there was no kinetic information on pentacyanoruthenates(II), $[Ru(CN)_5L]^{n-}$. This omission is now compensated for since a straightforward synthetic route to such complexes has now been described.[36]

8.2.2. Iron(II)-Diimine Complexes

8.2.2.1. Oxidation via Dissociation

The first step in reactions of iron(II)-diimine complexes with sluggish oxidants is sometimes rate-determining dissociation. This is the case for the reaction with peroxodiphosphate,[37] and for peroxodisulfate also when strongly electron-withdrawing substituents, such as nitro or sulfonato, make electron transfer relatively difficult, and dissociation relatively easy.[38] Now perbromate has been shown to react with $[Fe(bipy)_3]^{2+}$ and several analogous (substituted) 1,10-phenanthroline complexes by rate-limiting dissociation.[39]

8.2.2.2. Medium Effects and Activation Volumes

Many investigators have studied substitution at iron(II)–diimine complexes in binary aqueous mixed solvents and other investigators in aqueous salt solutions. Some years ago the results of addition of salts and a cosolvent were assessed, for $[Fe(5NO_2phen)_3]^{2+}$ in water, t-butyl alcohol, acetone, dimethyl sulfoxide, and acetonitrile mixtures containing added potassium bromide or tetra-n-butylammonium bromide.[40] Now the effects of added chloride, thiocyanate, and perchlorate on dissociation and racemization rates of $[Fe(phen)_3]^{2+}$ in water–methanol mixtures have been established.[41] The main explanation is in terms of increasing formation of ion pairs as the methanol content of the medium increases, but it is somewhat spoiled by the (unnecessary) assumption of a mechanism involving interchange within the ion pairs. K_{IP} values (molar scale) of 11, 18, and 25 were estimated for perchlorate, chloride, and thiocyanate in 80% (volume) methanol at 298.2 K. These values may be compared with values of 20, 7, and 4 for association between $[Fe(phen)_3]^{2+}$ and iodide,[42] $[Fe(bipy)_3]^{2+}$ and iodide,[43] and $[Fe(phen)_3]^{2+}$ and cyanide[44] in aqueous solution (at 298.2, 298.2, and 283.2 K, respectively).

Reaction of $[Fe(phen)_3]^{2+}$ or $[Fe(5NO_2phen)_3]^{2+}$ with hydroxide is more than 1000 times faster in microemulsions than in water, and reaction with cyanide shows an even greater acceleration; but aquation of these complexes shows an acceleration of only a few times in the microemulsions. The larger accelerations of the hydroxide and cyanide reactions are more likely to be due to activation via desolvation of these anions than simply to electrostatic effects, since hydroxide attack at 2,4–dinitrochlorobenzene is also accelerated by about 1000 times under similar conditions.[45] A much more dramatic change of medium is to the solid state, where it is of interest to note that the mechanisms of racemization of the very similar complexes $[Fe(bipy)_3]^{2+}$ and $[Fe(phen)_3]^{2+}$ appear to differ significantly.[46]

The increasing values of the rate constant (298.2 K) for racemization of the $[Fe(phen)_3]^{2+}$ cation:

water < glycol < methanol < acetone < formamide < dimethylformamide

correlate well with increasing stabilization of the cation $[\delta_m \mu^{\ominus}(Fe(phen)_3^{2+})$ is increasingly negative in the order given]. Thus the transition state must be progressively more stabilized. A greater effect of solvation on the transition state than on the initial state is reasonable, as intramolecular racemization here involves expansion of the initial state on going to the transition state, increasing interaction between the hydrophobic periphery of the complex and the solvent. It is instructive to contrast these trends with those for the hydrophilic $[Cr(ox)_3]^{3-}$ anion.[47] Transition-state effects are also slightly greater than initial-state effects in determining activation enthalpies for attack of hydroxide or cyanide at the $[Fe(phen)_3]^{2+}$ cation in 0–40% methanol. Both are dominated by transfer enthalpies for this cation, which show (cf. $[Fe(CN)_5(4CNpy)]^{3-}$ above) a marked extremum around 20% methanol.[48] The especial importance of ligand–solvent interactions here recalls that established for the nickel(II)–ethane-1,2-diamine complex formation reaction (enthalpies again), except that in this nickel(II) system the initial state dominates.[49]

Several years ago the activation volume for aquation of the $[Fe(phen)_3]^{2+}$ cation was found to be $+15.4$ cm^3 mol^{-1}.[50] Now the measurement of the overall volume change for this reaction, -4.7 cm^3 mol^{-1}, permits the construction of its complete volume profile.[51] Activation volumes for hydroxide and cyanide attack at $[Fe(bipy)_3]^{2+}$ and $[Fe(phen)_3]^{2+}$ in water are also large and positive, lying between $+19$ and $+22$ cm^3 mol^{-1}.[52] These unexpected values ($\Delta V^{\ddagger} \sim -10$ cm^3 mol^{-1} for bimolecular reactions in the gas phase or in inert solvents) can be explained in part by the contribution of the leaving bipy or phen, and in part by desolvation of the heavily hydrated hydroxide or cyanide ion, on formation of the transition state for bimolecular reaction. Further insight into the role of solvation in this type of reaction has now been provided by comparing $\Delta V^{\ddagger}$ trends for hydroxide attack at the Schiff-base complexes $[Fe(sb)_3]^{2+}$ and $[Fe(hxsb)]^{2+}$ [sb = (**19**); hxsb = (**20**)] with transfer chemical potentials for the respective cations over a range of methanol–water solvent mixtures. The activation volume for hydroxide attack as $[Fe(sb)_3]^{2+}$ in water is $+11$ cm^3 mol^{-1}, and *increases* to $+28$ cm^3 in 95% methanol, whereas the activation volume for hydroxide attack at $[Fe(hxsb)]^{2+}$ in water is $+13$ cm^3 mol^{-1} but *decreases* as the per cent methanol increases. The transfer chemical potentials show that the large and very hydrophobic $[Fe(sb)_3]^{2+}$ cation is greatly stabilized on adding

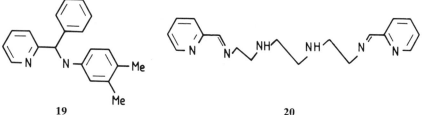

19   20

methanol, with $\delta_m\mu^{\ominus} = -45 \text{ kJ mol}^{-1}$ for transfer from water into methanol, but that the much smaller and less hydrophobic $[\text{Fe(hxsb)}]^{2+}$ cation shows little evidence for selective solvation in methanol–water mixtures. Thus, for the latter complex, $\Delta V^{\ddagger}$ decreases as the methanol content increases since the contribution of hydroxide desolvation to transition-state formation decreases. But for the former complex this trend is overwhelmed by increasing desolvation (methanol this time) of the complex as the methanol content and thus complex solvation increase.[53]

In principle, the transfer chemical potentials recently published[54] for $[\text{Fe(bipy)}_3]^{2+}$ and $[\text{Fe(phen)}_3]^{2+}$ for transfer from water into methanol–water mixtures should be valuable in the interpretation of solvent effects on reactivities of these cations, but unfortunately the authors' derivations are unreliable. There are errors in calculating solubility products and transfer chemical potentials for the salts studied; the transfer chemical potentials for the perchlorate anion are ignored (a serious error in methanol-rich mixtures); and there are transcription or printing errors in their published Table 3 for methanol contents of above 44.7%. Acceptable transfer chemical potentials for the $[\text{Fe(phen)}_3]^{2+}$ cation, for these and other mixed aqueous solvent media, are available from van Meter and Neumann.[55]

8.2.2.3. Reactions of Coordinated Ligands

Hydrazine reacts with cis-$[\text{Fe(bipy)}_2(\text{CNMe})_2]^{2+}$ by a second-order reaction, involving attack by nitrogen at the cyano-carbon as the rate-determining step. The second hydrazine-nitrogen then attacks the other cyano-carbon rapidly to give an "inverted diimine" chelate ring (**21**). Rate constants and Arrhenius parameters for this reaction were reported.[56] The ligand oxidation reaction sequence which converts ethane-1,2-diamine coordinated to iron(II) in the $[\text{Fe(CN)}_4(\text{en})]^{2-}$ anion into the parent diimine

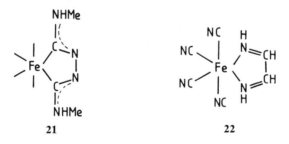

21   22

ligand coordinated to iron(II), in (**22**), has already been mentioned in Section 8.2.1.[34]

8.2.2.4. Covalent Hydration

Gillard's suggestion that nucleophilic attack at metal–diimine complexes may occur initially at the coordinated ligand continues to generate research and much discussion. One recent extensive review[57] is aggressively against mechanisms of this type, and has drawn a succinct retort from Gillard.[58] A more balanced review by Constable[59] deals less extensively with iron than with other metals. The relative lack of justice accorded to iron(II)–diimine complexes in these reviews is disappointing, for surely iron(II) is the key metal center, where expansion of coordination number at the metal is less likely than at ruthenium or at palladium or platinum. Interested readers will find further consideration of evidence for Gillardian intermediates in reactions of iron(II)–diimine complexes in a lengthy paper[60]. The voluminous organic chemistry which underlies these coordinated-ligand attacks has been excellently covered in a recent book written by four of the leading practitioners in this field[61] and in a comprehensive review.[62] There are also some relevant facts in another recent review less centrally concerned with this topic of the intermediacy of σ-anionic species.[63] One paper wholeheartedly in favor of the covalent hydration hypothesis in fact deals with iron(III) complexes of substituted 1,10-phenanthrolines.[64] Another recent paper, on hydroxide attack at $[Fe(bipy)_3]^{3+}$ and at $[Fe(phen)_3]^{3+}$, explores the possibilities of ligand deprotonation as an alternative to covalent hydration,[65] while a detailed kinetic and product study of these reactions implicates some attack by hydroxide at ligand-nitrogen rather than at ligand-carbon![66]

An interesting observation which may be of relevance is the ring-to-oxygen bond formation—as yet an undisclosed mechanism—in the reaction (23) → (24).[67] A diversionary topic connected with covalent hydration is the establishment of bipy C,N bonded to iridium, (25), based on X-ray diffraction studies of the perchlorate and on ^{1}H and ^{13}C NMR spectra of solutions containing the N-protonated form of (25).[68] Such a species has

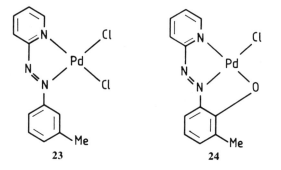

23   24

an interesting analogue or model in (26):[69] Is [Pt(C,N-bipy)$_2$] about to be prepared?

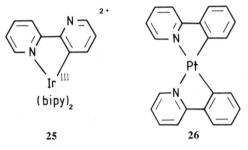

25   26

8.2.3. Other Low-Spin Iron(II) Complexes

The benzoquinonedioxime complexes (27) (L, L' from carbon monoxide, 1-methylimidazole (28), pyridine, benzyl isocyanide) link this section to the previous section, for although they contain two diimine moieties their kinetic and mechanistic interest lies in the axial ligands L and L'. Reactivities here are compared with those for analogous bis-dimethylglyoximatoiron(II) complexes. Rate constants are given for carbon monoxide loss from the complexes with L = CO and L' = 1-methylimidazole or pyridine; Arrhenius parameters are also given for the former ($\Delta H^{\ddagger}$ = 96.5 kJ mol^{-1} and $\Delta S^{\ddagger}$ = +42 J K^{-1} mol^{-1}). Benzyl isocyanide is replaced only very slowly by 1-methylimidazole under thermal conditions and much more rapidly photochemically. These reactions provide yet another set of illustrations of the importance of the *trans* effect in substitution kinetics for iron(II) complexes *trans*-[Fe(LLLL)LL'].[70] The operation of a *D* mechanism for the replacement of acetonitrile by 1-methylimidazole in [Fe(LLLL)(MeCN)$_2$]$^{2+}$, LLLL = (29) or (30) (another diimine-related ligand), has been demonstrated and kinetic parameters obtained (k, $\Delta H^{\ddagger}$, and $\Delta S^{\ddagger}$). The kinetic results refer to replacement of the second MeCN ligand; replacement of the first takes place very quickly. Unexpectedly rapid reaction of the complex of (30) was attributed to the presence of a very small but nonetheless significant concentration of a high-spin form in equilibrium with the predominant low-spin form.[71] Isocyanides RNC (eight compounds studied) generally react more rapidly than carbon monoxide or dioxygen with iron(II) in protohaem mono-3-(1-imidazoyl)-propylamide monomethyl ester (benzene solution or 2% myristyltrimethylammonium bromide in water). This is attributed to the polar nature of RNC. Rate constants vary very little with the nature of R (Me to 'Bu), indicating minimal steric effects.[72] Certain encapsulated iron(II) complexes react with dioxygen in a manner reminiscent of [Fe(CN)$_4$(en)]$^{2-}$ (cf. Sections

8.2.1 and 8.2.2.3), in that ligand oxidation occurs without oxidation of the iron(II). The net change is shown in the partial formulas (**31**) → (**32**).[73]

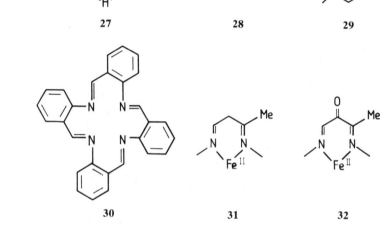

27   28   29

30   31   32

8.2.4. *Iron(III) Complexes*

Several references deal with slow or relatively slow substitution at iron(III), while a few are directly relevant to analogous iron(II) systems discussed in Sections 8.2.1 and 8.2.2. Indeed iron(III)–diimine complexes have already been mentioned in connection with covalent hydration (Section 8.2.2.4).[64-66] Substitution at $[Fe(phen)_3]^{3+}$ is a very much less popular area of study than that of its iron(II) analogue (Section 8.2.2). Dissociation of the iron(III) complex in aqueous acetone is claimed to be first order with respect to $[Fe(phen)_3]^{3+}$, second order with respect to acetone, and reciprocal first order with respect to H^+. This information is derived from observations at 620 nm; a more complicated picture emerged from kinetic studies carried out at 470 nm! The interpretation offered rather conceals the key role of acetone in solvating the leaving 1,10-phenanthroline, though this is mentioned in the final sentence.[74] It is regrettable that the authors do not report their primary experimental results, namely their rate constants, as they thus prevent readers from testing alternative hypotheses and assessing the contributions of initial state, transition state, and ligand solvation.

Rate constants for the forward and reverse reactions in the replacement of *iso*-propanol in $[Fe(LLLL)(^iPrOH)_2]^+$, LLLL = deuteroporphyrin (IX) dimethyl ester, by chloride have been obtained as a byproduct of pulse-radiolysis studies of this complex.[75] The tetraphenylporphyrin complex [Fe(tpp)Cl] is high spin and five coordinate. It reacts quickly with *N*-methylimidazole to give a low-spin product, with low concentrations of added trifluoroethanol or phenol providing marked acceleration by hydrogen bonding to the leaving chloride and thus facilitating its departure from the iron. Solvent effects on rate constants indicate that the rate-determining step is indeed ionization of the chloride from the metal; [Fe(tpp)(*N*-Me imid)Cl] is detectable as a transient intermediate, especially at low temperatures.[76]

Attack of cyanide at polyaminocarboxylatonickel(II) complexes has been much studied kinetically. Now the first report has appeared on the kinetics of cyanide reaction with an analogous iron(III) complex, $[Fe(edta)(OH)]^{2-}$.[77] The reaction sequence involves substitution steps giving $[Fe(CN)_5(OH)]^{3-}$ and then $[Fe(CN)_6]^{3-}$, which finally is reduced by the edta to $[Fe(CN)_6]^{4-}$. The initial reaction between $[Fe(edta)(OH)]^{2-}$ and cyanide has a complicated dependence on cyanide concentration, but the subsequent reaction of $[Fe(CN)_5(OH)]^{3-}$ with cyanide follows simple second-order kinetics. The kinetics of the reverse step, $[Fe(CN)_5(OH)]^{3-}$ with edta, were also examined.[78] Brief irradiation of $[Fe(CN)_5(OH_2)]^{2-}$ in aqueous solution in the presence of 1,10-phenanthroline yields $[Fe(phen)_3]^{3+}$ and $[Fe(CN)_6]^{3-}$. Longer irradiation results in some reduction to iron(II), producing $[Fe(phen)_3]^{2+}$ and $[Fe(phen)(CN)_4]^{2-}$.[79]

Finally, two studies of the kinetics of β-diketone replacement should be mentioned, involving the reactions of $[Fe(acac)_3]$ with tfacH and vice versa,[80] and of $[Fe(bzac)_3]$ with acacH.[81] Solvent effects in the latter system, especially the accelerating effect of *t*-butyl alcohol on the reaction in hexane, suggest a free-radical mechanism.

8.3. Ruthenium

8.3.1. Ruthenium(II)

8.3.1.1. Thermal Substitution

The rate constant for formation of the benzyl isocyanide complex $[Ru(NH_3)_5(BzNC)]^{2+}$ from $[Ru(NH_3)_5(OH_2)]^{2+}$, $0.1\ M^{-1}\ s^{-1}$ at 298.2 K in aqueous solution, is within the normal range for formation reactions from $[Ru(NH_3)_5(OH_2)]^{2+}$. The aquation rate constant for $[Ru(NH_3)_5(BzNC)]^{2+}$ indicates that benzyl isocyanide has about 40 times the *trans*-labilizing effect of ammonia, but comparison of the rate constants for reaction of $[Ru(NH_3)_4(OH_2)(BzNC)]^{2+}$ and of $[Ru(NH_3)_5(OH_2)]^{2+}$ with isonicotinamide indicates that benzyl isocyanide is about 10 times less labilizing than ammonia.[82] The rate constant for aquation of $[Ru(NH_3)_5(dmso)]^{2+}$ is approximately $10\,s^{-1}$ [cf. very much slower aquation of the ruthenium(III) analogue, Section 8.3.2 below] and for aquation of the methionine sulfonate analogue three times faster.[83] Much kinetic information is available for complexes $[Ru(NH_3)_5X]^+$, but none for $[Ru(CN)_5L]^{3-}$. The latter situation is about to be remedied, since a straightforward synthetic route to the latter class of complexes has just been reported.[36]

Kinetic parameters $(k, \Delta H^{\ddagger}, \Delta S^{\ddagger})$ for substitution at *trans*-$[RuL_4XY]$, where X and Y are variously chloride and bromide, are compared with those for analogous ruthenium(III) and cobalt(III) complexes, and greater similarities noted between d^6 ruthenium(II) and d^5 ruthenium(III) than between the two d^6 centers ruthenium(II) and cobalt(III). The ligands L_4 here are $(NH_3)_4$, three bidentate diamines, and five cyclic and garland four-nitrogen ligands.[84] Further qualitative information on the lability of ruthenium(II) complexes of this type is available in a paper primarily concerned with ruthenium(II)/(III) redox behavior, which gives full consideration to a variety of ligand effects on reactivity.[85] Rate constants are reported for reaction of $[Ru(NH_3)_4(N,O\text{-glycinamide})]^{2+}$ with acetonitrile and 4-cyanopyridine.[86] Kinetic parameters $(k, \Delta H^{\ddagger}, \Delta S^{\ddagger})$ for dissociation of axial ligands in ruthenium(II) complexes of phthalocyanine, octaethylporphyrin, and tetraphenylporphyrin have been determined, in toluene solution, and compared with those for iron(II) analogues. The axial ligands involved include benzyl isocyanide, 1-methylimidazole, 4-t-butylpyridine, and carbon monoxide. *Cis* and *trans* effects on reactivity were also discussed. Both for iron and ruthenium, axial ligands in porphyrin complexes are about $20\,kJ\,mol^{-1}$ more labile than in the corresponding phthalocyanine complexes.[87]

The diimine ligands 2-(phenylazo)pyridine and its p-methyl derivative [(**33**); R = H or Me] form bis complexes $[Ru(LL)_2XY]^{n+}$ with ruthenium(II). The complexes with $X = Y = Cl$, Br, or I $(n = 0)$ react with tertiary phosphines by a second-order process with k_2 values apparently controlled by the bulk of the entering ligand.[88] The complexes with $X = py$, $Y = OH_2$, and $X = Y = OH_2$ $(n = 2)$ react quite quickly with donor solvents such as acetonitrile or dimethyl sulfoxide; the bis-aquo complex also reacts rapidly with pyridine.[89] Complexes *cis*-$[Ru(LL)_2X_2]$, when LL = an arylazooxime [(**34**) with R = aryl], readily lose HX on addition of a base such as triethylamine. The proton for the eliminated HX is derived from the ligand LL, which has a proton in close proximity to coordinated X^-.[90]

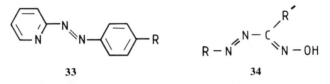

33   34

A preparative study of the *cis* and *trans* isomers of $[Ru(LL)_2Cl_2]$, where LL = diars or diphos, indicates that the *cis* isomers are, as usual, considerably more labile than the *trans* isomers. The *cis* isomers readily undergo sterespecific substitution by iodide or by carbon monoxide.[91] Some qualitative observations on lability of chloride coordinated to ruthenium(II) is also available for several μ-chlorodiruthenium species.[92] The Ru(II)Ru(II) form ($n = 4$) of the μ-dinitrogen series $[(H_3N)_5RuN_2Ru(NH_3)_5]^{n+}$ is very inert; the Ru(II)Ru(III) form ($n = 5$) has an aquation rate constant of $0.024^{(93)}$ or $0.1\ s^{-1}$,[94] while the Ru(III)Ru(III) form ($n = 6$) has an aquation rate constant $\geq 10^3\ s^{-1}$.[93]

Exchange of nitric oxide with $[Ru(NO)_2(PPh_3)_2]$ is fast, being complete in 15 min at room temperature in tetrahydrofuran solution. An associative mechanism is proposed; dissociative exchange of nitric oxide at $[Co(NO)_2(PPh_3)_2]^+$ has a half-life of about 4 h (in dichloromethane). Nitric oxide exchange at $[Fe(NO)_2(PPh_3)_2]$ is much slower and indeed is claimed to occur only in the presence of added triphenylphosphine.[95] Labeling experiments indicate that the reaction of $[Ru(py)_4(NO)Cl]^{2+}$ with azide, to give $[Ru(py)_4(N_2)Cl]^+$, proceeds through a cyclic intermediate.[96]

8.3.1.2. Isomerization

Isomerization of N-bonded ligands to C-bonded ligands in $[Ru(NH_3)_5L]^+$, L = imidazole, N-methylimidazole, 4,5-dimethylimidazole, or histidine, is acid catalyzed and accompanied by aquation.[97] The rate constants for linkage isomerization (O bonded to S bonded) in $[Ru(NH_3)_5(dmso)]^{2+}$ is $30\ s^{-1}$ at 298.2 K.[83] Slow epimerization has been reported for *cis*-$[Ru(CO)(LL)(dmso)Cl_2]$.[91] Isomerization of the NN'-glycinamide form of $[Ru(NH_3)_4(glycinamide)]^{2+}$ to the NO-bonded isomer has a half-life of only about 0.2 s in aqueous solution at room temperature.[86]

8.3.1.3. Photochemistry

$[Ru(bipy)_3]^{2+}$ continues to attract more attention than any other ruthenium(II) complex. A recent review is mainly concerned with redox processes, but does devote five pages to photosubstitution and photo-racemization of this complex.[98] The likelihood of separate excited states for photosubstitution and luminescence has been established,[99] and electronic structural models of $[Ru(bipy)_3]^{2+}$ and its osmium and iron analogues presented and discussed.[100] Photosubstitution mechanisms at $[Ru(bipy)_3]^{2+}$ in water and in dichloromethane have been compared; the relatively low reactivity of this complex in photosubstitution is ascribed to the ease of reclosing the chelate ring in $[Ru(bipy)_2(\text{monodentate bipy})]^{2+}$ rather than to any particularly great difficulty in breaking the first ruthenium–nitrogen bond.[101] A more extensive examination of solvent effects on the photochemistry of $[Ru(bipy)_3]^{2+}$ assesses radiative and nonradiative components of excited-state decay, though dealing mainly with charge transfer excited states.[102] Added silver(I) assists photochemical substitution at $[Ru(bipy)_3]^{2+}$ in acetonitrile. The reaction sequence involves a d–d excited state which decays to a ligand-labilized intermediate. The Ag^+ scavenges $[Ru(bipy)_2(\text{monodentate bipy})]^{2+}$ or $[Ru(bipy)_2(\text{monodentatebipy})\text{-}(MeCN)]^{2+}$. This promotion by Ag^+ is compared with anion-assisted photosubstitution, for example, by thiocyanate.[103] Direct attack by hydroxide may be involved in photodecomposition of $[Ru(bipy)_3]^{2+}$ in 0.3 M NaOH, where the quantum yield is 3.3×10^{-4}, but the main interest in this system appears to be its redox aspect[104] (cf. $[Ru(bipy)_3]^{3+}$ plus hydroxide below).

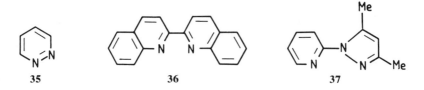

35   36   37

A photophysical study of *cis*-$[Ru(bipy)_2L_2]^{2+}$, where L = pyridine, pyridazine (**35**), *N*-methylimidazole (**28**), and related ligands is of relevance to photosubstitution at ruthenium(II) as well as to photoredox processes.[105] Irradiation of an aqueous solution of $[Ru(bipy)_2Cl(NO)]^{2+}$ at pH 5 leads to addition of OH^- to the nitrosyl ligand, albeit with a low quantum yield.[106] The mechanism of ligand substitution in $[Ru(pc)(py)_2]$ depends on the solvent (whether this is a potential ligand or not); substitution here is thought to be photoredox induced, via an $Ru(I)(pc\cdot)$ species.[107] Photochemical studies on substitution in related diimine complexes include those at $[Ru(LL)_3]^{2+}$, where LL = biquinolyl (**36**),[108] or the pyrazole derivative (**37**).[109] Both complexes are considerably more photolabile than $[Ru(bipy)_3]^{2+}$; the photodissociation excited state of the biquinolyl complex is easily populated at room temperature. On the other hand, the dinuclear complex $[(H_3N)_4Ru(bipym)Ru(NH_3)_4]^{4+}$, bipym = bipyrimidine (**38**), shows a very low photoreactivity on irradiation in the visible region.[110] Relative quantum yields for ruthenium(II)–nitrogen and rhodium(III)–nitrogen bond breaking in $[(H_3N)_5Ru(\mu\text{-}LL)Rh(NH_3)_5]^{5+}$ depend on LL, with Rh–N bond breaking the dominant pathway when LL = 4-cyanopyridine, Ru–N when LL = pyrazine (8), and both comparable when LL = 4,4'-bipyridyl (13).[30]

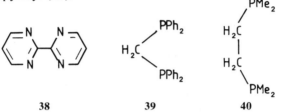

38   39   40

Irradiation in the ligand field bands of the diphosphine complexes $[Ru(LL)_2Cl_2]$, where LL = dppm, (39),[111] or dmpe, (40),[112] gives *trans* products both from *trans-* and *cis*-starting materials. In methanol or ethanol, *cis*-$[Ru(dmpe)_2Cl_2]$ gives *trans*-$[Ru(dmpe)_2Cl_2]$; in alcohol–water mixtures *trans*-$[Ru(dmpe)_2(OH_2)Cl]^+$ is produced; and in dimethyl sulfoxide *trans*-$[Ru(dmpe)_2(dmso)Cl]^+$ is produced. These results are all consistent with the intermediacy of a transient five-coordinate (square pyramidal) photo-intermediate $[Ru(dmpe)_2Cl]^+$, wherein the chloride ligand prefers to be apical.[112]

Photosubstitution at ruthenium(II)–bipyridyl complexes has been described for the polymer-bound complex $[Ru_n(bipy)_{2n}(pvp)_2]^{2n+}$, where pvp is poly(4-vinylpyridine). Reactivity is lower here than for equivalent monomeric complexes.[113] Photoisomerization (*cis* ⇌ *trans*) of $[Ru(bipy)_2(OH_2)_2]^{2+}$ has been studied at the surface of hectorite, a layer silicate, and the results compared with those for aqueous solution.[114]

8.3.1.4. Reactions of Coordinated Ligands

Stereoselective deuterium exchange at the coordinated amino group of L-alanine in $[Ru(bipy)_2(\text{L-ala})]^+$, monitored by NMR spectroscopy, reflects sterically induced asymmetry of solvation of this amino group.[115]

$[Ru(NH_3)_5(OH_2)]^{2+}$, generated *in situ* from $[Ru(NH_3)_5Cl]^+$, reacts with α-substituted ketoximes to give an acetonitrile complex and a hydroxyl derivative of the α-substituent, for example,

$$[Ru(NH_3)_5(OH_2)]^{2+} + HON{=}C\begin{array}{l} \diagup CH_3 \\ \diagdown COC_6H_5 \end{array} \rightarrow$$

$$[Ru(NH_3)_5(MeCN)]^{2+} + HO_2CC_6H_5 + H_2O \qquad (4)$$

Presumably here, as elsewhere, the ruthenium(II) is promoting hydrolysis of the coordinated ketoxime ligand. A lower limit of about 0.02 s^{-1} was suggested for cleavage of the $C-C$ bond in the intermediate complex.[116]

At pH ~ 5, hydroxide adds photochemically to NO coordinated to ruthenium in $[Ru(bipy)_2Cl(NO)]^{2+}$; the $-NO_2H$ ligand reverts to $-NO$ in the dark, following first-order kinetics.[106] In acid solution, $-ONSO_3$ is formed reversibly via attack of sulfite at coordinated $-NO$ in $[Ru(bipy)_2Cl(NO)]^{2+}$, behavior similar to that of nitroprusside. In neutral solution this reaction is not fully reversible, while in aqueous ammonia $-ONSO_3$ is converted into $-NO_2$, by attack of hydroxide at the nitrogen of the $-ONSO_3$. In fact the product of this last reaction is $[Ru(bipy)_2(NH_3)-(NO_2)]^+$, since the ammonia also displaces chloride from the original ruthenium(II) complex.[117]

Reaction of cis-$[PtCl_2(C_2H_4)(py)]$ with cis-$[Ru(bipy)_2(CN)_2]$ is first order in each compound, with a k_2 value of $10.6 \ M^{-1} \text{ s}^{-1}$ at room temperature in dimethylformamide solution. There is spectroscopic evidence for 1:1 and 1:2 binuclear products.[118] Two coalescence temperatures have been reported from variable-temperature NMR studies of the dithiahexane complex $[Ru(dth)_2Cl_2]$. Simple inversion at sulfur, a sufficient explanation for variable-temperature NMR spectra of, for instance, $[M(dth)Cl_4]^{n-}$ with $M = Rh(III)$, $Ir(III)$, $Pt(IV)$ (q.v.), is clearly not a sufficient explanation here.[119]

8.3.2. *Ruthenium(III)*

The rate constant for aquation of $[Ru(NH_3)_5(OSO_2CF_3)]^{2+}$ is $9.3 \times 10^{-2} \text{ s}^{-1}$ at 298 K[120] and for $[Ru(NH_3)_5(dmso)]^{3+}$ is $7.7 \times 10^{-5} \text{ s}^{-1}$,[83] the big difference reflecting the ease of departure of the good leaving-group $-OSO_2CF_3$. The rate constant for linkage isomerization, S-bonded to O-bonded dimethyl sulfoxide, in $[Ru(NH_3)_5(dmso)]^{3+}$ is $7 \times 10^{-2} \text{ s}^{-1}$, slightly faster for the analogous methionine sulfonate complex.[83] At pH 3, complexes $[Ru(NH_3)_5(LL)]^{3+}$, where LL, for example, = glycinamide, glycylglycine, or glycylglycinamide, acting as monodentate ligands, give $[Ru(NH_3)_4(LL)]^{3+}$ in which the ligands are N, O bonded. At higher acidities there is concurrent aquation, which can be catalyzed by the addition of ruthenium(II). Ruthenium(II) can also catalyze isomerization of $[Ru(NH_3)_4(NO\text{-bonded } LL)]^{3+}$ to the NN-bonded form.[121] The $trans$-$[Ru(NH_3)_4(OH_2)L]^{3+}$ cation, where L = C-bonded dimethylimidazole, reacts very reluctantly with chloride, pyridine, or imidazole, but fairly rapidly with thiocyanate. The rate law for this last reaction shows a complicated dependence on thiocyanate concentration, suggesting parallel paths with mono- and bis-thiocyanate outer-sphere intermediates.[97]

Kinetic parameters $(k, \Delta H^{\ddagger}, \Delta S^{\ddagger})$ are reported for reaction of *trans*-$[Ru(tet-a)Cl_2]^+$ with bromide and for *trans*-$[Ru(tet-a)Br_2]^+$ and its tet-b analogue [tet-a = (**41**); tet-b = (**42**)] with chloride. Comparison of steric effects here with those found in reactions of analogous cyclam (**43**) complexes suggests the operation of dissociative mechanisms for aquation and for base hydrolysis of all these complexes.[122] Steric effects have also been held to support a dissociative mechanism for substitution at $[Ru(NH_3)_4X_2]^+$, $[Ru(en)_2X_2]^+$, and their ruthenium(II) analogues[84] (cf. Section 8.3.1.1). Irradiation of complexes *trans*-$[RuL_4X_2]^+$, where $L_4 = (en)_2$, $(pn)_2$, tet-a, tet-b, or one of four N_4 macrocyclic ligands, at their lowest-energy CTTM band gives stereoretentive aquation. Irradiation at the second CTTM band also results in aquation, but now with some stereochemical change. The results are discussed in terms of a dissociative mechanism involving a quartet ligand field state[123] [cf. ruthenium(II), Section 8.3.1.3). Correlation diagrams for d^5 complexes of this type have been constructed and shown to differ significantly from those for d^6 centers.[124] The emerging pattern for ligand field excitation for d^5 ruthenium(III) complexes has been compared with the patterns established for d^3 and d^6 centers, with special features specific to d^5 photosubstitution processes identified.[124,125] Various mechanistic pathways have been considered for reaction of $[Ru(bipy)_3]^{3+}$ and $[Ru(phen)_3]^{3+}$ with hydroxide under photochemical conditions, with particular reference to the chemiluminescence perceived in the former system.[65]

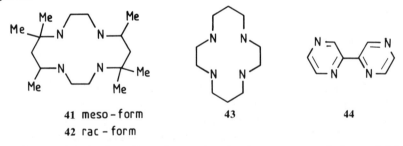

41 meso-form
42 rac-form   **43**   **44**

Electrochemical studies of several ruthenium compounds have yielded qualitative information on reactivities with respect to substitution. Indeed studies of the reversible one-electron reduction of compounds $[RuL_3Cl_3]$, with L, for example, = pyridine or tolunitrile, give information both on substitution and on isomerization. A repeat scan for *mer*-$[Ru(m$-$MeC_6H_4CN)_3Cl_3]$ going to *fac*-$[Ru(m$-$MeC_6H_4CN)_2(MeOH)Cl_3]$ plus released ligand is reproduced.[126] Ligand-exchange reactions involving tris-halogenoruthenium(III) complexes have also been investigated electrochemically in heterogeneous (solution/solid) systems.[127] Electrochemical oxidation behavior of $[Ru(bpz)_3]^{2+}$ in acetonitrile indicates fairly rapid solvolysis of the oxidation product $[Ru(bpz)_3]^{3+}$ [bpz = 2,2'-bipyrazine, (**44**)].[128] The complex $[Ru(bipy)_3]^{4+}$ can be generated electrochemically in liquid sulfur dioxide. This cation apparently contains ruthenium(III) rather than ruthenium(IV), with a bipy$^{+\cdot}$ radical cation ligand.[129] Ligand displacement from ruthenium truly in oxidation state IV is involved in heterogeneous reactions of Ru(IV) ternary oxides.[130]

The dinuclear complex $[(H_3N)_5Ru(\mu\text{-}N_2)Ru(NH_3)_5]^{6+}$ has an aquation rate constant $\geq 10^3 \, s^{-1}$ at room temperature, much faster than that for $[Ru(NH_3)_5(N_2)]^{3+}$.[93] The tetranuclear cation $[Ru_4(OH)_4]^{8+}$ undergoes depolymerization, presumably giving initially $[Ru_2(OH)_2]^{4+}$, much more slowly. At 298 K the depolymerization rate constant is $2.5 \times 10^{-3} \, s^{-1}$ ($E_A = 63 \, kJ \, mol^{-1}$).[131] Linkage isomerization of complexes $[Ru(NCS)_n\text{-}(SCN)_{6-n}]^{3-}$ ($n = 1$–4), produced by reacting ruthenium trichloride with potassium thiocyanate in aqueous solution, under various conditions has been discussed in qualitative terms.[132]

One of the most dramatic examples of catalysis of organic reactions by coordinating the substrate to a metal center was provided by the 10^8-fold acceleration of base hydrolysis of acetonitrile when coordinated to ruthenium(III) in $[Ru(NH_3)_5(MeCN)]^{3+}$.[133] Now the very much less dramatic acceleration of base-independent hydrolysis by the $[Ru(NH_3)_5]^{3+}$ moiety has been examined, for benzonitrile and 1-adamantyl cyanide as well as for acetonitrile. Rate constants and activation parameters $\Delta H^\ddagger$ and $\Delta S^\ddagger$ were determined. They are similar for all three compounds, indicating a close similarity of mechanism.[134]

8.4. Osmium

Photochemical reactions involving the osmium(II) complex $[Os(bipy)_3]^{2+}$ have been discussed in relation to electronic structural models for this and related d^6 complexes $[Ru(bipy)_3]^{2+}$ and $[Fe(bipy)_3]^{2+}$.[100] The mechanism of base hydrolysis of the osmium(III) equivalent $[Os(bipy)_3]^{3+}$ has been discussed, including consideration of ligand deprotonation as an initial step.[65] The dinuclear osmium(III) complex $[(H_3N)_5Os(\mu\text{-}N_2)\text{-}Os(NH_3)_5]^{6+}$ aquates readily, whereas the Os(III)Os(II) mixed-valence cation is indefinitely stable in water. This contrasts with the ruthenium situation, where the mixed-valence $[(H_3N)_5Ru(\mu\text{-}N_2)Ru(NH_3)_5]^{5+}$ is relatively labile.[93]

The complete reaction scheme for the stepwise conversion of the osmium(IV) complex $[Os(ox)I_4]^{2-}$ into $[Os(ox)Br_4]^{2-}$ has been elucidated, with product isomer ratios given for every stage where relevant. All seven possible intermediates were detected and characterized by ion exchange chromatography.[135] Reactions of the tetrakis(NN'-diethylimidazoline-2-thione)trans-dioxo-osmium(VI) cation, $[OsL_4O_2]^{2+}$ where L = (45) with R = R′ = Et and Y = S, with imidazolidine-2-selones, (45) with R,R′ = H,Me, Et and Y = Se, proceed by a fast substitution step followed by slower redox reactions, eventually to yield osmium(III) products.[136] A few qualitative observations on reactivities of related osmium(VI) complexes trans-$[OsO_2X_2(PPh_3)_2]$, where X = Cl or Br, have also been published.[137]

45

8.5 Rhodium

The organization of this section follows as closely as possible that of the sections on rhodium in the two previous volumes, though a few differences are dictated by changing emphasis on various areas of kinetics and mechanisms.

8.5.1. Aquation

In acid solution both O- and N-bonded isomers of $[Rh(NH_3)_5(urea)]^{3+}$ interconvert and aquate very slowly, giving $[Rh(NH_3)_5(OH_2)]^{3+}$ and $[Rh(NH_3)_6]^{3+}$. Production of the latter involves hydrolysis of coordinated N-bonded urea (cf. Section 8.5.10).[138] The rate constant for the aquation of $[Rh(NH_3)_5(OSO_2CF_3)]^{2+}$ is $0.019\,s^{-1}$ at 298 K.[120] Here, as for ruthenium(III), $OSO_2CF_3^-$ is an excellent leaving group. Aquation of trans-$[Rh(NH_3)_4(CN)Cl]^+$ is considerably slower ($k_{aq} = 3.6 \times 10^{-4}\,s^{-1}$ at 298 K), but is still 500 times faster than that of $[Rh(NH_3)_5Cl]^{2+}$.[139] Kinetic parameters have been established for loss of the first and second nitro-ligands from cis-$[Rh(biguanide)_2(NO_2)_2]^+$. They suggest an associative mechanism, favored by the electron-accepting nature of the nitro-ligands and consequent lower electron density in the vicinity of the rhodium center.[140] Kinetics and mechanisms of aquation, that is decarboxylation, of carbonato-rhodium(III) complexes have been reviewed and compared with those for cobalt(III) and chromium(III) analogues.[141]

Electrochemical reduction of $[Rh(bipy)_3]^{3+}$ gives $[Rh(bipy)_3]^{2+}$ and then $[Rh(bipy)_3]^+$. This last complex, octahedral d^8, readily and rapidly loses bipy ($k = 5 \times 10^4\,s^{-1}$ in water at 298 K).[142]

8.5.2. Base Hydrolysis

Base hydrolysis of $[Rh(NH_3)_5(NCS)]^{2+}$ has been studied in aqueous alcohol (10–30% by volume ethanol) over the rather restricted temperature range 323–333 K. The dependence of rate constants on solvent permittivity and of $\Delta H^{\ddagger}$ on solvent composition, the observed correlation of $\Delta H^{\ddagger}$ with $\Delta S^{\ddagger}$, and the relation between these results and those for analogous cobalt(III) and chromium(III) complexes are held to indicate a common mechanism. This is thought to be an ion-pair rather than a conjugate-base mechanism. It is suggested that the rhodium and cobalt complexes react by a somewhat more associative mechanism than the chromium(III) complexes.[143] These conclusions are somewhat surprising in some ways, but no one will deny the important role ascribed to solvation of the leaving thiocyanate group in determining the dependence of rate constants on ethanol content. Rate constants have been obtained for base hydrolysis of the O- and N-bonded isomers of $[Rh(NH_3)_5(urea)]^{3+}$, in aqueous solution.[138]

8.5.3. Catalyzed Aquation

The dependence of observed first-order rate constants for mercury(II)-catalyzed aquation of *trans*-$[Rh(en)_2Cl_2]^+$ on mercury(II) concentration [mercury(II) in large excess] indicates a two-stage mechanism in aqueous methanol as in water, as shown in equation (5):

$$trans\text{-}[Rh(en)_2Cl_2]^+ + Hg^{2+} \overset{K}{\rightleftharpoons} [Rh(en)_2Cl(ClHg)]^{3+}$$

$$\overset{k}{\longrightarrow} [Rh(en)_2(OH_2)Cl]^{2+} + HgCl^+ \quad (5)$$

Medium effects (methanol/water) on K and k have been established at 298.2 K.[144] Wells's single-ion assumptions were used in deriving transfer chemical potentials for reactants and the intermediate. The recent appearance of a full set of transfer chemical potentials for simple and complex ions for methanol/water,[145] supplementing earlier sparse data,[146] on the $Ph_4P^+(Ph_4As^+) = BPh_4^-$ assumption (TATB) permits a more satisfactory analysis of this *trans*-$[Rh(en)_2Cl_2]^+/Hg^{2+}$ reaction. This is shown in Figure 8.1, where comparison is made with the formally very similar redox reaction between $[Co(NH_3)_5Cl]^{2+}$ and Fe^{2+} [147] (these results also converted to the TATB assumption).

There are marked divergences between transfer chemical potential trends according to the Wells and TATB assumptions, and the conclusions

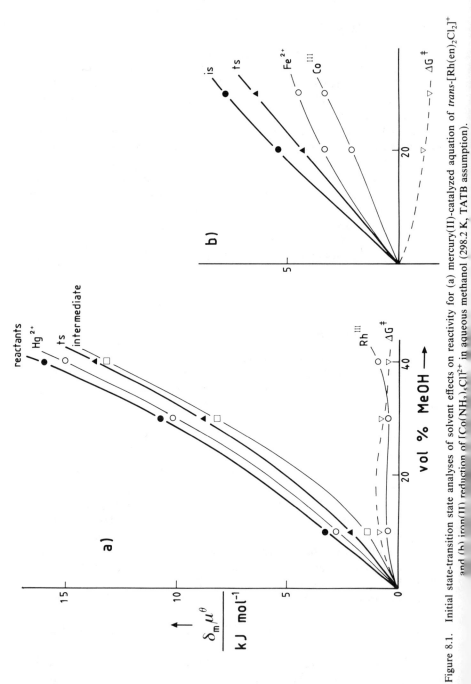

Figure 8.1.   Initial state–transition state analyses of solvent effects on reactivity for (a) mercury(II)-catalyzed aquation of *trans*-[Rh(en)$_2$Cl$_2$]$^+$ and (b) iron(II) reduction of [Co(NH$_3$)$_5$Cl]$^{2+}$ in aqueous methanol (298.2 K, TATB assumption).

of initial-state/transition-state analyses of solvent effects on reactivities can often, as here, depend critically on the assumption used. Thus for the $[Co(NH_3)_5Cl]^{2+}/Fe^{2+}$ reaction, use of the TATB or $K^+ = Cl^-$ assumptions leads to the conclusion that addition of methanol destabilizes the Fe^{2+}_{aq} more than the $[Co(NH_3)_5Cl]^{2+}$, and that the observed increase in rate constant is due to somewhat less destabilization of the $[(H_3N)_5CoClFe]^{4+\ddagger}$ transition state than of the initial state. Wells's assumptions lead to an altogether more complicated (though not therefore necessarily less correct!) pattern, in which, inter alia, the transition state is first destabilized then stabilized as the methanol content increases (contrast Figure 8.1b). Similarly, the complex pattern which follows from Wells's assumption for the *trans*-$[Rh(en)_2Cl_2]^+/Hg^{2+}$ reaction becomes simpler and markedly different on converting to the TATB basis. Figure 8.1a shows the dominant role of Hg^{2+} destabilization in determining transfer chemical potential trends for the initial state, dinuclear intermediate, and transition state—the small variation in rate constant results from almost complete compensation.

8.5.4. Formation

Kinetic and volume measurements have permitted the establishment of complete volume profiles for formation, and for decarboxylation, of $[M(NH_3)_5(CO_3)]^+$, M = Co and Ir as well as Rh. The volume profiles are compared in Figure 8.2, which shows their similarities; volume changes are biggest for the smallest metal center, Co(III). The partial molar volumes of the transition states are intermediate between reactants and products, lying slightly closer to the hydroxo complex plus carbon dioxide side.[148] Kinetics and mechanisms of formation of carbonatorhodium(III) complexes have been reviewed.[141]

The activation volume for reaction of the *meso*-tetrakis(*p*-sulfonatophenyl)porphyrin complex $[Rh(tpps)(OH_2)_2]^{3-}$ with thiocyanate is 8.8 cm^3 mol^{-1} (at 288 K), indicating a dissociative mechanism.[149]

8.5.5. Solvent Exchange

The determination of partial molar volumes for $[M(NH_3)_5(OH_2)]^{3+}$, M = Co, Rh, Ir, and Cr, permits an initial-state/transition-state split of the effects of the nature of the metal ion on activation volumes for water exchange.[150,151] The initial state dominates, which is surprising for a series in which the transition state is believed to range from associative to dissociative. In other words, the partial molar volume of the transition state appears

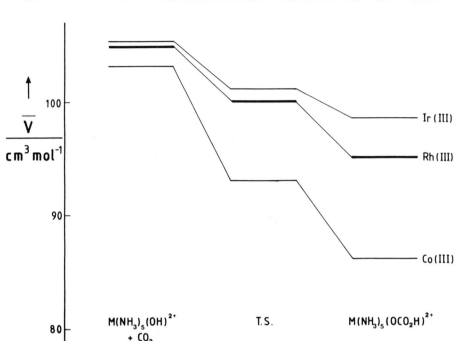

Figure 8.2. Volume profiles for reaction of $[M(NH_3)_5(OH)]^{2+}$ complex with carbon monoxide in aqueous solution at 298 K for M = cobalt(III), rhodium(III), and iridium(III).

remarkably insensitive to whether it is $[M(NH_3)_5(OH_2)_2]^{3+\ddagger}$ or $[M(NH_3)_5]^{3+\ddagger}$! Parallel conclusions emerge from a parallel analysis for the hexaaquoions $[M(OH_2)_6]^{2+}$, M = Mn, Fe, Co, or Ni,[150,151] and $[M(OH_2)_6]^{3+}$, M = Cr or Rh[151], where again mechanisms of water exchange are thought to span the range from associative to dissociative.

8.5.6. Ligand Replacement

There has been a preliminary report of stereochemical inversion in the reaction of $[Rh(LLLL)Cl_2]^-$, where LLLL = the optically active tetradentate ligand ethane-1,2-diamine-NN'-di-S-α-propionate (**46**), with amino acids,[152] and a superficial examination of the kinetics of reaction of rhodium trichloride solution with α-benzil monoxime in the presence of added acetate.[153] The mechanism of the latter substitution was discussed in terms of competition between acetate and the oxime for $[Rh(OH_2)_2Cl_4]^-$ and $[Rh(OH_2)Cl_5]^{2-}$.

$$^-O_2CCHNHCH_2CH_2NHCH-CO_2^-$$

Me   Me

46

8.5.7. Ring Opening and Closing

Solvent effects on the kinetics of ring closure and ring opening have been studied for reaction (6),

$$mer\text{-}[Rh(LL)LCl_2X]^+ \underset{k_b}{\overset{k_f}{\rightleftharpoons}} trans\text{-}[Rh(LL)_2Cl_2]^+ + X \qquad (6)$$

where $X = SCN^-$, $SeCN^-$, N_3^-, NO_2^-, or pyridine, and LL and L represent the ligand 1-dimethylamino-2-dimethylarsinobenzene, (**47**), in its bidentate (N,As bonded) and monodentate (As bonded) forms. The solvents used included methanol, ethanol, 1-propanol, 2-propanol, ethane-1,2-diol, and 2-methoxyethanol. For X = pyridine, neither k_f nor k_b are much affected by the solvent nature, but when X is anionic k_f is particularly affected by the solvent. Grunwald–Winstein and Kamlet–Taft analyses of these solvent effects support the operation of a dissociative mechanism for ring closure, with solvation of the charged leaving group having a major effect on reactivity trends. Change of solvent has an increasing effect on reactivity in the order $X = N_3^- \sim NCS^- < I^- < Br^- < Cl^-$ (Grunwald–Winstein m values are 0.88, 0.88, 0.95, 1.19, and 1.24 respectively), which is just that expected from the solvation characteristics of these anions.[154]

47

Ring closure of monodentate carbonatorhodium(III) complexes is dealt with in a general review of formation and dissociation of this class of complexes.[141]

8.5.8. Isomerization

In neutral or basic solution, the O-bonded isomer of $[Rh(NH_3)_5(urea)]^{3+}$ isomerizes to the N-bonded form and undergoes hydrolysis in parallel reaction pathways. Interconversion of the O-bonded and N-bonded isomers is also an important feature of hydrolysis in acid solution.[138]

8.5.9. Photochemistry

References in this section are arranged according to ligand type, starting with pentammine complexes and continuing through tetrammines and bis-ethane-1,2-diamine complexes to complexes containing N_4 macrocycles. A solitary tris-β-diketonate reference then precedes the final citations of relevant review articles.

Activation volumes (Table 8.3), reaction profiles, quantum yields, and pressure dependences of luminescence lifetimes have all been established for ligand field photoaquation of $[Rh(NH_3)_5X]^{2+}$ and $[Rh(ND_3)_5X]^{2+}$, X = Cl or Br. All the information is consistent with rate-determining dissociation; the difference in sign between $\Delta V^{\ddagger}$ values for X^- loss and for NH_3 loss may be ascribed to the large contribution from leaving-group solvation effects in the former case. The H/D isotope effect is, expectedly, proportionately larger for NH_3 (ND_3) loss than for halide loss, but deuteration does have a significant effect even on halide loss.[155] Photoaquation of $[Rh(NH_3)_5I]^{2+}$ is slower in zeolite cavities than in aqueous solution, presumably because of the lower availability of water under the former conditions.[156] Specifically equatorial photolabilization of ammonia in $[Rh(NH_3)_5(CN)]^{2+}$ has been demonstrated by using a sample containing $^{15}NH_3$ *trans* to the cyanide (readily synthesized since *trans*-$[Rh(NH_3)_4Cl(CN)]^+$ undergoes stereoretentive thermal substitution).[157] Both $[Rh(NH_3)_5(CN)]^{2+}$ and *trans*-$[Rh(NH_3)_4(OH_2)(CN)]^{2+}$ give *cis*-$[Rh(NH_3)_4(OH_2)(CN)]^{2+}$ on ligand field irradiation, but the quantum yield is much higher for the isomerization than for the aquation.[139] Irradiation of *cis*- and *trans*-$[Rh(NH_3)_4(OH_2)Cl]^{2+}$ results in both water exchange and isomerization. Quantum yields for these individual reactions, as well as for aquation of *cis*-$[Rh(NH_3)_4Cl_2]^+$ to *cis*- and *trans*-chloroaquo products, have been established.[158] Time-resolved luminescence studies on aqueous solutions of $[Rh(NH_3)_5Br]^{2+}$ and the *cis* and *trans* isomers of $[Rh(NH_3)_4X_2]^+$ and $[Rh(NH_3)_4(OH_2)X]^{2+}$ give important background information relevant to the mechanisms of photoaquation.[159]

Table 8.3. *Activation Volumes ($cm^3\ mol^{-1}$) for Photoaquation of $[Rh(NH_3)_5X]^{2+}$ and Deuterated Analogues, in Water at 298 K*

		NH_3 complex	ND_3 complex
Photoaquation: X^- loss	X = Cl	−5.2	−4.2
	X = Br	−10.3	−9.4
Photoaquation: NH_3 loss	X = Cl	12.7	9.5
	X = Br	4.6	3.4

Table 8.4. *Volumes of Reaction and Activation* $(cm^3\ mol^{-1})$ *for Photochemical Reactions of Rhodium(III) Complexes* $[Rh(NH_3)_4X_2]^+$, *in Aqueous Solution at 298 K (from Reference 160)*

Reaction	Complex	Main product	$\Delta V^{\ominus}$	$\Delta V^{\ddagger\ a}$
Aquation	cis-Cl	trans-(OH_2)Cl	2.2	-3.5
	cis-Br^b	trans-(OH_2)Br	6.4	-2.3
	trans-Cl	trans-(OH_2)Cl	-18.3	2.8
	trans-Br	trans-(OH_2)Br	-10.3	2.9
	trans-(OH)Cl	cis-$(OH)_2$		-8.8
Isomerization	cis-(OH_2)Cl	trans-(OH_2)Cl	2.5	0
	cis-(OH_2)Br	trans-(OH_2)Br	0.5	-1.0

^{a} Estimated activation volumes for a dissociative mechanism are also tabulated in Reference 160.
^{b} Data for parallel ammonia loss are also given in Reference 160.

Activation and reaction volumes (Table 8.4) and pressure dependences of quantum yields for isomerization and substitution reactions of *cis*- and *trans*-$[Rh(NH_3)_4XY]^{n+}$, with X and Y variously from Cl, Br, OH_2, are, as for $[Rh(NH_3)_5X]^{2+}$ complexes (Table 8.3), consistent with dissociative activation. The excited state is square-pyramidal $[Rh(NH_3)_4X]^{n+}$ with X apical.[160] Quantum yields have been determined for (stereoretentive) photoaquation of a series of complexes *trans*-$[RhL_4Cl_2]^+$, where L = a heterocyclic amine, such as pyrazine (8) or a picoline. The relative quantum yields for chloride loss and for heterocyclic ligand (L) loss vary with the nature of L, with a marked correlation with ligand pK_a values. Relatively little of the aquation goes by chloride loss here, in contrast to ammine analogues.[161]

Cis-$[Rh(en)_2(OH_2)_2]^{3+}$ photoisomerizes to the *trans* isomer, which is photoinert, but *trans*-$[Rh(en)_2(OH_2)(OH)]^{2+}$ photoisomerizes to its *cis* isomer. Both *cis*- and *trans*-$[Rh(en)_2(OH)_2]^+$ are photoinert. Differences between the behavior of these $(en)_2$ complexes and their $(NH_3)_4$ analogues are ascribed to the stronger σ-bonding of the en ligands.[162]

The strongest $d-d$ phosphorescence yet observed has been reported for the *trans*-$[Rh(cyclam)(CN)_2]^+$ cation [cyclam = (43)]. The emission lifetime here is 1000 times longer than for analogous tetrammine complexes. *Trans*-$[Rh(cyclam)(CN)_2]^+$ is photoinert at $d-d$ frequencies. The reason underlying the cyclam vs. tetrammine differences lies in the relative energies of the ^3E_g and $^3A_{2g}$ terms which derive from the $^3T_{1g}$ excited state as $O_h \rightarrow D_{4h}$. When the ^3E_g state is lower, the axial ligands are easily lost. However, when the $^3A_{2g}$ state is lower, as it is in *trans*-$[Rh(cyclam)(CN)_2]^+$, prompted by the particularly strong σ-donor and π-acceptor properties of the cyanide, then the equatorial ligands are, in principle, labilized. But here the macrocycle cyclam occupies all the equatorial positions and thus ligand loss is rendered extremely difficult from equatorial *and* from axial positions.[163]

Continuous, flash, and laser flash photolyses of phthalocyanine complexes [Rh(pc)(MeOH)X], with X = Cl, Br, or I, show that replacement of halide by methanol is a redox-induced process, via rhodium(III)-pc˙ and rhodium(II) species.[164] This may be compared with copper(II) analogues, where redox-induced substitution involves copper(III)-pc complexes.[165] Axial ligand substitution processes must also be involved in photoinduced hydrogen abstraction from coordinated solvent in rhodium(III)-phthalocyanine complexes.[166]

Comparisons of results from gas-phase and solution photochemical studies may prove valuable in elucidating mechanistic details, but unfortunately very few inorganic complexes are sufficiently volatile for gas-phase studies. One conveniently volatile complex is [Rh(tfac)$_3$], which undergoes *mer* → *fac* isomerization on irradiation in the gas phase, but not in benzene or mesitylene.[167]

Quantum yields for metal-bridging ligand bond breaking in complexes [(H$_3$N)$_5$Rh(μ-LL)Fe(CN)$_5$] and [(H$_3$N)$_5$Rh(μ-LL)Ru(NH$_3$)$_5$]$^{5+}$, whereas LL = 4-cyanopyridine, pyrazine, or 4,4′-bipyridyl, have already been mentioned (Sections 8.2.1 and 8.3.1).[30]

Two general reviews of photosubstitution and related reactions devote considerable space to the photochemistry of rhodium(III) complexes.[125,168]

8.5.10. Reactions of Coordinated Ligands

Urea hydrolyzes about 3×10^4 times faster coordinated to [Rh(NH$_3$)$_5$]$^{3+}$ than in the uncoordinated state.[138] This is a marked acceleration, but still very much less than the promotion of base hydrolysis of acetonitrile by [Ru(NH$_3$)$_5$]$^{3+}$ (cf. Section 8.3.2).[133] Formation and aquation reactions of carbonato complexes are often coordinated ligand reactions, involving C–O rather than Rh–O bond formation and breaking.[141,148] Inversion at coordinated sulfur has an activation barrier ($\Delta G^{\ddagger}$) of 87.3 and 85.5 kJ mol^{-1} in d_6-DMSO (at 383 and 378 K, respectively) in the dithiahexane complexes [Rh(dth)Cl$_4$]$^-$ and [Rh(dth)Br$_4$]$^-$. As one might well expect, these barriers, though high, are less than that for the platinum(IV) analogue [Pt(dth)Cl$_4$][119] (cf. Section 8.7.2, and see Sections 8.3.1 and 8.6).

8.6. Iridium

Aquation of [Ir(NH$_3$)$_5$(OSO$_2$CF$_3$)]$^{2+}$, $k = 1.6 \times 10^{-4}$ s^{-1} at 298 K, is fast by iridium(III) standards, but considerably slower than aquation of the cobalt(III), rhodium(III), and ruthenium(III) analogues (cf. Sections 7.1, 8.3.2, and 8.5.1).[120] Volume profiles for reaction of $[Ir(NH_3)_5(OH)]^{2+}$ plus carbon dioxide[120] and for water exchange at $[Ir(NH_3)_5(OH_2)]^{3+}$ [150,151] have been dealt with above, and in connection with their rhodium(III) analogues (Sections 8.5.4 and 8.5.5). Thermal aquation of cis-$[Ir(py)_2Cl_4]^-$ gives a mixture of the fac-cis- (48) and mer-cis- (49) isomers of $[Ir(py)_2(OH_2)Cl_3]$; none of the third isomer, the mer-trans form (50), was detected. (48) isomerizes to (49) on prolonged heating. The stereochemistry of the reaction of (48) and (49) with pyridine, pyrazine (8), pyrimidine (9),

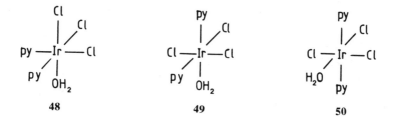

and pyridazine (35) has also been established, and the stereochemical course of several of these reactions under photochemical conditions investigated.[169] The first synthesis of an N_4-macrocyclic ligand complex of iridium(III), $[Ir(N_4)X_2]^+$,[170] will doubtless be followed by kinetic studies of aquation and substitution. Various solution interconversion and structural characterization studies of the chloro and bromo complexes $[MX_6]^{2-}$, $[MX_6]^{3-}$, $[MX_5(OH_2)]^-$, $[MX_5(OH_2)]^{2-}$, and $[M_2X_9]^{3-}$ are relevant to mechanistic studies in this area.[171] A few qualitative kinetic observations on aquation of $[IrCl_6]^{2-}$ and $[IrCl_6]^{3-}$ have been published,[172] and the effects of pH and concentration of added sodium acetate on rates of aquation of chloroaquo- and sulfatoaquoiridium(III) species investigated.[173] The rate constant for hydrolysis of $[Ir_3O(SO_4)_6(OH_2)_3]^{4-}$, an Ir(III)Ir(III)Ir(IV) mixed-valence anion which is an oxygen analogue of Delépine's salt's anion, has been measured as part of a study of catalytic properties of such solutions. The first-order rate constant, for what is apparently a one-step process kinetically, is $2.3 \times 10^{-3} s^{-1}$ in water at 296 K.[174]

The barrier to inversion at coordinated sulfur in the dithiahexane complex $[Ir(dth)Cl_4]^-$ is 63 kJ mol^{-1} at 283 K in $CDCl_3$/DMSO solution. Surprisingly, this is much lower than for the rhodium(III) analogue (Section 8.5.10); expectedly, the barrier is much lower than for the platinum(IV) analogue $[Pt(dth)Cl_4]$.[119]

The relevance of iridium(III)–bipy complexes to the topic of covalent hydration has been mentioned in Section 8.2.2.4.

8.7. Platinum(IV)

8.7.1. General

The rate constant for aquation of $[Pt(NH_3)_5(OSO_2CF_3)]^{3+}$ is 2.1×10^{-3} s^{-1} at 298 K.[120] $[Pt(NH_3)_5(OSO_2)]^{2+}$ reacts slowly with sulfite to give *cis*-$[Pt(NH_3)_4(SO_3)_2]$. This reaction can hardly be simple replacement, for the rate is independent of sulfite concentration and the O-bonded sulfite isomerizes to S-bonded during the course of the reaction. $[Pt(NH_3)_5(OSO_2)]^{2+}$ is formed from $[Pt(NH_3)_5(OH)]^{3+}$ by attack of S(IV), either in the form of SO_2 or HSO_3^-, at the coordinated oxygen.[175]

Kinetic parameters (k, $\Delta H^{\ddagger}$, $\Delta S^{\ddagger}$) for site exchange (effectively racemization) reactions in di- and trimethylplatinum(IV) glycinate complexes, such as $[PtMe_3(gly)(OD_2)]$ and isomers of $[PtMe_2Br(gly)(OD_2)]$, have been derived from NMR line-shape analysis. The mechanism suggested involves exchange of water (D_2O) with bulk solvent, which permits migration of glycinate donor atoms by competing dissociative and associative pathways. $[PtMe_3(gly)_2]^-$ contains mono- and bidentate glycinate ligands, with Pt–O bond dissociation to the latter determining its reactivity ($\Delta S^{\ddagger}$ is large and positive).[176]

Bond breaking is a key feature of transition-state formation in reduction of *trans*-$[Pt(CN)_4(OH_2)Br]^-$ or *trans*-$[Pt(CN)_4(OH)Br]^{2-}$ by iodide, thiocyanate, thiosulfate, cyanide, or bisulfite. Indeed rate constants for these reactions are almost independent of the nature of the reducing anion.[177] The fact that recrystallization of *trans, trans, trans*-$[Pt(OH)_2(NH_3)_2Cl_2]$ from water gives the *cis, trans, cis* isomer (**51**), but from water plus hydrogen peroxide gives the *trans, trans, trans* compound unchanged, suggests a key role for a platinum(II) intermediate in the former recrystallization.[178]

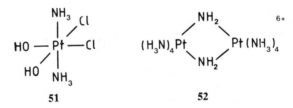

51   52

In continuation of recent studies on conjugate bases of ammine and amine complexes of metal(III) ions, the dinuclear species (**52**) has been fully characterized in the solid state (X-ray diffraction), and almost proved (^{15}N NMR) to exist in solution. A speculative mechanism for its formation in solution has been proposed.[179] The photoreactivity of $[Pt(NCS)_6]^{2-}$ has been ascribed to its lowest ligand field excited state. Quantum efficiency and intersystem crossing probability have been discussed.[180]

8.7.2. Isomerization at Coordinated Sulfur or Selenium

^{195}Pt, ^{77}Se, and ^{13}C NMR data on structures and stabilities of complexes [PtXMe$_3${MeE(CH$_2$)$_n$EMe}], where X = Cl, Br, or I; E = S or Se; n = 2 or 3, provide a useful and informative supplement[181] to extensive earlier studies of barriers to inversion at coordinated sulfur or selenium in platinum(IV) complexes of this type (cf. pp. 191–192 of Volume 1 and pp. 212–213 of Volume 2).

Barriers to inversion at coordinated sulfur in the dithiahexane complexes [Pt(dth)X$_4$] are $\Delta G^{\ddagger} = 90\ \text{kJ mol}^{-1}$ at 393 K for X = Br, and $>100\ \text{kJ mol}^{-1}$ at >423 K for X = Cl, in d_6-DMSO.[119]

Chapter 9

Substitution Reactions of Labile Metal Complexes

9.1. Complex Formation Involving Unsubstituted Metal Ions: Unidentate Ligands and Solvent Exchange

9.1.1. Bivalent Ions

The pressure-jump technique has been used[1] to study the kinetics of $BeSO_4$ formation in water/DMF mixtures. The rate of substitution depends on the composition of the solvated metal ion $[Be(H_2O)_i(DMF)_{4-i}]^{2+}$ and also on the solvent molecule which is being replaced.

It has been pointed out[2] that, in the exchange of solvent water on M_{aq}^{2+} (M = Mn, Fe, Co, Ni) [or, incidentally, on $M(NH_3)_5(OH_2)^{3+}$ (M = Cr, Co, Rh, Ir)], the partial molar volumes of the transition states display less than one-third of the variability of the partial molar volumes of the initial states as M is varied; in other words, trends in $\Delta V^{\ddagger}$ are largely determined by initial state, rather than transition state, properties.

In a high-pressure ^{17}O NMR study on solvent exchange at V_{aq}^{2+} (choride), it is suggested[3] that an I_a mechanism is operating; the following rate parameters were obtained: $k_{298} = 87\,s^{-1}$, $\Delta H^{\ddagger} = 61.8\,kJ\,mol^{-1}$, $\Delta S^{\ddagger} = -0.4\,J\,K^{-1}\,mol^{-1}$, and $\Delta V^{\ddagger} = -4.1\,cm^3\,mol^{-1}$. The rate constant for the exchange of acetic acid on Mn^{2+} in acetic acid/CH_2Cl_2 solutions has been determined[4] by proton NMR: $k\,(298\,K) = 1.4 \times 10^8\,s^{-1}$, $\Delta H^{\ddagger} = 42\,kJ\,mol^{-1}$, and $\Delta S^{\ddagger} = 50\,J\,K^{-1}\,mol^{-1}$. The activation parameters are independent of the solvent composition and concentration of perchloric acid added to prevent solvolysis of Mn(II). The same technique has been used[5] to correlate the kinetics of solvent exchange at Co(II) with the nature of the solvation of this ion in mixtures of methanol with DMF or hexamethyl-phosphoric triamide. A further study has been made[6] of the kinetics of solvent exchange at $[Co(DMSO)_6]^{2+}$ and $[Ni(DMSO)_6]^{2+}$.

A recent ^{17}O NMR investigation of water exchange at Ni(II) in a 41.5 wt % methanol solution finds[7] that the kinetic parameters are similar to those for $[Ni(H_2O)_6]^{2+}$ in water. On the assumption that $\bar{n} = 4.91$ in $[Ni(H_2O)_{\bar{n}}(CH_3OH)_{6-\bar{n}}]^{2+}$, k_1 (25°C) $= 2.77 \times 10^4 \, s^{-1}$, $\Delta H^{\ddagger} = 12.79 \, kcal \, mol^{-1}$, and $\Delta S^{\ddagger} = 4.8 \, cal \, K^{-1} \, mol^{-1}$, compared to, respectively, $3.2 \times 10^4 \, s^{-1}$, 13.6 kcal mol^{-1}, and 7.7 cal K^{-1} mol^{-1}, for the hexaqua species.

Evidence has been presented[8] for an I_d mechanism for complex formation at Ni(II) in acetic acid, as well as[9] for the reaction of this metal with isoquinoline in water, DMF, acetonitrile, methanol, and ethanol; in the latter case, activation volumes of, respectively, 7.4, 9.3, 12.8, and 12.6 cm^3 mol^{-1} are reported. The rate of dissociation of the Ni(II) isoquino-line complex in THF is strongly affected[10] by the free-water concentration in the medium, and comparison of the activation enthalpy for this reaction in pure THF with previous results in other solvents confirms the correlation with the Gutmann donor number.

Hunt and co-workers have measured[11] the activation parameters for the exchange of NH$_3$ with $[Ni(NH_3)_6]^{2+}$ in 15 M aqueous ammonia and have obtained values which are remarkably consistent with those found in other solvent-exchange studies on this metal (Table 9.1). In particular, the activation entropies and volumes correlate well and are consistent with a dissociative interchange mechanism in all cases, and $\Delta V^{\ddagger}$ is linearly related to the solvent parameter b_s, which was derived by Jordan et al.[12] in fitting $\Delta H^{\ddagger}$ to Dq for several metal-solvent pairs. It remains to be seen how generally useful and significant this parameter will turn out to be. Among recent reviews is an "update" by Wilkins[13] on mechanistic aspects of solvent exchange and complex formation at nickel(II).

Table 9.1.   *Kinetic Parameters for Solvent Exchange on Ni(II) (Reference 11)*

Solvent	k_1 at 298 K (s^{-1})	$\Delta H^{\ddagger}$ (kcal mol^{-1})	$\Delta S^{\ddagger}$ (cal K^{-1} mol^{-1})	$\Delta V^{\ddagger}$ (cm^3 mol^{-1})	b_s (kcal mol^{-1})
NH$_3$ (aq)	7.0×10^4	13.7	9.6	6	1.63
H$_2$O	3.1×10^4	13.6	7.7	7.2	3.58
DMF	3.8×10^4	15.0	8.0	9.1	4.64
CH$_3$CN	3.1×10^3	14.5	6.2	7.3	1.52
CH$_3$OH	1.0×10^3	15.8	8.0	11.4	5.44
NH$_3$ (aq)^a	6.5×10^3	10	−6	6.0	1.63

^{a} Substitution on $[Ni(H_2O)_6]^{2+}$ by NH$_3$ (aq) at 303 K. Units of k are dm^3 mol^{-1} s^{-1}.

The kinetics and mechanism of substitution of the quadruply bonded molybdenum(II) aqua dimer with thiocyanate and oxalate have been reported.[14]

9.1.2. Ions of Valency Three and Higher

The effect of pressure on the reaction between Fe(III) and thiocyanate has been reinvestigated[15] using the temperature-jump technique. The values of $\Delta V^{\ddagger}$ for the reaction of $[Fe(OH_2)_6]^{3+}$ and $[Fe(OH_2)_5OH]^{2+}$ with NCS⁻ are, respectively, 6.7 and ~0 cm³ mol⁻¹ at 20°C. While these data support the interchange nature of these complexation reactions of Fe(III), Doss *et al.* emphasize the danger in attempting to differentiate between the I_a and I_d reaction modes for anionic ligands on the basis of the sign of $\Delta V^{\ddagger}$ alone. For such a distinction to be possible, it is necessary to make allowance for the volume change associated with the formation of the outer-sphere complex which, in view of the partial charge neutralization involved, is likely to be fairly large and positive. Unfortunately, no experimental data are available which would allow even reasonable estimates of this contribution to be made. In another determination of activation volumes for these reactions, using the stopped-flow technique, values for low- and high-pH pathways of −6.1 and 8.5 cm³ mol⁻¹, respectively, were obtained.[16] Funahashi *et al.* also determined $\Delta V^{\ddagger}$ for the reaction of $[Fe(OH_2)_6]^{3+}$ and $[Fe(OH_2)_5(OH)]^{2+}$ with acetohydroxamic acid[16] (−10.0 and 7.7 cm³ mol⁻¹, respectively) and with 4-isopropyltropolone[17] (−8.7 and 4.1 cm³ mol⁻¹). The kinetics have also been reported for DMSO exchange[6] on $[Fe(DMSO)_6]^{3+}$ and for the dimerization of aquairon(III) species in ethanol/water mixtures.[18]

Variable-pressure and variable-temperature NMR studies have been reported[19,20] of solvent exchange on various lanthanide ions in DMF, while the exchange of 1,1,3,3-tetramethylurea (tmu) on $[Lu(tmu)_6]^{3+}$ in CD₃CN solution is apparently[21] dissociatively activated. The kinetics of DMF exchange and anation by SCN⁻ and N₃⁻ on $[UO_2(DMF)_5]^{2+}$ have been measured[22] by the NMR and stopped-flow techniques, respectively.

9.2. Complex Formation Involving Unsubstituted Metal Ions: Multidentate Ligands

9.2.1. Univalent Ions

Kinetics of the formation and dissociation of the cryptates Ag(2, 2, 2)⁺ and K(2, 2, 2)⁺ in acetonitrile/water mixtures have been reported.[23] The variation in k_f and k_d with solvent composition for the two metals is rather different and it appears that for Ag^+ the transition state is very close to that of the products. ^{39}K NMR line-width broadening has been used[24] to study the complexation kinetics of the K^+ ion by 18-crown-6 in several nonaqueous solvents, while ultrasonic absorption data have been published[25] for the Li^+, Na^+, and K^+ salts of the same crown ether in methanol.

9.2.2. Bivalent Ions

Rate and stability constants have been reported[26] for the formation of alkaline earth metal complexes with the diaza crown ethers (2, 2) and 2, 2-Me$_2$ in methanol. The values of K_s and k_d for the (2, 2) complexes are displaced in parallel to the corresponding values for the 2, 2-Me$_2$ complexes, and variations of K_s with metal ion radius are reflected in similar (inverse) variations of k_d.

It has been noted[27] that a positive activation volume ($8.0 \, cm^3 \, mol^{-1}$) for the incorporation of Co^{2+} into N-methyltetraphenylporphyrin in DMF is consistent with an I_d mechanism. The kinetics of the dissociation of the five-coordinate cobalt(II) complex of tris(2-dimethylaminoethyl)amine in solutions of chloroacetic acid in DMF and other solvents have been reported,[28] but no firm conclusions are drawn concerning the mechanism. The formation of the pentane-2,4-dione complex of Co(II) in aqueous solution has been studied,[29] and the values of k_f for the reaction of the enol and keto tautomers (respectively, 136 and $0.84 \, dm^3 \, mol^{-1} \, s^{-1}$; 25°C, $I = 0.5 \, mol \, dm^{-3}$) found to be considerably lower than those predicted on the basis of the solvent exchange rate for this metal.

Chopra and Jordan have studied the kinetics of complex formation between Ni_{aq}^{2+} and salicylic, 5-chlorosalicylic and 3,5-dinitrosalicylic acids in the pH range 5.5–7.0 and are able to resolve[30] the disagreement between previous studies on these systems. The chelation kinetics and equilibria have been reported[31] between Ni_{aq}^{2+} and two 2-(2-pyridylazo)-1-naphthol (α-PAN) dyes at various metal : ligand ratios and over the pH range 4.0–10.3. It is suggested that the large value of k_2 (rate constant for formation of the bis complex) and a large catalytic effect of the imidazole buffer (compared to those of nonaromatic amines of similar basicity and ligand strength) result from an enhanced value of the outer-sphere formation constant associated with a stacking interaction between bound and incoming ligands. The kinetics for the reaction of Ni_{aq}^{2+} with 1-nitroso-2-naphthol-6-sulfonate are similar[32] to those for the 3,6-disulfonate. The pH profiles have been reported of the formation rate constants for the 2,2'-bipyridyl,[33] 1,10-phenanthroline,[33] and 2-thenoyltrifluoroacetone[34] monocomplexes of

Ni_{aq}^{2+}, and the low values for the protonated species rationalized in terms of rate-limiting deprotonation. Whereas the activation volumes for the formation of nickel(II) glycolate and lactate in water (14.7 and 13.5 cm³ mol⁻¹, respectively) are consistent[35] with rate-determining ring closure, those for the succinate and maleate (14.1 and 13.5 cm³ mol⁻¹, respectively) suggest[36] the normal rate-limiting water loss at the metal ion. Further kinetic studies have been made[37] on the catalyzed chelate-ring opening of $[Ni(en)]^{2+}$.

Extrema in the enthalpy and entropy of activation for the substitution of 2,2'-bipyridyl into Ni_{aq}^{2+} in a series of water/acetonitrile mixtures correlate well[38] with the extrema in the physical properties of the mixture which are related to sharp changes in the solvent structure. Further reports have appeared on rate enhancements for the reaction of Ni^{2+} with pyridine-2-azo-p-dimethylaniline (pada) in micellar systems,[39,40] and the kinetics of complex formation with lasalocid (X-537A) in methanol[41] and with poly(4-vinylpyridine) in aqueous methanol[42] have been reported.

The rate constant for the reaction of Cu_{aq}^{2+} with the enolate ion of benzoylacetone $(1.1 \times 10^9 \ dm^3 \ mol^{-1} \ s^{-1})$[43] is typical for reactions of this metal, while the rate constant for its reaction with apo-transferrin is[44] some five orders of magnitude lower. Rapid incorporation of aquacopper(II) into porphyrins has been shown[45] to be dependent on the presence of nucleophiles such as Br^- and NO_2^- but, to a first approximation, the rate constant is independent of the nature of the porphyrin itself. Presumably[45] the nucleophile enters the inner coordination sphere of the metal and labilizes one or more of the remaining water molecules. The acid-decomposition kinetics of the monocopper(II) complexes of 20 linear polyamines have been studies[46] over a range of acid concentrations. The rates for complexes which contain only six-membered chelate rings are independent of pH, while complexes with five-membered rings have rate constants which show a nonlinear dependence on $[H^+]$. It is proposed that the rate-determining step is either the breaking of the first Cu–N bond or the subsequent protonation of the nitrogen atom.

Ultrasonic relaxation absorption has been observed[47] in aqueous solutions of Zn(II) and Cd(II) edta complexes. The absorption is ascribed to a rapid structural change of the complex itself involving penta- and hexacoordinated forms of the ligand, and in both cases the rate constant $(5 \rightarrow 6)$ is close to the water substitution rate constant of the metal ion. NMR spectroscopy has been used[48] to demonstrate that a similar dynamic equilibrium exists in the edta complexes of Co^{2+}, Ni^{2+}, and Cu^{2+}, although in the case of the Mn^{2+} complex the ligand is present only in the pentacoordinate form. In the reaction of Mg^{2+}, Ca^{2+}, Sr^{2+}, Ba^{2+}, Ni^{2+}, Cu^{2+}, or Zn^{2+}

with 1,4,7,10-tetraazacyclododecane-N,N',N'',N'''-tetraacetate (DOTA) or 1,4,8,11-tetraazacyclotetradecane-N,N',N'',N'''-tetraacetate (TETA), it has been found[49] that the reactive species is the monoprotonated form of the ligand. The kinetics of the reactions of 1,1,1-trifluoropentane-2,4-dione (Htfpd) with Co^{2+}, Ni^{2+}, and Cu^{2+} are consistent[50] with a mechanism in which the metal ion reacts exclusively with the enol tautomer of this β-diketone.

The rate constant for the incorporation of lead into the cryptand (2,1,1) in methanol is[51] unusually high and reflects the high selectivity of this ligand for Pb^{2+} over the alkaline-earth-metal ions. The kinetics of metalloporphyrin formation have been reviewed[52] with particular reference to the metal-ion-assisted mechanism, and the kinetics and mechanism of the recombination of Co^{2+}, Ni^{2+}, and Zn^{2+} with the metal-depleted catalytic site of horse liver alcohol dehydrogenase have been reported.[53]

9.2.3. Ions of Valency Three and Higher

The kinetics of aluminum citrate formation have been reported,[54] as have[55] those for the interaction between In(III) and the indicator 8-hydroxy-7-iodoquinolinium-5-sulfonate (ferron). Two kinetic studies of the formation of the iron(III) complex of Htfpd have been made,[50,56] and it has been shown[48] that the edta complex of Fe^{3+} contains a metal-bound water molecule with a pK_a of ~7. The NMR technique has been used to investigate the mechanism of the dissociation[57] of the edta complexes of Ce(III) and Y(III), and the kinetics[58] of the intramolecular acetate scrambling occurring in the complexes of this ligand with several metals-[including In(III), Sc(III), Y(III), and Lu(III)]. The rate constants and activation enthalpies and entropies for Fe(III)-complex formation with NCS^- and Br^- in DMF and with NCS^- in DMSO are consistent[59] with an I_a mechanism in all cases. On the other hand, the values of $\Delta V^{\ddagger}$ for $Fe^{3+} + NCS^-$ and other ligands in water, DMF, and DMSO suggest[60] a gradual changeover from I_a to I_d as the sizes of the coordinated solvent molecules and the entering ligand increase. Since it is sterically difficult for the donor atom in a bulky entering ligand to approach a central Fe^{3+} ion hexacoordinated by a bulky solvent such as DMF and DMSO, one of the metal–solvent bonds has to be lengthened in the activation process in order to accommodate the entering ligand whereas this is not necessary with a small solvent molecule such as H_2O.

The kinetics of the formation[61] of the (tris) dimethylglyoxime complex of Ni(IV) in aqueous alkaline media [and of its Cu(II) ion-mediated decomposition[62] in acid solution] have been measured.

9.3. The Effects of Bound Ligands

9.3.1. Reactions in Water

The addition of a second 1,10-phenanthroline ligand to the monophen complex of Cu(I) is a remarkably slow process ([equation (1), whose rate is, incidentally, little changed by the addition of sodium dodecyl sulfate (SDS) micelles]. The value of k_f (in water) for equation (1) (where X is presumbly water) is[63] $6.2 \times 10^4 \, dm^3 \, mol^{-1} \, s^{-1}$ ($25°C$; $I = 0.2 \, mol \, dm^{-3}$; $pH = 6.0$), some four to five orders of magnitude lower than might have been expected. The reason for this low rate is not altogether clear but it is presumably associated with the lack of flexibility of the ligand. A slightly smaller retarding effect (some three to four orders of magnitude) has been recorded[64] for the substitution of the second coordinating unit (presumably consisting of two adjacent carboxylate side groups on the polymer chain) of poly(acrylic acid) into Co_{aq}^{2+}. Evidence has been presented[65,66] for an $S_N 1$ CB mechanism in the dissociation of the bis complexes of Cu(II) with two amino acids.

$$[Cu(phen)X_2]^+ + phen \underset{}{\overset{k_f}{\rightleftharpoons}} [Cu(phen)_2]^+ + 2X \qquad (1)$$

The water-exchange rate constant at $[Ni(2,3,2\text{-tet})(H_2O)_2]^{2+}$ (2,3,2-tet = 1,4,8,11-tetraazaundecane) is[67] $4 \times 10^6 \, s^{-1}$, some 100-fold greater than for the hexaqua species. A stacking interaction is thought to be responsible[31] for the enhanced rate of addition of the second molecule of α-PAN dyes to Ni^{2+} as compared with the first, while the sizes of the chelate rings in NiL (L = dien, 2,3-tri, and dpt) have a significant effect[68] on the rate of ternary complex formation. The kinetics of substitution reactions of the type in equation (2) with L = dien and various ethyl-substituted derivatives, X = Cl, Br, or I, and Y = OH and I have been studied as a function of temperature and pressure.[69] A simple one-term rate law as in equation (3) has been observed[70] for the rapid cyanide exchange in the planar tetracyanometalate complexes of Ni(II), Pd(II), Pt(II), and Au(III), suggesting a single pathway involving a five-coordinate transition state or intermediate.

$$Pd(L)X^+ + Y^- \rightarrow Pd(L)Y^+ + X^- \qquad (2)$$

$$\text{rate} = k_2[M(CN)_4^{n-}][CN^-] \qquad (3)$$

9.3.2. Reactions in Nonaqueous Solvents

Rate constants have been measured[71] for the binding of carbon monoxide to a series of five-coordinate "capped" iron(II) porphyrin com-

Table 9.2. Rate Constants (k) for CO Binding to Five-Coordinate Hemes in Toluene (Reference 71)

Heme	Base	k (dm^3 mol^{-1} s^{-1})
Fe (C$_2$-Cap)	Pyridine	2.0×10^6
Fe (C$_3$-Cap)	Pyridine	5.5×10^6
Fe (C$_2$-Cap)	N-methylimidazole	9.5×10^5
Fe (C$_3$-Cap)	N-methylimidazole	6.2×10^6
Fe (C$_2$-Cap)	1,5-Dicyclohexylimidazole	3.0×10^6
Fe (C$_3$-Cap)	1,5-Dicyclohexylimidazole	4.1×10^6

plexes (Table 9.2). They are effectively independent of the cap size and, at $\sim 4 \times 10^6$ dm^3 mol^{-1} s^{-1}, are comparable to those for the reaction of carbon monoxide and other small molecules to hemoglobin in its high-affinity state. In the same study, a kinetic scheme is described which involves a novel seven-coordinate porphyrin species. The kinetics of dioxygen and carbon monoxide binding to "picket fence" and "pocket" porphyrins have been compared.[72] Substitution of axial acetonitrile by N-methylimidazole in low-spin iron(II) complexes of several synthetic macrocyclic ligands in acetone are consistent[73] with a D mechanism.

A high-pressure NMR technique has been used[74] to study solvent (acetonitrile) exchange in the five-coordinate Co(II) and Ni(II) complexes of tetramethylcyclam. While a $\Delta V^{\ddagger}$ of -9.6 cm^3 mol^{-1} supports an I_a mechanism for the cobalt system, the value of 2.3 cm^3 mol^{-1} for the nickel system suggests a dissociative mechanism (probably D). Substantial variations in the lability of the coordinated pyridine are found[75] in a series of bis-pyridine adducts of square-planar nickel(II) complexes containing dithiocarbamate, xanthate, dithiophosphate, and monothioacetylacetonate ligands, with rate constants at 298 K varying from about 10^3 to 10^7 s^{-1} (Table 9.3). The exchange of dimethylsulfide on [Pd(Me$_2$S)$_2$Cl$_2$] in various solvents is thought,[76] on the basis of its second-order rate law and the values of the activation volume, enthalpy, and entropy, to proceed via a trigonal bipyramidal transition state or intermediate.

Proton NMR studies show[77] that DMF exchange on five-coordinate [Cu(Me$_6$tren)(DMF)]$^{2+}$ has a rate constant (298 K) = 555 s^{-1}, $\Delta H^{\ddagger} = 43.3$ kJ mol^{-1}, $\Delta S^{\ddagger} = -47.0$ J K^{-1} mol^{-1}, and $\Delta V^{\ddagger}$ (365 K) = 6.1 cm^3 mol^{-1}, suggesting a dissociative mode of activation. Anation of this complex by NCS$^-$, N$_3^-$, and Br$^-$ is characterized by kinetic data consistent with an I_d mechanism. The observance[78] of comparatively slow solvent exchange on the zinc analog also emphasizes the importance of steric hindrance in these compounds.

Table 9.3. Kinetic Parameters for Exchange of Coordinated Pyridine in $[NiL_2(py)_2]$
(in Toluene) (Reference 75)

L^-	$\log k_{ex}$ at 298 K	$\Delta H^{\ddagger}$ (kJ mol^{-1})	$\Delta S^{\ddagger}$ (J K^{-1} mol^{-1})
$(PhCH_2)_2NCS_2^-$	6.11	44	21
$C_4H_4NCS_2^-$	4.06	63	45
$^iPrOCS_2^-$	4.99	57	41
$EtOCS_2^-$	4.75	58	40
$(^tBu-CO-CH-CS-^tBu)^-$	6.87	41	24
$(Me-CO-CH-CS-Me)^-$	6.64	38	11
$(Ph-CO-CH-CS-Me)^-$	6.48	39	9
$(Ph-CO-CH-CS-Ph)^-$	6.21	41	12
$(EtO-CO-CH-CS-Me)^-$	6.31	33	−14
$(EtO-CO-CH-CS-Ph)^-$	5.96	32	−25
$(CF_3-CO-CH-CS-Me)^-$	5.08	47	11
$(CF_3-CO-CH-CS-Ph)^-$	4.76	46	2
$(CF_3-CO-CH-CS-2\text{-thienyl})^-$	4.82	48	8

A NMR study of DMSO exchange in two $[UO_2L_2(DMSO)]$ complexes (L = 2,4-pentanedione and dibenzoylmethanate) suggests[79] the operation of a D mechanism in contrast to the A mechanism proposed for $[UO_2(DMSO)_5]^{2+}$.

Part 3

Reactions of
Organometallic Compounds

Chapter 10

Substitution and Insertion Reactions of Organometallic Compounds

10.1. Substitution Reactions

10.1.1. Introduction

A rather comprehensive review of the ligand substitution reactions of low-valent organometallic complexes has appeared[1] and is a most welcome addition to the literature. Another useful review deals with ligand substitution processes in metal carbonyls.[2] Brief accounts of ligand effects on organometallic substitution reactions[3] and of the mechanistic behavior of metal–metal bonded carbonyls[4] have also been published.

During the period July 1982 to December 1983 significant advances have been made in our understanding of ligand substitution at 17-electron centers and of electron transfer processes occurring in some ligand substitution reactions. Several recent papers provide striking illustrations of electron-transfer-induced stoichiometric and catalytic transformations. After a number of years of confusion, the mechanism of CO substitution in $[MM'(CO)_{10}]$ (M, M' = Mn or Re) now seems to be understood. Details of these and other studies are given below.

10.1.2. Substitution in 17-Electron Mononuclear Complexes

The vast majority of ligand substitution studies have been with diamagnetic 18-electron and 16-electron complexes. Because of their general scarcity and instability, only rarely have 17-electron complexes been investigated. Usually, photochemical or electrochemical means are used to generate 17-electron complexes, and qualitative observations with these transient species have indicated that they rapidly exchange or substitute their ligands. Until recently, the mechanism(s) of these reactions have been largely uncharacterized, although some authors have espoused dissociative pathways and some associative ones.

Basolo and co-workers[5,6] recently studied CO substitution in $[V(CO)_6]$, which is the only stable homoleptic metal carbonyl radical. With phosphine and phosphite nucleophiles (L) reaction (1) follows a strictly second-order

$$[V(CO)_6] + L \rightarrow [V(CO)_5L] + CO \qquad (1)$$

rate law. Table 10.1 gives some of the results obtained. The activation parameters and the dependence of the rate constants on the nature of the nucleophile clearly establish an associative mechanism. Substitution of CO in $[V(CO)_5L]$ also occurs by an associative mechanism and is at least 10^3 times slower than reaction (1). Similarly, replacement of phosphine by phosphite or phosphite exchange in $[V(CO)_5L]$ is associative. The rate of reaction (1) is strongly dependent on the basicity of the nucleophile as measured by ΔHNP values, which are a measure of basicity towards the proton. Only with the largest nucleophile, $P(i\text{-}Pr)_3$, is a marked steric retardation seen. This is surprising since the activated complex presumably is seven coordinate. However, the V–L bond order is formally $\frac{1}{2}$ in $[V(CO)_6L]$ and this implies a long V–L bond, thus reducing steric congestion. The lability of $[V(CO)_6]$ is indeed remarkable when compared to 18-electron metal carbonyls. For example, associative CO replacement by $P(n\text{-}Bu)_3$ is $\sim 10^{10}$ times faster in $[V(CO)_6]$ than that for the second-order (interchange)

Table 10.1. Rate Constants and Activation Parameters for the Reaction of $[V(CO)_6]$ and L According to Reaction (1) at 25°C in Hexane[a]

Nucleophile (L)	k ($M^{-1}s^{-1}$)	$\Delta H^{\ddagger}$ (kcal mol^{-1})	$\Delta S^{\ddagger}$ (cal deg^{-1} mol^{-1})
PMe$_3$	132	7.6 ± 0.7	−23 ± 3
P(n-Bu)$_3$	50.2	7.6 ± 0.4	−25 ± 2
PMePh$_2$	3.99	8.9 ± 0.3	−26 ± 1
P(O-i-Pr)$_3$	0.94		
P(OMe)$_3$	0.70	10.9 ± 0.2	−23 ± 1
PPh$_3$	0.25	10.0 ± 0.4	−28 ± 2
P(i-Pr)$_3$	0.11		
AsPh$_3$	0.018		

[a] Data from Reference 6.

pathway in $[Cr(CO)_6]$. Likewise, $[V(CO)_6]^-$ is essentially inert to thermal ligand substitution.

Hard bases also lead to associative CO substitution in $[V(CO)_6]$.[7,8] However, the product $[V(CO)_5L]$ cannot be characterized before it disproportionates to give $[V(L)_6][V(CO)_6]_2$. Nevertheless, kinetic data suggest that reaction (1) is the rate-limiting step in the overall reaction. The nucleophilicity order in CH_2Cl_2 is $L = py > Et_3N > MeCN > MeOH > acetone > THF > 2,5-Me_2THF > DME > MeNO_2 > Et_2O$.

The relatively stable 17-electron manganese radical $[Mn(CO)_3L_2]$, $L = P(i\text{-}Bu)_3$ or $P(n\text{-}Bu)_3$, follows an associative pathway for L replacement at 20°C in hexane, equation (2)[9]:

$$[Mn(CO)_3L_2] + CO \rightarrow [Mn(CO)_4L] + L \qquad (2)$$

This conclusion is indicated from the observation of a second-order rate law, a lack of rate retardation with excess L, and a greater rate with the smaller leaving group, $P(n\text{-}Bu)_3$. Carbon monoxide exchange in $[Mn(CO)_3L_2]$ is also associative, as is the reverse of equation (2) by microscopic reversibility.

The mechanism of CO substitution in $[Mn(CO)_5]$ has been disputed for a number of years. Recent studies suggest that in fact the mechanism is associative, in conformity with the manganese radical reactions discussed above. Thus, oxidation of $[Mn(CO)_5]^-$ with $[(Cp)_2Fe]^+$ in MeCN gives only $[Mn_2(CO)_{10}]$, but in the presence of 10^{-2} M PPh_3 the major product is $[Mn_2(CO)_8(PPh_3)_2]$.[10] This means that PPh_3 but not MeCN can replace a CO in $[Mn(CO)_5]$ in competition with radical coupling to give $[Mn_2(CO)_{10}]$. Such sensitivity to the nature of the nucleophile clearly points to an associative process. Similarly, laser photolysis of $[Mn_2(CO)_{10}]$ in cyclohexane

Table 10.2. Rate Constants for the Reaction of $[(MeCp)Mn(CO)_2L]^+$ and L' According to Equation (3) at 25°C [a]

L	L'	Solvent	k (M^{-1} s^{-1})
MeCN	PPh_3	MeCN	1.3×10^4
MeCN	PPh_3	Me_2CO	3.3×10^3
MeCN	$P(OPh)_3$	MeCN	1.2×10^1
MeCN	$PMePh_2$	MeCN	$>10^5$
MeCN	$P(OMe)Ph_2$	MeCN	2.5×10^4
MeCN	$P(p\text{-}ClC_6H_4)_3$	MeCN	9.5×10^2
MeCN	$P(p\text{-}MeC_6H_4)_3$	MeCN	3.0×10^4
py	PPh_3	Me_2CO	5.0
py	PEt_3	Me_2CO	1.3×10^3

[a] Data from Reference 12.

produces [Mn(CO)$_5$] radicals that can react with PBu$_3$ (before dimerization) to an extent dependent on the PBu$_3$ concentration.[11]

The electrochemical oxidation of [(MeCp)Mn(CO)$_2$L] in MeCN or acetone produces 17-electron radicals that are stable enough to allow an investigation of the mechanistic aspects of equation (3), in which L and L' are nitrogen or phosphorus donor ligands.[12,13] Equation (3) was found to

$$[(MeCp)Mn(CO)_2L]^+ + L' \rightarrow [(MeCp)Mn(CO)_2L']^+ + L \qquad (3)$$

be second order and the rate constants were obtained by digital simulation of cyclic voltammograms, and are given in Table 10.2. The second-order rate law and the sensitivity of the rate to variation in L and L' strongly suggest an associative mechanism. As found for the other 17-electron systems, [(MeCp)Mn(CO)$_2$L]$^+$ complexes are far more reactive than their neutral 18-electron analogues.

10.1.3. Substitution in Other Mononuclear Complexes

Substitution of CO by an alkyne in the 16-electron complex (1) shown in equation (4) is slow in toluene at 25°C.[14] Under an N$_2$ atmosphere k_{obs}

$$[Mo(CO)(RC_2R')(S_2CNMe_2)_2] + RC_2R'$$
$$\mathbf{1}$$
$$\rightarrow [Mo(RC_2R')_2(S_2CNMe_2)_2] + CO \quad (4)$$

is nucleophile independent and follows the order PhC$_2$Ph > PhC$_2$H > PhC$_2$Me > HC$_2$Bu > EtC$_2$Et, showing that CO loss is facilitated by electron-withdrawing substituents on the coordinated alkyne. Under a CO atmosphere k_{obs} follows equation (5) with $k_3 \approx 130k_2$. This implies a dissociative

$$k_{obs} = \frac{k_1 k_2 [RC_2R']}{k_3[CO] + k_2[RC_2R']} \qquad (5)$$

mechanism, which is somewhat surprising since (1) is a 16-electron complex and since seven-coordinate species such as [Mo(CO)$_3$(S$_2$CNMe$_2$)$_2$] are known. In contrast to this result, alkyne exchange in (1) is rapid and associative, as is alkyne displacement, equation (6), where L = CO, P(OMe)$_3$:

$$[Mo(CO)(RC_2R')(S_2CNMe_2)_2] + 2L \rightarrow [Mo(CO)L_2(S_2CNMe_2)_2] + RC_2R'$$
$$(6)$$

Evidence for a 14-electron [(phen)Mo(CO)$_2$] intermediate was obtained from a study of ^{13}CO isotopic enrichment of *fac*-[(MeCN)-(phen)Mo(CO)$_3$].[15] Pulsed laser flash photolysis of (2) (M = Cr or Mo) in 1,2-dichloroethane or chlorobenzene produces the five-coordinate (3) which can be trapped by P(O-*i*-Pr)$_3$ (L), equation (7).[16] A kinetic study of this reaction gave $k_{obs} = k_1 + k_2[L]$. Both the unimolecular chelate-ring closure

2   **3**

$$(3) + L \rightarrow (2) + [(\eta^1\text{-Et}_2\text{NCH}_2\text{CH}_2\text{PPh}_2)\text{M(CO)}_4\text{L}] \qquad (7)$$

(k_1) and the bimolecular term (k_2) are several orders of magnitude smaller than those found for the bimolecular addition of L to five-coordinate [(L$'$)M(CO)$_4$] (M = Cr or Mo), where L$'$ is unidentate. This suggests the existence of a barrier to ring reclosure and also a steric interaction between L and the free end of the bidentate in the k_2 path.

Photolysis of [M(CO)$_6$] (M = Cr, Mo, or W) followed by reaction with 4,4$'$-(*n*-C$_{19}$H$_{39}$)-2,2$'$-bipy, N–N, generates the monodentate complex (**4**).

Scheme 1

$$[M(CO)_6] \xrightarrow{h\nu} [M(CO)_5] + CO$$

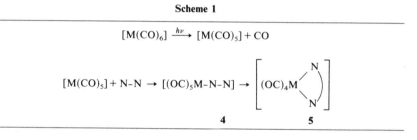

4   **5**

Rapid-scan FT IR spectroscopy was used to follow the conversion of (**4**) to the bidentate product (**5**) as shown in Scheme 1.[17] At 25°C in methylcyclohexane the rate constant k has the following values: Cr, 6×10^{-3} s^{-1}; Mo, 2×10^{-2} s^{-1}; W, 6×10^{-3} s^{-1}. The long alkyl side chains on N–N were present for solubility reasons and do not seem to affect the rate of CO loss from (**4**) since 4,4$'$-Me$_2$-2,2$'$-bipy reacts at approximately the same rate. The reactivity order Mo $\gg$ W $\approx$ Cr is the normal one for substitutions in this triad, but the absolute rate constants are much larger than usually observed for CO dissociation. This is no doubt due to the promixity of a dangling nitrogen in (**4**) and its assistance in CO loss via an interchange process.

The hydrosulfide ligand in [M(CO)$_5$SH]$^-$ (M = Cr or W) is a good *cis* labilizer, better than bromide and slightly weaker than acetate.[18] The *trans* effect for the dissociative ligand substitution shown in equation (8) spans a range of $\sim 10^4$ and follows the order $L = PPh_3 > PBu_3 > P(OPh)_3 \approx P(OMe)_3 > CO.$[19] It is suggested that this *trans* effect operates in the transition state and is primarily electronic (as opposed to steric) in origin.

$$\textit{trans-}[Cr(CO)_4LL'] + CO \rightarrow [Cr(CO)_5L] + L' \qquad (8)$$

$$\text{(L, L}' = PPh_3, AsPh_3, P(OPh)_3, P(OMe)_3, PBu_3)$$

The reaction of $[M(CO)_6]$ (M = Cr, Mo, or W) with aryl isonitriles in refluxing benzene to give $[M(CO)_{6-n}(RNC)_n]$ (n = 1-6) is catalyzed by PdO.[20,21] Little or no reaction occurs without the PdO. Similar catalysis obtains for isonitrile displacement of CO in $[(arene)Cr(CO)_3].$[22] The electrochemical oxidation of $[(C_6H_3Me_3)Cr(CO)_3]$ in the presence of P(OEt)$_3$, followed by reduction, yields a mixture of $[(C_6H_3Me_3)-Cr(CO)_{3-n}(P(OEt)_3)_n]$ (n = 1 or 2).[23] Scheme 2 shows an intramolecular

Scheme 2

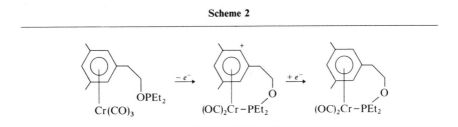

variation of this reaction. Undoubtedly, the mechanism of these reactions involves rapid CO substitution in the 17-electron cation radical. Unfortunately, the reaction cannot generally be made catalytic because the neutral product is more easily oxidized than the reactant. Although quite interesting, synthetic applications of these reactions are limited due to the instability of the cation radical intermediates, especially $[(arene)Cr(CO)_3]^+$. This often results in moderate to low product yields.

Thermal substitution of L in $[(MeCp)Mn(CO)_2L]$ is very slow. However, the application of a slight anodic current at room temperature in acetone or MeCN solvent induces rapid ligand substitution according to equation (9) [L = MeCN, py, NHMe$_2$; L' = PR$_3$, P(OR)$_3$, AsPh$_3$, SbPh$_3$,

$$[(MeCp)Mn(CO)_2L] + L' \rightarrow [(MeCp)Mn(CO)_2L'] + L \qquad (9)$$

t-BuNC, py, MeCN].[12,13] This is a good example of electrocatalytic ligand substitution. The current efficiencies are large (some > 1000) and the product yields are high, showing that the catalysis occurs with long kinetic chain

Scheme 3

$[MnL] - e^- \rightarrow [MnL]^+$ (initiation)

$[MnL]^+ + L' \rightarrow [MnL']^+ + L$ (fast substitution)

$[MnL']^+ + [MnL] \rightarrow [MnL]^+ + [MnL']$ (electron transfer)

---

lengths. The proposed ECE mechanism is given in Scheme 3. The electron transfer step can occur homogeneously (as shown) or $[MnL']^+$ can be reduced at the electrode surface. The chain length depends on the stability of $[MnL]^+$ and its rate of substitution, and is a function of the solvent and the nature and concentration of L and L'. Electrocatalysis works well for these reactions because the radical intermediates are fairly stable and because the reduction potential of $[MnL']^+$ is well positive of that for $[MnL]^+$. This means that application of a potential just positive enough to oxidize [MnL] will produce neutral [MnL']. If the product were more easily oxidized than the reactant, electrocatalysis would not, in general, be observed. Thus, there is a serious limitation on the feasibility of electrocatalyzed ligand substitutions initiated by oxidation, and the majority of reactions will not be subject to such catalysis, including almost all CO displacements. The best chance of success occurs when the incoming ligand is a better π-acid than the leaving ligand. This restriction on electrocatalysis does not apply if the reaction is initiated by reduction (see below).

The ease of the CO substitution shown in equation (10) is in the order cyclopentadienyl $\ll$ indenyl $<$ fluorenyl.[24] At 130°C in decalin the second-order rate constants ($M^{-1} s^{-1}$) are fluorenyl, 1.13×10^{-4}; indenyl, 1.68×10^{-6}; cyclopentadienyl, $<10^{-8}$. These associative reactions are thought to

$$[Mn(CO)_3(\eta\text{-dienyl})] + PPh_3 \rightarrow [Mn(CO)_2(PPh_3)(\eta\text{-dienyl})] + CO \quad (10)$$

have η^3 bonding of the dienyl in the activated complex (or intermediate) as shown for (6) and (7). The rate order is believed to reflect the extent of delocalization (stabilization) of the uncoordinated ene fragment in the allyl,

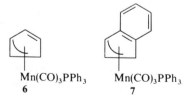

            $Mn(CO)_3PPh_3$ $Mn(CO)_3PPh_3$

                     6   **7**

ene intermediate. A similar η^3-Cp intermediate probably occurs during reaction (11), which is first order in PMe$_3$ in refluxing benzene.[25] The radical complex [(DTBQ)Re(CO)$_4$], where DTBQ is 3,5-di-t-butyl-o-quinone, undergoes CO substitution by phosphines and phosphites via a dissociative mechanism.[26,27]

$$[CpRe(CO)_3] + 2PMe_3 \rightarrow fac\text{-}[(\eta^1\text{-}Cp)Re(CO)_3(PMe_3)_2] \quad (11)$$

Reaction (12) is reported to follow a dissociative mechanism under a CO atmosphere in chlorobenzene or decane, with the rate order L = AsPh$_3$ > PPh$_3$ > P(OPh)$_3$ > CO.[28] Modi and Atwood note that the rate of L

$$[Fe(CO)_3L_2] + CO \rightarrow [Fe(CO)_4L] + L \quad (12)$$

dissociation spans a range of ~10^8 in the series [Ni(CO)$_2$L$_2$] > [Cr(CO)$_4$L$_2$] > [Fe(CO)$_3$L$_2$]. This reactivity order correlates with anticipated crystal-field activation energies, assuming that the [Fe(CO)$_3$L] intermediate is tetrahedral and not square planar. In any case, there is no correlation to M–L bond lengths. High-intensity ultrasound has been used to induce ligand substitution.[29] The ultrasound produces rapid growth and violent collapse of gas vacuoles in liquids; this generates short-lived localized hot spots with peak temperatures and pressures of ~3000 K and 300 atm. Sonolysis of [Fe(CO)$_5$] in the presence of phosphines or phosphites in alkane solvents produces [Fe(CO)$_4$L] and [Fe(CO)$_3$L$_2$]. In all cases k_{obs} is nucleophile independent, indicating that [Fe(CO)$_4$] is produced by the acoustic cavitation process.

The diazabutadiene (DAB) ligand acts as an electron sink in reaction (13) and the mechanism is strictly associative, with R$_1$ = aryl and R$_2$ = H or Me.[30] Electron-withdrawing substituents in the R$_1$ aryl groups increase the reaction rate, and the nucleophilic reactivity order is L = PMe$_3$ > PBu$_3$ > P(OMe)$_3$ > PPh$_3$. The rates are ~10^5 smaller than for the tetraazabutadiene complex [(R$_2$N$_4$)Fe(CO)$_3$], which is expected since the R$_2$N$_4$ ligand is a better π-acid than DAB. Large R$_1$ substituents (e.g., t-Bu) prevents the associative attack on (8) with CO loss; instead the DAB ligand is displaced to yield [Fe(CO)$_3$L$_2$].

$$(13)$$

[CpFe(CO)$_2$Me] is inert towards phosphorus nucleophiles at room temperature. In contrast, the σ-Cp complex in equation (14) reacts rapidly

$$[CpFe(CO)_2(\eta^1\text{-}Cp)] + P(OPh)_3 \rightarrow [CpFe(CO)(P(OPh)_3)(\eta^1\text{-}Cp)] + CO$$

$$(14)$$

with $P(OPh)_3$ to give CO substitution.[31] With $P(OMe)_3$ rapid substitution is followed by an Arbuzov elimination to yield $[CpFe(CO)(P(OMe)_3)-(PO(OMe)_2)]$ as the final product. A radical-chain mechanism applies for these reactions (Scheme 4). The $[CpFe(CO)_2]$ chain carrier is formed via an adventitious impurity (Q) or can be generated by irradiation with some $[CpFe(CO)_2]_2$ present. The last step in Scheme 4, which probably involves

---

Scheme 4

$$[CpFe(CO)_2(\eta^1\text{-}Cp)] + Q \rightarrow QCp + [CpFe(CO)_2]$$

$$[CpFe(CO)_2] + L \xrightarrow{\text{fast}} [CpFe(CO)L] + CO$$

$$[CpFe(CO)L] + [CpFe(CO)_2(\eta^1\text{-}Cp)] \longrightarrow [CpFe(CO)_2] + [CpFe(CO)(L)(\eta^1\text{-}Cp)]$$

---

attack on the free diene portion of the σ-Cp, is the crucial one that makes the radical-chain mechanism facile with σ-Cp but not with saturated alkyl ligands. A similar mechanism would be anticipated for reaction (15), but

$$[CpFe(CO)_2I] + nRNC \xrightarrow{[CpFe(CO)_2]_2} [CpFe(CO)_{2-n}(CNR)_nI] + nCO \tag{15}$$

crossover experiments using labeled Cp rings show that this is not the case, at least at 42°C in benzene solvent.[32,33] An alternate radical pathway is proposed, Scheme 5, that requires electron transfer but no atom transfer between substrate and catalyst. At higher temperatures, a second mechanism, probably analogous to Scheme 4, contributes to the overall reaction. It is suggested that $[CpFe(CO)_2]_2$ will prove to be of general utility as a catalyst for metal carbonyl substitution reactions.

Scheme 5

$$[CpFe(CO)_2]_2 \xrightleftharpoons{\Delta \text{ or } h\nu} [CpFe(CO)_2]$$

$$[CpFe(CO)_2] + [CpFe(CO)_2I] \rightarrow [CpFe(CO)_2]^- + [CpFe(CO)_2I]^+$$

$$[CpFe(CO)_2I]^+ + RNC \rightarrow [CpFe(CO)(CNR)I]^+ + CO$$

$$[CpFe(CO)(CNR)I]^+ + [CpFe(CO)_2]^- \rightarrow [CpFe(CO)_2] + [CpFe(CO)(CNR)I]$$

$$[CpFe(CO)(CNR)I]^+ + [CpFe(CO)_2I] \rightarrow [CpFe(CO)CNR)I] + [CpFe(CO)_2I]^+$$

Some years ago the CO substitution reactions of $[CpM(CO)_2]$ ($M = Co$ or Rh) were reported. This work has now been extended to the analogous

C_5Me_5 complexes, equation (16), with a variety of phosphines and phosphites as nucleophiles.[34] The mechanism is strictly associative and, as

$$[(C_5Me_5)M(CO)_2] + L \rightarrow [(C_5Me_5)M(CO)L] + CO \qquad (16)$$

expected, replacement of C_5H_5 by C_5Me_5 lowers the reactivity. The reaction rate dependence on nucleophile size (cone angle) is much more marked with the C_5Me_5 complexes, no doubt reflecting greater steric congestion caused by the methyl groups. The activation energies are several kcal higher and the entropies are somewhat less negative for the cobalt reactions, with the result that rhodium is moderately more reactive in the temperature range 40–80°C. For the reasons discussed above concerning equation (10), $[(C_9H_7)Rh(CO)_2]$ $(C_9H_7 = indenyl)$ is $\sim 10^8$ times more reactive towards PPh_3 than is $[CpRh(CO)_2]$.[24]

Oxidation of $[CpCo(CO)_2]$ and $[CpCo(CO)(PPh_3)]$ in the presence of PPh_3 rapidly produces the radical species $[CpCo(CO)(PPh_3)]^+$ and $[CpCo(PPh_3)_2]^+$. Neither reaction occurs thermally under the conditions used.[35] For reasons outlined previously, these reactions are not subject to electrocatalysis. $[Rh(CO)_2(\beta\text{-diketone})]$ suffers associative loss of CO in the presence of cyclooctadiene to give $[Rh(COD)(\beta\text{-diketone})]$.[36]

UV photolysis of $[Ni(CO)_4]$ in liquid Kr/N_2 mixtures generates $[Ni(CO)_3N_2]$, which reacts thermally with dissolved CO according to equation (17).[37] Experiments over the temperature range 112–127 K suggest

$$[Ni(CO)_3N_2] + CO \rightarrow [Ni(CO)_4] + N_2 \qquad (17)$$

a combination of dissociative and associative interchange pathways. From the dissociative path, an estimate of 10 kcal mol^{-1} was made for the $Ni\text{-}N_2$ bond strength. Turner et al. point out that the existence of a $D + I_a$ mechanism is not necessarily inconsistent with the apparently strict D mechanism for substitutions in $[Ni(CO)_4]$. This follows from activation energies obtained for reaction (17), which show that the D path would completely dominate at 300 K. Conversely, an I_a path too slow to observe at 300 K for $[Ni(CO)_4]$ could easily dominate at very low temperatures.

10.1.4. Substitution in Polynuclear Complexes

Electron-transfer-catalyzed (ETC) nucleophilic substitution in polynuclear metal carbonyls promises to become a useful synthetic procedure for a wide variety of compounds.[38] Scheme 6 shows the general reaction with a nucleophile L and Scheme 7 gives some mechanistic details (nonessential ligands omitted). For the ETC process to be efficient the radical-anion formation must be reversible, and the rate of electron transfer must be fast

Scheme 6

$$[M_n(CO)_m] + e^- \rightarrow [M_n(CO)_m]^-$$
$$[M_n(CO)_m]^- + L \rightarrow [M_n(CO)_{m-1}L]^- + CO \quad (fast)$$
$$[M_n(CO)_{m-1}L]^- + [M_n(CO)_m] \rightarrow [M_n(CO)_m]^- + [M_n(CO)_{m-1}L]$$

Scheme 7

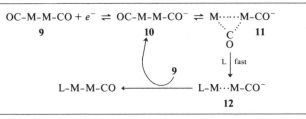

compared to radical decomposition (disproportionation or fragmentation). It is thought unlikely that direct nucleophilic attack occurs on the coordinatively saturated electron-rich cluster radical anions (**10**). Because the LUMO is usually M–M antibonding, it is relatively easy for (**10**) to form a reactive intermediate, for example, (**11**), in which the M–M bond is broken and one metal is formally a 17-electron center which rapidly adds nucleophile to give (**12**). The final step is electron transfer from (**12**) to reactant (**9**). For ETC this chain-propagating step must be spontaneous, and this will usually be the case since most ligands put more electron density on the metal than CO does. For the same reason, most oxidatively induced substitutions will not satisfy this requirement.

Substitution reactions under ETC conditions are orders of magnitude faster than thermal substitutions in the neutral substrate. For example, reaction (18) occurs slowly and in only 10% yield after 3 h of refluxing in

$$[(CF_3)_2C_2Co_2(CO)_6] + MeCN \rightarrow [(CF_3)_2C_2Co_2(CO)_5(MeCN)] + CO \quad (18)$$
$$\textbf{13}$$

MeCN. Application of a reducing potential, or addition of a small amount of benzophenone ketyl (BPK) gives 100% yield within 1 min. CO displacement from (**13**) by various phosphines and phosphites also occurs in high yield within 1 min when BPK is added.[38] Another example of ETC is given in equation (19), which proceeds readily when a slight cathodic current is applied.[39,40]

$$
\begin{array}{c}
\text{RO—C—S} \\
\quad\parallel\;\;\mid \\
\text{(OC)}_3\text{Fe—Fe(CO)}_3 \\
\qquad\backslash\diagup \\
\qquad\text{S} \\
\qquad\text{Me}
\end{array}
\;+\;\text{P(OMe)}_3\;\longrightarrow\;
\begin{array}{c}
\text{RO—C—S} \\
\quad\parallel\;\;\mid \\
\text{(MeO)}_3\text{P(OC)}_2\text{Fe—Fe(CO)}_3 \\
\qquad\qquad\backslash\diagup \\
\qquad\qquad\text{S} \\
\qquad\qquad\text{Me}
\end{array}
\tag{19}
$$

Reaction (20) is unusually rapid compared to complexes containing smaller phosphines.[41] It was shown that the rate-limiting step is not Co–Co

$$[\text{Co}_2(\text{CO})_8] + [\text{Co}_2(\text{CO})_6(\text{P}(t\text{-Bu})_3)_2] \;\rightarrow\; 2[\text{Co}_2(\text{CO})_7(\text{P}(t\text{-Bu})_3)] \tag{20}$$

bond rupture in $[\text{Co}_2(\text{CO})_8]$, nor dissociation of $\text{P}(t\text{-Bu})_3$ from $[\text{Co}_2(\text{CO})_6(\text{P}(t\text{-Bu})_3)_2]$. Based on kinetic studies, the mechanism shown in Scheme 8 was proposed, in which the first or second step is rate limiting.

Scheme 8

$$[\text{Co}_2(\text{CO})_6\text{L}_2] \;\rightleftharpoons\; 2[\text{Co}(\text{CO})_3\text{L}]$$

$$[\text{Co}(\text{CO})_3\text{L}] + [\text{Co}_2(\text{CO})_8] \;\rightarrow\; [\text{Co}_2(\text{CO})_8]^- + [\text{Co}(\text{CO})_3\text{L}]^+$$

$$[\text{Co}_2(\text{CO})_8]^- \;\rightarrow\; [\text{Co}(\text{CO})_4] + [\text{Co}(\text{CO})_4]^-$$

$$[\text{Co}(\text{CO})_4]^- + [\text{Co}(\text{CO})_3\text{L}]^+ \;\rightarrow\; [\text{Co}_2(\text{CO})_7\text{L}]$$

The rate of ligand addition shown in equation (21) is in the following order in cyclohexane: $\text{L} = \text{PBu}_3 \gg t\text{-BuNC} \approx \text{EtCN} \gg \text{CO}$.[11] The question of M–M bond cleavage $vs.$ CO dissociation in the reactions of $[\text{M}_2(\text{CO})_{10}]$ (M = Mn or Re) has been debated for some time. It is now known that

$$[\text{Mn}_2(\text{CO})_9] + \text{L} \;\rightarrow\; [\text{Mn}_2(\text{CO})_9\text{L}] \tag{21}$$

$$[\text{Re}_2(\text{CO})_{10}] + \text{L} \;\rightarrow\; [\text{Re}_2(\text{CO})_9\text{L}] + \text{CO} \tag{22}$$

$$[\text{Re}_2(\text{CO})_9\text{L}] + \text{L} \;\rightarrow\; [\text{Re}_2(\text{CO})_8\text{L}_2] + \text{CO} \tag{23}$$

$$[\text{Mn}_2(\text{CO})_{10}] + \text{L} \;\rightarrow\; [\text{Mn}_2(\text{CO})_9\text{L}] + \text{CO} \tag{24}$$

reactions (22)–(24) (L = CO, PPh$_3$) occur by simple CO dissociation and not by initial M–M cleavage.[42,43] This was established by performing crossover experiments with [185]Re-, [187]Re-, and [13]CO-labeled complexes in octane at 150 and 120°C. In contrast to these results, replacement of PCy$_3$ by P(OEt)$_3$ in $[\text{Mn}_2(\text{CO})_8(\text{PCy}_3)_2]$ does involve Mn–Mn cleavage in decalin at 40°C.[44] The rate increases with the P(OEt)$_3$ concentration and is not retarded by CO or PCy$_3$, in conformity with the postulated mechanism shown in Scheme 9. Another recent study of $[\text{Mn}_2(\text{CO})_{10}]$ demonstrates the feasibility of ultrasonically induced ligand substitution.[45]

Scheme 9

$$[Mn_2(CO)_8(PCy_3)_2] \rightleftharpoons 2[Mn(CO)_4(PCy_3)]$$

$$[Mn(CO)_4(PCy_3)] + P(OEt)_3 \rightarrow [Mn(CO)_4P(OEt)_3] + PCy_3$$

$$[Mn(CO)_4P(OEt)_3] + [Mn(CO)_4PCy_3] \xrightarrow{fast} [Mn_2(CO)_8(P(OEt)_3)(PCy_3)]$$

The substitution of E_2Ph_2 in (**14**) (E = Se or Te) by CO follows a dissociative pathway with Se > Te in rate.[46] Photosubstitution of CO by PPh_3 and $P(O\text{-}i\text{-}Pr)_3$ in $[Cp_2Fe_2(CO)_4]$ does not occur via $[CpFe(CO)_2]$ radicals as might be expected. Instead, (**15**) is postulated to be the reactive intermediate formed by photolysis that actually leads to observed products.[47]

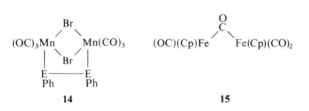

14   15

10.2. Insertion Reactions

10.2.1. Carbon Monoxide Insertion

Oxidation of the metal from Fe(II) to Fe(III) greatly increases both the rate and equilibrium constant (K_{eq}) for the CO insertion in equation (25).[48] Electrochemical studies show that K_{eq} is $\sim 10^{11}$ times greater for the Fe(III) analogues, and that this is largely due to an increase in the forward rate constant. This oxidatively induced insertion is not catalytic because the acyl product in equation (25) is more easily oxidized than the reactant. However, the carbonylation shown in equation (26) is subject to

$$[CpFe(CO)(L)Me] + MeCN \rightleftharpoons [CpFe(L)(MeCN)(COMe)] \quad (25)$$

$$(L = CO, PPh_3, P(i\text{-}OPr)_3)$$

$$[CpFe(CO)(PPh_3)Me] + CO \rightarrow [CpFe(CO)(PPh_3)(COMe)] \quad (26)$$

redox catalysis.[49] Thus, K_{eq} for equation (26) is quite large but the reaction does not progress detectably after five days at 0°C and 1 atm of CO. The addition of a few mole percent of a good oxidizing agent (Ag^+ or $[Cp_2Fe]^+$) causes complete conversion to product within 2 min. Cyclic voltammetry

Scheme 10

$$Fe\text{-}Me \xrightarrow{-e^-} Fe\text{-}Me^+$$

$$Fe\text{-}Me^+ + CO \rightarrow Fe\text{-}COMe^+$$

$$Fe\text{-}COMe^+ + Fe\text{-}Me \rightarrow Fe\text{-}Me^+ + Fe\text{-}COMe$$

---

studies indicate the mechanism in Scheme 10, in which the last step is spontaneous. The overall insertion rate for reaction (26) increases by $\sim 10^7$ upon oxidation. That oxidation facilitates CO insertions is probably due in part to the increased electrophilicity of the carbonyl carbon resulting from decreased M–CO π-backbonding. CO insertions can also be induced by reduction of the metal–alkyl substrate. Thus reaction (27) occurs electrocatalytically in THF when a reducing potential is applied.[50] Increased rates upon reduction are probably due to labilization of the Fe–Me bond in the 19-electron intermediate.

$$[CpFe(CO)_2Me] + PPh_3 \rightarrow [CpFe(CO)(PPh_3)(COMe)] \qquad (27)$$

Somewhat akin to oxidatively induced insertions are the ones promoted by electrophilic reagents that can bind to a carbonyl oxygen in the acyl intermediate. This was nicely illustrated in a recent study[51] in which (16) in methylene chloride was observed to react rapidly with simple alkali-metal and alkaline-earth salts to produce (17) in which the acyl oxygen is stabilized

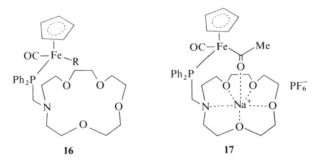

16   **17**

via bonding to the crown-ether-held metal cation. In a variation on this theme the amphoteric ligand $R_2PNR\,AlR_2$ (PNAl) was used to facilitate CO insertions by providing a Lewis-acid site (Al) *and* a Lewis-base site (P).[52,53] The acid increases the carbonyl carbon electrophilicity and the base assists methyl migration, perhaps by an I_a pathway. The normally inert $[CpFe(CO)_2Me]$ readily reacts with PNAl in benzene as shown in Scheme 11. A X-ray structure of (18) was obtained. An attempt to induce hydride migration to CO in $[HMn(CO)_5]$ using PNAl produces formyllike

Scheme 11

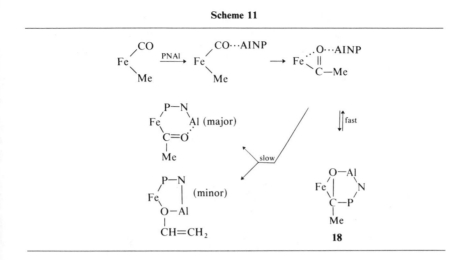

products, but a careful analysis showed that hydride migration does not occur; rather in the initial step the hydrogen is transferred as H^+ to the phosphorus end of PNAl.[54,55]

There is still debate concerning the bonding mode of the acyl in the coordinatively unsaturated intermediate usually assumed to occur during CO insertions. An IR study of $[Mn(CO)_4(COMe)]$ and $[CpFe(CO)(COMe)]$ in methane matrices at 12 K suggests that, at least under these conditions, the bonding is η^1-acetyl and not η^2-acetyl.[56]

Equation (28) gives the usual mechanism for CO insertion in a polar solvent (S). Reactions (29) and (30) with a variety of alkyl and aryl phosphines show clearly that increasing size of the nucleophile lowers the value of k_2.[57] Similarly, PPh_3-induced CO insertion in $[CpMo(CO)_3(CH_2C_6H_4X)]$ is accelerated by a bulky X group in the ortho position, reflecting relief of steric congestion in the transition state.[58]

$$[(OC)MR] \underset{k_{-1}}{\overset{k_1}{\rightleftharpoons}} [(S)M(COR)] \xrightarrow[L]{k_2} [(L)M(COR)] \qquad (28)$$

$$[CpFe(CO)(DMSO)(COCH_2Cy)] + PR_3$$

$$\xrightarrow{DMSO} [CpFe(CO)(PR_3)(COCH_2Cy)] + DMSO \qquad (29)$$

$$[CpMo(CO)_3(CH_2Ph)] + PR_3$$

$$\xrightarrow{MeCN} [CpMo(CO)_2(PR_3)(COCH_2Ph)] \qquad (30)$$

The rate of carbonylation of $[Mn(CO)_5CH_2X]$ in acetonitrile at 24°C is independent of CO pressure over the range 750–1500 psi, and is three times greater for $X = OSiMe_3$ than for $X = OMe$.[59] The decarbonylation in equation (31) occurs cleanly via allylic rearrangement of the migrating cyclopropenyl group.[60] The reaction does not proceed, as might be expected, by initial CO loss followed by 1,2-shifts of $Re(CO)_5$ around the cyclopropenyl ring.

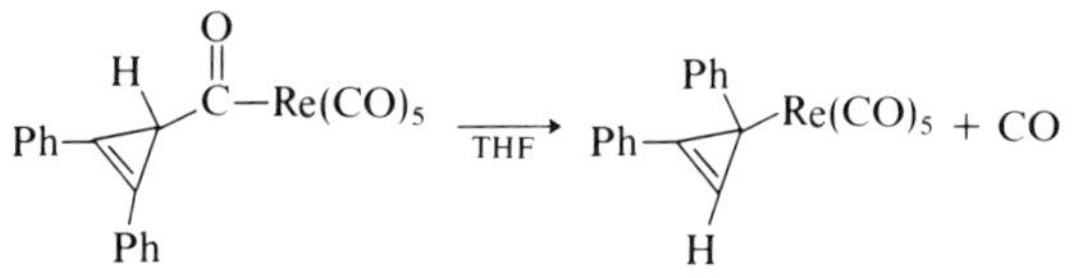

(31)

By using ^{13}CO-labeled complexes, the mechanistic details of reaction (32) have been determined (Scheme 12).[61] CO scrambling in the intermediate (**21**) is slow relative to methyl migration [(**21**) → (**22**)] but similar to the rate of phenyl migration [(**21**) → (**20**)].

$$cis\text{-}[(OC)_4Re(COMe)(COPh)]^- \rightarrow cis\text{-}[(OC)_4Re(COMe)(Ph)]^- + CO \quad (32)$$

19 → **20**

Scheme 12

(**19**) $\xrightarrow{\text{slow}}$ $[(OC)_3Re(COMe)(COPh)]^-$ $\underset{\text{slow}}{\overset{\text{slow}}{\rightleftharpoons}}$ (**20**)

21

fast ⇅ fast

$[(OC)_4Re(COPh)(Me)]^-$

22

The high-pressure carbonylation of [CpFe(CO)(L)Me] (L = PR_3) in toluene occurs readily at −30°C when the Lewis-acid catalyst BF_3 is present.[62] Below −20°C the reaction proceeds stereospecifically, with the methyl migrating to CO, and the incoming CO taking its place. At room temperature in the absence of BF_3 the reaction is not nearly so stereoselective. CO insertion into the Fe–C bond of optically active $[CpFe(CO)(PPh_3)Et]$ and $[CpFe(CO)(P(OCH_2)_3CCH_3)Et]$ in the presence of CO is stereospecific and corresponds to apparent ethyl migration in nitroethane.[63] In HMPA the reaction is stereoselective but gives mostly products corresponding to CO migration. Insertion induced by cyclohexyl isocyanide is only slightly stereoselective.

CO insertion in *fac*-$[(diars)Fe(CO)_3Me]^+$ is facile at 0°C in the presence of tertiary phosphines and phosphites (L).[64] In most cases a single kinetic product is obtained having a *trans*-L-Fe-COMe geometry. This could result from methyl migration to a *cis*-CO followed by rearrangement of the square pyramidal intermediate. However, Jablonski and Wang favor a concerted migration of CO and *cis*-Me to give a trigonal bipyramidal intermediate that allows entry of L *trans* to COMe (Scheme 13). In agreement with this,

Scheme 13

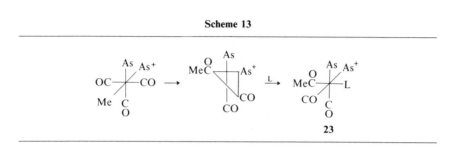

23

(**23**) (L = ^{13}CO) decarbonylates with exclusive loss of ^{13}CO *trans* to acetyl. The stronger *trans* effect of an acyl group compared to diars is proposed to account for these observations.

Insertion reactions of $[Fe(CO)_3(PMe_3)_2Me]^+$ and $[Fe(CO)_2(PMe_3)_2-(Me)X]$ (X = I or CN) have been studied.[65] It is claimed that CO insertion in *cis*-$[M(CO)_2(PMe_3)_2(Me)X]$ (X = I, Me, or CN; M = Fe or Ru) proceeds by CO migration to the *cis*-methyl.[66] An MO study of $[Pt(CO)(PH_3)(Me)F]$ (**24**) and $[Pt(COMe)(PH_3)F]$ (**25**) predicts that the Pt–Me bond in (**24**) is stronger when PH_3 is *trans*.[67] The most stable geometry for (**25**) is T-shaped with F and PH_3 *trans*. The calculations (*ab initio* Hartree–Fock) indicate that CO insertion [(**24**) → (**25**)] occurs most easily by methyl migration or a concerted motion of methyl and CO. Migration of CO is a high-energy process.

10.2.2. Alkene, Alkyne, and Carbene Insertion

Ethylene insertion into the Rh–H bond in *trans*-$[HRh(C_2H_4)(P(i\text{-}Pr_3)_2)]$ (**26**) has been studied using a magnetization transfer technique.[68] In toluene-d^8 the insertion rate is independent of added phosphine or ethylene, and therefore is truly intramolecular. The mechanism consists of very rapid *trans* $\rightleftharpoons$ *cis* equilibration of (**26**), followed by reversible hydride migration in the *cis* form. The migration step has the following activation parameters: $\Delta H^{\ddagger} = 13.0\ kcal\ mol^{-1}$ and $\Delta S^{\ddagger} = -2\ cal\ K^{-1}\ mol^{-1}$.

The hydride complexes $[CpM(CO)_3H]$ (M = Mo or W) react with cyanoacetylene and dicyanoacetylene in THF to give *trans* insertion products, $[CpM(CO)_3(CRCHCN)]$ (M = Mo, R = CN; M = W, R = H, CN).[69] The first example of methyl migration to a coordinated arene was recently reported, equation (33).[70] The rate of methylene insertion into the Pt–Pt bond of $[Pt_2XY(dppm)_2]^{n+}$ (X, Y = Cl$^-$, Br$^-$, I$^-$, CO, py, NH$_3$; n = 0, 1, 2) to give the molecular A-frame product (27) shown in equation (34) is first order in both reactants in dichloromethane.[71] It is proposed that in the rate-limiting step a pair of electrons are transferred from the metal–metal bond to the methylene group of diazomethane.

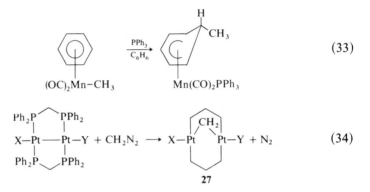

$$(33)$$

$$(34)$$

27

10.2.3. Insertion of Other Groups

Several examples of migrations to coordinated NO have recently appeared. The air-sensitive complex $[CpFe(NO)(Me)_2]$ decomposes via NO insertion.[72] Excess PMe$_3$ traps the insertion product as $[CpFe(PMe_3)(Me)(NOMe)]$. In the first detailed kinetic study of NO insertion it was shown that $[CpCo(NO)R]$ (R = Me, Et, i-Pr, p-CH$_2$C$_6$H$_4$Me) reacts according to the mechanism in Scheme 14.[73] Labeling studies indicate that the insertion is intramolecular and that β-elimination does not compete with the insertion. With PEt$_3$ as the incoming ligand there is rapid formation of an intermediate (28) that is believed to contain a bent nitrosyl. Complex (28) does not insert directly, but must first dissociate PEt$_3$, and accordingly the reaction rate decreases with increasing PEt$_3$ concentration. Estimates of rate constants at 28°C in THF with L = PEt$_3$ are $k_{-3} \approx 1 \times 10^{-3}\,s^{-1}$; $k_3/k_1 \approx 30\,M^{-1}$. Complex (28) is not formed with PPh$_3$ and the insertion rate is independent of PPh$_3$ concentration. The constant $k_1 = 1.6 \times 10^{-3}\,s^{-1}$ at 18°C in THF. In contrast to many CO insertions, the k_1 step in Scheme 14 is irreversible. It is also interesting that (28) must first dissociate L before NO insertion. Good nucleophiles often accelerate CO

Scheme 14

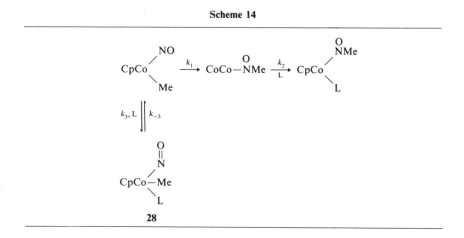

28

insertions, but this apparently is not the case with NO. One explanation, not advanced in Reference 73, is that the bent nitrosyl in (**28**) contains an electron-rich nitrogen and this retards alkyl-group migration in analogy to the reduced ease of alkyl to CO migration as the carbonyl carbon is made less electrophilic.

Formate complexes result from the rapid and quantitative insertion of CO_2 into the W–H bond in *cis*-$[HW(CO)_4(PR_3)]^-$ (R = Me, OMe, or Ph).[74] Insertion of thiocarbonyl into an osmium–aryl bond occurs with $[Os(CS)-(PPH_3)_2(C_6H_4Me)(X)(Y)]$ (X = Br, I, S_2CNEt_2, O_2CMe, or O_2CCF_3; Y = MeCN, CO, or CNR) to give products containing a η^2-thioacyl ligand, for example, $[Os(CO)(PPh_3)_2(\eta^1\text{-}O_2CCF_3)(\eta^2\text{-}CSC_6H_4Me)]$.[75] Carbonyl sulfide (COS) reacts with $[HRh(PPh_3)_4]$ to give a high yield of $[Rh(SH)(CO)-(PPh_3)_2]$.[76] This formal insertion of sulfur is believed to involve η^2 coordination of COS with subsequent hydride migration to the sulfur and C–S bond cleavage. Reversible migration of alkyl, aryl, or vinyl ligands from iron to a pyrrole nitrogen occurs when iron(III) porphyrins are treated with $FeCl_3$.[77] The products are the corresponding iron(II)(*N*-alkylporphyrin) complexes. The $FeCl_3$ serves to oxidize the reactant and transfer a chloride to the iron.

Chapter 11

Metal–Alkyl Bond Formation and Fission; Oxidative Addition and Reductive Elimination

11.1. Introduction

Once again an interesting 18 months. My thanks to those who have sent off-prints of their work. The same principles have been used for selection as were set out in the corresponding chapter in Volume 2: genuine evidence for a mechanism and measurement of rate constants and activation parameters, together with relevant work on equilibrium parameters and bond energies. However, in the hope that they will inspire some mechanistic work, a few references have been included to particularly novel and exciting reactions.

11.2. Metal–Alkyl Bonds

Several useful reviews have appeared on different aspects of the metal-carbon σ bond. Halpern[1] has collected data on bond dissociation energies of transition metal-alkyl bonds. An extensively referenced review on the breaking of organometallic bonds, R–M, in reactions in which R_2, RH, and

R^{-H} are formed has been written by Kashin and Beletskaya.[2] (Throughout R^{-H} is used to denote the olefin formed from R by loss of H.). Chapters 11, 12, and 16 of Volume 11 of *Organometallic Chemistry*[3] describe many new metal–alkyl systems, several crying out to have mechanistic and thermochemical studies made upon them, while the S_H2 reaction, involving homolytic displacement of a transition-metal atom from carbon, has been reviewed by Johnson.[4]

As in Volume 2, Cr–C and Co–C bonds come first in Sections 11.2.1 and 11.2.2, respectively, followed by other metal–carbon-containing systems usually running down groups from left to right across the Periodic Table. In general in each section bond breaking and bond making are followed by transalkylation.

11.2.1. Chromium

Two reviews on aspects of the Cr–C bond, very appropriately by Espenson, have appeared. One[5a] discusses areas including the synthesis of organochromium complexes and then analyzes different processes in which the Cr–C bond is broken: in general, unimolecular homolytic (S_H1), non-H^+ and H^+ heterolytic (acidolysis), S_E2, and S_H2.† One of the topics of the second review[5b] is the homolytic fission of the Cr–C bond in $[(H_2O)_5Cr\text{–}CMe_2OH]^{2+}$ as a source of $\cdot CMe_2OH$ radicals. This process goes easily in strongly acid solution; recently, it has been shown that this radical does not form chromium–alkyl bonds when reducing various complexes of the type $[Co(amine)_6]^{3+}$ and $[Cr(amine)_6]^{3+}$ which would be expected to react by outer-sphere mechanisms.[6] However, it reacts with $[V(OH_2)_6]^{2+}$ with a rate constant of $2.1 \times 10^5\ M^{-1}\ s^{-1}$ at 25°C possibly forming $[(H_2O)_6VR]^{2+}$ as an intermediate.[7]

Further rates of homolytic breaking, equation (1), of the Cr–C bond in $[(H_2O)_5Cr\text{–}R]^{2+}$ have been obtained.[8] They seem to depend not only on steric effects but also on the stabilization of $\cdot R$ by resonance. The rates of reaction of $\cdot R$ with $[(H_2O)_5CrR]^{2+}$ are equal to or greater than $10^8\ M^{-1}\ s^{-1}$.

$$[(H_2O)_5Cr\text{-}R]^{2+} \rightarrow [(H_2O)_5Cr\cdot]^{2+} + \cdot R \qquad (1)$$

The reaction of various alkylchromium complexes, $[(H_2O)_5CrR]^{2+}$ (or CrR^{2+}) and H_2O_2, have been studied (R = CH_2OMe, $CHMeOEt$, $CHMe_2$, and CH_2Ph).[9] A chain reaction occurs in which Cr^{2+} and $R\cdot$ probably both act as chain centers. The rate constant at 25°C for the attack of $\cdot OH$ on CrR^{2+}, when R is $CHMeOEt$, (i.e., $k_2 + k_3$) is about $1.6 \times 10^9\ M^{-1}\ s^{-1}$.

---

† In Volume 2 on p. 273, the reactant over the arrow in equation (3), the H^+ heterolytic process, should be H_3O^+ not H_2O.

Steps (2)–(5) are suggested:

$$CrR^{2+} + \cdot OH \rightarrow Cr^{2+} + R^{-H} + H_2O \quad \text{(or ROH)} \quad (2)$$

$$\text{or} \rightarrow CrOH^{2+} + R\cdot \quad (3)$$

$$Cr^{2+} + H_2O_2 \rightarrow CrOH^{2+} + \cdot OH \quad (4)$$

$$R + H_2O_2 \rightarrow R^{-H} + H_2O + \cdot OH \quad (5)$$

Acetate ions accelerate the rate of Cr–C bond cleavage in $[(H_2O)_5CrCH_2OH]^{2+}$. Kinetic studies point to reactions (6)–(9):[10]

$$[(H_2O)_5CrCH_2OH]^{2+} + OAc^- \rightleftharpoons [(H_2O)_4(AcO)CrCH_2OH]^+$$
$$+ H_2O \quad (6)$$

$$[(H_2O)_4(AcO)CrCH_2OH]^+ + 2H_2O \rightarrow [Cr(H_2O)_5(OAc)]^{2+}$$
$$+ \text{organic products} + OH^- \quad (7)$$

$$[(H_2O)_4(AcO)CrCH_2OH]^+ + H_3O^+ \rightarrow [Cr(H_2O)_5(OAc)]^{2+}$$
$$+ \text{organic products} \quad (8)$$

$$[(H_2O)_4(AcO)CrCH_2OH]^+ + HOAc + H_2O \rightarrow [Cr(H_2O)_5(OAc)]^{2+}$$
$$+ \text{organic products} + OAc^- \quad (9)$$

Reaction (7) is 1600 times faster than the corresponding one involving $[(H_2O)_5CrCH_2OH]^{2+}$, which is attributed to a labilizing effect by the acetate ligand in a trans position to the CH_2OH group.

The rate constant for the formation of the cyclopentylchromium complex, $[(H_2O)_5Cr(c\text{-}C_5H_9)]^{2+}$, in reaction (10) at 22°C is $8 \times 10^7 \ M^{-1} \ s^{-1}$.[11]

$$[Cr(OH_2)_6]^{2+} + \cdot C_5H_9 \rightarrow [(H_2O)_5Cr(c\text{-}C_5H_9)]^{2+} \quad (10)$$

Like many related organochromium compounds mentioned in Espenson's review[5a] and in Volume 2,[12] it decomposes by two unimolecular pathways, one heterolytic (11) and the other homolytic (1). Kinetic parameters are $k_{11}(298) = 4.86 \times 10^{-4} \ s^{-1}$, $\Delta H_{11}^{\ddagger} = 73.5 \ kJ \ mol^{-1}$, $\Delta S_{11}^{\ddagger} = -61.5 \ J \ K^{-1} \ mol^{-1}$, $k_1(298) = 1.07 \times 10^{-4} \ s^{-1}$, $\Delta H_1^{\ddagger} = 126 \ kJ \ mol^{-1}$, and $\Delta S_1^{\ddagger} = 102.5 \ J \ K^{-1} \ mol^{-1}$. The large values of $\Delta H_1^{\ddagger}$ and $\Delta S_1^{\ddagger}$ in the homolytic process are consistent with metal–carbon bond fission and are similar in size to those observed in other reactions involving Cr–C homolysis.[5a,12] It is suggested that the heterolytic process involves attack on the α-carbon of the $c\text{-}C_5H_9$ group by one of the cis-H_2O ligands.

$$CrC^{2+} \rightarrow CrOH^{2+} + HC \quad (11)$$

The rate constant for transalkylation from $[(H_2O)_5Cr(c\text{-}C_5H_9)]^{2+}$ to Hg^{2+} is 1.1 $M^{-1}s^{-1}$ at 298 K.[11]

11.2.2. Cobalt

11.2.2.1. R–Co(III) Bond Breaking and Making

An admirable two-volumed work *B12*, edited by Dolphin,[13] contains excellent reviews of many aspects of the Co–C bond in both the B12 coenzyme and its model compounds.

Halpern in a review article[14] has brought together his data on the effect of the basicity and size of L on the stability of the Co–C bond in *trans*-LCoR-containing systems. This includes an extension[15] of earlier work discussed in Volume 2[16] on reaction (12) and is designed to investigate steric effects associated with various phosphines L.

$$[(L(dmgH)_2CoCH(Me)Ph] \rightleftharpoons [L(dmgH)_2Co] + CH_2{=}CHPh + \tfrac{1}{2}H_2 \quad (12)$$

Just as before the rate-determining step in this overall process is reckoned to be (13) so that $\Delta H_{12}^{\ddagger} = \Delta H_{13}^{\ddagger}$. As the back reaction is diffusion controlled,

$$[L(dmgH)_2CoCH(Me)Ph] \rightleftharpoons [L(dmgH)_2Co] + \cdot CH(Me)Ph \quad (13)$$

$\Delta H_{-13}^{\ddagger} \simeq 2$ kcal mol^{-1}. Hence $D_{Co\text{-}R}$ can be calculated. k_{13} rises by a factor of 5000 when L is changed from $PhMe_2Ph$ through various other phosphines to $P(c\text{-}C_6H_{11})_3$ (see Table 11.1). In contrast a change from $PhMe_2Ph$ to 4-cyanopyridine increases this rate constant by only 13 which suggests that steric effects are much more important than electronic. The importance of the size of L as a factor in weakening the Co–C bond is demonstrated by the correlations between $D_{Co\text{-}R}$ and k_{13} with the cone angle of the phosphines.

Table 11.1. Data for the Decomposition of $[L(dmgH)_2CoCH(CH_3)C_6H_5]$ in Acetone at 25°C, Reaction (13)

L	$10^3 k_{13}$ (s^{-1})	$\Delta H_{13}^{\ddagger}$ (kcal mol^{-1})	$\Delta S_{13}^{\ddagger}$ (cal mol^{-1} K^{-1})	$D_{Co\text{-}R}$ (kcal mol^{-1})
$P(CH_3)_2C_6H_5$	0.1	25.9	10.5	24
$P(n\text{-}C_4H_9)_3$	1.2	22.8	5.1	21
$P(CH_3)(C_6H_5)_2$	1.4	—	—	—
$P(CH_2CH_2CN)_3$	3.1	22.1	4.5	20
$P(C_2H_5)(C_6H_5)_2$	3.3	21.3	1.9	19
$P(C_6H_5)_3$	19.2	19.3	−1.4	17
$P(cyclo\text{-}C_6H_{11})_3$	480	—	—	—

Another method of estimating the strength of a Co–CX bond is by study of the Co–C bond length, any drastic increase in the $Co\hat{C}X$ angle from tetrahedral probably also being relevant. In cobaloximes, $[Co(dmgH)_2(CH_2X)L]$, these parameters are dependent on the trans ligand L and on substituents X.[17-21]

Thermoanalytical studies show that organoiron(III) and cobalt(III) phthalocyanine THF complexes, $RMPc(THF)_x$, decompose thermally with increasing readiness: R, Ph < PhCO < Me < *n*-Pr.[22] For the decomposition of the *n*-propyl complexes in solution, the mechanism is suggested that is often proposed for the decomposition of alkylcobaloximes and cobalamines which is shown in equations (14)–(16) (HX = solvent):

$$CH_3CH_2CH_2M \rightarrow CH_3CH_2CH_2\cdot + M \tag{14}$$

$$CH_3CH_2CH_2\cdot + M \rightarrow CH_3CH{=}CH_2 + HM \tag{15}$$

$$CH_3CH_2CH_2\cdot + HX \rightarrow CH_3CH_2CH_3 + X\cdot \tag{16}$$

The ratio of propane to propene is dependent on solvent, and is favored by iron and the addition of hydrogen donors. In chinoline only propane is produced, though the presence of propyl radicals can be demonstrated by reduction of copper(II) to copper(I). It is proposed that the chinoline is intimately involved in reaction (14), a suggestion that is supported by $\Delta S^{\ddagger}$ for the overall processes being close to that for a S_N2 reaction, namely M = Fe, $\Delta H^{\ddagger} = 60.4\,kJ\,mol^{-1}$, and $\Delta S^{\ddagger} = -85.6\,J\,mol^{-1}\,K^{-1}$.

2,2,6,6-Tetramethylpiperidine-*N*-oxide(Tempo) traps R· but not the Co(II) unit in a Costa-type model B12 system, and has been used by Finke's group[23] to obtain kinetic parameters for reaction (17) for R = $PhCH_2$ and Me_3CH_2 (see Table 11.2). (Further data on the saloph, cobalamin, and

Table 11.2. *Activation Parameters for Reaction (17)*

	$\Delta H^{\ddagger}$ (kcal mol^{-1})	$\Delta S^{\ddagger}$ (cal mol^{-1})
$[PhCH_2Co\{C_2(DO)(DOH)_{pn}\}I]$	27.9	8
$[PhCH_2Co)SALOPH)py]$	23.6	1
$[PhCH_2Co(cobalamin)]$	24.6	12
$[PhCH_2Co(cobinamide)]$	26.9	9
$[(CH_3)_3CCH_2Co\{C_2(DO(DOH)_{pn}\}I]$	32.2	18
$[(CH_3)_3CCH_2Co(SALOPH)py]$	20.3	−6
$[(CH_3)_3CCH_2Co(cobalamin)]$	23.4	2.6
$[(CH_3)_3CCH_2Co(cobinamide)]$	32.1	17.3

cobinamide systems are given in Tables 11.4, p. 278 and 11.5, p. 279 of Volume 2.[24]

$$[R-Co\{C_2(DO)(DOH)_{pn}\}I] \rightarrow R\cdot + [Co^{II}\{C_2(DO)(DOH)_{pn}\}I] \quad (17)$$

Turning from S_H1 to a possible S_H2 process, it is observed that homolytic fission of a Co–C bond can be brought about by attack at the carbon atom by an alkyl radical.[25] The thermal decomposition in CCl_4 of hex-5-enyl(pyridine)cobaloxime and in particular its 5-methyl derivative leads to cyclopentyl products; reaction (18) is though to occur ($R = H$ or Me):

$$[CH_2 = CR(CH_2)_4Co(dmgH)_2py]$$

$$\xrightarrow{\cdot CCl_3} [Cl_3CCH_2\dot{C}R(CH_2)_4Co(dmgH)_2py]$$

$$\longrightarrow (Cl_3CCH_2)R\overline{C(CH_2)_3\dot{C}H_2} \quad (18)$$

Grate and Schrauzer have reinvestigated the products formed when alkylcobaloximes and other alkyl B12 model compounds containing β-hydrogen decompose on warming in water at various pHs.[26] In most cases the formation of alkene is increased by high basicity and if the alkyl group is branched, suggesting β-elimination of hydrogen as in equation (19):

$$Co-CH_2CHMe_2 + OH^- \rightarrow Co(I) + CH_2{=}CMe_2 + H_2O \quad (19)$$

The Co(I) species can bring about reductive cleavage of the Co–C bond giving alkane as in equation (20):

$$Co(I) + Co-R + H^+ \rightarrow 2Co(II) + RH \quad (20)$$

However, alkanes can also be formed by homolysis of the Co–R bond, which can be shown by the formation of methylcyclopentane from 5-hexenylcobaloxime as in reaction (21), and by the retardation of alkane formation by oxygen.

$$Co-(CH_2)_4CH{=}CH_2 \rightarrow Co(II) + \cdot CH_2(CH_2)_3CH{=}CH_2$$

$$\rightarrow \cdot CH_2\overline{\dot{C}H(CH_2)_3CH_2} \quad (21)$$

Fergusson and Baird[27] have observed that anhydrous HCl cleaves the R–Co bond in alkylcobaloximes, $[RCo(dmgH)_2py]$. In DCl the major product is RD. It is suggested that the reactions proceed by electrophilic attack of the hydrogen ion on R in one of the species produced by loss of one or both of the hydrogen-bonding protons which link the two dimethylglyoxime units, that is, in $[RCo(dmg)(dmgH)Cl]$ or $[RCo(dmg)_2Cl]^+$. When R is methyl or benzyl, some RCl is produced, perhaps by nucleophilic attack of chloride on the cobalt(IV) species, $[RCo(dmgH)_2py]^+$.

A preliminary report is published of work which, though incomplete mechanistically, is of sufficient interest in connection with the Co–C bond to justify inclusion here.[28] Hydrocarbons R^{-H}, RH, and R_2, are formed from the alkyl-*mer*-{*N*-(2-aminoethyl)-7-methylsalicylideneminato} (ethylenediamine)cobalt ion, $[RCo(7\text{-Mesalen})(en)]^+$, in aqueous perchloric acid. Use of $DClO_4$ in D_2O shows that RH is formed from the cobalt complex although $RN(\dot{O})(t\text{-Bu})$ radicals could be trapped on addition of t-BuNO. There is an induction period; this is decreased and the steady-state concentration of $RN(\dot{O})(t\text{-Bu})$ is increased as acidity rises. Two diol-dehydrase reactions catalyzed by B12 coenzyme can also be made to occur using this system.

Pulse-radiolysis studies by Mulac and Meyerstein[29] show that vitamin B12r, or Co(II), reacts with a variety of hydroxyalkyl radicals, R·, by the homolytic fusion process (22); there is no evidence for a redox reaction.

$$Co(II) + R\cdot \ \rightarrow\ Co\text{-}R \qquad (22)$$

Co(III)–CH_2CMe_2OH and Co(III)–CH_2CHO are stable for over 1 s between pH 3 and 10, the latter decomposing as in equation (23) in more acid solution:

$$Co(III)\text{-}CH_2CHO + H_3O^+ \ \rightarrow\ Co(III)(OH_2) + CH_3CHO \qquad (23)$$

Co(III)–CMe_2OH and Co(III)–$Ch(OH)CH_2OH$ decompose heterolytically over some 10 ms, as in (24), for example,

$$Co(III)\text{-}CH(OH)CH_2OH \ \rightarrow\ Co(I) + CH_2OHCHO + H^+ \qquad (24)$$

with some β-elimination probably taking place, equation (25):

$$Co(III)\text{-}CH(OH)CH_2OH \ \rightarrow\ Co\text{-}H + CHOH{=}CHOH \qquad (25)$$

k_{24} or $k_{24} + k_{25}$ at pH 5.9 is $\sim 1 \times 10^{-3}\,s^{-1}$. Pathway (24) contrasts with the homolytic fission process usually postulated for the B12-catalyzed diol-dehydrase reaction.

Cobalt(II) tetrasulfophthalocyanine, $[Co(II)(tspc)]^{4-}$, has been used by Meyerstein's group[30] as a model for coenzyme B12 and diol dehydrase. It reacts with $\cdot CH_2CMe_2OH$ to form both a mono- and dicobalt species, equations (26) and (27):

$$[Co(II)(tspc)]^{4-} + \cdot CH_2C(CH_3)_2OH$$

$$\rightarrow\ [(tspc)Co(III)CH_2C(CH_3)_2OH]^{4-} \qquad (26)$$

$$[Co(II)(tspc)]_2{}^{8-} + \cdot CH_2C(CH_3)_2OH$$

$$\rightarrow\ [Co(II)(tspc)(tspc)Co(III)CH_2C(CH_3)_2OH]^{8-}$$

$$\text{or } [Co(II)(tspc)]^{4-} + [Co(III)(tspc)CH_2C(CH_3)_2OH]^{4-} \qquad (27)$$

These transients, which have half-lifes of the order of milliseconds, eliminate water to give isobutenylcobalt intermediates as in (28):

$$[(tspc)Co(III)CH_2C(CH_3)_2OH]^{4-}$$

$$\rightarrow [(tspc)Co(III)CH{=}C(CH_3)_2]^{4-} + H_2O \quad (28)$$

Subsequent Co–C bond fission can lead to cobalt(I), II, and (III). Reactions (29) and (30) occur at a pH > 9.5 and ~6, respectively. At pH 3 reaction (30) is accompanied by reaction (31). (Similar equations are proposed for the dicobalt species.)

$$[(tspc)Co(III)CH{=}C(CH_3)_2]^{4-}$$

$$\xrightarrow{H_2O} [Co(I)(tspc)]^{5-} + OCH{-}CH(CH_3)_2 + H_3O^+ \quad (29)$$

$$[(tspc)Co(III)CH{=}C(CH_3)_2]^{4-}$$

$$\xrightarrow{H_2O} [Co(III)(tspc)]^{3-} + CH_2{=}C(CH_3)_2 + OH^- \quad (30)$$

$$[Co(III)(tspc)]^{3-} + [(tspc)Co(III)CH{=}C(CH_3)_2]^{4-}$$

$$\xrightarrow{H_2O} 2[Co(II)(tpsc)]^{4-} + OCH{-}CH(CH_3)_2 + H_3O^+ \quad (31)$$

Electrochemical oxidation of methylcobalamin in water yields aquocobalamin and methanol.[31] The two-electron process is considered to involve transfer out of the Co–C σ-orbital of two electrons yielding Co(III) and CH_3^+, the latter giving methanol.[32] Electrochemical reduction involves one electron giving vitamin B12s, the Co(I) form, and presumably ·CH_3, since ethane is also formed.[31] As reported in Volume 2,[33] the reduced species, $[py(dmgH)_2CoMe]^-$, which can be formed in matrix by electron capture, behaves similarly yielding ·CH_3 and $[Co(I)(dmgH)_2py]$. Ramakrishna Rao and Symons have shown that the electron enters the Co–C σ-antibonding orbital[34] in this system (in contrast to methylcobalamin where it enters a π^*-orbital on the corrin).

Further work by Fanchiang[35a] on dealkylation of alkylcobalamines by $[AuCl_4]^-$ helps to confirm that the reaction is not S_E2 but electron transfer in character as in equation (32) (see also Volume 2[36]):

$$Me{-}B_{12} + [AuX_4]^- \rightleftharpoons \{[Me{-}B_{12}], [AuX_4]^-\}$$

$$\rightarrow \{[Me{-}B_{12}]^+, [AuX_4]^{2-}\}$$

$$\rightarrow \cdots \quad (32)$$

The rates of reaction when R is Me and Et are comparable, which would not be expected if a simple substitution process occurred at the α-carbon atom. The oxidative demethylation of methylcobalamin by $[Ir(IV)Cl_6]^{2-}$ appears to be somewhat similar.[35b] The reaction is first order in each reactant, but the product distribution is affected by stoichiometry. As before an electron transfer process is postulated preceded by formation of two outer-sphere complexes only one of which, (**1**), is an intermediate [equations (33)–(38)]:

$$CH_3-B_{12} + [IrCl_6]^{2-} \rightleftharpoons \{[CH_3-B_{12}], [IrCl_6]^{2-}\} \qquad (33)$$
$$\mathbf{1}$$

$$CH_3-B_{12} + [IrCl_6]^{3-} \rightleftharpoons \{[CH_3-B_{12}], [IrCl_6]^{3-}\} \qquad (34)$$
$$\mathbf{2}$$

$$\{[CH_3-B_{12}], [IrCl_6]^{2-}\} \rightleftharpoons [CH_3-B_{12}]^+ + [IrCl_6]^{3-} \qquad (35)$$

$$[CH_3-B_{12}]^+ \rightarrow [H_2O-B_{12}]^+ + CH_3{\cdot} \qquad (36)$$

then either

$$CH_3{\cdot} + [Ir(IV)Cl_6]^{2-} \rightarrow CH_3Cl + [Ir(III)Cl_5]^{2-}$$
$$\rightarrow [IrCl_5(OH)_2]^{3-} \qquad (37)$$

or

$$CH_3{\cdot} + CH_3{\cdot} \rightarrow C_2H_6 \qquad (38)$$

However, methylcobalamin and tetracycloethylene (TCNE) behave differently forming a charge transfer complex which involves the π-system of the corrin rather than the Co–C bond.[37] (At high concentration of TCNE, a base-off species is also produced.) In the presence of a proton donor, demethylation occurs over a period of approximately seven days at 23°C.

The demethylation of methylcobalamin by various platinum compounds requires the presence of species in the II and IV oxidation states.[38] The initial step involves the platinum(II) complex, for example, $[PtCl_4]^{2-}$, which gives a binuclear intermediate, equation (39), that in turn reacts with platinum(IV) to give methylplatinum(IV) species such as $[CH_3PtCl_5]^{2-}$ either directly or via a trinuclear intermediate, as in equations (40) and (42), respectively:

$$CH_3-B_{12} + Pt(II) \rightleftharpoons CH_3-B_{12}, Pt(II) \qquad (39)$$

$$CH_3-B_{12}, Pt(II) + Pt(IV) \rightarrow products \qquad (40)$$

$$CH_3-B_{12}, Pt(II) + Pt(IV) \rightleftharpoons CH_3-B_{12}{\cdots}Pt(II){\cdots}Pt(IV) \qquad (41)$$

$$CH_3-B_{12}{\cdots}Pt(II){\cdots}Pt(IV) \rightarrow products \qquad (42)$$

There is kinetic and spectroscopic evidence for the intermediates. Rate and equilibrium constants vary by factors of less than 10 on changing the platinum species.

Bond dissociation energies, $D^0(Co^+-CH_3)$ and $D^0(Co^+-CH_2)$, are given later with those for the corresponding iron and nickel species.

11.2.2.2. The Vitamin B12 System

Many suggestions have been made on the mechanisms of the 1,2-shifts catalyzed by the B12 system; these have been reviewed by Golding.[39] Now in order to account for the very stereospecific hydrogen migration in the ethanolamine ammonia lyase reaction, Rooney's group[40] have suggested that the high CoĈR bond angle allows reactions (43) and (44) in Scheme 1 to occur (RH being the substrate).

Scheme 1

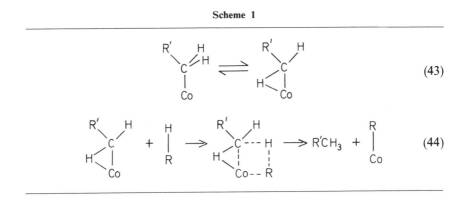

$$(43)$$

$$(44)$$

Using a Costa-type model B12 system, Finke's group[41] have prepared compounds of the type Co–CH(OH)CH$_2$OH and Co–CH$_2$CHO. Studies including the kinetics of decomposition of the first suggest that this type of compound is not produced during the diol-dehydrase reaction, and that Co-(π-C=C) intermediates are not formed either.[42]

MNDO SCF-MO calculations indicate three mechanisms whereby 1,2-shifts could occur in radicals implicated in B12 isomerization processes.[43] None involves the cobalt.

11.2.2.3. Dicobalacyclohydrocarbons

New ground has been broken by Hersh and Bergmann[44] studying the making and breaking of dicobalacyclic systems. The major pathway for the decomposition of the first dimetallacyclohexene to be discovered involves a sort of retro-Diels–Alder reaction in which two Co–C links are broken, *o*-xylylene is produced, and a Co=Co-containing species is formed initially as in Scheme 2. $\Delta H_{45}^{\ddagger}$ and $\Delta S_{45}^{\ddagger}$ are 24.3 kcal mol^{-1} and 12.1 cal deg^{-1} mol^{-1}, respectively. The first step in the decomposition of a somewhat similar dimetallacyclopentane may be the fission of the Co–Co bond accompanied by migration to a bridging CO to a terminal position.[45]

Scheme 2

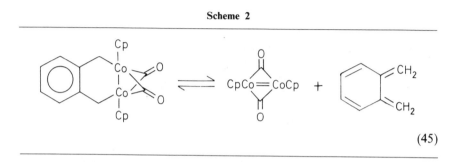

(45)

The dicobalt radical anion, (**3**), reacts with α,γ-diiodoalkanes to give a dicobalapentane, for example, (**4**).[46] Formation of the first σ-Co–C bond occurs with complete loss of chirality at the C atom and is thought to involve an electron transfer process. Some inversion takes place on formation of the second Co–C link which could involve either an S_N2 process accompanied by an electron transfer pathway or one of the recently identified electron transfer processes which occur with some inversion.

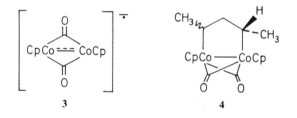

3   4

11.2.3. Lithium

Lithium intercalates are important in solid-state chemistry, and can be formed in heterogeneous reactions such as equation (46) [(M = Ti, Zr, Hf, V, Nb, Ta; X = S, Se, Te; R = *n*-Bu, *s*-Bu, *t*-Bu, *n*-Pr, $CH_2C(Me_2)Ph$]. Studies on product ratios of R^{-H} to R$_2$, cross-coupling reactions involving different R, and the influence of change of solvent, and comparisons with other systems, indicate the production of radicals, R·, in a process which

is not diffusion controlled. A one-electron oxidation of RLi(solid) to R·
and Li$^+$ is proposed.[47]

$$MX_2 \text{ (solid)} \xrightarrow[\text{solvent}]{\text{RLi}} LiMX_2 \text{ (solid)} \qquad (46)$$

By reacting lithium vapor with halomethanes, Lagow's group[48] have
prepared a variety of organolithium species such as $(CH_m Li_{4-m})_n$, in which
$0 < m < 3$ and $1 < n < 4$.

11.2.4. Germanium and Silicon

Transalkylation reactions, which may be S_E2 or electron transfer,
involving these elements are mentioned under tin.

11.2.5. Yttrium and Titanium

The kinetics of a reaction involving C–H activation of methane by
yttrium compounds are given under lutetium in Section 11.2.12.

Some transalkylation reactions of titanium compounds are discussed
under platinum, while a value for $D(\text{Ti–Ph})$ is given under thorium, Sections
11.2.10 and 11.2.13, respectively.

11.2.6. Tantalum

Chisholm's group[49] have prepared various moderately stable compounds of the sort $[Ta(NMe_2)_4R]$, where R are various alkyl groups which
can undergo α-, β-, or γ-elimination (or abstraction) of hydrogen. Thus
they strengthen the argument that σ-alkyl ligands attached to metal–alkyl
bonds involving early transition elements are stabilized by strong π-donors,
Me_2N^- in this case.

11.2.7. Manganese

Thermochemical studies by Connor's group[50] have enabled bond
dissociation energies, $D(\text{R–Mn})$, for reaction (47), to be estimated (see
Table 11.3).

$$[\text{R-Mn(CO)}_5](g) \rightarrow \text{R·}(g) + [\text{·Mn(CO)}_5](g) \qquad (47)$$

$D(\text{R–Mn})$ falls $R = C_6H_5 > CH_3 > CH_2C_6H_5$, thus paralleling the corresponding $D(\text{R–H})$ and $D(\text{R–CH}_3)$; the value for $F_3C\text{–Mn(CO)}_5$ is large in
comparison, though not unexpectedly so since replacement of H by F in
transition-metal alkyls often increases thermal stability. $D(C_6H_5C(=O)$–

Table 11.3. *Bond Energies and Enthalpies* $(/kcal\,mol^{-1})$

Mn–Me[a]	36	Fe^+–CH_3[b]	69	Th–Me[c]	77.2, 79.3
Ph	40	Co^+	61	Et	70.4, 72.9
CH_2Ph	21	Ni^+	48	n-Bu	71.6, 73.9
CF_3	41	Fe^+=CH_2	96	Ph	90.3, 95.1
COMe	31	Co^+	85	CH_2CMe_3	72.2, 77.0
COPh	21	Ni^+	86	CH_2SiMe_3	80.3, 82.7
$COCF_3$	35	Ni^+=CF_2	47		

[a] Bond dissociation enthalpies for reaction (47).[50]
[b] Bond dissociation energies for reaction (49).[51,52]
[c] Bond disruption energies for first and second Th–R bonds in $[\eta^5\text{-}Me_5C_5ThR_2]$, see reactions (71) and (72) and Section 11.2.13.

Mn) is much less than $D(CH_3C(=O)\text{-}Mn)$ although the analogous $D(R\text{-}H)$ and $D(R\text{-}CH_3)$ are approximately equal. Bond dissociation energies estimated for $R=CH_3$ and CF_3 for the cation show that reaction (48) is somewhat easier than (47):

$$[R\text{-}Mn(CO)_5]^+(g) \rightarrow R\cdot(g) + [Mn(CO)_5]^+(g) \qquad (48)$$

11.2.8. Iron, Cobalt, and Nickel

Bond dissociation energies, $D^0(M^+\text{-}R)$, have been estimated by Beauchamp's group[51,52] for various gas-phase iron, cobalt, and nickel alkyl and carbene positive ions, reaction (49) (see Table 11.3). The value for Ni^+=CF_2 is more like that for a single rather than a double bond.

$$M\text{-}R^+(g) \rightarrow M^+ + R(g) \qquad (49)$$

Of relevance to cytochrome P-450, it has been shown that methyl, phenyl, and 2,2-diphenylvinyl gropus can be transferred reversibly between Fe and N atoms in iron tetraphenyl porphyrin.[53]

11.2.9. Rhodium

An example of intramolecular transmethylation has been provided by Maitlis's group.[54] The CH_3Rh resonances of (5) and (6) are distinct at $-70°C$, but coalesce at higher temperatures. The process is retarded by free MeCN suggesting equilibria and the transition state (7) in Scheme 3. The ethyl group in the analogous system migrates faster than methyl.

Scheme 3

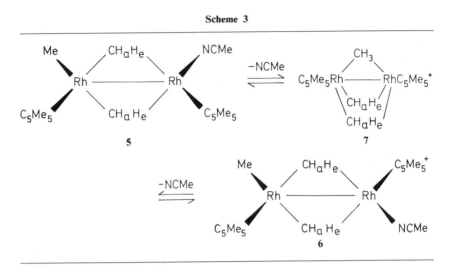

11.2.10. Palladium and Platinum

The rates of the transmetallation reaction (50) fall $R' = PhC{\equiv}C >$ $PrC{\equiv}C > PhCH{=}CH > CH_2{=}CH > Ph > PhCH_2 > CH_3OCH_2 > CH_3 >$ Bu. The *E-/Z-* geometry at the double bond is, on the whole, retained during the transfer of vinyl groups, while the α-C of alkyl groups inverts in polar solvents, suggesting an electrophilic role for the metals.[55]

$$[RCOPd(PPh_3)_2Cl] + R''_3SnR' \rightarrow R''_3SnCl + [RCOPd(PPh_3)_2R'] \quad (50)$$

Puddephatt's group[56,57] has investigated the mechanism of transalkylation or arylation between gold, platinum, and titanium complexes. Selective cleavage of aryl rather than methyl groups as in equations (51) and (52)

cis-$[AuMe_2(C_6H_4Me\text{-}p)(PPh_3)] + HgCl_2 \rightarrow$

$\quad$ *cis*-$[Au(Cl)(Me_2)(PPh_3)] + HgCl(C_6H_4Me\text{-}p)$ $\qquad$ (51)

$[PtMe(C_6H_4Me\text{-}p)(COD)] + [PtI_2(PMe_2Ph)] \rightarrow$

$\quad$ $[PtI(Me)(COD)] + trans$-$[PtI(C_6H_4Me\text{-}p)(PMe_2Ph)]$ $\qquad$ (52)

indicates an S_E2 mechanism[56]; increasing electron density on the aryl group lowers the rate of reaction as would be expected. In other reactions,[56] for example (53), the methyl group is transferred selectively. This is

cis-$[PtMe(C_6H_4Me\text{-}p)(PMe_2Ph)_2] + HCl$ or $HgCl_2 \rightarrow$

$\quad$ *cis*-$[PtCl(C_6H_4Me\text{-}p)(PMe_2Ph)_2] + CH_4$ or $HgClMe$ $\qquad$ (53)

attributed to the S_E(oxidative) mechanism in which oxidative addition and reductive elimination occur. The photoelectron spectra of the platinum complexes which react by the S_E(oxidative) mechanism reveal a HOMO (highest occupied molecular orbital) that is a $5d$ orbital. In contrast the HOMO of the gold complexes, where an S_E2 process occurs, corresponds to the Au–C σ-bond. In the titanium systems,[57] rates are similar in reaction (54) when R is Me and Ph; there is little selectivity when HCl is replaced by $HgCl_2$ in (55), and Hammett ρ values in (55) are abnormal. The reactions

$$[TiR_2Cp_2]^+[Ti(hal)_2Cp_2] \rightarrow 2[Ti(hal)(R)Cp] \qquad (54)$$

$$[TiMe(Ar)Cp_2] + HCl \rightarrow [TiCl(Ar)Cp_2] + CH_4$$

$$\text{or } [TiCl(Me)Cp_2] + ArH \qquad (55)$$

are thought not to be S_E2, and cannot be S_E(oxidative) since the titanium is in an oxidation state of IV.

Cross and Gemmill[58] have not only shown that transfer of methyl- and phenylethynyl groups between the platinum(II) and mercury(II) in Scheme 4 and between two platinum(II) centers in a somewhat similar system involves an S_E(oxidative) process; they have also demonstrated that the oxidative-addition and reductive-elimination processes must be cis. During these reactions it is the resident ethynyl group which moves, the mercury, for example, adding perpendicularly to the plane of the starting complex; the only plausible intermediate is (8) (Scheme 4, L = $PMePH_2$).

Scheme 4

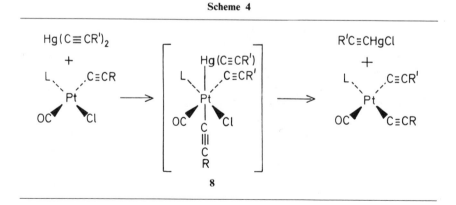

8

An S_E2(cyclic) mechanism would lead to a transition state (or intermediate), (9), from which rearrangement of the C≡CR and C≡CR′ would not occur.

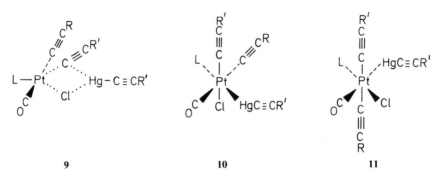

9   10   11

There is the same difficulty if there were a cis-oxidative addition with HgC≡CR′ adding trans to CO as in (10), while (11), which would be formed if HgC≡CR′ added trans to L, could only reach the observed product by a symmetry-forbidden rearrangement.

Both CpTl and Cp₂Hg react with (12) to produce only one isomer (13) (Y = C≡CR, L = *t*-phos).[59] CpTl and (13) (Y = Ph, L = *t*-phos) behave similarly. Treatment of (13) (Y = C≡CR or Ph) with HgCl₂ or *cis*-[PtCl₂(CO)L] also involves reaction of the group trans to L, the Cp ligand being transferred and (12) formed. These systems conform to the pattern expected for an S_E2(cyclic) mechanism. A reaction involving Hg(C≡CMe)₂ is more complicated.

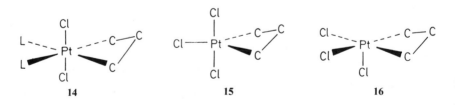

12   13

Calculations by Wilker and Hoffmann[60] suggest how skeletal rearrangements of platinacyclobutanes may occur. (Groups migrating between α- and β-carbon atoms remain on the same side of the ring.) It is assumed that loss of L from (14) leads not to (15), but a species with the geometry of (16). This is isolobal with $C_4H_7^+$, the nonclassical carbonium ion, and both can undergo five different types of rearrangements. One of these corresponds to the Puddephatt–Tipper mechanism as observed in platinum systems. Some of the other rearrangements have possible implications on

14   15   16

other metallacyclobutanes, one leading to migration with loss of stereochemistry, another corresponding to Herrisson–Chauvin metathesis.

11.2.11. *Gold, Mercury, Thallium, and Tin*

Reutov has reviewed the role of ion pairs in various types of S_E reactions involving organomercury and organotin compounds.[61] Some transalkylation reactions involving compounds of gold, mercury, and thallium were discussed in Section 11.2.10.

Various organothallium diacetates, $RTl(OAc)_2$ [$R = PhCH(OMe)CH_2$, for example], react with $Me_2CNO_2^-$ by an initial (reductive) electron transfer process (56) followed by radical, nonchain steps (57) and (58)[62]:

$$RTl(OAc)_2 + Me_2CNO_2^- \rightarrow RTl(OAc) + OAc^- + Me_2CNO_2 \cdot \quad (56)$$

$$RTl(OAc) \rightarrow R\cdot + TlOAc \quad (57)$$

$$R\cdot + Me_2CNO_2\cdot \rightarrow RMe_2CNO_2 \quad (58)$$

The Tl–C bond is cleaved more readily than the Hg–C link in $RHg(OAc)$ in the analogous reaction.

San Filippo and Silbermann[63] continue their studies on the stereochemistry of the reaction of metallate anions with alkyl halides. Me_3SnM (M=Li, Na, K) and 2-octyl tosylate, chloride, and bromide, $R'X$, react as in (59). While complete inversion

$$Me_3SnM + R'X \rightarrow Me_3SnR' + MX \quad (59)$$

at the carbon occurs with tosylate, considerable variation in stereoselectivity is observed for chloride and bromide which depends on factors such as order of addition of reagents, temperature, solvent, and gegenion. Ph_3SnM is more stereoselective than Me_3SnM. Both a free-radical pathway, equations (60)–(64), and an S_N2 process are implied. Somewhat similar results are obtained using Ph_3SiM and Ph_3GeM except that in some instances octenes are formed.

$$R'X + [R_3SnM]_n \rightarrow \{R'\cdot, X^-, [R_3SnM]_n^{\dagger}\} \quad (60)$$

$$\{R'\cdot, X^-, [R_3SnM]_n^{\dagger}\} \rightarrow R'\cdot + [R_3Sn]_n^{\dagger} + X^- \quad (61)$$

$$[R_3SnM]_n^{\dagger} + X^- \rightarrow [R_3SnM]_{n-1} + MX + R_3Sn\cdot \quad (62)$$

$$R'\cdot + R_3Sn\cdot \rightarrow R'SnR_3 \quad (63)$$

$$R'\cdot + [R_3SnM]_n^{\dagger} \rightarrow R'SnR_3 + [(R_3Sn)_{n-1}M_n]^+ \quad (64)$$

Allyl(tributyl)tin systems can be deallylated by polyhaloalkyl radicals.[64] The reaction is favored by electron-accepting substituents and is thought to be radical chain in nature involving an S'_H process[65]:

$$CH_3CH=CHCH_2SnBu_3 + \cdot CCl_3 \rightarrow CH_2=CHCHMeCCl_3 + SnBu_3 \quad (65)$$

Transmethylation reactions between various tin complexes have been studied by Petrosyan,[65] equations (66) to (68) (X = Cl or Br). Rates of reaction are dependent on solvent, rising with increasing polarity and with decrease in solvating power. Some of the reactions are catalyzed by products so that fractional orders may be observed.

$$Me_4Sn + SnX_4 \rightarrow Me_3SnX + MeSnX_3 \quad (66)$$

$$Me_3SnX + SnX_4 \rightarrow Me_2SnX_2 + MeSnX_3 \quad (67)$$

$$Me_4Sn + MeSnX_3 \rightarrow Me_3SnX + Me_2SnX_2 \quad (68)$$

11.2.12. Lutetium

Activation of a C–H bond frequently involves oxidative addition. However, in some examples discovered by Watson[66] there is no change of oxidation state. $[Lu(\eta^5\text{-}C_5Me_5)_2H]$ reacts with C_6H_6 and $SiMe_4$ forming $Lu–C_6H_5$ and $Lu–CH_2SiMe_3$ links. $[Lu(\eta^5\text{-}C_5Me_5)_2Me]$ behaves similarly with CH_4 (69), C_6H_6 (70), and $SiMe_4$, but with pyridine and $CH_2=PPh_3$ it gives Lu–C σ-bonds contained in ring systems.

$$[Lu(\eta^5\text{-}C_5Me_5)_2CH_3] + {}^{13}CH_4 \rightleftharpoons [Lu(\eta^5\text{-}C_5H_5)_2{}^{13}CH_3] + CH_4 \quad (69)$$

$$[Lu(\eta^5\text{-}C_5Me_5)_2CH_3] + C_6H_6 \rightleftharpoons [Lu(\eta^5\text{-}C_5H_5)_2C_6H_5] + CH_4 \quad (70)$$

The rate equations for (69) and (70) contain two terms, one first order in lutetium, the other first order in both lutetium and hydrocarbon. The yttrium analogue will also undergo reaction (69).[67]

11.2.13. Thorium

The enthalpies of reaction for (71) and (72) have been measured and used to calculate bond disruption enthalpies $\bar{D}$ for the Th–R bond.[68] The

$$[Th(\eta^2\text{-}C_5Me_5)R_2] + t\text{-BuOH} \rightarrow [Th(\eta^2\text{-}C_5Me_5)(R)(Ot\text{-Bu})] + RH \quad (71)$$

$$[Th(\eta^2\text{-}C_5Me_5)_2(R)(Ot\text{-Bu})] + t\text{-BuOH} \rightarrow [Th(\eta^2\text{-}C_5Me_5)_2(Ot\text{-Bu})_2]$$
$$+ RH \quad (72)$$

values of $\bar{D}$ are large compared with those for some other metal–carbon σ-bonds, namely 70–95 kcal mol^{-1}; related values for other (conventional) M–C σ-bonds vary from 17 to 41 kcal mol^{-1} (see Tables 11.1 to 11.3 in this chapter and Tables 11.3 and 11.4 of Volume 2[69]). However, the mean bond disruption enthalpy of the Ti–Ph bond in [Cp$_2$TiPh$_2$] is 74 kcal mol^{-1}.[70]

11.3. Oxidative Addition and Reductive Elimination

Oxidative addition is often quoted as proceeding by four recognized mechanisms: concerted, S_N2, radical chain, and radical nonchain. In principle reductive elimination can go by any of the mechanisms, but there are reactions for which designation is more complicated. For example, the change of a metal dialkyl to a metallacycloalkane and free alkane can involve an individual step in which reductive elimination occurs, see for example equation (75) in the thermal decomposition of various platinum compounds, as discussed in Volume 2, on p. 287[71] [equations (73)–(76)]. In (75) the hydrogen elimination can be β, γ, or even δ. However,

$$L_2PtR_2 \rightleftharpoons LPtR_2 + L \qquad (73)$$

$$LPtR_2 \rightleftharpoons H(L)(R)\overline{Pt\text{-}R}^{-H} \qquad (74)$$

$$H(L)(R)\overline{Pt\text{-}R}^{-H} \rightarrow L_2\overline{Pt\text{-}R}^{-H} + RH \qquad (75)$$

$$L\overline{Pt\text{-}R}^{-H} + L \rightarrow L_2\overline{Pt\text{-}R}^{-H} \qquad (76)$$

as Tulip and Thorn[72] have emphasized, the migration of the hydrogen can be either an elimination [as in (75)] or an abstraction, that is, A or B in Scheme 5. This scheme shows a γ process, but α migrations are possible

Scheme 5

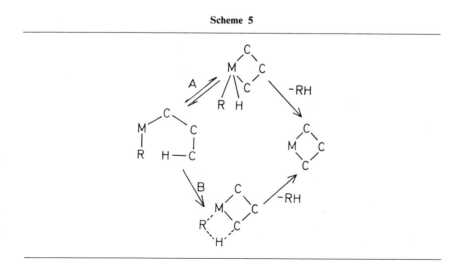

particularly if tantalum or tungsten are involved. Though hydrogen migrations are not reductive elimination, they have become such an important aspect of it that they are included in the review. Also considered is the question of whether migration is abstraction or elimination.

Closely related to H elimination is the phenomenon that the C–H group can coordinate to a metal as in C–H → M. Brookhart and Green[73] have gathered evidence for such bonding.

Several useful reviews have appeared. Mondal and Blake[74] have collected thermochemical data on oxidative addition, Halpern[75] has investigated the formation of C–H bonds by reductive elimination, while in a thought-provoking article on activation of $C[sp^3]$-X bonds, Chanon[76] stresses the importance of electron transfer in oxidative addition (among other topics). In a discussion of oxidation addition and reductive elimination involving two metal centers, Halpern[77] classifies and gives examples of three mechanisms whereby binuclear reductive elimination can occur: concerted two center (77), concerted one center (78), and free-radical [(79)–(81)] reactions given in Scheme 6.

Scheme 6

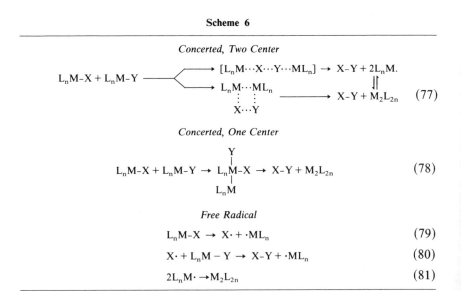

Concerted, Two Center

$$L_nM\text{-}X + L_nM\text{-}Y \longrightarrow \begin{cases} [L_nM\cdots X\cdots Y\cdots ML_n] \rightarrow X\text{-}Y + 2L_nM. \\ L_nM\cdots ML_n \\ \qquad \vdots \quad \vdots \\ \qquad X\cdots Y \end{cases}$$

$$\rightarrow X\text{-}Y + M_2L_{2n} \quad (77)$$

Concerted, One Center

$$L_nM\text{-}X + L_nM\text{-}Y \rightarrow L_nM\text{-}X \rightarrow X\text{-}Y + M_2L_{2n} \quad (78)$$
(with Y above and L_nM below the central M)

Free Radical

$$L_nM\text{-}X \rightarrow X\cdot + \cdot ML_n \quad (79)$$

$$X\cdot + L_nM - Y \rightarrow X\text{-}Y + \cdot ML_n \quad (80)$$

$$2L_nM\cdot \rightarrow M_2L_{2n} \quad (81)$$

11.3.1. Magnesium

Studies on the reaction of magnesium with phenyl and alkyl halides at low temperature in the solid phase to form Grignard reagents point to Scheme 7.[78]

Scheme 7

$$RX + Mg_n \rightarrow \{RX^-, Mg_n^+\} \rightarrow RMgX + Mg_{n-1}$$

$$\text{or} \quad \rightarrow Mg_nX\cdot + R\cdot \rightarrow RMgX + Mg_{n-1}$$

$$Mg_nX\cdot + RX \rightarrow Mg_{n-1} + MgX_2 + R\cdot$$

$$R\cdot + R\cdot \rightarrow R_2$$

11.3.2. Gallium and Indium

Reductive elimination does not occur in the reaction of $M(CH_2SiMe_3)_3$ with alkali-metal hydrides, MH. $MM(III)(CH_2SiMe_3)_3H$ is formed and not $MM(I)(CH_2SiMe_3)_2$ (M = Ga or In)[79] as previously thought.

11.3.3. Silicon

Calculations show that in reaction (82) in the transition state the incoming hydrogen atoms are still rather H_2-like in character:[80]

$$SiH_2 + H_2 \rightarrow SiH_4 \tag{82}$$

Recoil ^{31}Si atoms abstract F from PF_3 two to three times more quickly than H from PH_3. A stepwise mechanism is thought to occur.[81]

Some work on hydrosilylation is reported in Section 11.3.8.

11.3.4. Titanium, Zirconium, and Hafnium

$[(\eta^5\text{-}C_5Me_5)_2Ti(CH_3)_2]$ decomposes thermally in toluene following first-order kinetics giving $[(\eta^5\text{-}C_5Me_5)(C_5Me_4CH_2)Ti(CH_3)]$ and methane. For the overall process, $\Delta H^{\ddagger}$ and $\Delta S^{\ddagger}$ are 27.6 kcal mol^{-1} and -2.9 cal deg^{-1} mol^{-1}, respectively.[82] H/D-labeling experiments show that it is the Ti–Me hydrogen atoms that are involved in the formation of the methane, and that either α-hydrogen abstraction or elimination occurs; comparison with $[(\eta^5\text{-}C_5Me_5)_2Ti(CD_3)_2]$ shows k_H/k_D to be 2.92 and independent of temperature, points which, respectively, emphasize the role of the TiMe hydrogen and hint perhaps at an abstraction rather than an elimination process.

Zirconium dialkyls, $[R_2Zr\{N(SiMe_3)_2\}_2]$, decompose thermally to give the carbene dimer and RH, the latter involving either γ-elimination (or abstraction) of hydrogen (R = Me, Et, or CH_2SiMe_3).[83] The analogous hafnium compounds decompose less readily but probably in the same way. In contrast $[Me_2Ti\{N(SiMe_3)_2\}_2]$ is stable thermally, perhaps for kinetic rather than thermodynamic reasons associated with the smaller size of titanium.

Two examples of reductive elimination[84] from bimetallic systems are shown in Scheme 8. Although $[Cp_2ZrMe_2]$ is coordinatively unsaturated, elimination takes place more rapidly from $[Cp_2Zr\{\eta^2\text{-}C(=O)Me\}Me]$ (perhaps owing to rapid transfer of the proton from HMo to the oxygen of the $Zr\{C(=O)Me\}$ group), that is reaction (84) is faster than reaction (83). It is notable that acetaldehyde is not eliminated in the second reaction.

Scheme 8

$$[Cp_2ZrMe_2] + [CpMo(CO)_3H] \xrightarrow{-CH_4} [Cp_2ZrMeMo(CO)_3Cp] \qquad (83)$$

$$-CO \Big\|\Big. +CO \qquad\qquad\qquad\qquad\qquad\qquad \Big\downarrow +CO$$

$$[Cp_2Zr(\eta^2\text{-}C(=O)Me)Me] + [CpMo(CO)_3H] \xrightarrow{-CH_4} [Cp_2Zr(\eta_2\text{-}C(=O)Me)Mo(CO)_3Cp]$$

$$(84)$$

11.3.5. Tantalum

$[\eta^5\text{-}C_5Me_5)Ta(NMe_2)Me_3]$ decomposes to $[(\eta^5\text{-}C_5Me_5)Ta(\eta^2\text{-}CH_2NMe)Me_2]$ and methane by first-order kinetics.[85] While it is not possible to say whether hydrogen abstraction or elimination takes place, the very high kinetic isotope effect of 9.7 suggests that the migrating hydrogen atom is very loosely bonded in the transition state. Hoffmann's group have carried out extended Hückel-type calculations to test the feasibility of α-elimination.[86] They have shown that the formation of a hydride from the five-coordinate 14-electron species as in reaction (85) is symmetry

$$[H_4TaCH_2]^{3-} \rightarrow [H_5TaCH]^{3-} \qquad (85)$$

forbidden. They have also shown that hydrogen transfer from methylidene to methyl to give methane is reasonable energetically in reaction (86) and

$$[H_3Ta(CH_3)(CH_2)]^{3-} \rightarrow [H_3Ta(CH)]^{3-} + CH_4 \qquad (86)$$

in a system containing two fewer electrons, but not in ones with two or four more.

11.3.6. Chromium, Molybdenum, and Tungsten

Matrix isolation has been used to demonstrate a reversible α-hydrogen elimination in a chromium-containing system, reaction (87):[87]

$$[CpCrCH_3(CO)_2] \rightleftharpoons [CpCrH(CH_2)(CO_2] \qquad (87)$$

The same technique has been used by Perutz's group[88] to show that $[Cp_2W]$ is very probably formed during the ultraviolet photolysis of $[Cp_2W(CH_3)H]$ to give methane.

Alkenes are catalytically hydrogenated by photoexcited $[H_4M(dppe)_2]$ in toluene (M = Mo or W). Several oxidative addition/reductive elimination steps are postulated.[89]

The first example of oxidative addition, namely of HCl, to a $W \equiv W$ quadruple bond has been observed by Cotton and Mott.[90] Both the H and the Cl bridge the two W atoms.

Addition of strong acid to

$$[(\eta^5\text{-}C_6H_5)Cr(CO)_2[P(Ph)_2CH_2P(Ph)_2](OC)_2Cr(\eta^5\text{-}C_6H_5)]$$

and its arsenic analogue leads to rapid evolution of H_2. It is proposed that reductive elimination takes place through a diprotonated species.[91]

Kinetic data for an α-hydrogen abstraction are provided very appropriately by Pedersen and Schrock.[92] $[W(NPh)(CH_2SiMe_3)_4]$ decomposes thermally in toluene to give an alkylidene, as opposed to a ring, reaction (88). The activation parameters, $\Delta H^{\ddagger} = 22 \, \text{kcal mol}^{-1}$ and $\Delta S^{\ddagger} = -8 \, \text{cal K}^{-1} \text{mol}^{-1}$, are not dissimilar to those observed for the thorium system in Section 11.3.17.

$$[W(NPh)(CH_2SiMe_3)_4] \rightarrow [W(NPh)(CHSiMe_3)(CH_2SiMe_3)_2] + SiMe_4$$

(88)

11.3.7. Manganese and Rhenium

The thermochemical studies by Connor's group[50] have enabled them to make estimates of the standard enthalpies of reactions involving reductive elimination from various bimanganese processes, the kinetics of some of which were reported in Volume 2.[93] Reactions (89)–(91) for the R or COR given in Table 11.3 are all exothermic, while (92) is thermoneutral (X = Cl, Br, and I), with two exceptions, namely CF_3I and C_6H_5COBr:

$$[Mn(CO)_5R] + [Mn(CO)_5H] \rightarrow [Mn_2(CO)_{10}] + RH \qquad (89)$$

$$[Mn(CO)_5COR] + [Mn(CO)_5H] \rightarrow [Mn_2(CO)_{10}] + RCHO \qquad (90)$$

$$[Mn(CO)_5COR] + 3[Mn(CO)_5H] \rightarrow 2[Mn_2(CO)_{10}] + RCH_2OH \qquad (91)$$

$$[Mn_2(CO)_{10}](g) + RX(g) \rightarrow [Mn(CO)_5R](g) + [Mn(CO)_5X](g) \qquad (92)$$

Rhenium arouses interest as an element whose compounds can activate C–H bonds. The η^2-species, $[(\eta^2\text{-}C_6H_6)Re(CO)(Cp)NO]^+$, has been identified and proposed as an intermediate in C–H activation,[94] since it takes part in the type of acid–base equilibrium depicted in (93):

$$(\eta^1\text{-}C_6H_5)\text{Re} \underset{Et_3N}{\overset{HX}{\rightleftharpoons}} (\eta^2\text{-}C_6H_6)\text{Re}^+ \tag{93}$$

Activation of C–H bonds of a methyl group occurs when rhenium atoms generated from an electron-gun furnace are co-condensed with various methylbenzenes.[95]

11.3.8. Iron, Ruthenium, and Osmium

CH_3FeH provides the first example of a ligand-free system undergoing reductive elimination. Ozin and McCaffrey[96] have observed that it takes part in a photoreversible reaction as in equation (94):

$$CH_3FeH \underset{300\ nm}{\overset{420\ nm}{\rightleftharpoons}} Fe + CH_4 \tag{94}$$

Monoatomic iron will also photoinsert into the C–H bonds of ethane and n-propane and into the C–C bond of cyclopropane.[97] Irradiation (in either the visible or ultraviolet) of cis-[FeH$_2$(dppe)$_2$] gives H$_2$, the other product being [HFe(C$_6$H$_4$P(Ph)CH$_2$PPh$_2$)(dppe)]. Treatment of the latter with H$_2$ gives the starting material.[98]

Reductive elimination of HSiPh$_3$ from [Fe(CO)$_4$(H)(SiPh$_3$)], which is an intermediate in the [Fe(CO)$_5$]-catalyzed hydrosilylation of alkenes, has been studied by Bellachioma and Cardaci.[99] Elimination is from the tricarbonyl not the tetracarbonyl but requires the presence of CO or PPh$_3$.

$$Fe(CO)_4(H)(SiPh_3) \rightleftharpoons Fe(CO)_3(H)(SiPh_3) + CO \tag{95}$$

A heterobimetallic system has been observed by Geoffroy's group to oxidatively add and reductively eliminate H$_2$ in two ways, reactions (96) and (97):[100]

$$[(OC)_3(Ph_3P)\overline{Fe(\mu\text{-}PPh_2)Ir}(PPh_3)(CO)_2] + H$$
$$\rightleftharpoons [(OC)_3(Ph_3P)Fe(\mu\text{-}PPh_2)Ir(H)_2(PPh_3)(CO)_2] \tag{96}$$

$$[(OC)_3(Ph_3P)\overline{Fe(\mu\text{-}PPh_2)Ir}(PPh_3)(CO)] + 2H_2$$
$$\rightleftharpoons [(OC)_3(Ph_3P)(H)_2\overline{Fe(\mu\text{-}PPh_2)Ir}(H)_2(PPh_3)(CO)] \tag{97}$$

Gaffney and Ibers have prepared [Ru(η^2-SCO)(CO)$_2$(PPh$_3$)$_2$] by addition of OCS to [Ru(CO$_2$(PPh$_3$)$_2$]; there is evidence for the iron analog.[101]

Ruthenium clusters are active towards hydrogen. The overall exchange reaction (98) has a rate-determining step (99) which involves reductive

$$[Ru_3(\mu\text{-}H)_3(\mu\text{-}COMe)(CO)_9] + CO \rightleftharpoons [Ru_3(\mu\text{-}H)(\mu\text{-}COMe)(CO)_{10}] + H_2 \tag{98}$$

$$[Ru_3(\mu\text{-}H)_3(\mu_3\text{-}COMe)(CO)_9] \rightleftharpoons [Ru_3(H)(COMe)(CO)_9] + H_2 \quad (99)$$

elimination of H_2 apparently from a $(\mu\text{-}H)_3$ system.[102] However, $\Delta H^{\ddagger}$ and $\Delta S^{\ddagger}$ are 31.0 kcal mol^{-1} and 8 cal deg^{-1} mol^{-1}, respectively, values which are not dissimilar to those involving reductive elimination of H_2 from monometal complexes. The H/D kinetic isotope effect is also similar. It is suggested that a terminally bonded hydride could be formed in reaction (99) prior to reductive elimination.

The tetrauthenium cluster $[H_4Ru_4(CO)_{12}]$ catalyzes the hydrogenation of ethene. Deuterium labeling shows that HD and C_2H_3D are formed more quickly than ethane and that the solvent, heptane, is not involved.[103] Reactions (100)–(103) are proposed:

$$[H_4Ru_4(CO)_{12}] \rightleftharpoons [H_4Ru_4(CO)_{11}] + CO \quad (100)$$

$$[H_4Ru(CO)_{11}] + H_2 \rightleftharpoons [H_6Ru_4(CO)_{11}] \quad (101)$$

$$[H_4Ru_4(CO)_{11}] + C_2H_4 \rightleftharpoons [H_3Ru_4(CO)_{11}(C_2H_5)] \quad (102)$$

$$[H_3Ru_4(CO)_{11}(C_2H_5)] + H_2 \rightarrow [H_4Ru_4(CO)_{11}] + C_2H_6 \quad (103)$$

Relative values are quoted for K_{100}, K_{101}, K_{102}, and k_{103} at 72°C, showing $K_{101} > K_{102} > K_{100}$.

H_2 adds reversibly to $[Os_3Pt(\mu\text{-}H)_2(CO)_{10}\{P(c\text{-}Hex)_3\}]$ to give a $(\mu\text{-}H)_4$-containing compound with a closo structure.[104]

11.3.9. Gas-Phase Reactions of Fe^+, Co^+, Ni^+, and Rh^+

A considerable amount of effort has been devoted to the reactions in the gas phase of alkanes and metal ions, particularly, Fe^+, Co^+, and Ni^+, by Beauchamp's[51,52,105,106] and Freiser's[107–110] groups. Individual steps and species seem to be similar to those involved and postulated for more conventional processes, for example, insertion into C–H and C–C bonds (namely oxidative addition), β-hydrogen elimination, π-alkene complexes, and metallacycloalkanes. While methane may add oxidatively to Fe^+, Co^+, and Ni^+ to give $HMCH_3^+$ from which MCH_3^+ and MCH_2^+ are formed, the rather large production of MH^+ suggests that direct abstraction of H from the methane can also occur. All three ions can insert into C–H and C–C bonds of higher alkanes (butane and beyond). Thus in the reaction of Ni^+ and *n*-butane, the following species are postulated: $[n\text{-}PrNiMe]^+$, $[EtNiEt]^+$, $[NiH(Et)(\eta^2\text{-}C_2H_4)]^+$, $[Ni(\eta^2\text{-}CH_2{=}CHCH_3)]^+$, $[Ni(\eta^2\text{-}C_2H_4)]^+$, $[H_2Ni(\eta^2\text{-}C_2H_4)_2]^+$, and $[Ni(\eta^2\text{-}C_2H_4)_2]^+$. The last species is quite distinct from an isomer produced from Ni^+ and cyclopentanone which could be either [nickelacyclopentane]$^+$ or a [nickel(η^2-butene)]$^+$. In longer-chain molecules there is an increase in selectivity, $Fe^+ < Co^+ < Ni^+$, toward internal as opposed to terminal C–C bonds.

All three metal ions react with cyclobutane to give ring-cleaved products. Dehydrogenation occurs with cyclopentane and cyclohexane, the formation of cycloalkene and cyclodiene metal complexes being postulated. It is suggested that in both cases insertion into the C–C bond occurs, but that in the latter instance only α-hydrogen abstraction then takes place.

In the case of Rh^+ and alkanes there is evidence for C–H but not C–C insertion.[109] For n-propane it is suggested that species such as $[HRhCHCMe_2]^+$ and $[H_2Rh(\eta^3\text{-}CH_2CHCH_2)(CH_2CH_2Me)]^+$ are formed as intermediates prior to the production of $[RhC_3H_6]^+$ and $[RhC_6H_{10}]^+$.

11.3.10. Cobalt

Halpern[111] has analysed the data of Mooiman and Pratt, discussed in Volume 2[112] on the mechanism of reaction of H_2 with $[Co(CN)_5]^{3-}$, which was thought to prove a heterolytic splitting mechanism. He concludes that the data are more compatible with the older homolytic process (104)

$$H_2 + 2[Co(CN)_5]^{3-} \rightarrow 2[Co(CN)_5H]^{3-} \tag{104}$$

(which could be intermolecular as shown or proceed through rapidly formed $[Co_2(CN)_{10}]^{6-}$; however, the latter is symmetry forbidden[77]).

The reaction of cobaloximes, $[Co(dmgH)_2L]$, with polyhalomethanes has been studied by Espenson and McDowell (L = PPh_3 or py).[113] Product analysis and use of scavengers demonstrate the same steps occur as for alkyl halides, reactions (105) and (106), the first being rate determining. In

$$[Co(dmgH)_2L] + RX \rightarrow [XCo(dmgH)_2L] + R\cdot \tag{105}$$

$$[Co(dmgH)_2L] + R\cdot \rightarrow [RCo(dmgH)_2L] \tag{106}$$

reaction (105) the C–Br bond is broken much more readily than that of C–Cl, while negative entropies of activation suggest an ordered transition state. An estimate is made for k_{106} at 25°C of $8 \times 10^7 \, M^{-1} \, s^{-1}$, which is reasonable since the rate constants for reactions of radicals with $[Co(Me_6[14]4,11\text{-diene-}N_4)(OH_2)_2]^{2+}$ lie between 1×10^7 and 7×10^8 $M^{-1} \, s^{-1}$.

The rates of reaction of various cobalt(I) salen-type complexes and n-PrBr, n-BuBr, and t-BuBr have been measured by Puxeddi and Costa.[114] A Co–alkyl bond is formed in the first two cases, for example reaction (107), but in the last β-elimination of hydrogen takes place giving isobutene.

$$[Co(I)(salen)]^- + n\text{-PrBr} \rightarrow [n\text{-Pr-Co(salen)}] + Br^- \tag{107}$$

For related unhindered cobalt systems there is a linear relationship between the log of the rate constant and the Co(II) → Co(I) half-wave potential; introduction of bulky groups cause deviations. It is suggested that the rate-determining step involves electron transfer to give an intermediate, such as $\{[Co(II)(salen)]\cdots R\cdots Br^-\}$, so that the reaction, while having some S_N2 character, is not of conventional type.

Cobaloxime(I) reacts with 3-oxa-5-hexenyl tosylate to form the organocobalt compound expected in an S_N2 reaction (108).

$$[Co(I)]^- + RX \rightarrow Co-R + X^- \tag{108}$$

However, the bulky 2,2-diphenyl derivative, Co-R, forms (3,3-diphenyl-4-oxa-cyclopentyl)methylcobaloxime, Co-R', suggesting the formation of the radical, R·, which cyclizes to R·' before bonding to cobalt, reactions (109) and (110)[115]:

$$[Co(I)]^- + RX \rightarrow Co(II) + RX^- \rightarrow Co(II) + X^- + R\cdot \tag{109}$$

$$R\cdot \rightarrow R\cdot' \rightarrow Co-R' \tag{110}$$

11.3.11. Rhodium

H_2 adds oxidatively to $[((4\text{-tolyl})_3P)_2Rh(Cl)B]$, where B is $P(4\text{-tolyl})_3$, pyridine, or tetrahydrothiophene, to give six coordinate species in which the hydrides are cis to each other. $\Delta H°$ for the first and last B are 11.0 and 11.6 kcal mol^{-1}, respectively.[116]

Milstein[117,118] has shown that $cis\text{-}[RhH(COR)(PMe_3)_3Cl]$ (R = Me, Ph, 4-C_6H_4F, OMe) can undergo either reductive elimination of aldehyde or decarbonylation to the alkane. A common intermediate appears to be involved, $[HRh(COR)(PMe_3)_2Cl]$ [see reactions (111)–(113)]. He has also isolated and characterized the intermediate believed to be formed in several

$$cis\text{-}[RhH(COR)(PMe_3)_3Cl] \rightleftharpoons [RhH(COR)(PMe_3)_2Cl] + PMe_3 \tag{111}$$

$$[RhH(COR)(PMe_3)_2Cl] + PMe_3 \rightleftharpoons [Rh(PMe_3)_2Cl] + RCHO \tag{112}$$

$$[RhH(COR)(PMe_3)_2Cl] + PMe_3 \rightleftharpoons [RhH(R)(CO)(Cl)(PMe_3)_2]$$

$$\rightarrow [Rh(CO)(PMe_3)_2] + RH \tag{113}$$

catalytic processes, namely $[RhH(CH_2COR)(PMe_3)_3Cl]$. (R = Me or Ph; H and CH_2COR are cis; H and the mer P_3 are in plane.) Both of these complexes undergo reductive elimination to give the expected ketones. For the first, the process is accelerated by the phosphine sponge, $[Rh(acac)(C_2H_4)_2]$, and retarded by free PMe_3 as expected for a dissociative process, reactions (114) and (115):

$$[RhH(CH_2COCH_3)(PMe_3)_3Cl] \rightleftharpoons [RhH(CH_2COCH_3)(PMe_3)_2Cl] + PMe_3$$

(114)

$$[RhH(CH_2COCH_3)(PMe_3)_2Cl] \rightarrow [Rh(PMe_3)_3Cl] + CH_3COCH_3 \qquad (115)$$

At 31°C, $k_{114} = 1.02 \times 10^{-4}\,s^{-1}$. Use of the $[RhD(CD_2COCD_3)]$ isotopomer shows that the process is intramolecular. This system provides the first example of a reductive elimination process involving formation of a C–H bond taking place from a complex of reduced coordination number. For example, the corresponding reaction involving cis-$[PtH(Me)(PPh_3)_2]$ is not retarded by free PPh_3.

Reductive elimination of RCl from $[R(X)RhCl(CO)L_2]$ (which occurs during the decarbonylation of acid halides using rhodium catalysts) has been assumed to involve a *trans* stereochemistry; the fact that the yields of chlorobenzene and bromobenzene from benzoyl halides rise as the cone angle of the phosphine, L, is decreased, confirms this point of view.[119]

Arene C–H bonds can be activated easily by octaethylprophyrinato-rhodium(III) chloride, [(OEP)RhCl], and $AgClO_4$ or $AgBF_4$. The rate increases as substituents on the benzene become more donor: Cl < H < Me < OMe, while Hammett ρ-values suggest that an ionic electrophilic substitution reaction occurs.[120] Although it is not stated, presumably the formation of AgCl causes $[(OEP)Rh]^+$ to be produced as the reactive intermediate.

Jones and Feher[121] have shown that the photoactivated reaction of d_6-benzene with $[Rh(\eta^5\text{-}C_5Me_5)(PMe_3)(H)_2]$ to give $[Rh(\eta^5\text{-}C_5Me_5)\text{-}D(C_6D_5)(PMe_3)]$ may involve oxidation addition of arene to $Rh(C_5Me_5)$-(PMe_3) to form a η^2-complex, $[Rh(\eta^2\text{-}C_6H_6)(\eta^5\text{-}C_5Me_5)(PMe_3)]$. Alkane activation is also possible.[122] Warming $[Rh(\eta^5\text{-}C_5Me_5)(PMe_3)(H)(Me)]$ to −17°C in C_6D_6 gives $[Rh(\eta^5\text{-}C_5Me_5)(PMe_3)(D)(C_6H_5)]$ and CH_4 ($k = 6.5 \times 10^{-5}\,s^{-1}$), while irradiation of the dihydride, $[Rh(\eta^5\text{-}C_5Me_5)(PMe_3)\text{-}(H)_2]$, in (liquid) propane at −55°C gives $[Rh(\eta^5\text{-}C_5Me_5)(PMe_3)(H)(n\text{-}Pr)]$. Presumably $(\eta^5\text{-}C_5Me_5)(Me_3P)Rh$ is also an intermediate in these reactions.

The photoinduced oxidation addition to $[Rh_2(\mu\text{-dicp})_4]^{2+}$ (dicp = 1,3-diisocyanopropane) of n-butyl iodide in MeCN yields chiefly $[n\text{-}BuRh(\mu\text{-dicp})_4RhI]^{2+}$. High quantum yields, the fact that the reaction is inhibited by reagents such as benzoquinone and oxygen, and the formation of 1-butanol and butyraldehyde in the latter case, indicate a radical-chain process (116) and (117)[123]:

$$n\text{-}Bu\cdot + [Rh_2(\mu\text{-dicp})_4]^{2+} \rightarrow [n\text{-}BuRh(\mu\text{-dicp})_4Rh]^{2+}_2 \qquad (116)$$

$$[n\text{-}BuRh(\mu\text{-dicp})_4Rh]^{2+}_2 + n\text{-}BuI \rightarrow [n\text{-}BuRh(\mu\text{-dicp})RhI]^{2+}_2 + n\text{-}Bu\cdot$$

(117)

Enthalpies of oxidative addition of I_2 to [RhCl(CO)dppe], [Rh(dppen)$_2$]BF$_4$, [Rh(dppe)$_2$]BF$_4$, and [RhCl(CO)dppm$_2$] are found to be -143, -118, -110, and -147 kJ mol^{-1}, respectively.[124]

Normally in reactions such as (118), the equilibrium lies to the right. Rhodium octaethylporphyrin hydride appears to be the first such hydride for which the reverse process is thermodynamically favorable, reacting with aldehydes to give α-hydroxyalkylrhodium species.[125]

$$\text{MCH(R)OH} \rightleftharpoons \text{MH} + \text{RCHO} \qquad (118)$$

The first example of oxidative addition to a heterobimetallic compound has been reported.[126] MeI adds to [CpRh(μ-PMe$_2$)$_2$Mo(CO)$_4$] to yield (19). Two points of interest are the addition of the methyl group only to rhodium and the relative positions of it and the iodide ligand. (Incidentally, the article in which this work is reported contains a very good bibliography of oxidative addition to homobimetallic sytems.)

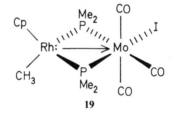

19

11.3.12. Iridium

The oxidative addition of H_2 to [IrX(CO)(dppe)] is highly stereoselective, (20) being formed initially (X = Cl, Br or I).[127] The more stable isomer (21) is produced over several hours. Similarly, D_2 adds to [IrH(CO)-(dppe)] to give (22) before other isotopomers are formed. It appears that it is easier for H_2 to approach parallel to the OC–Ir–P axis than to the X (or H)–Ir–P.

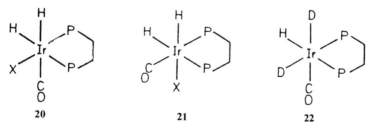

20   21   22

Anton and Crabtree[128] have shown that H_2 adds to [Ir(dct)(PPh$_3$)$_2$]$^+$, in CH$_2$Cl at reduced temperatures, to give *cis, cis*-[IrH$_2$(dct)(PPh$_3$)$_2$] (dct = dibenzo[a,e]cyclotetraene). However, at 30°C in the presence of MeOH, the

thermodynamically more stable isomer *cis,trans*-[IrH₂(dct)(PPh₃)₂] is
formed. Isomerization probably proceeds by loss of H⁺ through
[Ir(dct)H(PPh₃)₂], which can be prepared by the action of *t*-BuOK on the
cis,trans-dihydride. Crabtree and Uriate[129] have shown that only one
isomer (kinetically and thermodynamically favored) is formed in the
analogous reaction of [Ir(cod)(Ph₂POCH₂CH₂PPh₂)]⁺. The H add cis to
each other and trans to one of the C=C bonds and to the P=O, as opposed
to the P, group; the latter suggests that the reaction is directed by electronic
rather than steric effects.

First examples have been quoted of the oxidative addition of methane.
In methane under pressure, [Ir(η^5-C₅Me₅)(CO)₂] in perfluoroexane slowly
reacts to give [Ir(η^5-C₅Me₅(CO)(H)(Me)] (the C₅H₅ analogue behaving
similarly),[130] while heating [Ir(η^5-C₅Me₅)(PMe₃)(H)(c-C₆H₁₁)] in cyclooctane yields [Ir(η^5-C₅Me₅)(PMe₃)(H)(Me)].[131]

Irradiation of [Ir(η^5-C₅Me₅)(PMe₃)(H)₂] in a hydrocarbon, RH, leads
to extrusion of H₂ and formation of [Ir(η^5-C₅Me₅)(PMe₃)(H)(R)].[132] It is
suggested that [Ir(η^5-C₅Me₅)(PMe₃)] is formed as an intermediate and that
the addition of RH to it proceeds by a concerted process. Reaction rates
rise by a factor of about 40 in the sequence: R is cyclooctane <
cyclodecane < cyclohexane < neopentane < cyclopentane < cyclopropane
< benzene. At elevated temperatures reductive elimination of RH occurs
from [Ir(η^5-C₅Me₅)(PMe₃)(H)(R)] to give the intermediate which can then
react with hydrocarbon solvent as before.

CS₂ and OCS can add oxidatively[133] to *trans*-[IrCl(CO)(phos)]₂ as in
(119). The stability of the adduct is increased as the phosphine becomes

$$[IrCl(CO)(phos)_2] + SCY \rightleftharpoons [Ir(\eta^2\text{-}SCY)Cl(CO)(phos)_2] \quad (119)$$

more basic (and/or smaller), and when Y is changed from O to S suggesting
donation from the Ir center to the SCY unit. (Other η^2-SCY complexes are
mentioned under ruthenium and platinum.)

Iodine adds oxidatively to the di-iridium complex, [Ir₂(μ-*t*-
BuS)₂(CO)₂(phos)₂] to give an iridium(II) system, [Ir₂(μ-*t*-BuS)₂(phos)₂I₂].
The reaction is fast and maybe radical in nature.[134]

11.3.13. Nickel

Siegbahn's group[135,136] have calculated energy barriers to reaction and
the energies and shapes of the products formed when H–H, H₃C–H, and
H₃C–CH₃ add oxidatively to nickel. The addition of H₂ is estimated to
occur very readily, and the formation of HNiH is exothermic; this is ascribed
to some H–H bonding which persists even in the product as illustrated by
the low HNiH angle of 50°. (The incoming H₂ in the reaction, H₂ + SiH₂ →
SiH₄, are also relatively close in the transition state—Section 13.3.3.)

Preparation of the first *cis* (aryl–alkyl) nickel(II) complexes, *cis*-[NiMe(C$_6$H$_4$X)(dmpe)] (X = *p*-MeO, *p*-Me, H, *p*-F, and *o*-Me), is reported.[137] They decompose in benzene at room temperature (which *trans*-[NiMe(C$_6$H$_4$X)(PEt$_3$)$_2$] does not) to give chiefly MeC$_6$H$_4$X. Kinetics are first order. The rate of decomposition of the cis complexes is accelerated by free phosphine in contrast to that of the trans systems where the opposite occurs. It is suggested that in both instances a trigonal bipyramidal intermediate is formed, but that in the cis system the Me and Ar groups are in an apical and an equatorial position (or vice versa), while in the trans case both are in the same. Reductive elimination is only allowed by symmetry in the first instance. Presumably the trans complex decomposes through the T-shaped three-coordinate intermediate as discussed in Volume 2.[138].

As reported in Volume 1,[139] Tsou and Kochi have shown that [Ni(I)X(PEt$_3$)$_3$] is formed as well as [Ni(II)X(R)(PEt$_3$)$_2$] as an intrinsic feature of the reaction of aryl halides and [Ni(PEt$_3$)$_4$]. However, Morvillo and Turco consider that the reaction of alkyl halides yields directly only nickel(II) complexes, which can decompose slowly to give nickel(I) species, (see Volume 2[140]—lines 23 and 24 on page 294 should read "10 times slower"). Sales's group have now observed that polychlorobenzenes, like tha alkyl halides, appear to form only nickel(II) compounds directly.[141].

Kinetic studies[142] suggest that the oxidative addition of PhCH$_2$CN and PhCH(CH$_3$)CN, BN, to [Ni(PCy$_3$)$_2$] and [{Ni(PCy$_3$)$_2$}$_2$N$_2$] is preceded by formation of σ- and π-bonded intermediates.

$$[Ni(PCy_3)_2 + BN \rightleftharpoons [Ni(PCy_3)_2(\sigma\text{-}BN)] \rightleftharpoons [Ni(PCy_3)(\pi\text{-}BN)]$$
$$\xrightarrow{\text{oxidative addition}} \text{products} \qquad (120)$$

A short mention is made of the mechanism at the end of a review on C–C bond formation by palladium- or nickel-catalyzed cross-coupling.[143]

11.3.14. Palladium

Grubbs' group has joined Yamamoto's to extend the latter's thermolytic studies on dialkyl(bisphosphine)palladium complexes reported in Volume 2[144] into the field of photochemistry.[145] The photochemical pathway for the decomposition of *cis*- and *trans*-[PdEt$_2$(PMe$_2$Ph)$_2$] and of [(Me$_2$PhP)$_2${*cyclo*-Pd(CH$_2$)$_4$}] is quite different from the thermolytic. The *cis* and *trans* isomers decompose to C$_2$H$_4$, C$_2$H$_6$, and C$_4$H$_{10}$ through common intermediates, one of which may be tetrahedral [R$_2$PtP$_2$] and the other [R$_2$PtP]. Lack of H/D scrambling excludes radical pathways.

Oxidative addition of an optically active allyl acetate to [Pd(dppe)-(PPh$_3$)] to give a [(η^3-allyl)Pd(dppe)]$^+$-type ion proceeds with inversion of configuration.[146]

11.3.15. Platinum

Ab initio calculations have been carried out by Noell and Hay[147] on the oxidative addition of H_2 to $[Pt(PH_3)_2]$ and $[Pt(PMe_3)_2]$, the symmetry-allowed products being the *cis* isomers. The activation energy in the first (hypothetical) system is 17.4 kcal mol^{-1}, the preferred approach of the H_2 unit being broadside-on giving a transition state of C_{2v} symmetry. The main increases in the distance between the H atoms and in their anionic character take place after the transition state. *cis*-$[Pt(PH_3)_2H_2]$ and the *trans* isomer whose formation is symmetry forbidden are calculated to be more stable than the reactants by 6.7 and 10.6 kcal mol^{-1}, respectively. Somewhat similar results are obtained for the $[Pt(PMe_3)_2]$ system. *Cis* to *trans* isomerism of $[Pt(PMe_3)_2H_2]$ is suggested to occur through three- or five-coordinate intermediates since calculations point to an activation energy of at least 60 kcal-mol^{-1} for an intramolecular process. Paonessa and Trogler[148] have shown that in practice both isomers containing PMe_3 (or PEt_3) coexist in equilibrium in solution. The existence of cisisomers of this sort of compound is thus not dependent on the presence of bulky phosphines as had been supposed.

Oxidative addition of MeI, EtI, and *n*-PrI to $[PtMe_2(phen)]$ gives only the expected product $[PtMe_2(phen)(I)(R)]$ (cf. *i*-PrI). Second-order kinetics and a fall in rate, MeI $\gg$ EtI $>$ *n*-PrI, support an S_N2 mechanism (121) and (122).[149] Further studies show that α,ω-diodoalkanes add oxidatively

$$Pt + RX \rightarrow PtR^+ + X^- \qquad (121)$$

$$PtR^+ + X^- \rightarrow PtXR \qquad (122)$$

to $[PtMe_2(phen)]$ in two steps (123) and (124) with second-order kinetics when *n* is 2, 4, or 5, all six rate constants being approximately equal. In

$$[PtMe_2(phen)] + I(CH_2)_nI \rightarrow \textit{cis,trans-}[PtMe_2(phen)(I)\{(CH_2)_nI\}] \quad (123)$$

$$[PtMe_2(phen)] + \textit{cis,trans-}[PtMe_2(phen)(I)\{(CH_2)_nI\}] \rightarrow$$

$$[I(phen)Me_2Pt(\mu\text{-}CH_2)_nPtMe_2(phen)I] \qquad (124)$$

contrast when *n* is 1, (123) has an induction period, does not obey second-order kinetics, and is accompanied by the formation of *cis,cis*-product, which suggest a radical process, while (124) does not occur. [CH_2I_2 is unreactive towards nucleophiles, and attack on the CH_2I unit in (124) would be difficult sterically.] When *n* is 2, ethene is also formed in (123).

The oxidative addition of *i*-PrI to $[PtMe_2(phen)]$ appears to proceed by both the radical-chain and the nonchain mechanisms. In addition to the expected $[PtMe_2(phen)(I)(i\text{-}Pr)]$, $[PtMe_2(phen)(I)_2]$ and $[PtMe_2(phen)(I)-(OO\text{-}i\text{-}Pr)]$ are also formed in air.[150] The yield of the last depends on the pressure of oxygen but the rate of reaction of [PtMe$_2$(phen)] and i-PrI does not, which suggests that there is a radical-chain mechanism (125) and (126) and also a fast step (127):

$$Pt + i\text{-}Pr\cdot \rightarrow Pt(i\text{-}Pr) \tag{125}$$

$$Pt(i\text{-}Pr) + i\text{-}PrX \rightarrow PtX(i\text{-}Pr) + i\text{-}Pr\cdot \tag{126}$$

$$i\text{-}Pr\cdot + O_2 \rightarrow i\text{-}PrO_2\cdot \tag{127}$$

(In contrast, the primary alkyl iodides, MeI, EtI, and n-PrI, give only the expected product and no alkylperoxo complex.) Further, isopropyliodide reacts with [Me$_2$Pt(phen)] in the presence of CH$_2$=CHZ [Z = CN, CHO, or C(= O)Me] to give [PtMe$_2$(phen)(I)(i-Pr)] and [PtMe$_2$(phen)(I)-(CHZCH$_2$R)].[151] The second compound is not formed in the presence of alkenes, styrene, or methyl methacrylate, polymer being produced instead. A radical-chain process is proposed (125), (126), (128)–(130):

$$i\text{-}Pr\cdot + CH_2=CHZ \rightarrow i\text{-}PrCH_2CHZ\cdot \tag{128}$$

$$i\text{-}PrCH_2CHZ\cdot + Pt \rightarrow i\text{-}PrCH_2CHZPt\cdot \tag{129}$$

$$i\text{-}PrCH_2CHZPt\cdot + i\text{-}PrI \rightarrow i\text{-}PrCH_2CHZPtI + i\text{-}Pr\cdot \tag{130}$$

k_{125} is estimated to be 4×10^6 M^{-1} s^{-1} in acetone at 20°C. (In contrast, ethyl iodide gives only [PtMe$_2$(phen)(I)(Et)] even in the presence of high concentrations of CH$_2$=CHCN.) However, in its early stages the reaction of i-PrI is also first order in each reactant (though it accelerates later) and is not retarded by galvinoxyl, the rate being much slower than that for n-PrI. Ferguson *et al.*[150] suggest tentatively the nonchain radical process at this stage, (131) and (132), step (131) perhaps proceeding through the electron

$$Pt + i\text{-}PrI \rightarrow PtI + i\text{-}Pr\cdot \tag{131}$$

$$PtI + i\text{-}Pr\cdot \rightarrow PtI(i\text{-}Pr) \tag{132}$$

transfer intermediate, $\{[Pt\cdot]^+[i\text{-}PrI]^-\}$. However, the evidence does not seem to exclude an S_N2 mechanism.

Alcohols can add oxidatively to [PtMe$_2$(bipy)] and [PtMe$_2$(phen)] to give cations, [PtMe$_2$(OR)(NN)]$^+$. Preliminary studies indicate that rates fall: MeOH > EtOH > i-PrOH.[152]

The importance of neighboring-group participation in oxidative addition has been shown by Shaw's group.[153] Rates of addition of MeBr to compounds such as *cis*-[PtMe$_2$R)$_2$], *cis*-[PtMe$_2$(AsMe$_2$R)$_2$], and *trans*-[RhBr(CO)(PEt$_2$R)$_2$] were found to rise by factors of 145–250 when R was changed from phenyl to o-methoxyphenyls. Since it is already known that introduction of p-methoxy groups only accelerates the reaction by a small factor and since a purely steric effect of an o-methoxy group would lower the rate, neighboring-group participation is proposed.

[Pt(PPh₃)₃] reacts[154] with OCS to form [Pt(η^2-SCO)(PPh₃)₂]. [Pd(PPh₃)₃] behaves similarly but the product decomposes more readily. Thus within the triad stability falls: Pt > Pd > Ni, since [Ni(SCO)(PPh₃)₂] has not been observed.

Some transalkylation reactions of platinum compounds which proceed through the S_E(oxidative) mechanism are mentioned in Section 11.2.10.

Reductive elimination of CH_4 from cis-[PtH(CH₃)L₂] has been the subject of extended Hückel calculations similar to those reported in Volume 2[156] for more symmetric complexes cis-[MR₂L₂]. As found earlier if L is made more electron withdrawing, the activation energy is lowered as is observed experimentally. The calculations confirm the statement that the reverse oxidation addition is precluded on thermodynamic rather than kinetic grounds.[157]

A full account has now appeared of the studies of Yamamoto's group on the thermal decomposition of unsymmetrical dialkylplatinum compounds, [L₂PtR'R''], where L = various phosphines.[158] These complement the work of Whitesides' group mentioned in Volume 2[159] and reach similar conclusions. Decomposition involves β-hydrogen elimination and can occur by either a dissociative or nondissociative pathway, (133) or (134), as before.

$$[L_2PtR_2] \rightleftharpoons [LPtR_2] \rightleftharpoons [LPtH(R)(\eta^2\text{-}R^{-H})]$$

$$\longrightarrow R^{-H} + [LPtH(R)] \rightarrow RH \qquad (133)$$

$$[L_2PtR_2] \rightleftharpoons [L_2PtH(R)(\eta^2\text{-}R^{-H})] \rightarrow RH + R^{-H} \qquad (134)$$

However, the Yamamoto scheme for the first of the two pathways shows the alkene eliminating before the alkane. Evidence for the Pt-R $\rightleftharpoons$ PtH(η^2-R^{-H}) rearrangements is provided by the fact that alkane and alkene products show that H/D scrambling can occur within one R (but not between R' and R''). In the dissociative pathway, for alkyl systems the ratio of alkenes to alkanes in the products show that all β-hydrogen atoms have an equal tendency to be eliminated. They are, however, activated by phenyl groups. An isopropyl group is isomerized to an n-propyl ligand before decomposition occurs. In the nondissociative pathway, ease of β-elimination from groups appears to rise: PtEt < Pt^{n}Pr ≤ Pt^{n}Bu. Rate constants for overall process (134) vary by a factor of only 4 for various R.

Photolysis of [L₂Pt(CH₂)₄][L=PPh₃ or P(C₆H₄Me-p)₃] in CH_2Cl_2 gives alkanes, alkenes, and L₂Pt; the latter reacts rapidly with solvent to form trans-[L₂Pt(CH₂CH₂Cl)Cl].[160]

The attack of acids on trans-[PtH(σ-CH₂CN)(PPh₃)₂] can lead to a variety of products depending on the acid and the solvent, (135)–(138):[161]

$$\text{HBF}_4/\text{Et}_2\text{O:} \rightarrow \textit{trans-}[\text{Pt}(\text{CH}_2\text{CN})(\text{PPh}_3)_2]_m(\text{BF}_4)_m + \text{H}_2 \quad (135)$$

$$\text{HCl}/\text{Et}_2\text{O, toluene:} \rightarrow \textit{trans-}[\text{PtH}(\text{Cl})(\text{PPh}_3)_2] + \text{CH}_3\text{CN} \quad (136)$$

$$\text{excess HCl}/\text{Et}_2\text{O, toluene:} \rightarrow \textit{cis-} \text{and}$$

$$\textit{trans-}[\text{PtCl}_2(\text{PPh}_3)_2] + \text{H}_2 + \text{CH}_3\text{CN} \quad (137)$$

$$\text{HCl}/\text{MeOH, CH}_2\text{ClCH}_2\text{Cl:} \rightarrow \textit{trans-}[\text{PtH}(\text{Cl})(\text{PPh}_3)_2] + \text{CH}_3\text{CN} \quad (138)$$

Equation (138) is the reaction of interest here. Its rate follows a two-term rate law, one first order in $[\text{H}^+]$, the other first order in $[\text{H}^+]$ and $[\text{Cl}^-]$. It is proposed that rapid oxidative addition occurs (139) followed by slower steps involving reductive elimination, (140) and (142). [(140) is followed by a fast step (141)]:

$$[\text{PtH}(\text{CH}_2\text{CN})(\text{PPh}_3)_2] + \text{H}^+ + \text{S} \rightleftharpoons [\text{PtH}_2(\text{CH}_2\text{CN})(\text{PPh}_3)_2\text{S}]^+ \quad (139)$$

$$[\text{PtH}_2(\text{CH}_2\text{CN})(\text{PPh}_3)_2\text{S}]^+ \rightarrow [\text{PtH}(\text{PPh}_3)_2\text{S}]^+ + \text{RH} \quad (140)$$

$$[\text{PtH}(\text{PPh}_3)_2\text{S}]^+ + \text{Cl}^- \rightarrow [\text{PtH}(\text{Cl})(\text{PPh}_3)_2] + \text{S} \quad (141)$$

$$[\text{PtH}_2(\text{CH}_2\text{CN})(\text{PPh}_3)_2\text{S}]^+ + \text{Cl}^- \rightarrow [\text{PtH}(\text{Cl})(\text{PPh}_3)_2] + \text{RH} + \text{S} \quad (142)$$

Kinetic parameters: $\Delta H^\circ_{139} + \Delta H^\ddagger_{140} = 17.5 \text{ kcal mol}^{-1}$, $\Delta S^\circ_{139} + \Delta S^\ddagger_{140} = -15 \text{ cal deg}^{-1} \text{ mol}^{-1}$, $\Delta H^\circ_{139} + \Delta H^\ddagger_{142} = 20.5 \text{ kcal mol}^{-1}$, and $\Delta S^\circ_{139} + \Delta S^\ddagger_{142} = 0.5 \text{ cal deg}^{-1} \text{ mol}^{-1}$. The CH_2CN stabilizes the system towards oxidative addition; $\textit{trans-}[\text{PtH}(\text{Me})(\text{PPh}_3)_2]$ undergoes reductive elimination to Pt(0) and methane even at low temperatures.

The photoinduced reductive elimination of cyanoalkanes from complexes such as $\textit{trans-}[\text{PtH}(\text{CH}_2\text{CN})(\text{PPh}_3)_2]$ proceeds through the *cis* isomer, reaction (143) and incidentally provides a useful way of forming $[\text{Pt}(\text{PPh}_3)_2]$.[162]

$$\textit{trans-}[\text{PtH}(\text{CH}_2\text{CN})(\text{PPh}_3)_2] \rightarrow \textit{cis-}[\text{PtH}(\text{CH}_2\text{CN})(\text{PPh}_3)_2]$$

$$\rightarrow [\text{PtH}(\text{PPh}_3)_2] + \text{CH}_3\text{CN} \quad (143)$$

$\textit{cis-}[\text{PtH}(\text{CH}_2\text{CF}_3)(\text{PPh}_3)_2]$ decomposes thermally in solution by a first-order process, which appears to be a straightforward concerted reductive elimination (144):[163]

$$\textit{cis-}[\text{PtH}(\text{CH}_2\text{CF}_3)(\text{PPh}_3)_2] \rightarrow \text{CH}_3\text{CF}_3 + \text{``}[\text{Pt}(\text{PPh}_3)_2]\text{''} \quad (144)$$

In benzene, $\Delta H^\ddagger$ and $\Delta S^\ddagger$ are 24.6 kcal mol^{-1} and 5 cal K^{-1} mol^{-1}, respectively, the second parameter being slightly dependent on solvent. Use of the deuterido complex gives an H/D kinetic isotope effect of 2.2 at 40°C. Unlike many systems, the rate is not dependent on the concentration of free phosphine, which shows that there is no initial dissociative step. What is formulated as "[Pt(PPh$_3$)$_2$]" does not seem to have exactly the same properties as this compound has in other systems.

Morvillo and Turco have shown by H/D-labeling studies that platinum monoalkyl complexes trans-[PtMe(I)(L)$_2$] can decompose thermally by two pathways (L = phosphine). The initial step in the more important one is homolytic (145), the methyl radical so produced abstracting a hydrogen

$$Pt-Me \rightarrow Pt + Me \cdot \qquad (145)$$

atom from a phosphine ligand or solvent.[164] The less important path involves elimination of methane from coordinated methyl groups presumably in a bimolecular process.

H/D-labeling experiments on the pyridine-induced rearrangement of platinacyclobutanes to platinum alkenes show that the reaction is intramolecular and proceeds through α- rather than β-elimination.[165] For example, in the E-CHMeCHMeCD$_2$ units in {Cl$_2$PtCHMeCHMeCD$_2$}$_4$ it is D, as opposed to H, which migrates giving MeCHDMe=CHD rather than MeCH$_2$Me=CD$_2$.

Studies on cations containing the biplatinum unit, Pt(μ-dmpm)$_2$Pt, have been carried out by Puddephatt's group. The oxidative addition of I$_2$ to [Me$_2$Pt(μ-dmpm)$_2$PtMe$_2$] involves transfer of a methyl group from one platinum to the other, the ion [Me$_3$Pt(μ-dmpm)$_2$(μ-I)PtMe]$^+$ being formed.[166]

Thermal reductive elimination of H$_2$ from the Pt$_2$H$_2$(μ-H) system in equation (146) is intramolecular,[167] as was reported in Volume 2[168]

$$[HPt(\mu-H)(\mu-dppm)_2PtH]^+ + L \rightarrow [HPt(\mu-dppm)_2PtL]^+ + H_2 \quad (146)$$

(L = η^1-dppm or PPh$_3$). It is now shown the same is true of the photochemical process (L = solvent = MeCN or pyridine); the reaction is efficient, the quantum yield being 0.81 for the first solvent (λ = 366 nm). Photolytic reductive elimination of ethane occurs from [Me$_2$Pt(μ-dppm)$_2$PtMe]$^+$, [MePt(μ-dppm)$_2$Pt(O=CMe$_2$)]$^+$ or [MePt(μ-Cl)(μ-dppm)$_2$PtMe]$^+$ also being formed in acetone and dichloromethane, respectively.[169] H/D-labeling experiments show that elimination is both intramolecular and intermolecular. The Pt$_2$(H)(Me)(μ-H) cation, [MePt(μ-H)(μdppm)$_2$PtH]$^+$, will not reductively eliminate H$_2$ or CH$_4$ thermolytically, presumably because the departing groups are trans to each other. However, the former and latter are eliminated by the action of phosphines and CF$_3$C≡CCF$_3$, respectively.

Returning to the Pt$_2$H$_2$(μ-H) system (146), kinetic studies now confirm that elimination occurs from a phosphine-containing intermediate, equation (147) (L = phosphine):[170]

$$[Pt_2H_2(\mu-H)(\mu-dppm)_2]^+ + L \rightleftharpoons [Pt_2H_3L(\mu-dppm)_2]^+ \qquad (147)$$

$$[Pt_2H_3L(\mu\text{-dppm})_2]^+ \rightarrow [\overline{HPt(\mu\text{-dppm})_2Pt}L]^+ + H_2 \qquad (148)$$

For L = PPh$_3$, $\Delta H^{\ddagger}_{148} = 92$ kJ mol^{-1}, $\Delta S^{\ddagger}_{148} = 45$ J K^{-1} mol^{-1} and for L = P(p-ClC$_6$H$_4$)$_3$, $\Delta H^{\ddagger}_{148} = 85$ kJ mol^{-1} and $\Delta S^{\ddagger}_{148} = 17$ J K^{-1} mol^{-1}. The reaction is not accelerated significantly when the bulky phosphine, PPh$_2$(2-MeC$_6$H$_4$), is used which suggests that H elimination occurs somewhat remotely from L, possibly from an intermediate containing no μ-H group, as in equation (149). Although the activation enthalpies appear to be comparable with that of 100 kJ mol^{-1} calculated for reductive elimination of H$_2$ from *cis*-[PtH$_2$(PMe$_3$)$_2$] (see the calculations of Noell and Hay at the beginning of this section), the true value associated with loss of H$_2$ is probably less since some energy will probably be associated with rearrangement of the binuclear system as in the first step of reaction (149):

$$[HPt(\mu\text{-H})(\mu\text{-dppm})_2PtHL]^+ \rightarrow [H_2Pt(\mu\text{-dppm})_2PtHL]^+$$

$$\xrightarrow{-H_2} [Pt(\mu\text{-dppm})_2PtHL]^+$$

$$\rightarrow [\overline{HPt(\mu\text{-dppm})_2Pt}L]^+ \qquad (149)$$

11.3.16. Gold and Mercury

Dyadchenko has reviewed oxidative addition and reductive elimination reactions of gold complexes.[171] Some transalkylation reactions of mercury compounds which are S_E(oxidative) are mentioned in Section 11.2.10.

11.3.17. Actinides

The thermolysis[172] of [Cp$_2$Th(CH$_2$CMe$_3$)$_2$] and [Cp$_2$Th(CH$_2$SiMe$_3$)$_2$] to give thoriacyclobutanes and neopentane is first order, the first decomposing about 20 times faster at 70°C, equation (150); $\Delta H^{\ddagger}$ and $\Delta S^{\ddagger}$ are 20.6 kcal-mol^{-1} and -17 cal mol^{-1} K^{-1}, and 25.9 kcal mol^{-1} and -9 cal mol^{-1} K^{-1}, respectively. Use of CD$_2$CMe$_3$ ligands and C$_6$D$_6$ as solvent shows that the reaction is nonradical and involves transfer of a γ-hydrogen atom as in the decomposition of (Et$_3$P)$_2$Pt(CH$_2$CMe$_3$)$_2$, which was discussed in Volume 2,[173] and for which the overall $\Delta S^{\ddagger}$ is 35 cal mol^{-1} K^{-1}. In the platinum system the γ process is elimination.

$$[Cp_2Th(CH_2XMe_3)_2] \rightarrow [Cp_2\overline{ThCH_2XMe_2C}H_2] + CMe_4 \qquad (150)$$

The lower values of $\Delta S^{\ddagger}$ here suggest a different mechanism and γ-abstraction of hydrogen is proposed as in pathway B in Scheme 5 (at the beginning of Section 11.3). For the α-abstraction process, (88), $\Delta S^{\ddagger}$ is -8 cal mol^{-1} K^{-1} (see Section 11.3.6.)

A provocative paper by Evans' group[174] reexamines the reaction of uranium tetrachloride and alkyllithium reagents to form UR_x, and the interpretation of the reactions often written as in (151):

$$UCl_4 + 4\text{-}n\text{-BuLi} \xrightarrow{\text{hexane}} U(n\text{-Bu})_4 \xrightarrow{120 \text{ h}} \sim 2\text{-}n\text{-BuH}$$

$$+ \sim 2EtCH\!=\!CH_2 + U \qquad\qquad (151)$$

However, the major reaction appears to require only two moles of RLi. $RUCl_3$ is postulated as an intermediate which can undergo β-elimination to give R^{-H} or react with UCl_3H to form RH.

Chapter 12
Reactivity of Coordinated Hydrocarbons

12.1. Introduction

As in previous volumes, this chapter reviews kinetic and mechanistic studies on the stoichiometric reactions of coordinated hydrocarbons with nucleophiles and electrophiles, together with some related processes. There has been a marked increase of interest in the stereochemistry of these reactions, in particular with chiral complexes, because of their potential in asymmetric synthesis.

Related catalytic processes are discussed elsewhere (Chapter 14) as are intramolecular ligand rearrangements (Chapter 13). Quantitative studies of nucleophilic attack on carbonyl ligands are also generally outside the scope of this chapter, and are discussed only where attack at carbon monoxide occurs competitively with reaction at a coordinated hydrocarbon. A short section on nucleophilic attack at coordinated isocyanides is included for the first time.

12.2. Nucleophilic Addition and Substitution

12.2.1. σ-Bonded Hydrocarbons

There has been considerable recent interest in the reactivity of transition-metal carbene complexes and other species containing metal–carbon multiple bonds because of their involvement in olefin metathesis and the possible role of methylene coupling in chain growth during heterogeneous Fischer–Tropsch reactions.

Nucleophilic addition to the carbene ligand of $[Re(\eta\text{-}C_5H_5)(NO)\text{-}(PPh_3)(=CHR)]^+$ species is generally highly stereospecific.[1,2] For example, the reaction of the ethylidene cation (1) with a range of nucleophiles $(LiEt_3BD, PhCH_2MgBr, PhMgBr, PMe_3)$ gives the $[Re(\eta\text{-}C_5H_5)(NO)\text{-}(PPh_3)\{CH(Me)Y\}]$ adducts in $\geq 99:1$ diastereomer ratios, as in equation (1) with $Nu = D^-, PhCH_2^-, Ph^-, PMe_3$. Slightly lower selectivity was noted

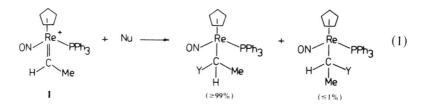

$$(1)$$

in addition to the analogous benzylidene cation. All nucleophiles therefore prefer to attack the carbene carbons of type (1) cations anti to the PPh_3 ligand. These observations indicate broad potential for asymmetric organic synthesis.

The reactions of electrophilic carbene complexes with alkenes have attracted particular attention. For example, enantioselective synthesis of cyclopropanes has been achieved[3] via the reaction of $(1SS)\text{-}(2)$ with styrene

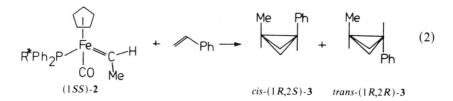

$$(2)$$

[equation (2); $R^* = (S)\text{-}2\text{-methylbutyl}$]. In analogy with the rhenium-carbene chemistry discussed above, the high enantiomeric excess $(>84\%)$ of the *cis*- and *trans*-cyclopropane adducts (3) results from selective attack of styrene on the *si* face of the prochiral ethylidene ligand in $(1SS)\text{-}(2)$, controlled by a preferred orientation of the carbene ligand and large steric differences in the ancillary ligands.

Preliminary studies[4] of the coupling of alkynes with the alkylidyne ligand in the clusters $H_3Ru_3(\mu_3\text{-}CX)(CO)_9$ {equation (3)} reveal the same relative reactivities as those found for substitution by $AsPh_3$ (X = MeO > Me > Ph). This implies that CO dissociation is the rate-limiting step. Hydrogen transfer to alkyne probably precedes carbon–carbon coupling.

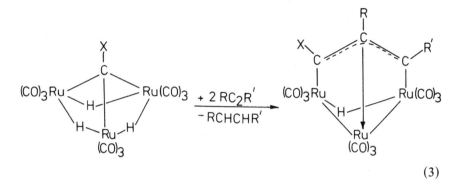

(3)

Nonparametrized MO calculations[5] on $[Fe(\eta\text{-}C_5H_5)(PH_3)_2(=CH_2)]^+$ and related acetylide and vinylidene complexes indicate that nucleophilic additions are directed by the character and localization of the LUMOs (lowest unoccupied molecular orbital) in the substrate molecules. While the overall positive charge of the complex enhances its reactivity towards nucleophiles, overall charge is probably not critically important for regioselectivity. On the other hand, a spectroscopic (^1H, ^{13}C, and ^{31}P NMR and Mössbauer) study[6] of the additions of amines to the $\mu\text{-}\eta^2$-acetylide complexes [(**4**); R = Ph, $p\text{-}MeOC_6H_4$, $p\text{-}BrC_6H_4$, C_6H_{11}, CMe_3] indicates that the site of nucleophilic addition is controlled mainly by the charge polarization in the μ-triple bond, the steric bulk of the substituent R, and the nature of the incoming amine.

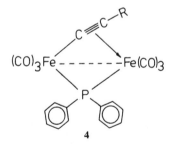

4

Significant asymmetric induction (18–60% e.e.) is observed in the alkylation of the chiral alkyl complexes $[Fe(\eta\text{-}C_5H_5)(CO)(L)(CH_2Cl)]$ [(**5**); L = PPh$_3$ or tri(o-biphenyl)phosphite] by the prochiral nucleophiles sodium t-butyl acetoacetate and pyrrolidine cyclohexanone enamine.[7] The product diastereomer ratio in the former case was shown to be thermodynamically controlled, while kinetic control was assumed in the latter reaction.

12.2.2. π-*Bonded Hydrocarbons*

12.2.2.1. *Addition at Mono-Olefins*

Analogous reactions with the chiral ethylene cations [(**6**); L = PPh$_3$ or tri(*o*-biphenyl)phosphite)] also proceed[7] with appreciable asymmetric induction (10–64% e.e.) (Scheme 1). As with the alkyl substrate (**5**), increased steric bulk of the phosphorus ligand L increases the extent of induction.

Scheme 1

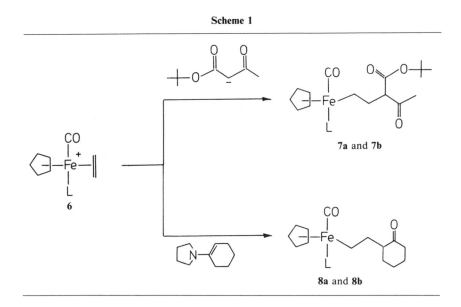

7a and 7b

6

8a and 8b

Support for the recent suggestion[8] that the enhanced electrophilicity of coordinated olefins is due to slippage of the metal fragment along the double-bond axis comes from studies of the reactivity of the complex [PtCl(tmen)(η^2-C$_2$H$_4$)]$^+$ (tmen = *N,N,N',N'*-tetramethylenediamine) towards nucleophiles.[9] An X-ray analysis of this cation shows a marked tilting (10%) of the C$_2$H$_4$ molecule from the expected position orthogonal to the coordination plane, and different Pt–C distances (2.176 and 2.242 Å, respectively). Exceptional activation of this unsymmetrically coordinated ethylene is confirmed[9] by the reactions in Scheme 2 with the nucleophiles NO$_2^-$, NCO$^-$, and N$_3^-$, which normally attack the Pt(II) center in other Pt(II)-olefin complexes. A plausible mechanism for the novel formation of

Scheme 2

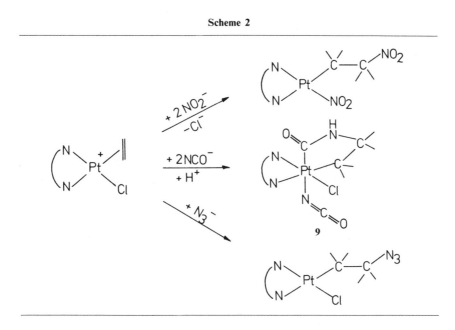

the carbamoyl product (**9**) involves initial attack by NCO^- on C_2H_4 to form a *N*-ethyleneisocyanate ligand, protonation of this ligand, and its subsequent addition to the Pt(II) center.[10]

12.2.2.2. Addition at Alkenes

The first definite proof of *trans* addition of nucleophiles to η^2-alkyne complexes has been obtained[11] by the isolation and characterization of the alkenyl products from the reactions of $[Fe(\eta\text{-}C_5H_5)(CO)\{P(OPh)_3\}\text{-}(\eta^2\text{-}MeC\equiv CMe)]^+$ with $Ph_2Cu(CN)Li_2$ and of $[Fe(\eta\text{-}C_5H_5)(CO)\text{-}\{P(OPh)_3\}(\eta^2\text{-}PhC\equiv CMe)]^+$ with $Me_2Cu(CN)Li_2$. Unlike earlier studies of related reactions, the stereochemistry of the alkenyl products could be directly related to the addition mechanism since these products were incapable of undergoing *cis/trans* interconversions under the reaction conditions. For the reaction with $\eta^2\text{-}PhC\equiv CMe$ the incoming nucleophile adds to the alkyne carbon bearing the phenyl substituent.

12.2.2.3. Addition at η^3-Enyls

Addition of dialkylamines (R = Me, Et) to the π-allyl complex (**10**) has been shown[12] to occur exclusively with *trans* stereochemistry (>98%)

in the presence of PPh₃ {equation (4)}:

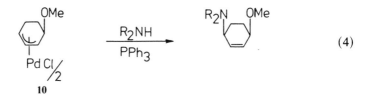

$$(4)$$

This is in accordance with previous observations[13] that stabilized car-banions and acetate ions in the presence of chloride ligands also add trans to (π-allyl) palladium complexes. In the presence of AgBF₄, reaction (4) also yields up to 14% of the *cis*-addition product (**11**). Possible mechanisms for the formation of (**11**) are via cis migration of an initially coordinated amine or via formation of a (σ-allyl) palladium complex (**12**) followed by a syn S_N2' attack by a free amine.

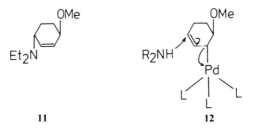

Nucleophilic attack on mixtures of exo and endo isomers of [Mo(η^5-NMCp)(NO)(CO)(η^3-1,3-dimethylallyl)]⁺ (NM = neomenthyl) provides facile routes to optically pure allylically substituted olefins.[14] A given configuration at the metal center controls the configuration at the allylic center because the exo isomer is attacked preferentially and addition occurs *cis* to the NO ligand. That is, in Scheme 3 the relationships $k_{2x}[\text{Nuc}] > k_{1n} > k_{1x} \gg k_{2n}[\text{Nuc}]$ hold (Nuc = enamine).

An attempt has been made to predict the sites of nucleophilic attack on [M(CO)₃(π-hydrocarbon)] complexes using the perturbation theory of reactivity.[15] For the model allyl substrate [Co(CO)₃(η^3-C₃H₅)] the site preference CO > M > C₃H₅ was predicted for reaction with hard nucleophiles in polar solvents. On the other hand, with soft nucleophiles initial attack at the π-allyl ligand was favored. Mechanistic studies have suggested only a small energy difference between attack by alkoxide ions on the allyl ligand and the metal in related (π-allyl) palladium(II) complexes.[16]

Scheme 3

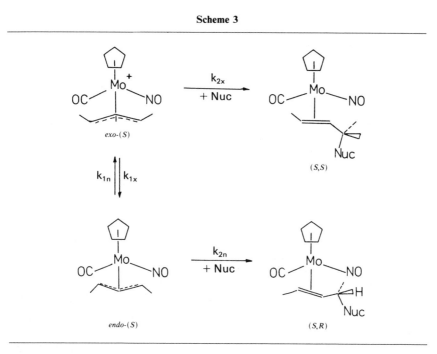

12.2.2.4. Addition at η^4-Dienes

Spectroscopic and kinetic studies indicate that reaction (5) can be regarded as an unusual electrophilic aromatic substitution in which the

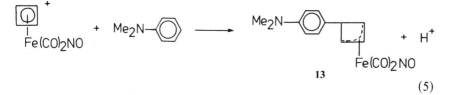

$$(5)$$

electrophile $[Fe(CO)_2(NO)(\eta\text{-}C_4H_4)]^+$ attacks the N,N-dimethylaniline substrate.[17] ^{1}H NMR and X-ray structural data confirm that substitution occurs exclusively at the carbon para to the NMe_2 group and that the cyclobutenyl product (**13**) has an exo configuration. The second-order rate law (6) was rationalized in terms of a mechanism involving initial π-complex formation followed by rate-determining conversion to a σ-complex:

$$\text{rate} = k[Fe][Me_2NC_6H_5] \qquad (6)$$

12.2.2.5. *Addition at η^5-Dienyls*

An interesting change in regioselectivity with temperature occurs in the reaction of $[Fe(\eta^5\text{-}C_5H_5)(Ph_2PC_2H_4PPh_2)(CO)]PF_6$ with $LiAlH_4$ in THF.[18] At $-78°C$ the iron hydride complex $[Fe(\eta^5\text{-}C_5H_5)(H)\text{-}(Ph_2PC_2H_4PPh_2)(CO)]$ is the exclusive product, whereas at 70°C hydride attack occurs on both the cyclopentadienyl and CO ligands to give (**14**) (X = H) and (**15**), respectively (2/3 ratio) {equation (7)}. Exo addition to

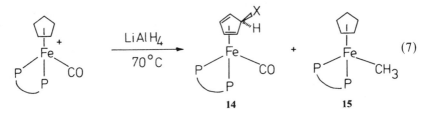

$$(7)$$

the cyclopentadienyl ligand was established from a ^1H and ^2H NMR study of the product (**14**) (X = D) obtained with $LiAlD_4$. This report provides one of the very few examples of nucleophilic attack upon a cyclopentadienyl ligand. In general, as predicted by molecular-orbital calculations,[15] nucleophilic attack at the metal or CO is preferred for such complexes.

Kinetic studies of reaction (8) in acetone, where PR_3 is a wide range of tertiary phosphines and phosphites, have established the general rate law (9) where Nuc = PR_3.[19] A good correlation ($r = 0.98$) is found between

$$\left[\text{⬡—Fe(CO)}_3 \right]^+ \ + \ PR_3 \ \xrightarrow{k_1} \ \left[\text{R}_3\text{P} \text{⬡—Fe(CO)}_3 \right]^+ \qquad (8)$$

16

$$\text{Rate} = k_1[\text{Fe}][\text{Nuc}] \qquad (9)$$

$\log k_1$ and the Tolman ΣX values of the phosphorus nucleophiles. For the triarylphosphines a Brønsted slope of 0.5 is derived from a plot of $\log k_1$ vs. pK_a, demonstrating the importance of basicity in controlling nucleophilicity towards (**16**). In contrast, the very slow or nonexistent reactions of $SbPh_3$ and $AsPh_3$ with (**16**) indicate that the polarizability of the donor group in EPh_3 (E = P, As, or Sb) has little influence on their nucleophilicity. Steric factors, quantified by negative deviations from the above Tolman and Brønsted plots by $P(p\text{-}MeC_6H_4)_3$ and $P(C_6H_{11})_3$, are much more important than in related reactions with EtI. The Hammett slope ρ of -1.3 observed for reaction (8) with triarylphosphines suggests moderate, but far from complete, phosphorus–carbon bond formation in the transition states.

Exo addition of PPh_3 to cation (**16**) has been established by an X-ray study,[20] and is assumed for each of the other reactions (8). Similarly, the rapid additions of PEt_3, PPr_3^n, PBu_3^n, PMe_2Pr^n, $PEtPh_2$, and PPh_3 to the cycloheptadienyl cation $[Fe(CO)_3(1-5-\eta-C_7H_9)]^+$ (**17**) have been shown[21] to give $[Fe(CO)_3(1-4-\eta-5-exo-R_3P \cdot C_7H_9)]^+$ products (**18**). Particularly interesting is the observation[21] that some of these exo products convert more slowly ($\sim$30 min) to their endo analogues (**19**). This rearrangement or less facile kinetic pathway for phosphine attack on (**17**) is not open to tertiary phosphines containing more than one phenyl substituent, presumably due to steric inhibition.

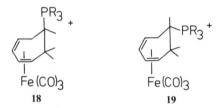

18   19

Rate law (9) (Nuc = imidazole) is also observed[22] for the additions of imidazole to (**16**) and the related cation $[Fe(CO)_3(1-5-\eta-2-MeOC_6H_6)]^+$ (**20**). The mechanism in Scheme 4 is proposed for these reactions, giving

Scheme 4

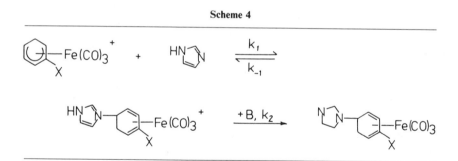

rise to the general rate law (10). ^1H NMR evidence supports the condition $k_2 \gg k_{-1}$, resulting in the simplification of equation (10) to the observed

$$\text{Rate} = \frac{k_1 k_2 [\text{Fe}][\text{imidazole}]^2}{k_{-1} + k_2[\text{imidazole}]} \qquad (10)$$

rate law (9). Direct addition of imidazole to the dienyl rings is indicated by the rate trend $C_6H_7 > 2-MeOC_6H_6$.

Kinetic results have also been reported for the reactions of cations (**16**) and (**20**) with a wide range of aryltrimethylsilanes and aryltrimethylstannanes of the type $XC_6H_4MMe_3$ (M = Si, Sn).[23] Rate law (9) (Nuc = $XC_6H_4MMe_3$) was uniformly observed, being rationalized in terms of electrophilic attack by the organometallic cations on the $XC_6H_4MMe_3$ substrates (Scheme 5). Linear free-energy relationships reveal close similarities with

Scheme 5

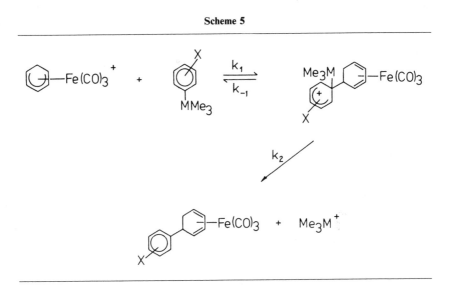

related protodemetalations. As expected, rates vary with the nature of the metal M in the order Pb ≫ Sn ≫ Ge > Si. In general the $XC_6H_4MMe_3$ substrates are considerably more reactive towards cations (**16**) and (**20**) than are the corresponding XC_6H_5 compounds, providing novel and often unique routes to diene-substituted arenes. Less extensive kinetic studies were reported for analogous reactions with 2-trimethylsilyl-furan or 2-trimethylsilyl-thiophen.

Direct addition of the above phosphorus and $XC_6H_4MMe_3$ nucleophiles to the dienyl ring of (**16**) is supported by perturbation theory molecular-orbital calculations which predict[15] the site-preference ring > CO > Fe for soft nucleophiles in solvents of moderate to high polarity. On the other hand, initial attack at a carbonyl ligand is predicted for hard nucleophiles such as OH^-. In keeping with this prediction, the carboxylic acid intermediates $[Fe(CO)_2(COOH)(1-5-\eta\text{-dienyl})]$ [(**21**); dienyl = C_6H_7 or C_7H_9] could be isolated[24] from the reactions of cations (**16**) and (**17**) with hydroxide ion. The kinetics observed,[24] shown in equation (11) for

$$k_{obs} = \frac{k_2[OH^-]}{1 + K_1[OH^-]} \tag{11}$$

the overall formation of the exo ring-addition product $[Fe(CO)_3(1\text{-}4\text{-}\eta\text{-}HOC_6H_7)]$ (22), supported the mechanism shown in Scheme 6. A value of 65 was estimated for the pre-equilibrium constant K_1.

Scheme 6

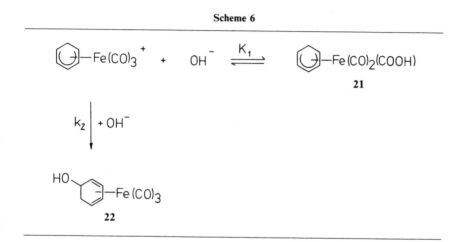

21

22

Similar initial formation of $[Fe(CO)_2(COOR)(1\text{-}5\text{-}\eta\text{-dienyl})]$ (R = Me or Et; dienyl = C_6H_7 or C_7H_9) species has been recently described[21,25] during attack by hard alkoxide nucleophiles on cations (16) and (17). In contrast, there is no evidence for initial attack at a CO ligand during the reaction of such cations with hard amine nucleophiles (RNH_2, R_2NH, R_3N).[21,22,26] However, hydrazine does react with (17) to form an intermediate carbazoyl complex $[Fe(CO)_2(CONHNH_2)(1\text{-}5\text{-}\eta\text{-}C_7H_9)]$, which undergoes a subsequent Curtius rearrangement to give the isocyanate derivative $[Fe(CO)_2(NCO)(1\text{-}5\text{-}\eta\text{-}C_7H_9)]$.[21]

There is considerable current interest in the preparation of optically active $[Fe(CO)_3(1\text{-}5\text{-}\eta\text{-dienyl})]^+$ cations because of their potential for enantioselective synthesis via reaction with nucleophiles. Of closest direct concern to this chapter are ^1H NMR[27] studies of chiral discrimination in the reactions of racemic (20) with (−)-neomenthyldiphenylphosphine, (S, S)-(−)-chiraphos and (+)-diop. Addition of these optically active tertiary phosphines to the dienyl ring of (20) is somewhat less diastereoselective than that previously observed with (S, S)-$(-)$-o-phenylenebis(methyl-phenylphosphine). Also of particular significance is the appreciable asymmetric induction (~33% e.e.) observed[28] in the reaction of CN$^-$ with the related cation $[Fe(CO)_2(NMDPP)(1\text{-}5\text{-}\eta\text{-}C_6H_7)]^+$ [NMDPP = $(+)$-neomenthyldiphenylphosphine] shown in equation (12). The two diastereomers (23a) and (23b) are formed in a ratio of ~2:1.

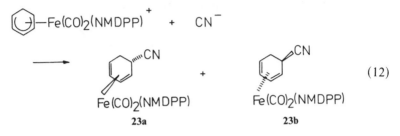

The first full resolution of (20) has been reported via its reaction with $(-)_{589}$-menthol and separation of the two diastereomers using medium-pressure liquid chromatography.[29] Reaction of $(2S)$-$(+)$-(20) with PPh_2Me to give a phosphonium salt followed by Wittig reaction with formaldehyde and protonation with HPF_6 has also provided[30] a novel route to optically pure $(2S,2R)$-$(+)$-$[Fe(CO)_3(2\text{-MeO-5-MeC}_6H_5)]^+$.

A surprising result was the first report of stereospecific kinetic *endo*-hydride addition to a coordinated hydrocarbon ligand.[31] The $^2H\{^1H\}$ NMR spectra of the species (24) (R = Me or Ph) obtained from reaction (13a) indicated 100% endo addition. Since at equilibrium the distribution of *exo*- and *endo*-deuteride should be near 1:1, it follows that the observed results refer to the kinetic product. Conclusive confirmation of endo attack came from an X-ray structural study of the product (25) from reaction (13b), in

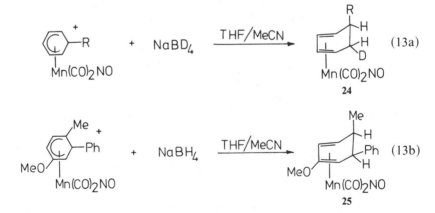

which the exo position of the C-5 methyl was clearly evident.[32] In contrast, phosphine nucleophiles are believed to add exo to $[Mn(CO)_2(NO)-(\eta^5\text{-dienyl})]^+$ substrates. The different stereospecificity for H^- addition may arise from initial attack at a CO ligand to give a Mn–CHO intermediate followed by H^- migration to the dienyl ring.

Few studies of nucleophilic attack on cyclooctadienyl ligands have been reported to date. In one such study, the tricarbonyl(1-3:5,6-η-cyclooctadienyl)iron cation (**26**) has been shown to exhibit different site preferences

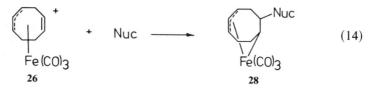

$$\tag{14}$$

for nucleophilic attack to the previously examined tricarbonyl (1-5-η-cyclooctadienyl) iron cation (**27**).[33] Addition with a range of nucleophiles (Nuc = MeO^-, CN^-, N_3^-) occurs on the olefinic portion of the dienyl ring to give stable σ, allyl products (**28**) as shown in equation (14).

In contrast, with cation (**27**) preferential attack occurs at the metal or a CO ligand. For example, with MeO^- initial addition at a CO group gives an ester complex prior to migration to the organic ligand. With PPh_3 or I^- as nucleophile, cation (**26**) exhibits a mixture of ring addition and metal attack to give products (**28**) and $[Fe(CO)_2(Nuc)(dienyl)]^{n+}$ ($n = 1$ or 0), respectively. Less extensive studies have also been reported for the ruthenium analogue of (**26**), revealing a greater preference for attack at the metal.[34]

12.2.2.6. Addition at η^6-Arenes

Detailed kinetic data have been reported for the equilibrium reactions (15) (M = Fe, Ru, Os) and (16) (M = Mn, Re; R = H, Me)[35]:

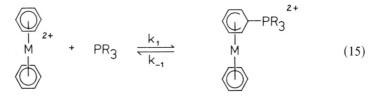

$$\tag{15}$$

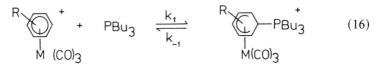

$$\text{(16)}$$

The k_1 values for ring addition varied with the nature of the metal in the order Fe ≫ Ru > Os (65:3:1) and Mn ≥ Re (2:1). Nucleophilic attack at the metal with ring displacement shows[36,37] very different reactivities (Ru ≫ Fe, Re ≫ Mn), supporting direct bimolecular addition to the rings in reactions (15) and (16). The trend in k_1 down the iron triad parallels the usual changes in redox potentials of iron-triad organometallics (Fe(II) ≫ Ru(II) > Os(II)). Extending this reasoning, Sweigart[35] has reported the linear relationship (17) between log k_{rel} and the reduction potential (E_p) for phosphine attack on a wide range of (π-hydrocarbon) metal complexes. This remarkable relationship suggests that reduction potential may be used to predict reactivities. The relatively low slope in equation (17) indicates

$$\log k_{rel} = 11.7 E_p + 12.9 \qquad (17)$$

that electron transfer is not occurring in the rate-determining step in such reactions, that is, they are of the Lewis-acid/Lewis-base type.

The reaction of $[\text{Mn(CO)}_3(\eta^6\text{-arene})]^+$ cations with nucleophiles normally results in exo attack at the ring to give 6-exo-substituted cyclohexadienyl products, as in equation (16). However, treatment of $[\text{Mn(CO)}_3(\eta^6\text{-C}_6\text{H}_6)]^+$ with Me_2CuLi at 0°C in diethylether gives the methyl complex (**29**) in 42% yield.[38] Surprisingly, treatment of (**29**) (0.04 M) with one equivalent of PPh_3 in refluxing benzene results in intramolecular migration ($t_{1/2} \sim 30$ h) of the methyl group to the arene and formation of the

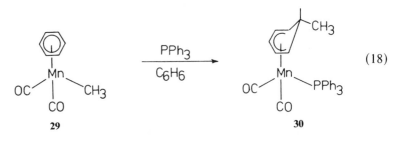

$$\text{(18)}$$

endo-methylcyclohexadienyl derivative (**30**) as in equation (18). This unprecedented reaction provides a rare demonstration of a pathway frequently postulated for endo-product formation.

Several papers have analyzed the regioselectivity of nucleophilic attack upon coordinated arene ligands. Considerable selectivity occurs in the attack of MeMgCl and PhMgBr on $[\text{Mn(CO)}_3(\eta^6\text{-XC}_6\text{H}_5)]^+$ (X = Me, MeO, Cl,

1-Cl, 4-Me), with ortho and meta addition being strongly favored over para. The degree of regioselectivity is nucleophile dependent,[39] since while H^- is known to add to $[Mn(CO)_3(\eta^6\text{-MeOC}_6H_5)]^+$ with a meta/ortho ratio of 2:1, RMgY (R = Me, Ph; Y = Cl, Br) adds exclusively meta.

Theoretical analyses indicate that there is a general correlation between the site of nucleophilic addition to $[Cr(CO)_3(\eta^6\text{-arene})]$ complexes and the magnitude of the coefficients in the lowest arene-centered unoccupied molecular orbital in the complex.[40,41] Other important factors are charge polarization induced by the conformation adopted by the $Cr(CO)_3$ group, the reactivity of the nucleophile, and the steric demands of arene substituents and the nucleophile. For example, with $[Cr(CO)_3(\eta^6\text{-alkylbenzene})]$ substrates and carbon nucleophiles, the generally unfavored para addition becomes more important with large alkyl groups (e.g., Bu^t) and more stabilized carbanions.[40] Extension of these studies to (1,1-dimethylindane)-tricarbonyl and (1,4-dimethoxynaphthalene)-tricarbonyl chromium complexes has been reported.[42,43]

Electrochemical studies in DMSO solvent show that the mechanisms of the one-electron reductions of bis(benzalacetophenone) chromium (**31**), as in equation (19), and free benzalacetophenone are virtually identical[44]:

$$[Cr(\eta^6\text{-}C_6H_5CH=CHCOPh)_2]$$
$$\textbf{31}$$

$$+ e^- \rightleftharpoons [Cr(\eta^6\text{-}C_6H_5CH=CHCOPh)_2]^{\overline{}} \qquad (19)$$

Comparison of the respective half-wave potentials confirms that π-coordination of the arene decreases its reactivity (thermodynamic) towards reduction by an order of magnitude.

12.2.2.7. Addition at the Tropylium Ligand

Kinetic studies of reaction (20) show[45] that π-complexation of the tropylium cation to the $Cr(CO)_3$ moiety reduces its reactivity towards MeOH

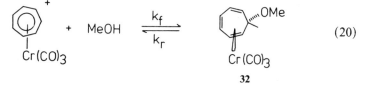

$$\qquad (20)$$

$$\textbf{32}$$

by a factor of 2×10^3. This may be due to an increase in the π-electron density at the ring carbon atoms of the complex[46] and/or the extra energy needed to break a metal–carbon bond accompanied by folding of the ring into an envelope conformation. Combination of the k_f (0.07 s^{-1}) and k_r

$(2.2 \times 10^5 \text{ litres mol}^{-1} \text{s}^{-1})$ values in MeOH at 25°C gives a pK_{R^+} of ~ 6.5. This compares with a pK_{R^+} value of ~ 2.2 for free $C_7H_7^+$ in MeOH, confirming a parallel increase in thermodynamic stability of the tropylium ligand upon coordination.

A somewhat smaller decrease in reactivity of $C_7H_7^+$ upon coordination to $M(CO)_3^+$ (M = Cr, Mo, W) groups was found in its methoxide-exchange reactions (21).[47] The relative second-order rate constants at 25°C in $MeNO_2/MeCOEt$ (2/3 v/v) decreased in the order $C_7H_7^+ > [(C_7H_7)Mo(CO)_3]^+ > [(C_7H_7)W(CO)_3]^+ > [(C_7H_7)Cr(CO)_3]^+$ (110:10:6:1). Given the statistical factor that reaction of $C_7H_7^+$ can occur from either side of the planar ring whereas only the exo face of the ring in the complexes is open to reaction, the effect of complexation on reactivity is quite small in free-energy terms for the Mo and W complexes. However, reaction (21) is mechanistically different to the nucleophilic addition process

$$R^+ + (MG)OMe \rightarrow ROMe + (MG)^+ \tag{21}$$

$$\{(MG)^+ = \text{malachite green} = (p\text{-}Me_2NC_6H_4)_2\overset{+}{C}Ph\}$$

(20) since it involves S_E displacement of $(MG)^+$ by attack of a carbocation on (MG)OMe. Finally, it should be noted that the metal dependence of reaction (21) (Mo > W > Cr) is somewhat larger than that previously found[48] for addition of PBu_3^n to these $[M(CO)_3(\eta\text{-}C_7H_7)]^+$ cations (Cr > Mo $\sim$ W).

12.2.3. Reactions at Side Chains and Exocyclic Carbocations

An X-ray structure of the phosphonium adduct (33) obtained from the reaction of $[Cr(CO)_3(\eta^6\text{-}6,6\text{-dimethylfulvene})]$ with PEt_3, shown in equation (22), confirms exo addition of the phosphorus nucleophile to the exocyclic carbon center.[49]

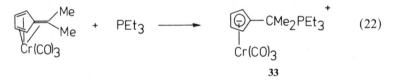

$$\tag{22}$$

33

12.2.4. Attack at Isocyanide Ligands

Mechanistic studies of aromatic amine addition to isocyanide ligands in $[PdCl_2(L)(CNAr)]$ complexes have been systematically reviewed.[50] These reactions, which yield coordinated carbene species, follow the general

Scheme 7

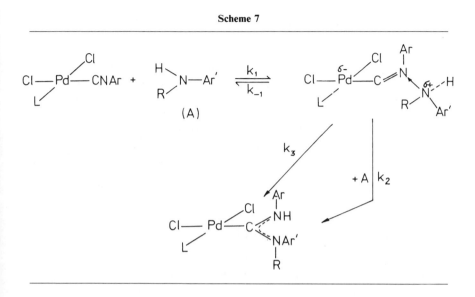

mechanism outlined in Scheme 7 [Ar, Ar' = ortho- or para-substituted aryl; R = H, Me, Et; L = PR$_3$, P(OR)$_3$, CNR]. The general rate law (23) holds.

$$k_{obs}/[A] = k_1 \frac{k_3 + k_2[A]}{k_{-1} + k_3 + k_2[A]} \tag{23}$$

The initial step (k_1) involves nucleophilic attack of the amine on the carbon of the coordinated isocyanide to give an intermediate iminopalladium(II) species. k_1 is sensitive to the basicity of the amine, the electrophilicity of the isocyanide carbon, and the degree of steric crowding about the reacting centers. The predominant importance of electronic factors is borne out by the fact that cis-[PdCl$_2$(PPh$_3$)(CNAr)] is much more reactive than the PEt$_3$ analogue in spite of the greater steric bulk of the triphenylphosphine complex.

12.3. *Electrophilic Attack*

Kinetic and mechanistic studies of electrophilic attack upon coordinated hydrocarbons continue to be far less numerous than reports of nucleophilic addition, despite the importance of the former processes in organometallic chemistry.

A second-order rate constant, k_r, of 2.2×10^5 liter mol s^{-1} has been determined at 25°C for the acid heterolysis of the 7-*exo*-methoxycycloheptatriene complex (**32**) in methanol (0.5–2.0 mM HClO$_4$), that is, for the reverse of reaction (20).[45] The 7-*endo*-methoxy isomer of (**32**) is very much less reactive. Heterolysis of (**32**) is general-acid catalyzed, and the k_r values for cyanoacetic and dichloroacetic acid buffers in water/MeCN (1/1 w/w) correspond to a Brønsted α-value of 0.8–0.9 (based on pK_a values in aqueous MeCN).

The exchange of protium for D$^+$ in a series of terminal alkynes RC≡CH is catalyzed by silver(I) trifluoromethanesulfonate, confirming activation of the *sp* C–H bond by π-coordination of RC≡CH to Ag(I) as in equation (24).[51] The rate law (25) is observed in CD$_3$NO$_2$. Deuterated alkynes react

$$RC≡CH + CD_3COOD \rightarrow RC≡CD + CD_3COOH \tag{24}$$
$$\underset{Ag^+}{|} \qquad\qquad\qquad \underset{Ag^+}{|}$$

$$\text{Rate} = k_1[RC≡CH \cdot AgOT_f] + k_2[RC≡CH \cdot AgOT_f][CD_3COOD] \tag{25}$$

faster than protiated analogues, consistent with an increase in p character of the alkyne terminal carbon in the rate-determining step. For the series of RC≡CH alkynes, a correlation was also noted between the rate of deuterium exchange and the decrease in the terminal *sp* C–H bond stretching frequency upon π-coordination to Ag(I).

Scheme 8

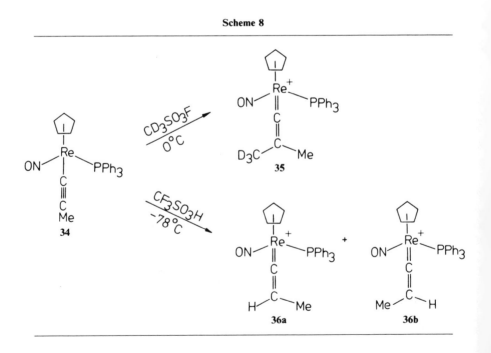

High stereospecificity has been observed in the reactions of the chiral acetylide complexes (**34**) (R = Me, Ph) with the electrophiles CD_3^+ and H^+.[52] Thus treatment of (**34**) (R = Me) with CD_3SO_3F at 0°C gives initially only the one isomeric vinylidene product (**35**) (Scheme 8). Similarly, protonation of (**34**) (R = Me) with CF_3SO_3H at −78°C generates a 90/10 mixture of (**36a**) and (**36b**). Even greater stereospecificity is found with the analogous phenyl acetylide (**34**) (R = Ph). The preferential formation of one isomeric product indicates that the rhenium chirality is transmitted through the formally cylindrically symmetrical C≡C triple bond, an unprecedented stereospecificity.

Similarly, chiral vinyl rhenium complexes such as (**37**) undergo electrophilic attack with appreciable 1,3-asymmetric induction.[53] For example, reaction of *E*-(**37**) with CD_3SO_3F in CD_2Cl_2 at −25°C gives a 92/8 mixture of diastereomers (**38a**) and (**38b**) [equation (26)]. Preferential attack of CF_3^+

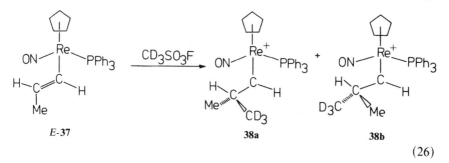

$$(26)$$

occurs on the C=C face anti to the PPh₃ ligand. The striking ability of the [Re(η-C₅H₅)(NO)(PPh₃)] system to participate in highly stereoselective reactions has also been demonstrated in studies of hydride abstraction[2] from [Re(η-C₅H₅)(NO)(PPh₃)(alkyl)] complexes with Ph₃C⁺BF₄⁻.

12.4. Miscellaneous Reactions

Upon standing at room temperature in CD₂Cl₂ the methylidene complex [Re(η-C₅H₅)(CO)(PPh₃)(=CH₂)]⁺ couples with itself to yield the η^2-ethylene cation [Re(η-C₅H₅)(CO)(PPh₃)(C₂H₄)]⁺.[54] Crossover experiments show that no PPh₃ dissociation or intermolecular =CH₂ scrambling occurs before the rate-determining step, which is considered from the second-order kinetics and retention of configuration at rhenium to be mutual front-side attack of two Re=CH₂ moieties. Remarkable enantiomer self-recognition occurs in this reaction, that is, *RR* and *SS* transition states are greatly preferred over the *RS* transition state.

The decomposition of $[Cr(CO)_3(\eta^6\text{-}2\text{-lithiochlorobenzene})]$ (**39**) in diethyl ether to give 1,2-dehydrobenzene has been found to follow first-order kinetics with a rate constant of $8.5 \times 10^{-5} \text{ s}^{-1}$ at $0°C$.[55] Complex (**39**) is thus much more stable than free o-chlorolithiobenzene. The simple first-order decomposition is consistent with the intermediacy of the benzyne complex (**40**). However, trapping experiments were successful in only low yield.

The triphenylplumbate anion (Ph_3Pb^-) adds to the carbyne carbon atom of $[Cr(CO)_5(CNEt_2)]BF_4$ to give the carbene complex (**41**).[56] Sub-

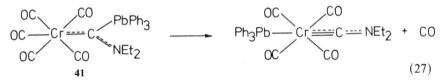

(27)

sequent rearrangement of (**41**) in 1,1,2-trichloroethane [equation (27)] followed first-order kinetics and gave activation parameters ($\Delta H^{\ddagger} = 103 \text{ kJ mol}^{-1}$, $\Delta S^{\ddagger} = 40 \text{ J K}^{-1} \text{ mol}^{-1}$) consistent with CO dissociation followed by $PbPh_3$ migration to the chromium atom. Similar kinetic behavior and activation parameters have been observed for the analogous carbon to chromium migration of TePh in the related phenyltellurocarbene complex (**42**).[57]

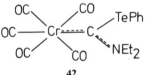

A kinetic study of the reaction of the carbene complex $[Cr(CO)_5\{C(Ph)OMe\}]$ (**43**) with arylalkynes to give naphthol complexes, as in equation (28), revealed[58] the rate law (29). This was rationalized in

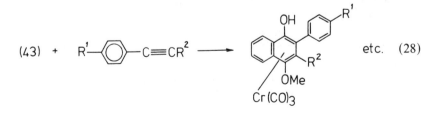

etc.   (28)

$$\text{Rate} = \frac{k_1 k_2 [\text{Cr}][\text{alkyne}]}{k_{-1}[\text{CO}] + k_2[\text{alkyne}]} \tag{29}$$

terms of the sequence shown in Scheme 9, in which initial CO dissociation (k_1) and coordination of the arylalkyne (k_2) was followed by rapid coupling (k_3) of the alkyne and carbene ligands to give the tricarbonyl (η^6-naphthol) chromium products.

Scheme 9

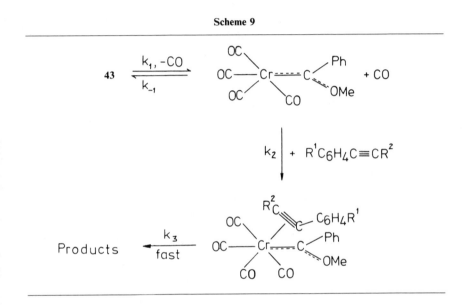

Chapter 13

Rearrangements, Intramolecular Exchanges, and Isomerizations of Organometallic Compounds

13.1. Mononuclear Compounds

13.1.1. Isomerizations and Ligand Site Exchange

Cis/trans isomerizations of square-planar d^8 complexes normally involve consecutive displacement of coordinated ligands by solvent or other ligand molecules. Stereochemical nonrigidity of a five-coordinate transient adduct or dissociative mechanisms have been less clearly defined.

For compound (1) (R = Ph, etc.) the spontaneous site exchange of the PPh$_3$ ligands occurs with retention of hydride ^{31}P nuclear spin–spin correlation showing that no Pt–H or Pt–P bonds are broken. Rates are little affected by complex concentration, by change of solvent, or by added ligands, except that added PPh$_3$ leads to a new reaction: the rapid intermolecular exchange of PPh$_3$ *trans* to SiR$_3$ with free PPh$_3$.[1] Rates of PPh$_3$ site-exchange in compound (2) are independent of complex concentration, solvent, and the presence of free PPh$_3$.[2] These are criteria normally used to characterize intramolecular exchange and direct geometry changes such as a digonal twist are possible. If this is correct, these are rare examples of this behavior for Pt(II) which is common for Ni(II). Dissociative mechanisms would require the recombination of the fragments without intermolecular exchange.

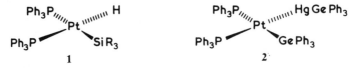

1

2

The kinetics of *cis/trans* isomerization of the octahedral complexes $[ML_2(RCO_2)_2]$ (M = Co or Ni, L = substituted pyridine) have been measured by NMR coalescence methods and the rate dependence upon the nature of RCO_2 studied. The reaction appears to become more dissociative in the order for R of $CF_3 < CF_2H < CFH_2 < C_2F_5$.[3] The isomerization of octahedral Ir(III) complexes can occur by different mechanisms. Compound (3) (L = PPh₃ or PMePh₂), formed by *cis* oxidative addition of H₂, does not isomerize in dichloromethane. However, at −30°C the addition of methanol leads rapidly to the isomer with ligands L *trans* and hydrides *cis*. A deprotonation/reprotonation mechanism seems likely especially as the supposed neutral intermediate was obtained by deprotonation with Bu'OK.[4] In contrast, the dihydrido cation (4), where the diphosphine is $(2\text{-MeOC}_6H_4PPhCH_2)_2$, which was also obtained by H₂ oxidative addition, isomerizes to the *trans*-dihydride but in this case without intermolecular scrambling either between complex molecules or with H₂, D₂, or methanol. An intramolecular mechanism is therefore more likely in this case.[5]

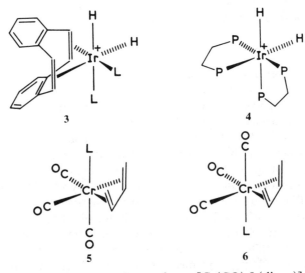

3

4

5

6

The chromium diene complexes of type $[Cr(CO)_3L(\text{diene})]$ [L = PMe₃ or P(OMe)₃] exist in some cases as mixtures of isomers (5) and (6). These compounds show rapid intramolecular CO scrambling and, where isomers are observed at low temperatures, these interchange rapidly at higher temperatures. Rearrangements involving trigonal prismatic geometries have been used to explain these observations.[6] Related fluxionality has also been observed for $[M(CO)_2\{P(OMe)_3\}_2(diene)]$ (M = Mo or W)[7] and for *cis*-$[W(CO)_2(diene)_2]$.[8] The rapid interconversion of the enantiomers (7) and (8) that are present for the bis(diene) tungsten complexes when using unsymmetrical dienes was observed by the exchange of the nonequivalent diene ligands and this occurs without exchange with a diastereomer of C_2 symmetry which has equivalent diene ligands. The mechanism shown in Scheme 1, a trigonal twist, is consistent with these observations ($\Delta G_{250\,K}^{\ddagger} = 47\ kJ\ mol^{-1}$).

Scheme 1

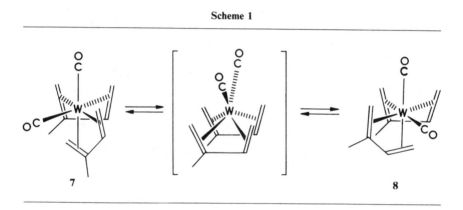

7   8

There has been an analysis of the molecular and lattice motions necessary for the *trans* to *cis* isomerization of compound (9) within a single crystal at 133°C and of the interrelation between these motions.[9] The seven-coordinate capped octahedral compound $[TaMe_3Cl\{H_2B(Me_2pz)\}_2]$ (10), where Me_2pz is the 3,5-dimethyl-1-pyrazolyl group, contains a tridentate ligand bound at an octahedral face opposite the capped one through two N atoms and a Ta–H–B bridge. The molecule is fluxional but the Ta–H–B bridge remains intact. The low-temperature NMR spectrum shows the presence of the three possible isomers derived by permutation of the three Me ligands and the Cl ligands, while changes in the spectra on raising the temperature involve three separate processes involving motions of the chloro and methyl ligands.[10]

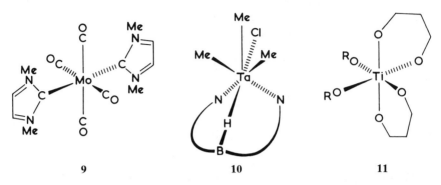

9   10   11

Other metal-centered rearrangements have been observed for bis (bidentate) tetrahedral gold(I) complexes,[11] for dialkylthiocarbamato complexes of titanium(IV) and zirconium(IV),[12] and for β-diketonato complexes of titanium(IV) such as $[Ti(acac)_2(OPr^i)_2]$.[13] For the latter complexes (11), inversion as probed by the exchange of diastereotopic groups of the alkoxy ligands occurs at the same rate as the exchange of the ends of the β-diketonate ligands. Various criteria (the lack of solvent effects, the negative $\Delta S^{\ddagger}$, and the inability to correlate rate with bond strengths) are consistent with an intramolecular twist mechanism.

13.1.2. Simple Ligand Rotation about the Metal–Ligand Axis

Enantiomers formed by coordination of *trans*-RCH=CHR in complex (12) do not interconvert since four PMe_2 ^1H NMR doublets are observed over a wide range of temperatures. Rotation about the C=C axis but not about the metal–alkene axis would lead to methyl exchange. However, alkene rotation about the metal–alkene axis occurs readily and rotational barriers have been measured for a range of complexes of types (12)–(14).[14,15] For compounds (14), these may be measured simply from ^{31}P NMR coalescences. For *trans*-$MeO_2CCH=CXCO_2Me$, $\Delta G^{\ddagger}$ (rotation, 303 K) values increase in the order X = H < Me < F < Cl < Br and correlate linearly with CO stretching force constants; the higher the force constant, the higher is $\Delta G^{\ddagger}$. This relation probably reflects competition between CO and the alkene for π-bonding orbitals of tungsten, differences in the rotational barriers being controlled largely by π-bonding changes. C_2F_4 is invariably more strongly coordinated than C_2H_4 and in the case of $[Pt(C_2H_4)(C_2F_4)(PCy_3)]$ the structural features of coordinated C_2H_4 compared with C_2F_4 support this. Replacing one C_2H_4 of $[Pt(C_2H_4)_2(PCy_3)]$ by C_2F_4 reduces the rotational barrier of the remaining C_2H_4 by 12 kJ mol^{-1}, reflecting mainly the weaker π-bonding for this remaining ethylene ligand.[16]

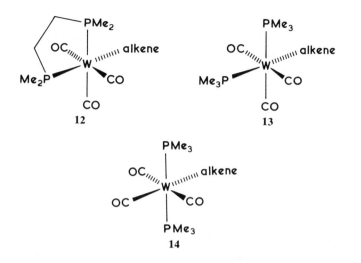

12

13

14

In compounds of type [M(CO)$_4$(alkene)] (M = Fe, Ru, or Os) alkene rotation cannot easily be separated from angular deformations of the CO ligands. Indeed, in some cases it has been shown that a Berry axial-equatorial exchange, which occurs as the pivotal alkene rotates by 90°, fits observations on the exchange. However, this cannot represent the lowest-energy motions in [Fe(CO)$_4$(cis-alkene)] where the alkene is *cis*-cycloheptene through to *cis*-cyclodecene. In these cases, a different exchange rate constant was determined for each of the axial CO ligands. Turnstile mechanisms involving an axial CO and the two equatorial ones are proposed and because of the symmetry of the alkene these are not equivalent. These processes must be faster than any motion that could be identified as alkene rotation.[17] As might be expected for [M(CO)$_4$(alkene)], exchange rates are in the order Fe > Ru > Os.[18]

Alkene rotation barriers, mass spectroscopic appearance potentials, and enthalpies of formation have all been determined for a series of rhodium and iridium alkene complexes of type [M(η-C$_5$H$_4$R)(C$_2$H$_4$)$_2$]. Again changes in $\Delta G^\ddagger$ (rotation) may be related to the expected changes in the π-contribution to the metal–alkene bond. Thus $\Delta G^\ddagger$(Ir) > $\Delta G^\ddagger$(Rh) and $\Delta G^\ddagger$ depends upon R in the order alkyl ~ H > CO$_2$R ~ CN > CHO > COCO$_2$R which would mean that the higher $\Delta G^\ddagger$ values are found when the cyclopentadienyl ligand is most donating.[19]

Alkene rotations and exchanges were studied for a range of complexes of type [Rh(acac)(alkene)$_2$] and for substituted styrenes the influence of the ring substituent on the rates of alkene exchange was established. These are 16-electron compounds and it is interesting that the *o*-MeOC$_6$H$_4$CH=CH$_2$ complex is non-dynamic which may be interpreted in

terms of MeO coordination.[20] Alkene rotation has been studied for
$[Ru(C_5H_5)H(PPh_3)(alkene)]$,[21] $[Ni(C_5H_5)(alkyl)(alkene)]$,[22] and for
$[Pt(\eta\text{-}C_5H_5)(\eta^1\text{-}C_5H_5)(C_2H_4)]$[23] and rotational barriers determined in
some cases. In one case the mechanism of transfer of coordination from
one face of an alkene to the other has been studied. For a simple η^2-alkene
complex this would require alkene dissociation, but in the case examined,
$[Rh(diphos)(PhCH=CCO_2EtNHCOMe)]^+$, the alkene is part of a chelate
ring so that ring opening rather than total dissociation is sufficient. NMR
methods gave the rate coefficient of $0.65\ s^{-1}$ for the process shown in Scheme
2 {diphos = $[(2\text{-}MeOC_6H_4)PhPCH_2]_2$} and require that the process is
intramolecular. This is another, still rather rare, example of a 14-electron
intermediate in the chemistry of square-planar d^8 metal complexes.[24]

Scheme 2

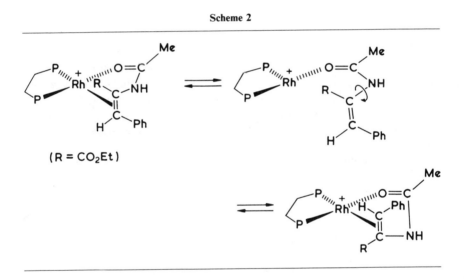

(R = CO$_2$Et)

Although mononuclear alkyne complexes are less commonly found
than alkene ones, there is a range of Mo and W alkyne complexes which
generally show alkyne rotation unlike, for example, platinum(0) alkyne
complexes. The structure of the fluxional complex $[MoBr_2(CO)$
$(PEt_3)_2(PhC_2H)]$ (**15**) has been determined. The corresponding MeC$_2$Me
complex shows nonequivalent Me groups which exchange with
$\Delta G^{\ddagger} = 13.0\ kcal\ mol^{-1}$.[25] Likewise the rotation of alkynes in
$[Mo(C_4H_4NCS_2)_2(alkyne)_2]$ (**16**) occurs with the rotational barrier of
$13.7\ kcal\ mol^{-1}$ for but-2-yne and $13.8\ kcal\ mol^{-1}$ for ethyne. In these cases

the fluxionality is a little complicated by a faster process (10.7 kcal mol^{-1}) which has been identified as intraligand rotation about the C–N bonds of the dithiocarbamates. No evidence for sulfur–ligand chelate-ring opening was found.[26]

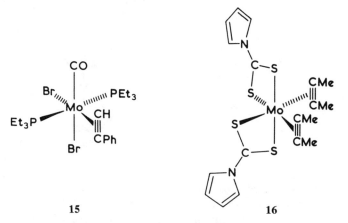

15   **16**

The but-2-yne ligand in [W$_2$(OPr^i)$_6$(MeC$_2$Me)(C$_4$Me$_4$)] is η^2 coordinated at one of the W atoms throughout two dynamic processes (W-coupling evidence). One of these processes is MeC$_2$Me rotation; the other is a slower process involving stereochemical nonrigidity localized at the W atom to which the alkyne is coordinated.[27] The barrier to alkyne rotation in [W(C$_5$H$_5$)(CO)(NO)(C$_2$H$_2$)] is >90 kJ mol^{-1}, and while the corresponding Mo compound has a high rotational barrier (146 kJ mol^{-1}), that for Cr is much lower (55.9 kJ mol^{-1}).[28]

Rotational barriers for the cyclic ligands, cyclopentadienyl and arene, in various complexes have been measured. In the solid state, rotation of the C$_5$H$_5$ rings in [Mn(C$_5$H$_5$)(CO)$_3$] (7.24 kJ mol^{-1}), [Re(C$_5$H$_5$)(CO)$_3$] (7.15 kJ mol^{-1}), and [V(C$_5$H$_5$)(CO)$_4$] (7.07 kJ mol^{-1}) were determined from measured spin-lattice relaxation times. Nonbonded atom–atom potential calculations for these compounds indicate that crystal-packing effects largely control the molecular conformations.[29] Mechanical spectroscopy has been extended to [Cr(C$_6$H$_6$)(CO)$_3$] and related compounds. The dispersion of an oscillating mechanical stress is maximized when its frequency corresponds with the rotational frequency of the C$_6$H$_6$ ring.[30] The temperature dependence of spin-lattice relaxation times has been used to obtain barriers for C$_5$H$_5$ ligand rotation in [TiCl$_3$(C$_5$H$_5$)] (9.6 kJ mol^{-1}), [Mo$_2$(CO)$_6$(C$_5$H$_5$)$_2$] (13.9 kJ mol^{-1}) and [TiS$_5$(C$_5$H$_5$)$_2$] (8.9 axial and 7.7 equatorial kJ mol^{-1}).[31]

A variable-temperature ^{13}C NMR study of [Cr(CO)$_2$(CS)(η-C$_6$Et$_6$)] has shown that at 163 K the spectrum is consistent with the C$_s$ structure

(17) (as in the crystal) but that at 231 K rotation of the $Cr(CO)_2(CS)$ unit with respect to the ring (but not exchange between proximal and distal Et groups) is occurring.[32] This is the first example of the freezing out of tripodal rotation in an arene–chromium compound.

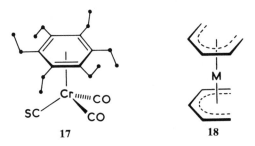

17   18

The pentadienyl complexes of iron and ruthenium $[M(C_5H_7)_2]$ (18), the open-ring analogues of ferrocene and ruthenocene, adopt gauche-eclipsed conformations with the ligand termini being different. Oscillation barriers for processes exchanging the termini increase slightly as the C_5 chains are methylated and are somewhat higher for Ru than for Fe. Compare, for example, $\Delta G^{\ddagger}$ of 9.1 kcal mol^{-1} for $[Fe(\eta^5\text{-}2,4\text{-}Me_2C_5H_5)_2]$ with 9.73 kcal mol^{-1} for its Ru analogue.[33] The variable-temperature NMR spectra of the complexes $[Re(\eta^3\text{-}C_3H_4R)(CO)_4])$ ($\Delta G^{\ddagger} = 61.5$ kJ mol^{-1} for R = H; $\Delta G^{\ddagger} > 69.5$ kJ mol^{-1} for R = Me) have been interpreted in terms of an allyl rotation mechanism.[34] However, for the mixed η^3-allyl-carbene complex $[PdCl(CNHRNR_2)(\eta^3\text{-}CH_2CHMeCH_2)]$, either allyl rotation or carbene rotation could be used to explain the observed NMR behavior, since a $\eta^3\text{-}\eta^1\text{-}\eta^3$ mechanism was ruled out by the lack of *syn/anti*-hydrogen exchange. A simple allyl rotation has never been observed for Pd(II) so the carbene rotation is favored.[35]

No coalescence of NMR signals was observed for the pseudotetrahedral alkylidene and vinylidene complexes $[Re(C_5H_5)(NO)(PPh_3)(=CHPh]^+$ and the related CH_2 and $C=CH_2$ complexes which sets lower limits for rotation barriers about the Re = C bonds of 18–19 kcal mol^{-1}.[36-38] However, slow isomerizations of kinetically formed isomers of the CHPh and C=CMePh complexes to thermodynamic mixtures could be followed by NMR methods and this has allowed kinetic activation parameters for the ligand rotation to be determined in these cases.

The tungsten complex (19) and its derivatives (20) and (21) formed by treatment with sulfur and diazomethane, respectively, are all fluxional. Rotation about the W = As bond of (19) is associated with $\Delta G^{\ddagger}_{227\,K}$ of 11.1 kcal mol^{-1} compared with a value of 10.9 kcal mol^{-1} for the Mo analogue. Rotations of the η^2 ligands of (20) and (21) are also fast enough to give NMR coalescence but it has not been established whether these are simple rotations or whether ring-opened intermediates are involved.[39]

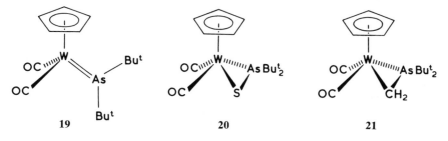

19   **20**   **21**

13.1.3. Migration of Metal Atoms between Different Ligand Sites

For the compound (**22**) to give a ^{1}H NMR singlet for the η^2-C_6H_6 ligand, it was shown that exchange between free and coordinated benzene was not the cause and that a ring-whizzing mechanism seemed likely. Substituted rings give rapidly interconverting isomers. The intermediate (**23**) has been discussed (Scheme 3) and, because firm experimental evidence

Scheme 3

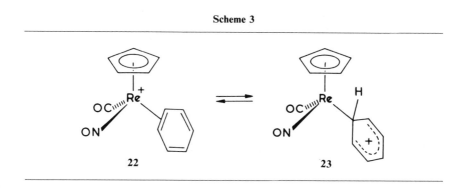

22   **23**

for this has not been given, this would justify further study.[40] Reversible C–H bond cleavage was also considered and this must be taken seriously in view of the observation that isomerization in tolyl–rhodium complexes occurs faster than exchange with free toluene (Scheme 4). The intramolecular mechanism shown is very likely.[41]

Magnetization transfer ^{13}C NMR experiments on the two isomers (**24**) and (**25**) of $[Fe(\eta^4\text{-}C_8H_8)(CO)_2(CNPr^i)]$ have been used to demonstrate a specific transfer of the iron about the cyclooctatetraene ring in a way

Scheme 4

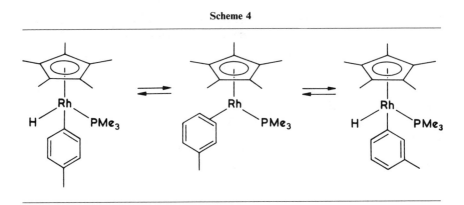

completely consistent with a mechanism based on Woodward–Hoffmann symmetry considerations and quite inconsistent with the least-motion mechanism.[42] A theoretical treatment of haptotropic rearrangements of

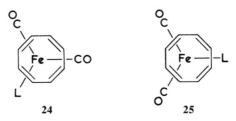

24   25

complexes with bicyclic polyene ligands coordinated to $M(C_5H_5)$ and $M(CO)_3$ groups has been made,[43] and the fluxionality of $SiMe_3$ and $GeMe_3$ cycloheptatriene complexes of $Fe(CO)_3$ also studied.[44] Complex (**26**) exists as the *anti*-isomer, which is illustrated, in mixtures with the *syn*-isomer. Magnetization transfer has been used to establish that, although the ends of the acyclic ligand are different, these are exchanging.[45] The compound $[Pd(\eta^5\text{-}C_5H_5)(\eta^1\text{-}C_5H_5)(PR_3)]$ undergoes a 1,2-shift process for the η^1-cyclopentadienyl ligand and a slower exchange of the η^1 and η^5-C_5H_5 ligands (^1H and ^{13}C NMR evidence).[46]

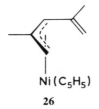

$Ni(C_5H_5)$

26

In most chelating BH_4 complexes exchange of terminal and bridging hydrogen atoms is very rapid. Exchange of the ^{31}P nuclei is observed for $[FeH(\eta^2\text{-}BH_4)\{MeC(CH_2PPh_2)_3\}]$ although the details of the process are unclear.[47] Bridge–terminal hydrogen exchange is very rapid in $[Cr(CO)_4(\eta^2\text{-}BH_4)]^-$,[48] and in $[Ln\{\eta\text{-}C_5H_3(SiMe_3)_2\}(THF)_n(BH_4)]$ for the bigger lanthanide metal atoms.[49] With smaller trivalent metal ions (e.g., Sc), no THF is coordinated and the BH_4 ligand is nonfluxional at room temperature, only showing rapid hydrogen atom exchange at 100°C. Two mechanistic possibilities arise in most of these cases. Either the $\eta^2\text{-}BH_4$ ligand passes through a η^1-form with any one of the three terminal hydrogen atoms bridging when the ligand returns to its η^2-form, or the ligand becomes η^3 in the intermediate. These mechanisms and even that involving complete dissociation have rarely been defined. The compound $[Cu(BH_4)\{P(OMe)_3\}_2]$ probably contains $\eta^3\text{-}BH_4$ which undergoes hydrogen exchange with an estimated $\Delta G^\ddagger = 4.3 \pm 0.5$ kcal mol^{-1}.[50]

When a potentially multidentate ligand has unused coordinative capacity, the exchange between different ligand sites could be regarded as an intramolecular model for ligand exchange. Where the metal center is unsaturated, associative reactions are likely as in the ^{31}P exchange within the bidentate ligand in $[PtCl(PEt_3)(\eta^2\text{-}SPPh_2CHPPh_2S)]$ in which the sulfur atoms are successively coordinated.[51] On the other hand sulfur atom exchange in $[Cr(CO)_5(\eta^1\text{-}SCH_2SCH_2SCH_2)]$ and related complexes occurs intramolecularly; but here an associative reaction would require a 20-electron intermediate and it seems likely that W–S bond breaking accompanies bond making.[52,52a] Another related and interesting system not only requires metal atom transfer between sulfur atoms but also a rearrangement of rings in the noncoordinated part of the ligand; $\Delta G^\ddagger(T_c) = 14.7$ kcal mol^{-1} (Scheme 5).[53]

Scheme 5

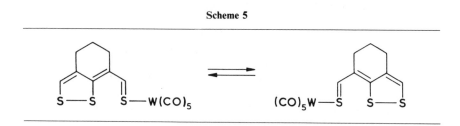

Another possibility that arises is that the other ligands reorganize to facilitate an associative-type reaction. An example is shown in Scheme 6.

Compound (**27**), formed at −78°C by hydride addition at the corresponding cationic carbonyl precursor, converts to the hydride (**28**), which contains η^1-diphosphine, above −50°C. It is believed that this conversion is fast enough to account for the exchange of phosphorus atoms in the hydride with $\Delta G^{\ddagger}_{297\,K} = 18$ kcal mol^{-1}.[54] It is a pity that the mechanistic details of many such interchanges are more speculative than experimentally established, since the study of intramolecular reactions of this sort provides access in principle to the mechanisms of important intermolecular reactions.

Scheme 6

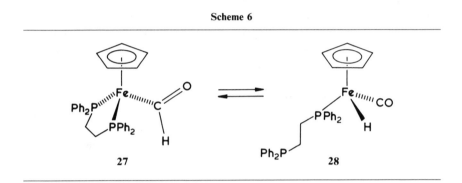

27 28

Other examples of the rapid exchanges of coordinated and noncoordinated sites in the same ligand are provided by [AuMe$_2${η^2-BH(C$_3$H$_3$N$_2$)$_3$}][55] and [RuCl$_2$(η^1-L)$_2$(η^2-L)], where L is Ph$_2$PCH$_2$CO$_2$Et. In the latter case the coordinated CO$_2$Et group of one ligand L exchanges with the free CO$_2$Et groups of the other two.[56]

13.1.4. Hydrogen Atom Transfer Reactions

As reported in Volume 2 of this series, the dynamic behavior of the compound [Co(C$_5$Me$_5$)(C$_2$H$_4$)$_2$H]$^+$ was interpreted in terms of hydride exchange with the terminal hydrogen atoms at only one end of each ethene. A curious feature is the proton–proton coupling between the hydride and these terminal CH$_2$ protons.[57] This coupling has allowed a reinterpretation of the structure and dynamic behavior of this compound, which was reformulated as [Co(C$_5$Me$_5$)(C$_2$H$_5$)(C$_2$H$_4$)]$^+$ in which the η^2-ethyl group contains one agostic hydrogen atom, the one which was previously assigned as a hydride.[58] Indeed various earlier observations in the literature may be reinterpreted in terms of agostic hydrogen bridges rather than hydrides. The dynamic behavior of the cobalt compound may now be understood in

Scheme 7

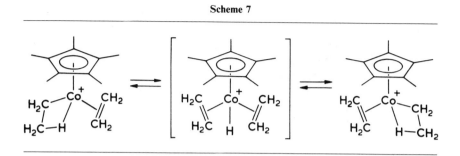

terms of the equilibrium in Scheme 7. This is not very different from the mechanism originally proposed except that the ground state is close to the original transition state and the higher-energy intermediate corresponds with the original ground state. These results have considerably developed our understanding of mechanisms of hydrogenations and related reactions.

In many ways the system $[Mn(CO)_3(cyclohexenyl)]$ (**29**) is closely related and involves three separate dynamic processes.[59,60] One is the exchange between the two corresponding hydrogen atoms adjacent to the η^3-allyl group that are available for agostic bonding, another is CO site exchange; the third is a hydrogen atom transfer reaction akin to that in Scheme 7. This process (Scheme 8) allows the exchange of all the ring carbon atoms via the hydride (**30**). The dynamic behavior observed for (**29**) has been extended to $[Mn(CO)_2(PMe_3)(cyclohexenyl)]$ and to alkyl-substituted cyclohexenyl compounds.

Scheme 8

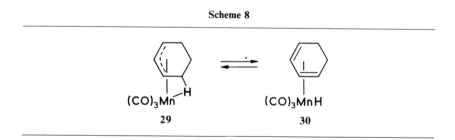

Ultraviolet photolysis of $[MR(CO)_3(C_5H_5)]$ (M = Mo or W; R = Et, etc.) generates the 16-electron alkyl compound $[MR(CO)_2(C_5H_5)]$ which was characterized by its infrared spectrum at 77 K. Warming in the presence of free ligands L gave $[MR(CO)_2L(C_5H_5)]$, but in the absence of added ligands the thermal β-hydrogen transfer in Scheme 9 occurs. The kinetics are consistent either with the mechanism shown in the scheme or with a fast pre-equilibrium between $[MoEt(CO)_2(C_5H_5)]$ and *cis*-$[MoH(C_2H_4)$-$(CO)_2(C_5H_5)]$ which slowly converts to the *trans*-isomer.[61] In the light of other results,[58–60] agostic bonding in $[MoEt(CO)_2(C_5H_5)]$ may be present.

Scheme 9

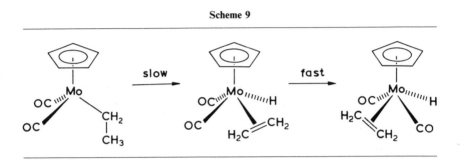

A kinetic examination of alkene insertion by magnetization transfer methods has been made for the reaction of C_2H_4 with *trans*-$[RhH(N_2)$-$(PPr_3^i)_2]$. The initially formed compound, *trans*-$[RhH(C_2H_4)(PPr_3^i)_2]$ isomerizes to the *cis*-isomer prior to an insertion to give $[RhEt(PPr_3^i)_2]$. The interconversion of the last three (isomeric) compounds was studied by magnetization transfer and this indicates an entirely intramolecular insertion. It may be that, for less bulky phosphines, the hydride migration may be induced by the association of another ligand but that is not the case here.[62]

Detailed kinetic and stereochemical experiments have been made on the isomerization of an alkylidene to an alkene complex (Scheme 10). The reactions as shown are highly stereospecific and the kinetics of the last step have given $\Delta H^{\ddagger} = 20.4 \pm 1.4 \, kcal \, mol^{-1}$ and $\Delta S^{\ddagger} = -27 \pm 0.3 \, cal \, K^{-1} \, mol^{-1}$. The 1,2-hydrogen shift seems to occur spontaneously but probably not via a single transition state.[63]

Magnetization transfer experiments have been applied to the α-hydrogen atom transfer equilibrium between $[TaH(=CHBu^t)I_2(PMe_3)_3]$ and $[Ta(CH_2Bu^t)I_2(PMe_3)_3]$,[64] while an even faster exchange giving coalescence occurs for $[WH(=CH)Cl_2(PMe_3)_3]$.[65] The CH and hydride ^1H NMR signals coalesce below 280 K but the data are better interpreted if some other form has some population at higher temperatures. This may be $[W(=CH_2)Cl_2(PMe_3)_3]$ which would be the intermediate for the exchange.

Scheme 10

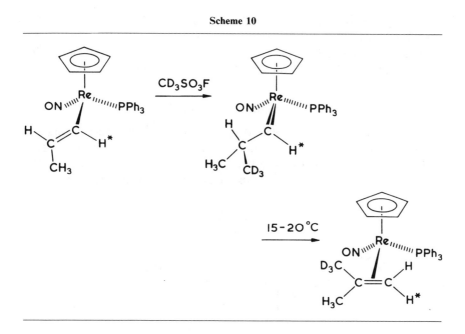

13.2. Dinuclear Compounds

The fluxional behavior of dinuclear compounds can involve the transfer of ligands between the metal atoms, rearrangements of a bridging ligand, the equilibration of the metal atoms by an oscillation of an unsymmetrically bridging ligand, exchange of bridging and terminal ligands, and other processes requiring the combined motion of two connected coordination spheres.

13.2.1. Carbonyl Ligand Exchange

The complex $[MoW(C_5H_5)_2(CO)_4]$ shows separate NMR resonances for the C_5H_5 ligands bound at each metal atom but only one ^{13}CO NMR signal at room temperature. At low temperatures two ^{13}CO signals are obtained and only one of these shows coupling to tungsten, so that CO transfer between the metal atoms is occurring ($E_a = 10.4 \pm 0.3$ kcal mol^{-1}).[66] However, site exchange of CO ligand does not necessarily require their movement. For example, the $[Rh_2(\mu\text{-}CO)_3L]^{2+}$, where L is a 24-membered ring $(-OCH_2CH_2NRCH_2CH_2NRCH_2CH_2NRCH_2CH_2-)_2$ which is coordinated through three N atoms at each Rh atom, contains three nonequivalent

CO bridges. These give three ^{13}CO NMR triplets (^{103}Rh coupling) at 170 K but there is a time-averaged plane of symmetry through the metal atoms relating two CO at 260 K. This originates from stereochemical nonrigidity at the octahedral Rh atoms and not from CO migration.[67]

13.2.2. Transfer of Other Ligands between Metal Atoms

Intramolecular transfer of a Me group between Rh atoms during the *cis/trans* isomerization of the cation [Rh$_2$(μ-CH$_2$)$_2$(C$_5$Me$_5$)Me(L)]$^+$ (L = MeCN) was confirmed by the observation of a ^{1}H NMR triplet (^{103}Rh coupling) for the Me group above the coalescence temperature. Scheme 11

Scheme 11

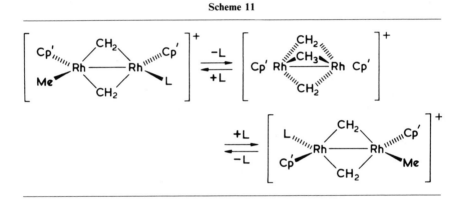

is consistent with the nondependence of rate on complex concentration and the lack of exchange of the diastereotopic methylene protons. Dissociation of MeCN is further supported by the complex's rigidity when L is more strongly coordinated (CO or py).[68]

The mechanisms of inversion of A-frame molecules (Scheme 12) have been considered and in some cases these involve the transfer of ligands

Scheme 12

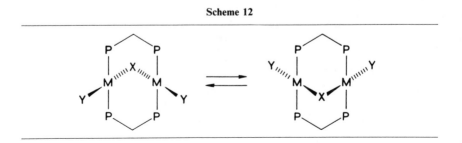

between metal atoms.[69-71] A-frame inversion on an NMR time scale normally only occurs for Pt_2 compounds when there is a bridgng hydride or where one is readily accessible. Then the mechanism is via a linear Pt–H–Pt system. The compounds $[Pd_2MeX_2(dppm)_2]^+$ have different behavior depending upon whether X is hydride or iodide. When X = I, the exchange of the two I ligands of compound (**31**) via compound (**32**) is at the same rate as A-frame inversion, whereas the dihydride analogue $[Pd_2MeH_2(dppm)_2]^+$ undergoes inversion without exchange of the non-equivalent hydrides. The bridging hydride must pass between the Pd atoms and out to the other side and the mechanistic differences relate in part to the size constraints imposed by the dppm bridges. The related cation $[Pt_2H_2Cl(dppm)_2]^+$ adopts structure (**33**) like the high-energy form (**32**), but this compound undergoes exchange involving Pt atom interconversion via the chloro-bridged species (**34**) (H, ^{31}P, and ^{195}Pt NMR evidence). No rapid hydride ligand transfer between Pt atoms occurs.

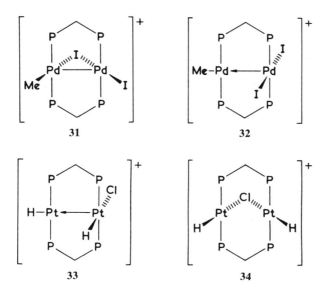

A form of ligand transfer between metal atoms is where an unsymmetrically bridging ligand oscillates between the metal atoms, such as in a range of compounds with $\mu\text{-}\eta^1,\eta^2$ ligands such as vinyl, alkynyl, cyanide, CO, or RNC. The rate of oscillation of the bridging CO in $[Mn_2(\mu\text{-}\eta^1, \eta^2\text{-}CO)(CO)_4(dppm)_2]$ is too slow to give NMR coalescence, whereas the cyanide oscillation in the related compound $[Mn_2H(\mu\text{-}\eta^1, \eta^2\text{-}CN)(CO)_4(dppm)_2]$ is fast. Using 90%-enriched ^{13}CN and ^{13}C NMR, the quintet at room temperature changes to a triplet at −90°C, confirming the intramolecular nature of the oscillation and that coupling to ^{31}P nuclei only occurs through the η^1-Mn–CN contact.[72] The *cis* and *trans* isomers of $[M_2(\mu\text{-}\eta^1, \eta^2\text{-}CH=CH_2)(CO)_3(C_5H_5)_2]^+$ (M = Fe or Ru) interconvert via a terminal vinyl intermediate in which there is rotation about the M–M bond, but only in the *cis* isomer does vinyl oscillation occur more rapidly than *cis-trans* isomerization.[73]

Some other examples of $\mu\text{-}\eta^1$, η^2 ligands have been found to be fluxional and some others rigid. The compounds $[Re_2H(\mu\text{-}C\equiv CPh)(CO)_8]$ and $[Re_2H(\mu\text{-vinyl})(CO)_8]$ have both been shown to have very rapidly oscillating ligands,[74,74a] whereas the compound $[Mo_2(C_5H_5)_2(CO)_4(\mu\text{-}CNPh)]$ is rigid. Separate C_5H_5 NMR resonances were found at room temperature for the latter compound which is in contrast to observations on the fluxional compound $[Mo_2(C_5H_5)_2(CO)_4(\mu\text{-}CN)]^-$ which is closely related.[75]

The interchange of the bonding of an unsymmetrically bridging ligand between two nonequivalent metal atoms also occurs for compound (**35**) (Scheme 13) but here this occurs by a specific *endo*-hydrogen atom 1,5-shift.[76]

Scheme 13

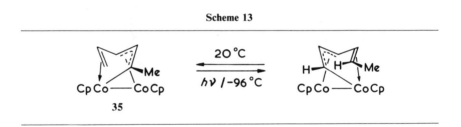

35

The kinetics of isomerization between the head-to-head and the head-to-tail isomer of $[Pt_2(\mu\text{-pyridonato})_2(en)_2]^{2+}$ have been measured by ^{195}Pt NMR methods.[77] This involves the turning round of an unsymmetrical bridge by bridge opening, O for N exchange of the nonbridging pyridonato ligand followed by bridge closure.

13.2.3. Bridge–Terminal Ligand Exchange

The cation $[Pt_2(\mu\text{-}H)H_2(dppe)_2]^+$ (dppe = $Ph_2PCH_2CH_2PPh_2$) undergoes the exchange of bridging and terminal hydride ligands too fast for the NMR spectrum to be frozen out.[78]

In $\mu\text{-}CH_2$ compounds there is evidence for $\mu\text{-}CH_2$ to $\eta^1\text{-}CH_2$ transformations.[79,80] Thus the rates of interconversion of *cis*- and *trans*-

[Co$_2$(MeC$_5$H$_4$)$_2$(μ-CH$_2$)(CO)$_2$] and other alkylidene analogues have been measured [$\Delta G^{\ddagger}$(*cis*-to-*trans*) = 17.0 and $\Delta G^{\ddagger}$(*trans*-to-*cis*) = 18.2 kcal mol^{-1} for the neohexylidene complex]. The mixed Co/Rh species show two processes: a slow process ($\Delta G^{\ddagger}$ = 13.0 kcal mol^{-1}) which interconverts isomers by a η^1-CH$_2$ intermediate (Process A in Scheme 14) and a

Scheme 14

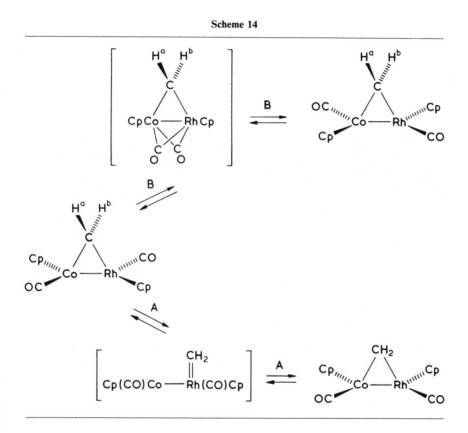

faster process B which leads to the exchange of the diastereotopic CH$_2$ protons of the *trans* isomer.[79] The complexes [Co$_2$(μ-CH$_2$)(μ-CO)(CO)$_4$(dppm)] and [Co$_2$(η-CH$_2$)$_2$(CO)$_4$(dppm)$_2$] (36) both undergo reversible CH$_2$-bridge opening. In the case of compound (36), the bis(η^1-CH$_2$) intermediate is required to undergo pseudorotation at each trigonal bipyramidal Co prior to readopting the bridged form.[80] A reaction related to (A) in Scheme 14 was proposed for the *cis*/*trans* isomerization of

[Ru$_2$(μ-CMe$_2$)(μ-CO)(CO)$_2$(C$_5$H$_5$)$_2$] but this also involves the simultaneous opening of the μ-CO and μ-CMe$_2$ bridges which leads to the exchange of bridging and terminal CO ligands.[81]

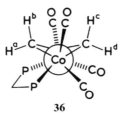

36

13.2.4. Bridging Alkyne Rotation

The compound [Co$_2$(CO)$_4$(dppm)(μ-alkyne)] is a derivative of [Co$_2$(CO)$_6$(alkyne)] and has been shown to undergo the oscillatory rotation shown in Scheme 15 which generates a time-averaged mirror plane.[82] There is, however, no evidence for the rotation of the alkyne through a form with the C–C bond parallel to the Co–Co bond. Calculations give the rotational barrier in [Co$_2$(CO)$_6$(C$_2$H$_2$)] as 58 kcal mol^{-1}, and in general this is not a likely process for perpendicularly coordinated acetylene compounds.[83] However, the barrier is lowered to 34 kcal mol^{-1} in [Co$_2$(C$_5$H$_5$)(CO)$_3$(C$_2$H$_2$)]$^-$ and to 26 kcal mol^{-1} in [Co$_2$(C$_5$H$_5$)$_2$(C$_2$H$_2$)]$^{2-}$. These calculated rotational barriers may be compared with the experimentally determined value of ~21 kcal mol^{-1} for the isostructural and isolobal compound [NiCo(C$_5$H$_5$)(CO)$_3$(PhC$_2$CO$_2$CHMe$_2$)]; coalescence of the ^{1}H NMR doublets of the diastereotopic isopropyl methyl groups was observed.[84] Alkyne orientations in complexes of type [L$_2$M(alkyne)ML$_2$], where M is Ni or Pt and L is a tertiary phosphine, and alkyne rotations have been considered theoretically.[85]

Scheme 15

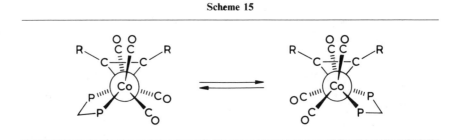

13.3. Cluster Compounds

13.3.1. Rearrangements Involving the Relative Motion of Metal Atoms in a Cluster

The ability of clusters to rearrange has been a feature of many large ones but there is only fragmentary evidence that this can be rapid enough for the cluster skeleton to be considered fluxional. A series of hexanuclear clusters formed by the addition of two Cu, Ag, or Au atoms to a Ru_4 cluster has the incoming atoms in nonequivalent sites. Thus $[Cu_2Ru_4H_2(CO)_{12}(PPh_3)_2]$ (37) has the structure shown (CO and H ligands are omitted). Site exchange of the two Cu atoms was deduced from the coalescence of the ^{31}P signals assuming no Cu–P bond cleavage. The barriers for rearrangement are 43 for Cu, 40 for Ag, and <35 kJ mol^{-1} for Au, whereas the mixed Cu, Ag, and Au compounds showed no dynamic effects presumably because of specific site preferences of different metal atoms.[86]

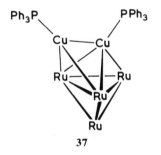

37

This type of behavior is more common, even becoming characteristic, for high clusters. The clusters $[Rh_9P(CO)_{21}]^{2-}$ and $[Rh_{10}P(CO)_{22}]^{3-}$ have structures based on a square antiprism of eight metal atoms with one or two Rh atoms respectively capping the square faces. The ^{31}P NMR signal for the encapsulated atom appears as a doublet of quintet of quintets (Rh_9) or as a triplet of nonets (Rh_{10}) at $-80°C$ consistent with these symmetries; but a dynamic process exchanging all the Rh atoms is observed at higher temperatures to give a symmetrical decet (Rh_9) or undecet (Rh_{10}) at room temperature or above. The precise relative motions of the metal atoms leading to these observations are still to be elucidated.[87]

13.3.2. Localized Carbonyl Exchange

Localized exchange at $M(CO)_3$ units containing axial and equatorial CO groups is commonly observed but will sometimes be obscured by faster exchange over several metal atoms (delocalized exchange). The compounds

$[Fe_3(CO)_9(PhC_2Ph)]$[88] and $[Fe_3(CO)_9(PPh_2)_2]$[89] each contain one $Fe(CO)_3$ unit different from the other two and while there is localized exchange at different rates at these units there is no CO scrambling between them. Likewise $[WOs_3(CO)_{11}(C_5H_5)(COCH_2C_6H_4Me)]$ contains one $Os(CO)_4$ and two nonequivalent $Os(CO)_3$ units. The rates of localized scrambling at each are different.[90] A rather different situation arises for $[RhCoMo(CO)_7(C_5Me_5)_2]$ (**38**) where there is exchange of the four CO ligands in a plane but not with the axial CO. These four ligands rotate in a windmill fashion which is similar to alkene rotation in compounds of type $[Mo(CO)_5(alkene)]$ where $RhCo(CO)_2(C_5Me_5)_2$ corresponds to the alkene.[91]

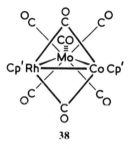

38

13.3.3. Delocalized Carbonyl Exchange

The accepted mechanism for CO exchange in $[Co_4(CO)_{12}]$ and $[Rh_4(CO)_{12}]$ is that given by Cotton: the reversible opening of CO bridges to convert the C_{3v} to the higher-energy T_d form. $[Ir_4(CO)_{11}(PEt_3)]$, unlike $[Ir_4(CO)_{12}]$, also adopts the bridging carbonyl structure and exists in solution as two isomers (abundances 7 : 1) with the PEt_3 at the basal plane in axial and equatorial sites, respectively. These isomers give separate NMR signals at $-90°C$. The dynamic behavior of the major isomer as studied by magnetization transfer is consistent with the Cotton mechanism but the minor isomer is in equilibrium with the major and plays a kinetically important role.[92] The complexes $[HFeCo_3(CO)_{12}]$ and $[FeCo_3(CO)_{12}]^-$ are structurally related and show two distinct processes: the interconversion of the nine CO ligands about the Co_3 basal plane which occurs faster than the exchange of these ligands with the remaining three at iron.[93]

An interesting and specific exchange has been found for $[Mo_2FeCo(S)(CO)_8(C_5H_5)_2]$ which leads to a mirror plane through the molecule (Scheme 16).[94] Some Fe/Rh clusters including $[FeRh_4(CO)_{15}]^{2-}$ have been studied by NMR from a structural and dynamic point of view. This anion undergoes a rapid oscillatory process specifically exchanging some of the bridging CO ligands (marked X) with some of the terminal

Scheme 16

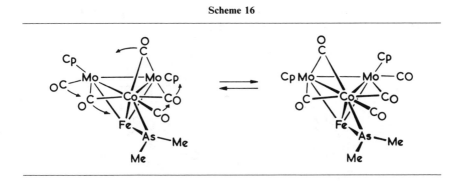

ones and a slower process leading to further exchange. These processes have been indicated in (**39**) and (**40**).[95]

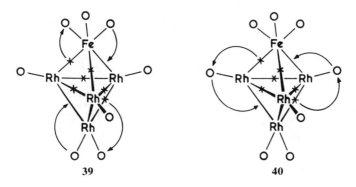

39   40

13.3.4. Hydride Exchange

A detailed examination of the exchange of terminal and bridging hydrides in $[Os_3H_2(CO)_{10}L]$ (L = CO, tertiary phosphine, or RNC), extending earlier work, has been made.[96] ^{13}C NMR spectra show that hydride exchange occurs as shown in Scheme 17. When the ligand L is equatorial (L = CO or PR_3) only one isomer is possible, but when L = RNC the ligand is axial and two isomers are present. These then interconvert as in Scheme 18. Very specifically the terminal hydride of one isomer exchanges with the bridging hydride of the other and vice versa. The values of $\Delta G^{\ddagger}$ are 13.2 and 12.8 kcal mol^{-1} for the major and minor isomers, respectively, which is consistent with $\Delta G° = 0.4$ kcal mol^{-1} determined from the position of equilibrium relating the isomers. Unlike in the equatorial forms of $[Os_3H_2(CO)_{10}L]$, exchange in the axial form does not lead to the exchange of the axial CO ligands of the $Os(CO)_4$ unit and hence a plane of symmetry is not generated through the metal triangle.[96]

Scheme 17

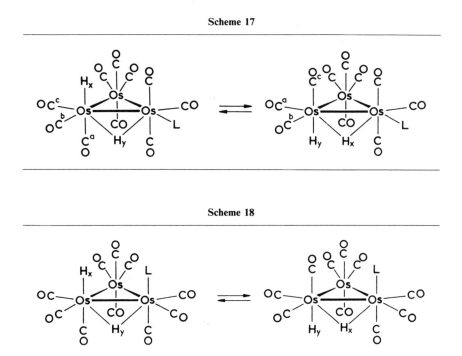

Scheme 18

Hydride exchange also occurs for $[Os_3H_3(CO)_9(SiPh_3)]$. Two hydride ligands bridge the same Os–Os pair and these exchange more rapidly with each other than they do with the other hydride which bridges another Os–Os pair (magnetization transfer experiments).[97] The cluster $[Ru_3(\mu-H)_3(CO)_3(C_5H_5)_3]$ does not have trigonal symmetry since two Cp ligands are above and the third below the Ru_3 plane. Fluxionality leading to C_5H_5 exchange is at the same rate as hydride exchange and this results from the C_5H_5 ligands moving between the two sides of the Ru_3 ring.[98]

13.3.5. Organic Ligand Mobility

There are now many examples of dynamic motion of μ^3-alkynes with respect to the metal triangle for "parallel" alkynes (**41**), but there are few reports of dynamic behavior for the "perpendicular" orientation (**42**) because these are much rarer systems. $[Fe_3(CO)_9(PhC_2Ph)]$ is of type (42) and has a rigid Fe_3C_2 framework,[8] whereas $[FeW_2(RC_2R)(CO)_6(C_5H_5)]$

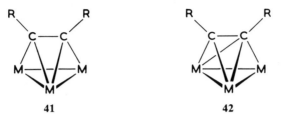

41   42

(R = *p*-tolyl) shows the rapid oscillation shown in Scheme 19[99] so it is not clear why the Fe_3 compound is not dynamic. The process in Scheme 19 equilibrates the W atoms but not the different ends of the alkyne ligand. μ^3-Alkyne compounds of type (41) are commonly fluxional.[100,101] Rotation of the alkyne with respect to the M_3 triangle leads to a plane of symmetry through the R–C–C–R set of atoms; the ligand must flip as it rotates.[101]

Scheme 19

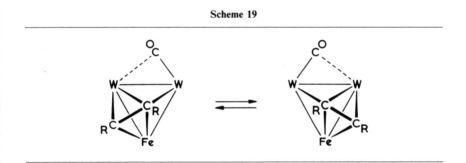

Chapter 14

Homogeneous Catalysis of Organic Reactions by Complexes of Metal Ions

14.1. Introduction

This chapter concentrates on reports of kinetic studies on organic reactions which are homogeneously catalyzed by transition–metal complexes. References to papers which provide an insight into the mechanism of a particular reaction or which report significant advances in the area under discussion are also included.

14.1.1. General Reviews and Elementary Steps in Homogeneous Catalysis

Applications of transient rather than steady-state kinetic methods have been reported; such methods allow a quantitative comparison between heterogeneous and homogeneous catalytic pathways.[1] Methods of distinguishing between homogeneous and heterogeneous catalysis are invaluable and one such test involves the use of dibenzo[a,e]cyclooctatetraene, a potent selective poison for homogeneous catalysts.[2]

A review on insertion reactions involved in catalytic processes will be welcomed.[3] Carbonyl "insertion" into a metal hydride bond and the possible involvement of formyl complexes in carbonylation reactions are topics of particular interest. It should be noted, therefore, that a thermochemical study of various manganese carbonyl species suggested that CO insertion into a metal hydride bond is unlikely to be significant unless the metal center is both coordinatively unsaturated and has a good affinity for oxygen so that the formyl ligand can be stabilized in the η^2-(C, O) form.[4] However, it has been demonstrated that deuteride transfer from $[RuD(CO)_4]^-$ to an electrophilic carbonyl ligand in $[Re(C_5H_5)(CO)_2NO]^+$ occurs to give $[Re(C_5H_5)(CO)(NO)CDO]$.[5] Also, it has been shown for the first time that carbonylation of a metal hydride may lead to a formaldehyde complex: carbonylation of $[Zr(C_5H_5)_2H_2]_x$ yields $[Zr(C_5H_5)_2(\eta^2-CH_2O)]_3$.[6] It is also pertinent to note that formaldehyde "inserts" into mesitylcopper(I) in the presence of triphenylphosphine to give mesityl alcohol $(C_6H_2Me_3)CH_2OH$.[7]

Reviews on transition-metal–alkyl bond dissociation energies[8a] and the formation of C–H bonds by reductive elimination[8b] are also relevant to the mechanisms of catalytic processes. It has been proposed that $[Ti(Me_2PCH_2CH_2PMe_2)(Et)Cl_3]$ contains a direct-bonding interaction between the titanium atom and the β-C–H system[9a]; the evidence for such agostic C–H–M bonds in other systems has been reviewed.[9b] Obviously, such bonds have important implications in alkane activation and it is particularly exciting that kinetic studies of the exchange reaction shown in equation (1) indicate that a bimolecular pathway predominates.[10]

$$[Lu(C_5Me_5)_2CH_3] + {}^{13}CH_4 \rightarrow [Lu(C_5Me_5){}^{13}CH_3] + CH_4 \qquad (1)$$

Related studies have shown that $[Ir(C_5Me_5)(CO)_2]$ undergoes oxidative addition with methane[11a] and 2,2-dimethylpropane.[11b] In recent years α-H elimination has been increasingly postulated in catalytic pathways and therefore it is interesting to note that $[Cr(C_5H_5)(CO)_2CH_3]$ in a gas matrix at 12 K undergoes reversible α-H elimination which may be observed directly using IR spectroscopy.[12] Reviews on catalysis by molecular metal clusters,[13] catalytic asymmetric synthesis,[14a-14c] applications of transition-metal nitrosyl complexes (particularly in the control of pollution by oxides of nitrogen),[15] and the role of palladium(I) complexes in catalysis[16] also include discussions on mechanistic aspects.

14.2. Reactions Involving Carbon Monoxide

14.2.1. Hydroformylation and Hydrocarboxylation of Olefins

Recent developments in hydroformylation have been reviewed.[17] Hydroformylation of $CD_3CH_2CH_2CH{=}CH_2$ using $[H_4Ru_4(CO)_8L_2]$ (L = diop = (**1a**)) yields methyl hexanoate in which all the deuterium is retained but is only found in the α,β- and ω-positions relative to the aldehyde group. Evidently, migration of deuterium within the substrate occurs without exchange with gaseous hydrogen but the selective redistribution of deuterium is curious.[18] Despite previous claims, complexes of the type $[Co_3(CO)_9CR]$ have been shown to break down under hydroformylation conditions to give mono- and dicobalt compounds which are the active catalytic species.[19] Insight into the mechanism of the hydrogen activation step in hydroformylation comes from the reaction of ethylene with approximately equimolar H_2/D_2 mixtures and $[Co_2(CO)_8]$ or $[Rh_4(CO)_{12}]$. Very little H_2/D_2 scrambling was observed and $\sim 50\%$ propionaldehyde d_1 was formed with a propionaldehyde d_0/d_2 ratio of ~ 3 and ~ 2.6 for Rh and Co, respectively. This would seem to exclude hydrogen activation through M$(H)_2$ or $M(H)_2$(olefin) complexes and indicate that a monohydride is formed, for example, equation (2)[20]:

$$[Rh(CO)_3(COR)] + H_2 \rightarrow [RhH(CO)_3] + RCHO \qquad (2)$$

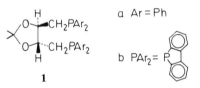

The hydroformylation activity of rhodium(I) complexes bound to silica via silylated phosphines decreases as the sites become isolated and this suggests that a dinuclear reductive-elimination step occurs as in equation (3)[21]:

$$[M]-H + [M]-C(O)R \rightarrow [M]-[M] + RCHO \qquad (3)$$

Further support for this proposal comes from the observation that $[Fe(Et)-(CO)_4]$ (**2**), which is assumed to be in equilibrium with $[Fe\{C(O)Et\}(CO)_3]$, reacts with $[FeH(CO)_4]^-$ (**3**) to give propionaldehyde; **2** and **3** are isolable intermediates in Reppe-modified hydroformylation of ethylene using $CO + H_2O$ in the presence of $Fe(CO)_5$.[22] A comparison of the $[Co_2(CO)_8]$- and $[RhCl(CO)_2]_2$-catalyzed hydroformylation of $Ph_2C{=}CH_2$ has shown that the cobalt reaction yields $\leq 5\%$ aldehyde and that the major product (95%) is 1,1-diphenylethane formed via a free-radical reaction; the rhodium reaction yields 85% aldehyde and 11% hydrogenation product via conventional olefin "insertion" into a rhodium hydride bond.[23] Further details have been reported of the photochemical hydroformylation using phosphine-modified cobalt compounds in methanol. The rate is nearly independent of the nature of the cobalt compound but increases with increasing synthesis gas pressure; an activation energy of 5–6 kcal mol^{-1}

was found.[24a] Similar UV irradiation of RhCl$_3$ and norbornadiene in methanol under hydroformylation conditions generates [Rh(nbd)$_2$H] which catalyzes the photochemical hydroformylation of 1-octene at room temperature.[24b] NMR spectroscopy has been used to measure rate constants and equilibrium constants of intermediates involved in the hydroformylation of styrene starting with [RhH(CO)(PPh$_3$)$_3$]. Significantly, it has been shown that although trans-[RhH(CO)(PPh$_3$)$_2$] has in the past been proposed as the key intermediate which reacts with the olefin, it is actually intercepted by PPh$_3$ and CO too efficiently for the olefin to compete.[25a] In contrast, the cis-[RhH(CO)$_2$PPh$_3$] present reacts rapidly with styrene to give [Rh(CO)$_2$(PPh$_3$)$_2${C(O)R}], where R = CH(Me)Ph and R = CH$_2$CH$_2$Ph in the ratio 91:9; this is similar to the ratio of 2-phenylpropanal and 3-phenylpropanal finally formed and it is therefore suggested that the normal:branched isomer ratio in hydroformylation is controlled by the kinetic liability of the two acyl isomers.[25b]

Sterically hindered olefins, for example, BuC(Me)=CH$_2$, may be hydroformylated under mild conditions (90°C, 10 bar) with [Rh(COD)OAc]$_2$ in the presence of excess bulky electron-withdrawing ligands, for example P{OCH(CF$_3$)$_2$}$_3$. The high rates have been attributed to the fact that the number of coordinated bulky phosphites is severely restricted and therefore the resulting unsaturated rhodium complexes are very reactive towards olefins.[26] Similarly, the addition of an equimolar amount of Ph$_2$P(CH$_2$)$_n$PPh (n = 2–4) to [RhH(CO)(PPh$_3$)$_3$] + 25PPh$_3$ activates this system to hydroformylate-substituted terminal olefins CH$_2$=CH(CH$_2$)$_m$OR (R = H or Ac, m = 0–2). The explanation for this intriguing observation is not really known although it is suggested that the chelating phosphine suppresses the formation of a meta-stable acyl complex in which a six-membered ring is formed by coordination to rhodium of the oxygen atom in a carbonyl or hydroxyl group of the substrate.[27a] By blocking coordination sites in this manner the α,ω-bis(diphenylphosphino)alkane also suppresses competing olefin isomerization reactions.[27b] 1-Hexene has been shown to react slowly with **4** to give **5** and this is believed to be the slow step in the hydroformylation of olefins using the precursor [Rh(acac){P(OPh)$_3$}$_2$]; subsequent CO insertion into **5** and reaction of the acyl with hydrogen gives the aldehyde and regenerates **4**.[28]

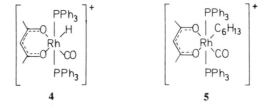

4   5

A detailed investigation of the *cis*-[PtCl$_2$(CO)PR$_3$] + SnCl$_2$ system has failed to identify the active hydroformylation catalyst.[29] It is interesting that hydroformylation of α-olefins in ethanol with [PtCl$_2$(PPh$_3$)$_2$] + SnCl$_2$ occurs without any significant hydrocarboalkoxylation.[30] An unusual "hydrocarboalkoxylation" reaction has been reported in which HCO$_2$Me replaces MeOH and CO; thus [RuCl$_2$(PPh$_3$)$_3$] catalyzes the formation of methyl propionate from ethylene and methyl formate. The mechanism is believed to proceed via initial fragmentation of the formate to give a formyl species Ru–C(O)H followed by ethylene insertion and methoxide attack.[31] Carbomethoxylation of butadiene in the presence of [Co$_2$(CO)$_8$] + pyridine is proposed to proceed as shown in Scheme 1; steps 2 and 3 have been demonstrated.[32]

Scheme 1

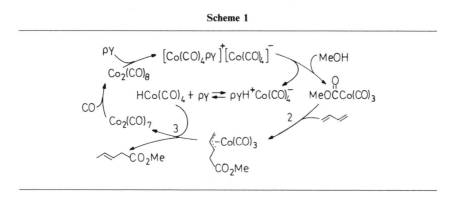

Asymmetric hydroformylation and hydrocarboxylation reactions have been reviewed.[33] From a comparison of the X-ray structures, it has been pointed out that the chiral coordination structure of diop (**1a**) is very different from that of diphol (**1b**) and the results obtained in rhodium-catalyzed hydroformylations with these ligands have been rationalized.[34]

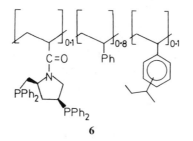

6

Previous estimated optical yields ($\leq 95\%$) in the asymmetric hydroformylation of styrene using $PtCl_2 + SnCl_2 +$ Diphol (**1b**) have now been corrected;
although considerably lower ($\leq 73\%$) they are still impressive.[35] Comparable optical yields have also been obtained from styrene using $SnCl_2$ together
with $PtCl_2$ supported on chiral polymers such as **6**.[36]

14.2.2. Carbonylation and Homologation of Alcohols, Ethers, Alkyl Halides, and Carboxylic Esters

The homologation of methanol [equation (4)] is an area of increasing
interest. A range of metal carbonyls (e.g., Rh, Fe, Mn, or Ru) will, in the
presence of amine, for example, Me_3N, catalyze this reaction; the amine B
promotes the formation of carbonyl anions and activates the methanol
[equations (5) and (6)]. The catalytic pathway is shown below with methyl
transfer [equation (9)] being the slow step. It is suggested that ethanol does
not undergo further homologation because attack of Me_3N upon ethyl
formate is sterically less favorable than attack upon methyl formate
[equation (7)].[37]

$$CH_3OH + 2CO + H_2 \rightarrow CH_3CH_2OH + CO_2 \qquad (4)$$

$$[Fe(CO)_5] + H_2 + B \rightleftharpoons [FeH(CO)_4]^- + CO + BH^+ \qquad (5)$$

$$MeOH + CO \overset{B}{\rightleftharpoons} HCO_2Me \qquad (6)$$

$$HCO_2Me + B \rightleftharpoons MeB^+ + HCO_2^- \qquad (7)$$

$$BH^+ + HCO_2^- \rightleftharpoons H_2 + CO_2 + B \qquad (8)$$

$$MeB^+ + [FeH(CO)_4]^- \rightarrow [FeH(Me)(CO)_4] + B \qquad (9)$$

$$[FeH(Me)(CO)_4] + CO \rightarrow [MeC(O)FeH(CO)_4] \qquad (10)$$

$$[MeC(O)FeH(CO)_4] + CO \rightarrow MeC(O)H + [Fe(CO)_5] \qquad (11)$$

$$MeC(O)H + H_2 \xrightarrow{[FeH(CO)_4]^-} EtOH \qquad (12)$$

Stoichiometric hydrogenation of $[Co(COMe)(CO)_3(PMePh_2)]$ in
methanol yields a product distribution similar to that obtained from
$[Co(CO)_3(PMePh_2)]_2$-catalyzed homologation of methanol; hence, it is
argued that the acyl species is an intermediate in the catalytic reaction.[38]
The rate of methanol to ethanol homologation with $[Co_2(CO)_8] + PBu_3 + I_2$
is first order with respect to methanol and cobalt concentrations and CO
partial pressure.[39] It has been demonstrated using such a catalyst that
homologation of CD_3OD with $CO + H_2$ gives CD_3CHO, CD_3CH_2OH, and
$CD_3CH(OCD_3)_2$; this confirms that methyl rather than methylene intermediates are involved.[40] If a rhodium iodide catalyst is used, then CD_3OD

yields CD_3H in addition to acetic acid; the proposed mechanism involves a reaction between CD_3I and $[RhH(CO)_2I_3]^-$.[41] A kinetic study of the $Bu_4N[CoCO)_4]$-catalyzed carbonylation of $PhCH_2Br$ to $PhCH_2CO_2H$ under phase-transfer conditions has shown that the rate-limiting step is cleavage of the acyl cobalt bond in $[Co(COCH_2Ph)(CO)_4]$ by Bu_4NOH at the liquid–liquid interface.[42] High pressures (140 MPa) and low temperatures (160–170°C] are reported to favor the formation of acetaldehyde from the reaction of methanol with $CO + H_2$ mixtures in the presence of $Co(OAc)_2 + I_2$.[43] A synergistic effect has been observed using MeI and an ionic iodide $A^+I^-(A^+ = Li^+, K^+,$ or $MePBu_3^+)$ in this reaction. This, it is suggested, is because MeI is required to act with the catalyst [equation (14)], whereas A^+I^- aids catalyst formation and regeneration [equations (13) and (18)].[44]

$$2[Co_2(CO)_8] + 4A^+I^- + 2H_2 \rightleftharpoons 4A^+[Co(CO)_4]^- + 4HI \qquad (13)$$

$$MeI + [Co(CO)_4]^- \rightleftharpoons [CoMe(CO)_4] + I^- \qquad (14)$$

$$[CoMe(CO)_4] + CO \rightleftharpoons [Co(COMe)(CO)_4] \qquad (15)$$

$$[Co(COMe)(CO)_4] + H_2 \rightleftharpoons [Co(COMe)H_2(CO)_3] \qquad (16)$$

$$[Co(COMe)H_2(CO)_3] + CO \rightarrow MeCHO + [CoH(CO)_4] \qquad (17)$$

$$[CoH(CO)_4] + I^- \rightarrow HI + [Co(CO)_4]^- \qquad (18)$$

Transition-metal halides $MCl_3 \cdot 3H_2O$ (M = Rh, Ir, or Ru) also promote the $(Co(OAc)_2 \cdot 4H_2O + LiI)$-catalyzed hydrocarbonylation of methanol to acetaldehyde. However, this is not due to metal cooperativity but arises from the ability of the promoting metal hydrate to generate HI [equation (19)]:

$$RhCl_3 + H_2O + CO + 2LiI \rightarrow Li[Rh(CO)_2Cl_2] + CO_2 + HI + HCl \qquad (19)$$

The HI reacts with methanol and leads to rapid formation of MeI with a concomitant increase in rate.[45] In contrast, it has been suggested that mixed-metal clusters may be involved in methanol homologation with $M[RuCo_3(CO)_{12}] + I^-$ (M = H, Na, Et_4N, or Ph_4N) since this catalyst system gives much higher yields of ethanol than $[Co_2(CO)_8] + I^-$ under comparable conditions.[46] In the presence of an iodide promoter and a proton donor, ruthenium complexes, for example $[Ru(acac)_3]$, react with $CO + H_2$ to give $[RuHI_y(CO)_x]$ which catalyzes the reductive carbonylation of dimethyl ether (to MeOH and EtOH) amd methyl acetate (to ethyl acetate) by a common mechanism (Scheme 2)[47a]; addition of cobalt[47b] or an alkali-metal iodide[47c] enhances the formation of ethyl acetate. These investigations have been extended to similar homologation of carboxylic methyl esters to the corresponding ethyl esters.[47d]

Scheme 2

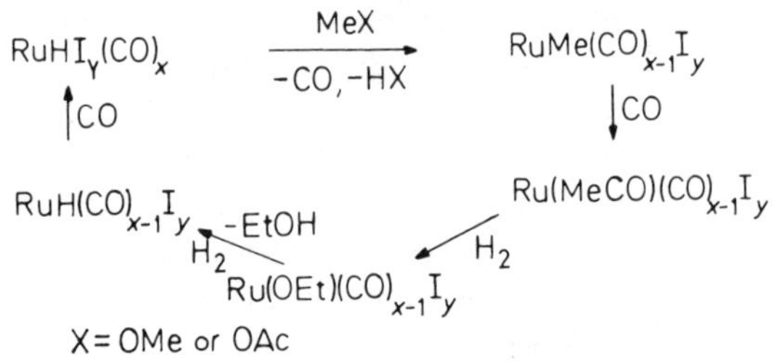

$$RuHI_y(CO)_x \xrightarrow[-CO,-HX]{MeX} RuMe(CO)_{x-1}I_y$$

with vertical arrows labeled $\uparrow CO$ (left) and $\downarrow CO$ (right), leading to:

$$RuH(CO)_{x-1}I_y \qquad Ru(MeCO)(CO)_{x-1}I_y$$

$$\xrightarrow{H_2} \xleftarrow{-EtOH} Ru(OEt)(CO)_{x-1}I_y \xleftarrow{H_2}$$

X = OMe or OAc

The mechanisms of the cobalt- rhodium- and iridium-catalyzed carbonylation of methanol in the presence of iodide have been reviewed. A new mechanism for the cobalt-catalyzed reaction is proposed which does not involve addition of MeI [equations (20)–(23)]. Thus it is proposed that the role of the iodide is to attack the intermediate acyl cobalt species to give acetyl iodide which is rapidly solvolyzed.[48]

$$[HCo(CO)_4] + MeOH \rightarrow [MeCo(CO)_4] + H_2O \qquad (20)$$

$$[MeCo(CO)_4] + CO \rightarrow [MeC(O)Co(CO)_4] \qquad (21)$$

$$I^- + [MeC(O)Co(CO)_4] \rightarrow [Co(CO)_4]^- + MeCOI \qquad (22)$$

$$H^+ + [Co(CO)_4]^- \rightarrow [HCo(CO)_4] \qquad (23)$$

Kinetic studies have allowed the optimum conditions to be defined for the synthesis of acetic anhydride by the carbonylation of methyl acetate using a variety of Group VIII metal catalysts.[49a] Such studies, complemented by IR and UV spectroscopic studies, have helped to elucidate the main catalytic pathways for the rhodium- and iridium-catalyzed reactions in the presence of iodide. Although complex, both mechanisms essentially involve oxidative addition of MeI to $[M(CO)_2I_2]^-$ (M = Rh or Ir) followed by CO insertion into the metal–methyl bond and subsequent reductive elimination of MeCOI; the latter reacts with acetate ion to give acetic anhydride and regenerate iodide.[49b]

14.2.3. Fischer–Tropsch Reactions

The proceedings of a symposium on Catalytic Reactions of One Carbon Molecules[50] and a review which includes homogeneous catalysis with syn gas (i.e., $H_2 + CO$) and metal clusters summarize recent developments in this area.[51]

An increasingly favored mechanism for the Fischer–Tropsch reaction is the so-called carbide mechanism which involves initial cleavage of CO and subsequent hydrogenation of the carbide to give carbenes which polymerize by inserting into an alkyl chain. It has, however, been argued that such a mechanism does not explain all the experimental observations and that a mechanism involving CO insertion into the growing chain with subsequent reduction of the acyl group is more satisfactory.[52a] Such arguments have been refuted.[52b] Circumstantial evidence for a carbide mechanism comes from the observations that CO cleavage may occur under mild conditions [equations (24)[53] and (25)[54]] and the fact that olefins may be formed by insertion of a methylene group into a metal–alkyl bond followed by β-hydrogen elimination {equation (26), $[Ru] = Ru(\eta\text{-}C_6Me_6)\text{-}(PPh_3)$}.[55]

$$[Fe_3(^{13}CO)_{10}(\mu\text{-}CCH_3)]^{2-} + MeOH \rightarrow [Fe_3(^{13}CO)_9(\mu\text{-}^{13}C\equiv CCH_3)]^- \tag{24}$$

$$3[((Me_3Si)_2N)_2ZrMe_2] + 2CO \rightarrow [\{((Me_3Si)_2N)_2ZrMe\}_2O]$$
$$+ [((Me_3Si)_2N)_2Zr(OC(Me)=CMe_2)Me] \tag{25}$$

$$[Ru](CH_3)_2 + [CPh_3]PF_6 \rightarrow \text{``}[Ru](CH_2)CH_3\text{''} \rightarrow [Ru]H(C_2H_4)^+ \tag{26}$$

Alternatively, a mechanism involving cyclopropylidene intermediates has been proposed,[56] whereas kinetic and thermochemical data have been interpreted as evidence that metal hydroxycarbyne species, $M\equiv C-OH$, are important intermediates in such CO reductions.[57] Ruthenium complexes with iodide promoters effect the hydrogenation of CO to give mainly MeOH, EtOH, and $HOCH_2CH_2OH$; spectroscopic and reaction studies have identified $[HRu_3(CO)_{11}]^-$ and $[Ru(CO)_3I_3]^-$ as being essential for optimum activity and it has been suggested that they may interact to generate $[Ru(CO)_4]^-$, a possible active intermediate.[58] Mixed ruthenium–rhodium catalyst systems dispersed in a low-melting quaternary phosphonium salt catalyze conversion of $CO + H_2$ mixtures to mainly MeOH and EtOH together with ethylene glycol. It is proposed that the active catalyst is a mixed rhodium–ruthenium species since maximum glycol production was achieved at a Ru : Rh ratio of 1 : 1; also under such conditions $[RhRu_2(CO)_{12}]$ was identified as the major species present.[59a] In contrast, no evidence for mixed ruthenium–rhodium species was obtained when this reaction was carried out in acetic acid with nitrogen-containing bases or alkali-metal cations. A synergistic effect was, however, noted with an optimum Ru : Rh ratio of 10 : 1; it appears as though the minor component of the catalyst(Rh) determined the selectivity towards C_2-oxygenated products, whereas the major component (Ru) enhanced the activity.[59b] ^{14}C-tracer studies have demonstrated that methanol and ethylene glycol are primary products of the homogeneous [$Rh_4(CO)_{12}$]-catalyzed conversion of $CO + H_2$ mixtures and are not derived from the solvent mixture N-methylpyrrolidin-2-one and tetraethyleneglycol dimethyl ether (1:4). Further, paraformaldehyde is readily converted under the reaction conditions into the typical mixture of reaction products lending support to the proposal that formaldehyde may be a transient intermediate.[60] Since formaldehyde is readily available, it has been suggested that if it is indeed an intermediate in such reactions, using it as a starting material may avoid the severe conditions of syn-gas chemistry and lead to more selective reactions. Therefore the hydroformylation of formaldehyde to glycolaldehyde has been investigated using rhodium complexes such as [$RhCl(CO)(PPh_3)_2$] combined with triphenylphosphine and triethylamine (1:30:3-4). The proposed role of the base is to activate the catalyst precursor via deprotonation; the ammonium cation thus formed may also serve as a proton source necessary for the formation of (7) [Scheme 3]. A problem which becomes severe under more basic conditions is the condensation of the glycolaldehyde.[61]

Scheme 3

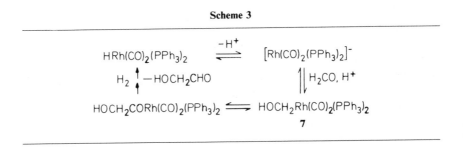

7

Kinetic studies have shown that the previously reported "homogeneous Fischer–Tropsch catalysis" by [$Ir_4(CO)_{12}$] in molten $AlCl_3$–NaCl(2:1) involves the homogeneous reduction of CO to MeCl followed by homologation and/or hydrogenation reactions leading to CH_4, C_2H_6, and other hydrocarbon products.[62]

14.2.4. *The Homogeneous Water-Gas Shift Reaction (WGSR)*

Although it is agreed that the ([$M(CO)_6$] + KOH)-catalyzed (M = Cr, Mo, W) WGSR proceeds via [$M(CO)_5H$]$^-$, the process by which this is formed is controversial. The observation that the activity is inhibited by carbon monoxide has been interpreted as evidence for the formation of [M(CO)$_5$].[63a] In contrast it has been argued that nucleophilic attack of hydroxide upon [M(CO)$_6$] occurs [step a in Scheme 4]; the decarboxylation process (step b) has been shown to proceed via CO dissociation which could account for the observed CO inhibition.[63b]

Scheme 4

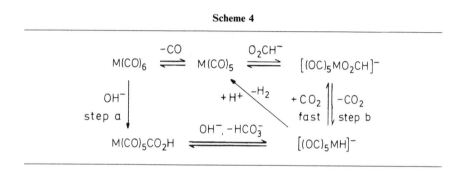

Nucleophilic attack upon a carbonyl ligand has been proposed in other WGSR mechanisms. For example, [Ru$_3$(CO)$_{12}$] in aqueous acidic diglyme effects the WGSR at $P_{CO} < 1$ atm: Based on kinetic and *in situ* spectroscopic

Scheme 5

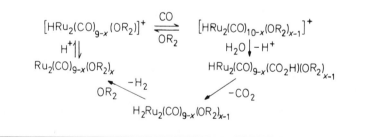

studies the above mechanism in Scheme 5 has been proposed [OR$_2$ = diglyme or H$_2$O].[64] Similarly, under white light irradiation [RuCl(CO)L$_2$]Cl (L = 2,2'-bipyridyl or 1,10-phenanthroline) is believed to catalyze the WGSR by the mechanism outlined in Scheme 6.[65]

Scheme 6

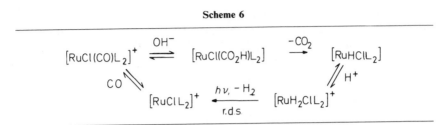

The reactions shown in equations (27)–(29) have been demonstrated; combined together they constitute a catalytic cycle for the WGSR and although there is no direct evidence that it operates in this manner it is significant that $[Rh_2(\mu\text{-}H)(\mu\text{-}CO)(\mu\text{-}dpm)_2(CO)_2]^+$ (dpm $=$ $Ph_2PCH_2PPh_2$) in 1-propanol is a WGSR catalyst.[66]

$[Rh_2(\mu\text{-}H)(\mu\text{-}CO)(\mu\text{-}dpm)_2(CO)_2]^+ + H^+$

$$\rightarrow H_2 + [Rh_2(\mu\text{-}dpm)_2(CO)_2(solv)_2]^{2+} \quad (27)$$

$[Rh_2(\mu\text{-}dpm)_2(CO)_2(solv)_2]^{2+} + HCO_2^-(CO + OH^-)$

$$\rightarrow [Rh_2(\mu\text{-}dpm)_2(\mu\text{-}O_2CH)(CO)_2]^+ \quad (28)$$

$[Rh_2(\mu\text{-}dpm)_2(\mu\text{-}O_2CH)(CO)_2]^+ + CO$

$$\rightarrow [Rh_2(\mu\text{-}dpm)_2(\mu\text{-}H)(\mu\text{-}CO)(CO)_2]^+ + CO_2 \quad (29)$$

The WGSR may be used to effect selective reductions. For example, $[Mn_2(CO)_8(PBu_3)_2]$ catalyzes the reduction of anthracene by $CO + D_2O$ to give 9,10-dideuterioanthracene[67]; similarly with $CO + H_2O$, $[Fe(CO)_5] +$ Et_3N catalyzes reduction of $PhCH{=}NPh$ to $PhCH_2NHPh$[68] and $[Rh_6(CO)_{16}] + Me_2NCH_2CH_2NMe_2$ catalyzes the reduction of unsaturated aldehydes to unsaturated alcohols.[69]

14.3. Oxidation

Two reviews—one covering the mechanisms of catalytic oxygen transfer processes[70] and the other comparing the mechanisms of oxygen transfer from inorganic and organic peroxides to organic substrates[71]—are particularly recommended to readers who wish to follow developments in this area. Other useful reviews cover the mechanisms of metal–porphine– and metal–azaporphine-catalyzed oxidations by molecular oxygen,[72a] oxidations using heteropolyacids,[72b] the mechanisms of palladium(II)-catalyzed oxidative coupling of unsaturated hydrocarbons,[72c] and the mechanism of the titanium-tartrate-catalyzed asymmetric epoxidations by *tert*-butyl hydroperoxide.[72d]

It has been shown that the course of Mo(VI)-catalyzed epoxidation of olefins by hydroperoxides depends upon the oxygen source. With H_2O_2 or Ph_3COOH, stable reactive peroxo complexes (**8**) have been isolated; these react as shown in Scheme 7 (path a, [Mo] = $MoL_2Cl(^{18}O)$, where L = DMF or HMPT). In the case of other alkyl peroxides ^{18}O-labeling experiments rule out such intermediates and an alkyl-peroxides species (**9**) is believed to be formed (Scheme 7, path b).[73]

Scheme 7

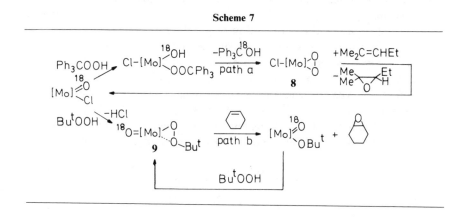

Kinetic studies of the epoxidation of olefins using $[MoO_2(acac)_2]$ + H_2O_2 or $[MoO_5(HMPT)]$ (HMPT = hexamethylphosphorotriamide) are also consistent with a mechanism involving Mo(VI) peroxo complexes which act as electrophilic oxidants towards the uncomplexed olefin.[74] It is also noteworthy that added acid significantly enhances the rate of the $[W(CO)_6]$- or $[WO_5(HMPT)H_2O]$-catalyzed epoxidation of olefins by H_2O_2 since protonation of such a W(VI) peroxo intermediate would certainly enhance electrophilic attack upon an olefin.[75] Chemical and kinetic studies of the epoxidation of cyclohexene by cumene hydroperoxide using molybdenum(V) bis(diphenylphosphino)ethane complexes have been reported; the active catalyst isolated from such systems is $[MoO_2(C_6H_{10}O_2)$-$(dpeO_2)]$.[76] The rate of the $[Na_3PMo_{12}O_{40}]$-catalyzed epoxidation of hex-1-ene by isopropylbenzene hydroperoxide shows a dependence upon $[Na_3PMo_{12}O_{40}]^{0.5-0.9}$ and a mechanism involving partial decomposition of the heteropoly compound and formation of a glycol complex has been proposed.[77] In a kinetic study of the $[C_{16}H_{33}NMe]^+[Mo(O)(O_2)_2$-$(NC_5H_4COO)]^-$-catalyzed H_2O_2 oxidation of alcohols to ketones, it was noted that added $[C_{16}H_{33}NMe_3]_2SO_4$ significantly enhanced the rate, indicating that substrate micellization is an important factor in determining the oxidizing ability of the system.[78]

The proposed mechanism of the [FeCl(TTP)] (TPP = 5,10,15,20-tetraphenyl porphyrinato)-catalyzed epoxidation of olefins by iodosylbenzene involves the olefin approaching the oxygen in the intermediate Fe(=O)(TPP)Cl side-on parallel to the porphyrin plane, due to a stereoelectronic effect involving partially filled oxygen–iron p_π–d_π antibonding orbitals; such a mechanism would account for the observation that *cis* olefins are more reactive than *trans* olefins.[79a] If the porphyrin system is modified with optically active functionalities, then this system may be used for asymmetric epoxidations, for example, [Fe(T($\alpha,\beta,\alpha,\beta$-binaphthyl)PP)]Cl catalyzes the epoxidation of *p*-chlorostyrene with 51% e.e.[79b] Oxidation of sulfides to sulfoxides using Bu^{t}OOH or H_2O_2 and a [Ti(O)(acac)$_2$] catalyst is reported to proceed via coordination of the sulfide to the titanium,[80a] whereas with [V(O)(acac)$_2$] or [MoO$_2$(acac)$_2$] catalysts the reaction is believed to proceed via attack of a metal peroxo species upon the uncoordinated sulfide.[80b] The rate of the [RuCl$_2$(Me$_2$SO)$_4$]-catalyzed selective oxidation of alkyl sulfides in alcohols depends upon the nature of the sulfide but is zero order in sulfide. The proposed mechanism is shown in equations (30)–(32)[81]:

$$[\text{Ru(II)}] + O_2 \rightleftharpoons [\text{Ru(IV)}] + O_2^{2-} \qquad (30)$$

$$SR_2 + H_2O_2 + ROH \rightleftharpoons R_2SO + H_2O + ROH \qquad (31)$$

$$[\text{Ru(IV)}] + R_2CHOH \rightleftharpoons [\text{Ru(II)}] + 2H^+ + R_2C{=}O \qquad (32)$$

Under aerobic conditions [RhCl(PPh$_3$)$_3$^{18}O$_2$] catalyzes the dioxygenation of cyclooctadiene to give cyclo-octane-1,4-dione, C$_8$H$_{12}$(^{18}O)$_2$. Kinetic studies are consistent with a pathway involving slow initial coordination of cyclooctadiene to [RhCl(PPh$_3$)$_2$^{18}O$_2$].[82] However, it has been proposed that the Rh(III)/Cu(II)-catalyzed oxidation by molecular oxygen of 1-alkenes to 2-ketones in alcohols actually involves oxidation by hydrogen peroxide generated *in situ* from the reduction of O$_2$ by the solvent. This suggestion is based upon the fact that such oxidations may be carried out with RhCl$_3$·3H$_2$O + H$_2$O$_2$ and the observation that dioxygen reacts with [Rh(CO)$_2$Cl]$_2$ in alcohols to give H$_2$[Rh(CO)Cl$_2$(OOH)].[83] Rhodium complexes such as [RhCl(PPh$_3$)$_3$] catalyze the unusual oxidation of Me$_2$C=CMe$_2$ by dioxygen to give acetone. The reaction is believed to involve the reaction of the alkene with a peroxo complex to give a 1,2-dioxa-3-rhodacyclopentane intermediate, [Rh]OOC(Me)$_2$C(Me)$_2$; elimination of a molecule of acetone generates [Rh]OCMe$_2$ which reacts with dioxygen to eliminate acetone and regenerate the initial peroxo complex.[84] In the presence of [Rh$_6$(CO)$_{16}$] in benzene, molecular oxygen will cooxidize CO and PPh$_3$ to CO$_2$ and OPPh$_3$, respectively. Addition of H$_2$^{18}O increases the yield of CO$_2$ which contains the bulk of the ^{18}O label; although the detailed mechanism is uncertain, a pathway involving nucleophilic attack of H_2O, or more probably OH^-, upon a carbonyl ligand would seem to be involved.[85] Further details of a kinetic study of the $[IrHCl_2(C_8H_{12})]_2$-catalyzed cooxidation of H_2 and cyclooctene have been reported providing evidence that this reaction proceeds via an iridium–hydroproxide intermediate.[86]

Controversy continues over the mechanism of the Wacker process. For example, it has been pointed out that allyl alcohol is oxidized by $PdCl_2$ in aqueous solution to β-hydroxypropanal $HOCH_2CH_2CHO$, a product which must be derived from a symmetrical hydroxypalladation adduct. Therefore, if hydroxypalladation is an equilibrium process [i.e., equation (33), $k_{-1} \gg k_2$], then $CH_2{=}CHCD_2OH$ will be isomerized into a 1:1 mixture of $CH_2{=}CHCD_2OH$ and $CD_2{=}CHCH_2OH$ before appreciable oxidation occurs. However, very little isomerization of $CH_2{=}CHCD_2OH$ occurs, supporting a mechanism having *cis* addition of coordinated hydroxyl to the olefin as the rate-limiting step.[87]

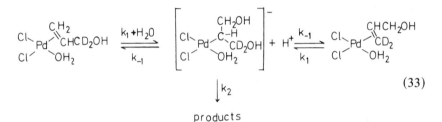

$$(33)$$

products

In contrast, oxidation of E- or Z-$Bu^tCH{=}CHD$ by $Pd(OAc)_2$ in chloride-free acetic acid produces the corresponding vinyl acetate via an acetoxypalladation β-H-elimination process, and analysis of the deuteriated products indicates that nucleophilic attack by acetate occurs with *trans* addition to the coordinated olefin. Hence it is argued that *cis* migration of an oxygen nucleophile from palladium to a coordinated olefin is an unfavorable process.[88] A kinetic study of the $(PdCl_2 + CuCl_2)$-catalyzed oxidation of 1-butene to 2-butanone has been reported[89] and the mechanisms and factors controlling the regioselectivity of palladium(II)-catalyzed oxidation of 1,3-dienes in acetic acid have been discussed.[90] Insight into the mechanism of formation of ethylene glycol monoacetate from ethylene in acetic acid solution containing $Pd(OAc)_2 + MNO_3$, comes from a kinetic study (M = H)[91a] and an ^{17}O NMR study which has demonstrated that oxygen atom transfer occurs from the oxidant (M = Li) to the carbonyl group of the product in equation (34).[91b]

$$3C_2H_4 + 2LiN^{17}O_3 + 5MeCO_2H$$

$$\rightarrow 3HOCH_2CH_2OC(^{17}O)Me + 2LiOAc + H_2^{17}O + 2N^{17}O \quad (34)$$

Oxygen transfer from a nitro group is proposed for the [Pd(NCMe)$_2$Cl(NO$_2$)]-catalyzed epoxidation of cyclic alkenes; with norbornene the intermediate (**10**) was isolated and shown to decompose to give **11** and [PdCl(NO)]$_n$ which reacts with molecular oxygen in acetonitrile to regenerate [Pd(NCMe)$_2$Cl(NO$_2$)]. Alkenes with β-H atoms preferentially undergo β-hydrogen elimination to give ketones rather than epoxides.[92]

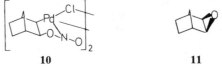

10   **11**

Kinetic studies of the oxidative coupling of 2,6-dialkylphenols using cobalt–dioxygen complexes[93] or CuCl$_2$–KOH–pyridine[94] catalysts have been reported.

14.4. Hydrogenation

14.4.1. Hydrogenation of Alkenes

A general test for the reversibility of the hydrogen-addition step has been applied to a number of rhodium-catalyzed hydrogenations and involves using para-enriched hydrogen. If addition is reversible then equilibration of o- and p-H occurs and this can be detected using Raman spectroscopy.[95] The results of a kinetic study of the [Cr(CO)$_4$(nbd)]-catalyzed photochemical hydrogenation of dienes have been rationalized by a mechanism in which 1,4-hydrogenation of conjugated dienes or norbornadiene proceeds via initial dissociation of a carbonyl ligand, whereas any 1,2-hydrogenation of norbornadiene to norbornene proceeds via [Cr(CO)$_4$(η^2-nbd)].[96] Kinetic studies of the thermal[97a] and photoinduced[97b] hydrogenation of ethylene by [H$_4$Ru$_4$(CO)$_{12}$] are consistent with the initial dissociation of CO followed by addition of ethylene, hydrogen transfer, and the rate-determining step—the reaction of hydrogen gas with [H$_3$Ru$_4$(CO)$_{11}$C$_2$H$_5$]. The [Ru(η^4-cod)(η^6-C$_8$H$_{10}$)]-catalyzed hydrogenation of cycloheptatriene or cyclooctadienes to monoenes has been shown to proceed via the corresponding conjugated diene.[98] The kinetics of the [RuCl$_2$(PPh$_3$)$_3$]-catalyzed hydrogenation of allyl bromide are consistent with the rate-limiting step being dissociation of triphenylphosphine from the intermediate [RuHCl(PPh$_3$)$_3$].[99] The recent claim that hydrogen addition to [Co(CN)$_5$]$^{3-}$ proceeds through heterolytic cleavage has been refuted.[100] Hydrogenation of cinnamate ion in aqueous NaCl solution using [Co(CN)$_5$H]$^{3-}$ has been followed by IR spectroscopy and an activation energy of 14.5 ± 0.5 kcal mol^{-1} determined.[101]

A review includes a discussion of the kinetics and mechanisms of rhodium(I)-catalyzed hydrogenations[102] and general electron and atom group transfer theory has been applied to such reactions to calculate activation energies and pre-exponential nuclear tunneling factors for the various steps including oxidative-addition and hydrogen transfer.[103] Kinetic studies suggest that the $[RhCl(PPh_3)_3]$-catalyzed hydrogenation of allyl alcohol proceeds via initial formation of $[RhCl(PPh_3)_2(CH_2=CHCH_2OH)]$ followed by the rate-determining step—hydrogen addition. A plot of the rate *vs.* allyl alcohol concentration shows a maximum; this substrate inhibition at high concentrations is ascribed to the formation of $[RhCl(PPh_3)_2(CH_2=CHCH_2OH)_2]$.[104] The effect of ring size on the $[Rh(Ph_2P(CH_2)_nPPh_2)(MeOH)_2]^+$-catalyzed hydrogenation of olefins has been investigated. For hex-1-ene the binding constant K_{eq} of the olefin to the catalyst is relatively independent of n, whereas the rate constant k_2 for the reaction of H_2 with $[Rh(Ph_2P(CH_2)_nPPh_3)(olefin)]^+$ increases by a factor of 75 as n increases from 2 to 4. For methyl-(Z)-α-acetamidocinnamate, as n increases from 2 to 4 the binding constant decreases but the rate constant k_2 becomes too fast to measure. It is suggested that the increasing overall catalytic rate with increasing ring size may reflect the ability of the larger more flexible chelates to adopt the favored transition-state geometries.[105] Kinetic and chemical studies of the $[(\mu\text{-}H)RhL_2]_2$-catalyzed [12; $L = P(OPr^i)_3$] reductions of alkynes to *trans*-alkenes are consistent with initial alkyne addition to give 13, conversion to 14 followed by the rate-determining step—hydrogen addition to give *trans*-alkene and 12.[106]

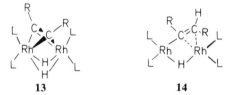

13   14

A stopped-flow kinetic study has revealed a marked *inverse* kinetic isotope effect for the reduction of norbornadiene to norbornene by $[Ir(PPh_3)_2(OCMe_2)_2R_2]^+$ (R = H or D). The faster rate of the deuteride complex has been attributed to the large isotope ratio ($k_{-1}^H/k_{-1}^D = 5.73 \pm 0.32$) in the equilibrium between 15 and 16 in equation (35)[107]:

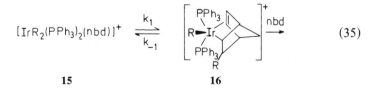

$$[IrR_2(PPh_3)_2(nbd)]^+ \underset{k_{-1}}{\overset{k_1}{\rightleftharpoons}} \qquad (35)$$

15   16

$[Ir(cod)\{P(C_6H_{11})_3\}py]PF_6$ catalyzes the stereoselective reduction of unsaturated alcohols with hydrogen addition taking place almost exclusively at the face of the substrate bearing the alcohol; this high stereoselectivity has been attributed to the substrates chelating to the catalyst *via* the olefin and the hydroxyl group.[108] In the $([MCl_2(PR_3)_2] + SnCl_2)$-catalyzed (R = *n*-alkyl or *p-n*-alkylphenyl) hydrogenation of methyl linoleate it is proposed that when M = Pd, metal hydride formation is the rate-limiting step, whereas when M = Pt, metal–alkene formation is rate limiting.[109] The rate-limiting step in the $[Pt(\eta^3\text{-}C_3H_5)HL]$-catalyzed ($[L = P(C_6H_{11})_3$ or $PBu_3^i]$ hydrogenation of conjugated dienes to olefins is believed to be addition of hydrogen to [Pt(diene)L] since there is a linear dependence of the rate upon the hydrogen pressure.[110]

14.4.2. Hydrogenation of Arenes and Ketones

Reaction of $[\overline{Ru(PPh_2C_6H_4)}H_2(PPh_3)_2]^-$ with hydrogen and anthracene yields (17) via *fac*-$[RuH_3(PPh_3)_3]^-$; further reaction of (17) with hydrogen gives (18) which reacts with anthracene to regenerate (17). These reactions constitute a catalytic cycle for the reduction of anthracene to 1,2,3,4-anthracene as in equation (36)[111]:

$$[RuH(\eta^2\text{-}C_{14}H_{10})(PPh_3)_2]^- \underset{2C_{14}H_{10}}{\overset{4H_2,-C_{14}H_{14}}{\rightleftharpoons}} [RuH_5(PPh_3)_2]^- \qquad (36)$$

$$\mathbf{17} \hspace{6cm} \mathbf{18}$$

When the above reduction is carried out using $[Rh(dpe)(anthracene)]^+$, the proposed rate-limiting step is the initial oxidative addition of hydrogen to give $[RhH_2(dpe)C_{14}H_{10}]^+$; the observation that the rate is given by the expression, rate = $k[Rh(dpe)(C_{14}H_{10})^+][H_2]$ is consistent with this proposal.[112] $[Rh(acac)\{P(OPh)_3\}_2]$ catalyzes the reduction of arenes to cyclohexanes.[113]

A kinetic study of the ($[Rh(nbd)Cl]_2 + PPh_3 + Et_3N$)-catalyzed reduction of acetophenone has been reported and the proposed rate-limiting step is the reaction between the ketone and the intermediate $[RhH(PPh_3)_3]$.[114a] Interestingly, this system hydrogenates 4-*t*-butylcyclohexanone to give predominantly the *cis*-alcohol, whereas if PBu_3 is used the *trans*-alcohol is obtained. It is suggested that the *cis*-alcohol results from coordination of the C=O group side-on to a soft Rh(I)–H species and the *trans*-alcohol results from coordination of the carbonyl oxygen atom to a hard Rh(III)–H_2 species.[114b] The proposed mechanism for the reduction of R^1R^2CO (R^1 = alkyl or aryl; R^2 = H, alkyl, or aryl) using $[Fe(CO)_5]$ in tertiary amine solvent is shown in Scheme 8.[115]

Scheme 8

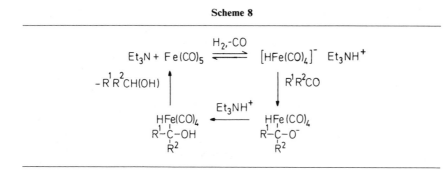

14.4.3. Asymmetric Hydrogenation

Several reviews on this topic have been published.[116a-116c] It has been demonstrated by ^{31}P and ^{13}C NMR spectroscopy using the DANTE spin-excitation transfer technique that the enamide complexes **19** and **20** interconvert via an intramolecular mechanism involving an olefin dissociated species. Such interconversion is essential since it is the minor isomer **19** which selectively reacts with hydrogen.[117] Reduction of CH_2=CHX (X = Ph, NHCOMe, or CO_2H) with either D_2 or HD has been investigated with a variety of rhodium(I)-chiral-bis(phosphine) complexes to gain an insight into the factors other than steric effects which influence the asymmetric hydrogenation of PhCH=C(NHCOMe)CO_2H.[118] It has been shown that the Ruch-Ugi model[119a] does not allow the enantioselectivities in the (rhodium(I) + chiral-bis(phosphine))-catalyzed reduction of PhCH=C(CO_2R^1)NHCOR^2 to be calculated.[119b]

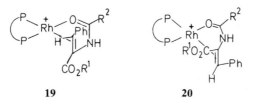

      19   **20**

14.4.4. Hydrogen Transfer and Dehydrogenation Reactions

Undoubtedly, the most exciting recent development in this area is the [ReH$_7$(PPh$_3$)$_2$]-catalyzed dehydrogenation of cycloalkanes C_nH_{2n} (n = 6, 7, or 8) in the presence of a hydrogen acceptor, for example, 3,3-dimethylbut-1-ene. With methylcyclohexane the major product is 4-methylcyclohexene, consistent with a pathway involving coordination of the least sterically hindered alkene rather than a radical pathway.[120a] Further insight into the mechanism comes from the observation that treatment of *n*-pentane with $[ReH_7(PPh_3)_2]$ and $Bu^tCH=CH_2$ gives $[ReH_3(trans$-penta-1,3-diene)-$(PPh_3)_2]$ which reacts with trimethyl phosphite to give pent-1-ene.[120b] It is also of interest that irradiation of hexane or heptane in a solution of $[PtCl_6]^{2-}$ in acetic acid yields the corresponding olefin complex $[Pt(Me[CH_2]_nCH=CH_2)Cl_2]_2$; the reaction is assumed to proceed via a Pt(IV)-alkyl complex which undergoes β-elimination of a proton.[121] Asymmetric transfer hydrogenation from chiral alcohols to α,β-unsaturated acids [e.g., $MeCH=C(Me)CO_2H]^{[122]}$ or prochiral ketones (e.g., PhCOMe),[123] is catalyzed by $[RuCl_2(PPh_3)_3]$ and $[H_4Ru_4(CO)_8(PBu_3)_4$, respectively. The optical yields, although low (i.e., $\leq 9\%$), are influenced by the structure of the chiral alcohol suggesting that the reaction proceeds via a ruthenium intermediate containing both the hydrogen donor and hydrogen acceptor. Kinetic data on the $[RuHCl(CO)(PPh_3)_3]$-catalyzed disproportionation of α-glucose, as in equation (37), are compatible with coordination of the acyclic form of glucose (G) to $[RuHCl(CO)(PPh_3)_2]$ being the rate-limiting step, followed by rapid hydrogen transfer to G and coordination of the D-glucopyranose which readily transfers two hydrogens to complete the catalytic cycle.[124]

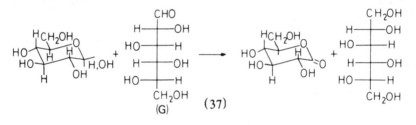

A kinetic study of the $[RuCl_2(PPh_3)_3]$-catalyzed hydrogen transfer from 1-phenylethanol to cyclohexanone has shown that both reactants may add to the active catalyst $[RuCl_2(PPh_3)_2]$ although only the ketone adduct may react further; hence, the order of mixing the reactants is important.[125] Reactions of the type (38) are catalyzed by $[\{(p$-$XC_6H_4)_3P\}RuCl_2]$ (X = H, Cl, Me, or OMe) and are believed to proceed via the intermediate $[RuCl(OCH_2C_6H_4X)(P\{C_6H_4X\}_3)_2]$; however, the rate is only influenced by the nature of the X group in either the catalyst or the hydrogen acceptor and not by that in the hydrogen donor which suggests that the formation of this alkoxide intermediate and its subsequent rearrangement to give a hydride cannot be rate determining.[126]

$$p\text{-}XC_6H_4CH_2OH + p\text{-}XC_6H_4CH(OH)CCl_3$$

$$\rightarrow p\text{-}XC_6H_4CHO + p\text{-}XC_6H_4CH(OH)CHCl_2 + HCl \quad (38)$$

Hydrogen transfer from 9,10-dihydroanthracene or 9-fluorenol to 1,1-diphenylethylene occurs in the presence of $[Co_2(CO)_8]$ or $[Co_4(CO)_{14}]$; the reaction is probably radical in character with initial hydrogen abstraction by $Co(CO)_4^{\cdot}$ being the rate-limiting step.[127] Detailed kinetic, ^{1}H NMR and IR studies have been carried out on the transfer hydrogenation of α,β-unsaturated ketones by alcohols in the presence of $[RhH(PPh_3)_4]$.[128] Of particular interest is the fact that it has been demonstrated that in such reactions the hydroxylic hydrogen is selectively transferred to the α-carbon of the ketone as in equation (39).[128c] Surprisingly, the rate-determining step appears to be the cleavage of the O–H bond since in a similar reduction of benzalacetone with $PhCD(OH)CH_3$ a kinetic isotope effect of 1.20 was observed which increased to 2.55 when $PhCD(OD)CH_3$ was used.[128d]

$$CH_3CH_2OD + PhCH{=}CHCOCH_3$$

$$\rightarrow PhCH_2CHDCOCH_3 + CH_3CHO \quad (39)$$

$[Ru(O)(py)L_2]^{2+}$ (**21**) (L = 2-(phenylazo)pyridine) dehydrogenates water to dioxygen and gives $[Ru(OH_2)(py)L_2]^{2+}$ (**22**); in the presence of Ce(IV) to oxidize **22** back to **21** this dehydrogenation becomes catalytic.[129] Rate parameters have been determined for the corresponding $[Pt_{12}(CO)_{24}]^{2-}$-catalyzed dehydrogenation of water in the presence of p-benzoquinone to reoxidize the platinum complex, equations (40) and (41)[130]:

$$3[Pt_{12}(CO)_{24}]^{2-} + H_2O \rightarrow 4[Pt_9(CO)_{18}]^{2-} + 2H^+ + \tfrac{1}{2}O_2 \quad (40)$$

$$4[Pt_9(CO)_{18}]^{2-} + C_6H_4O_2 + 2H^+ \rightarrow 3[Pt_{12}(CO)_{24}]^{2-} + C_6H_4(OH)_2 \quad (41)$$

14.5. Skeletal Rearrangements

Isomerization of 1,5-hexadiene to $CH_2{=}\overline{C(CH_2)_3CH_2}$, $CH_3C{=}CH(CH_2)_2CH_2$, and linear hexadienes is effected by $[Ti(C_5H_5)_2Cl]$ or $[Ti(C_5H_5)_2Cl_2]$ in the presence of Pr^iMgCl. The active catalyst is believed to be $[Ti(C_5H_5)_2H]$ which reacts with 1,5-hexadiene to give an alkenyl titanium intermediate.[131] Coenzyme B_{12} serves as a cofactor for a variety of isomerization reactions of the type $H{-}C_1{-}C_2{-}X \rightarrow X{-}C_1{-}C_2{-}H$ [X = OH, NH_2, $CH(NH_2)CO_2H$, $C(C{=}CH_2)CO_2H$, etc.] and it has been argued that the principal, if not the only, role of coenzyme B_{12} is to serve as a precursor for an organic free radical generated by homolytic dissociation of the cobalt–carbon bond.[132] A free-radical mechanism has also been proposed for the (*meso*-tetraphenylporphine)cobalt-catalyzed isomerization of cycloheptatriene-1,4-endoperoxide as in equation $(42)^{(133)}$:

$$\text{(42)}$$

[RhCl(PMe$_3$)$_3$] catalyzes the isomerization of $\overline{\text{RCHCH}_2\text{O}}$ (R = Ph, H, or Me) to RCOCH$_3$; the initial step involves the formation of *cis*-[RhH(CH$_2$COR)Cl(PMe$_3$)$_3$] which may be isolated (R = Me or Ph). Kinetic studies support a mechanism involving rate-limiting loss of PMe$_3$ from this intermediate followed by reductive elimination of RCOCH$_3$.$^{(134)}$ Cyclopropane moieties such as **23** are converted primarily to exocyclic methylene compounds **25** by [IrCl(CO)(PPh$_3$)$_2$] in the presence of oxygen; the proposed reaction mechanism involves the formation of an iridacyclobutane intermediate (**24**).$^{(135)}$

$$\text{(43)}$$

23   **24**   **25**

In a carboxylic acid solvent, [Ir(cod)Cl]$_2$ with a MeI promotor effects the isomerization of methyl formate to acetic acid. Kinetic and chemical studies indicate that the reaction proceeds in the carboxylic acid by transesterification to give formic acid which then reacts with iridium (Scheme 9).$^{(136)}$ $^{(37)}$

Scheme 9

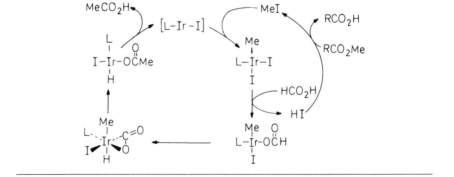

[Ni(cod)$_2$] + (R,R)-Bu$^t\overline{\text{POCH(R)CH(R)O}}$ (R = CO$_2$CH$_2$CH$_2$OMe) catalyzes the asymmetric isomerization reaction (44) with 22% e.e.$^{(137)}$

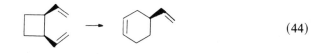

(44)

14.6. Metathesis Reactions

14.6.1. Alkene and Alkyne Metathesis

A significant development would be the ability to metathesize a wide range of functionalized olefins. Progress towards this goal has been reviewed.[138]

It has been proposed that if a metathesis catalyst contains groups which vibrate with frequencies close to each other, for example, $\nu_{M=O}$ and δ_{CH}, vibrational resonance may provide a low-energy pathway by allowing energy to be transferred to the part of the catalyst where it is needed. This model not only accounts for the activating effect of an oxo ligand but can also explain the product distribution in various metathesis reactions.[139] The accepted mechanism of the degenerate or nonproductive metathesis of terminal alkenes assumes that a metal-bound fragment readily exchanges with the terminal methylene group of the olefin. Such a process has now been observed directly. Thus, when $[Cp_2TiCD_2Al(CD_3)_2Cl]$ (26) reacts with methylene cyclohexane $CH_2{=}C_6H_{10}$ the initial broad singlet of 26 in the ESR spectrum is replaced by a triplet characteristic of a $Ti-CH_2$ containing species.[140] Insight into the mechanism of metathesis reactions comes from the observation that the reaction of 27 [R = H or D] with diphenylacetylene is first order in 27 and zero order in $PhC{\equiv}CPh$. This together with the observed large secondary isotope effect suggests that the reaction proceeds via rate-limiting opening of the metallacycle to give a carbene which is rapidly trapped by an incoming olefin or acetylene as in equation (45)[141]:

$$Cp_2Ti\overset{R\,R}{\underset{R\,R}{\diamond}}Bu^t \overset{r.d.s}{\rightleftharpoons} R_2C{=}CHBu^t + Cp_2Ti{=}CR_2 \xrightarrow{PhC{\equiv}CPh} Cp_2Ti\overset{R\,R}{\underset{Ph}{\diamond}}-Ph \quad (45)$$

27

Conductivity measurements show that there is no correlation between the activity of the metathesis catalyst $WCl_6 + Ph_4Sn$ and the formation of ionic species.[142] In contrast, conductivity and NMR experiments indicate that the active intermediate in the $[W(CHBu')(OCH_2Bu')_2Br_2] + nGaBr_3$ system is the chiral ionic salt $[W(CHBu')(OCH_2Bu')_2Br]^+Ga_2Br_7^-$. Further, it is argued that the principle of microscopic reversibility requires the metallacycle intermediate to have a plane of symmetry and hence ligand rearrangement must accompany metallacycle formation. Thus, a *cis*-olefin reacting to give a *cis*-olefin product should occur with inversion of chirality at the carbene which inverts the chirality of the catalyst (Scheme 10); a *cis*-olefin giving a *trans*-olefin product occurs with retention of configuration.[143] Such a mechanism could account for the observed alternating copolymerization of enantiomers of 1-methybicyclo[2.2.1]hept-2-ene by ReCl$_5$.[144]

Scheme 10

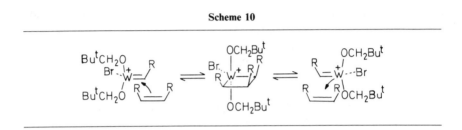

There is an inverse correlation between the stereoselectivity of the metathesis of cyclic olefins by [Mo(NO)$_2$Cl$_2$(PPh$_3$)$_2$] + EtAlCl$_2$ and the ring strain in the olefin. This lends further support to the proposal that if the coordinated olefin has an energy greater than that of either the *cis*- or *trans*-directing metallacyclobutanes then the system will lose its stereoselectivity.[145] Extended Hückel calculations on [Cl$_4$MCH(Me)CHR^{1}CHR^2] (M = Cr or W, R^1 = H or Me; R^2 = H, Me, or Et) support the proposal that for unstrained olefins the stereochemistries of the products correlate with the favored conformation of the substituted metallacyclobutane intermediate.[146] Another theoretical study based upon the carbene fragment X≡M(CHR) (M = Mo, W; X = O, N, C–R) has attempted to define the essential requirements of an active metathesis catalyst.[147] The first alkylation step in the metathesis catalyst WCl$_6$ + SnMe$_4$ shows second-order kinetics.[148]

Metathesis of functionalized alkynes is catalyzed by [MoO$_2$(acac)$_2$] + AlEt$_3$ or [Mo(CO)$_6$] + PhOH.[149] The observed reaction of phenylacetylene with the alkyne metathesis catalyst [W(≡CPh)(OBu$'$)$_3$] [equation (46)] has led to the suggestion that terminal alkynes are not metathesized because the metallacyclobutadiene intermediate readily loses a proton to give a stable complex of the type (28).[150]

[W(CPh)(OBu$'$)$_3$] + PhC≡CH

$$\xrightarrow[-\text{Bu}'\text{OH}]{+2\text{py}} [\overline{\text{WC(Ph)C}}=\overline{\text{C}}\text{Ph}(\text{OBu}')_2(\text{py})_2] \qquad (46)$$
                                                  28

14.6.2. Amine Disproportionation

Disproportionation of primary amines RCH_2NH_2 to ammonia and secondary amines occurs in the presence of $RuCl_3 + PBu_3$. The reaction involves the initial formation of $[RuCl_2(PBu_3)_3(H_2NCH_2R)]$ which eliminates HCl to generate the active catalyst—the imine complex $[Ru]NH=CHR$, where $[Ru] = RuHCl(PBu_3)_x$; the catalytic cycle is outlined in Scheme 11.[151]

Scheme 11

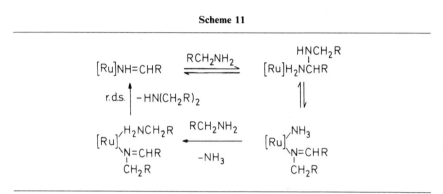

14.7. Oligomerization and Polymerization of Alkenes and Alkynes

14.7.1. Reactions of Alkenes

The β-alkyl elimination reaction (47) suggests that such a process may be the chain-terminating step in the $[Lu(\eta^5\text{-}C_5Me_5)_2Me]$-catalyzed oligomerization of propene.[152]

$$[Lu(\eta^5\text{-}C_5Me_5)_2CH_2CHMe_2]$$

$$\rightleftharpoons [Lu(\eta^5\text{-}C_5Me_5)_2Me] + CH_2=CHMe \quad (47)$$

In the $([Ti(C_5H_5)_2(Et)Cl] + EtAlCl_2)$-catalyzed polymerization of C_2H_4 and C_2D_4, the rate of propagation k_p shows only a small primary isotope effect $(k_{pH}/k_{pD} = 1.04 \pm 0.03)$. This is consistent with an olefin insertion mechanism (path a in Scheme 12) rather than a carbene-to-metallacycle mechanism which involves a hydrogen migration step (path b in Scheme 12.)[153]

In contrast the $[M(NO)_2(NCMe)_4]^{2+}$-catalyzed (M = Mo or W)[154a] and $[Pd(\eta^3\text{-}C_3H_5)(NCMe)_2]^+$-catalyzed[154b] oligomerization of olefins is postulated to proceed via the formation of incipient carbonium ions through

Scheme 12

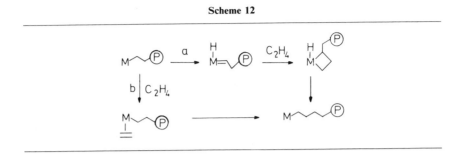

the interaction of the olefin with the electrophilic transition-metal center [equation (48)]. The palladium catalyst may be quantitatively recovered suggesting that the allyl group is retained throughout the reaction.

$$\begin{array}{c} [M^{2+}]-\| \\ \downarrow \end{array} \qquad (48)$$

NCCHDCH$_2$CH=CHCN is an initial product when the [RuCl$_2$(CH$_2$=CHCN)$_3$]-catalyzed linear dimerization of acrylonitrile is carried out under deuterium; this is consistent with a mechanism involving initial olefin "insertion" into a Ru–D bond followed by olefin "insertion" into the Ru–CH$_2$CHDCN bond.[155] A model system for the Shell–Higher Olefin Process is the [Ni(PPh$_2$CH$_2$COO)(η^3-C$_8$H$_{13}$)]-catalyzed linear oligomerization of ethylene to α-olefins; the kinetics of the reaction are consistent with the initial formation of [Ni(PPh$_2$CH$_2$COO)H] followed by subsequent olefin "insertions."[156] Related complexes [Ni(cyclooctenyl)L$_2$] (HL$_2$ = cyclic 1,2-diketones, α-acyl cycloalkanones, or substituted 1,3-propanediones) catalyze the linear oligomerization of 1-butene and the catalytic activity increases almost linearly with increasing acidity of HL$_2$. Complexes containing the η^1:η^2-cyclooctenyl ligand readily eliminate cyclooctadiene to form the nickel-hydride catalyst; consequently, they are more active than the corresponding η^3 complexes.[157] For systems of this type, an attempt has been made to correlate the kinetics of linear olefin oligomerization with the electron density at the nickel as measured by K-edge XAS and ESCA spectra; however, the correlation is not simple.[158] To explain the structures of the trimers obtained from the (Ni(O$_2$CC$_7$H$_{15}$)$_2$ + AlEtCl$_2$)-catalyzed or (NiCl$_2$ + 2P(C$_6$H$_{11}$)$_3$ + AlEtCl$_2$)-catalyzed oligomerization of propene, it has been argued that the nickel alkyl intermediates rapidly equilibrate by a process involving the Lewis-acid co-catalyst A as in equation (49) rather than via the usual metal–olefin–hydride intermediate.[159]

$$[Ni]\text{-}CH_2CH_2Pr^i \underset{}{\overset{+A}{\rightleftharpoons}} [Ni](\eta^2\text{-}CH_2\text{=}CHCH_2Pr^i\cdot HA)$$

$$\overset{-A}{\rightleftharpoons} [Ni]\text{-}CH(Me)CH_2Pr^i \qquad (49)$$

14.7.2. Reactions of Alkynes

An ingenious method has been described for differentiating between a metallacycle mechanism (path a in Scheme 13) and the classical Cossee-Arlman mechanism (path b) for the $[Ti(OBu^n)_4] + Et_3Al)$-catalyzed copolymerization of $H^{13}C\equiv^{13}CH + H^{12}C\equiv^{12}CH$ (4:96). The proton-coupled ^{13}C mutation NMR spectrum shows that the polymer contains the ^{13}C atoms linked by a double bond as predicted by path b and not a single bond as predicted by path a.[160]

Scheme 13

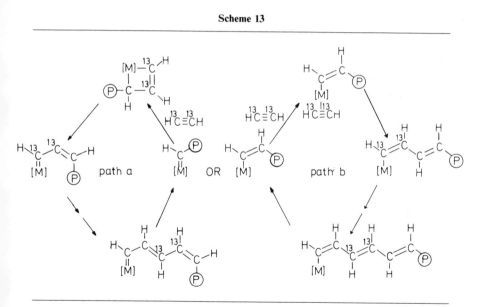

Circumstantial evidence that μ-carbene complexes may initiate alkyne oligomerization comes from the observation that diruthenium μ-carbene complexes do cause linking of alkynes as in equation (50); the stereochemistry of the growing carbon chain is controlled by the nature of the carbene substituents.[161]

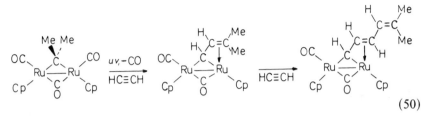

$$(50)$$

[RhClL$_3$] (L = PPh$_3$, AsAr$_3$, etc.) catalyzes the dimerization and trimerization of R^{1}R^{2}C(OH)C≡CH and the influence of L, R^1, and R^2 upon the selectivity of the reaction has been discussed.[162]

14.8. Reactions of Dinitrogen

The reactions of dinitrogen in its metal complexes have been reviewed.[163] Unfortunately, however, such complexes are still far removed from a viable homogeneous catalyst capable of fixing atmospheric nitrogen. A MO study of the complexes [ML$_4$Cl(N$_2$)]$^{n+}$ (M = Ti, V, Cr, Mn, or Fe) has attempted to define the factors which activate the coordinated dinitrogen as measured by its π-acceptor ability. One conclusion was that the degree of dinitrogen activation decreases as the equatorial ligands L change along the series NH$_3$ > H$_2$O ≫ H$_2$S, PH$_3$.[164] A ^{13}C and ^{1}H NMR spectroscopic investigation of the mechanism of reduction of dinitrogen by [Cp$_2$TiCl$_2$] + Mg suggested that N$_2$ is reduced to nitride N^{3-} which then forms M$_3$N bridges (M = Ti or Mg); the nitride may then be oxidized to N$^-$ which may react further, for example, with CO to produce isocyanates.[165] Dinitrogen coordinated end-on in a mononuclear molybdenum complex has been converted to an imido complex with splitting of the N–N bond and generation of an amine, equations (51)–(53). By analogy it has been suggested that reaction (54) may occur in nitrogenase.[166]

$$[Mo(N_2)_2(dpe)_2] + 2RBr \rightarrow [MoBr(N_2R_2)(dpe)_2]^+ \tag{51}$$

$$[MoBr(N_2R_2)(dpe)_2]^+ + 2e^- \rightarrow [Mo(N_2R_2)(dpe)_2] \tag{52}$$

$$[Mo(N_2R_2)(dpe)_2] + 2HBr \rightarrow [MoBr(NH)(dpe)_2]^+ + R_2NH \tag{53}$$

$$[Mo]{=}NNH_2 + 2H^+ \rightarrow [Mo]{=}NH^{2+} + NH_3 \tag{54}$$

Part 4

Compilations of Numerical Data

Chapter 15

Volumes of Activation for Inorganic and Organometallic Reactions: A Tabulated Compilation

15.1. Introduction

This chapter is devoted to a compilation of activation volume data for inorganic and organometallic reactions published during the period January 1980 to December 1983. Data published prior to 1980 are adequately covered by a series of review papers[1-4] and readers are advised to consult these for further information. No major review has covered the data published since January 1980, and this compilation is intended to serve as a continuation of the data summarized for inorganic reactions by Asano and le Noble in 1978.[1] A number of specialized review articles, which also contain volume-of-activation data, have appeared in recent years.[5-14] However, these data are included in the present compilation to make it as complete as possible. Readers are advised to consult the following articles for detailed interpretation and mechanistic discussions dealing with the following topics: dependence of equilibrium and rate constants on pressure;[5] chemistry in compressed solutions;[6] solvent and pressure effects on kinetics of reactions of complexes;[7] kinetics and mechanisms of solvation, solvolysis, and substitution in nonaqueous media;[8] elucidation of solvent exchange mechanisms by high-pressure NMR studies;[9] operationism, mechanism, and high-pressure,[10] pressure effects and substitution mechanisms;[11] substitution reactions of divalent and trivalent metal ions;[12] high-pressure photochemistry in gas–liquid systems;[13] and biophysical chemistry at high pressure.[14]

The significant increase in activity in this area during recent years, as reflected by the number of entries in this compilation (more than 360) as compared to those published before 1978 (about 170), is due mainly to the general availability of specialized high-pressure equipment. In addition to the conventional equipment used to study the kinetics of slow reactions under pressures up to a few kilobar, instruments to handle fast reactions in the milli-, and micro-, and nanosecond time range have been developed and are being employed by various groups. Such instrumentation includes stopped flow,[15-19] T-jump,[20,21] NMR,[22,23] and photochemical techniques.[24,25] A special apparatus for high-pressure studies in liquid ammonia has also been developed.[26]

The mechanistic interpretation of volume of activation and reaction volume data according to a volume profile analysis cannot be treated in this compilation. However, we would like to direct the readers' attention to some interesting papers of general interest to this topic in the following references: a method to estimate the intrinsic term in activation and reaction volumes;[27] the ionic-strength dependence of volumes of activation;[28] the correction for the compressibility of the solvent;[29] the role of nonlabile ligands in the interpretation of volumes of activation;[30,32] and the theoretical prediction of partial molar volumes and volumes of activation.[33-35]

15.2. Data in Tabular Form

The data in tabular form have been arranged according to the presentation by Asano and le Noble.[1] The sequence in which different types of reactions are presented is solvent exchange, substitution, isomerization, electron transfer, and photochemical. The reactions are ordered according to the atomic number of the central metal atom. Reaction volume data ($\Delta \bar{V}$) are only reported for systems where appropriate volume-of-activation data ($\Delta V^{\ddagger}$) is available, and this is included in the tables. The methods employed to determine $\Delta \bar{V}$ are: a—the pressure dependence of the equilibrium constant; b—the dilatometric or partial molar volume (density) measurements, and; c—theoretical prediction. Other general remarks are $\Delta V^{\ddagger}$ data are quoted at ambient pressure—in case of significant curvature in the ln k vs. P plots the compressibility coefficient ($\Delta \beta^{\ddagger}$) is also given; no. of data refers to the different pressures at which kinetic measurements (usually more than one experiment) were performed—the maximum applied pressure being quoted in the fourth column; concentration is given in mol liter^{-1} (M) or mol kg^{-1} (m); anions quoted in brackets refer to the ionic-strength controlling medium employed, namely ClO_4^- usually refers to a mixture of $HClO_4$ and $NaClO_4$.

Abbreviations Used in the Tables

acac	acetylacetonate
bipy	bipyridine
bsb	
cod	cycloocta-1,5-diene
cyclam	1,4,8,11-tetraazacyclotetradecane
dien	diethylenetriamine
DMF	dimethylformamide
dmg	dimethylglyoximate
DMSO	dimethylsulfoxide
dtcd	*meso*-5,12-dimethyl-1,4,8,11-tetraazacyclotetradeca-4,11-diene
edda	ethylenediaminediacetate
edta	ethylenediaminetetraacetate
en	ethylenediamine
Eten	*N*-ethylethylenediamine
fz	ferrozine(3-(2-pyridyl)-5,6-bis-(4-sulfophenyl)-1,2,4-triazine)
gly	glycolate
Hahx	acetohydroxamic acid
Hame	2-aminoethanol
Hipt	4-isopropyltropolone
hxsb	
isoq	isoquinoline
lac	lactate
mal	malonate
male	maleate

4-Medien	4-methyldiethylenetriamine
Meen	*N*-methylethylenediamine
MFA	*N*-methylformamide
mtpp	*N*-methyltetraphenylporphine
nta	nitrilotriacetate
pada	pyridine-2-azo-4-dimethylaniline
pan	1-(2-pyridylazo)-2-naphthol
PC	propylene carbonate
PES	poly(ethylenesulfonate)
Ph$_2$-dtc	diphenyldithiocarbamate
phen	1,10-phenanthroline
Pren	*N*-propylethylenediamine
PSS	poly(styrenesulfonate)

Table 15.1. Volumes of Activation

Reaction: Solvent exchange	Solvent	T (°C)
$Al(DMSO)_6^{3+} + DMSO$	CH_3NO_2	85
$Al(DMF)_6^{3+} + DMF$	CH_3NO_2	82
$Al(tmp)_6^{3+} + tmp$	CD_3NO_2	68
$[Sc(tmp)_6]^{3+} + tmp$	CD_3NO_2	26
	CD_3NO_2	−33 to 77
$[V(H_2O)_6]^{3+} + H_2O$	H_2O	
	H_2O	
$[V(H_2O)_6]^{2+} + H_2O$	H_2O	77
$[Mn(H_2O)_6]^{2+} + H_2O$	H_2O	5 to 6
	H_2O	5
	H_2O	6
	H_2O	103
$[Mn(CH_3OH)_6]^{2+} + CH_3OH$	CH_3OH	4
$[Mn(CH_3CN)_6]^{2+} + CH_3CN$	CH_3CN	−20 to −13
$[Fe(H_2O)_6]^{3+} + H_2O$	H_2O	110
$[Fe(H_2O)_5OH]^{2+} + H_2O$	H_2O	110
$[Fe(DMSO)_6]^{3+} + DMSO$	DMSO	5 to 35
$[Fe(DMF)_6]^{3+} + DMF$	DMF	−18 to 127
$[Fe(CH_3OH)_5(OCH_3)]^{2+} + CH_3OH$	CH_3OH	42 to 43
$[Fe(H_2O)_6]^{2+} + H_2O$	H_2O	−5 to 19
$[Fe(CH_3OH)_6]^{2+} + CH_3OH$	CH_3OH	−23
$[Fe(CH_3CN)_6]^{2+} + CH_3CN$	CH_3CN	−15 to −10
$[Co(H_2O)_6]^{2+} + H_2O$	H_2O	−10 to 34
$[Co(CH_3CN)_6]^{2+} + CH_3CN$	CH_3CN	13
	CH_3CN	−0.8
$[Co(tmc)(CH_3CN)]^{2+} + CH_3CN$	CH_3CN	−25
$[Ni(H_2O)_6]^{2+} + H_2O$	H_2O	36 to 41
	H_2O	35 to 44

continued

pyrdtc	pyrrolidinecarbodithionate
Q	quencher
SFL	sulfolane
suc	succinate
TCE	1,1,2,2-tetrachloroethane
2,3,2-tet	1,3-bis-(2'-aminoethylamino)propane
3,2,3-tet	N,N'-bis-(3-aminopropyl)
tmc	1,4,8,11-tetramethylcyclam
tmp	trimethylphosphate
tmpp	*meso*-tetrakis(4-N-methylpyridyl)porphine
tpps	*meso*-tetrakis(*p*-sulfonatophenyl)porphine
tpyp	5,10,15,20-tetra-4-pyridylporphine
tu	thiourea

for Solvent Exchange Reactions

P (kbar)	No. of data	$\Delta V^{\ddagger}$ (cm^3 mol^{-1})	$\Delta\beta^{\ddagger}$ (cm^3 mol^{-1} kbar^{-1})	$\Delta\bar{V}$ (cm^3 mol^{-1})	Ref.	Remarks
2.0	12	+15.6 ± 1.4	+4.8 ± 1.4		36	
2.0	12	+13.7 ± 1.2	+5.4 ± 1.2		36	
		+22.5 ± 0.6			9	
		−18.7 ± 1.1	−2.4 ± 1.2		9	
		−23.8 ± 2.7			9	
2.5		−9.3 ± 0.5			37	Cl$^-$ medium
2.5		−10.1 ± 0.5			37	Br$^-$ medium
2.5	13	−4.1 ± 0.1			37, 38	$\mu \sim 2\,M$ (Cl$^-$)
2.5	20	−6.2 ± 0.2			39	0.06–0.1 m HClO$_4$
2.5	>10	−6.0 ± 0.6			40	0.06–0.1 m HClO$_4$
2.5	>10	−6.3 ± 0.5			40	0.06–0.1 m HClO$_4$
2.5	>10	−4.3 ± 0.4			40	0.06–0.1 m MClO$_4$
2.0	9	−5.0 ± 0.2			41	
1.1	~10	−7.0 ± 0.4			42	
2.4	13	−5.4 ± 0.4			43	$\mu = 6\,m$ (ClO$_4^-$)
2.4	13	+7.0 ± 0.2			43	$\mu = 6\,m$ (ClO$_4^-$)
2.0	10	−3.1 ± 0.3			44	
2.0	10	−0.9 ± 0.2			44	
2.0	6	+6.4 ± 0.2			44	
2.5	>15	+3.8 ± 0.2			40	$\mu = 0.2$–1 m (ClO$_4^-$)
2.0	9	+0.4 ± 0.3			41	
1.4	~10	+3.0 ± 0.5			42	
2.5	>15	+6.1 ± 0.2			40	$\mu = 0.1$–0.8 m (ClO$_4^-$)
2.0	11	+7.7 ± 1.7			45	
1.5	~12	+6.7 ± 0.4			46	
1.5	9	−9.6 ± 0.5			47	
2.5		+7.2 ± 0.3			40	
2.25	10	+7.1 ± 0.2			48	$\mu = 0.1$–0.3 m (ClO$_4^-$)

continued

Table 15.1.

Reaction: Solvent exchange	Solvent	T (°C)
$[Ni(CH_3CN)_6]^{2+} + CH_3CN$	CH_3CN	35
$[Ni(tmc)(CH_3CN)]^{2+} + CH_3CN$	CH_3CN	-25
$[Ni(NH_3)_6]^{2+} + NH_3$	NH_3	10 to 30
$Ga(DMSO)_6^{3+} + DMSO$	CH_3NO_2	62
$Ga(DMF)_6^{3+} + DMF$	CH_3NO_2	40
$Ga(tmp)_6^{3+} + tmp$	CD_3NO_2	46
trans-$[Pd(DMSO)_2Cl_2] + DMSO$	$CHCl_3$	35
	CH_2Cl_2	23
	o-$C_6H_4Cl_2$	14
	C_6H_5CN	25
$In(tmp)_6^{3+} + tmp$	CD_3NO_2	50
		62
$SbCl_5 \cdot MeCN + MeCN$	CH_2Cl_2	-20
$SbCl_5 \cdot Me_3CCN + Me_3CCN$	CH_2Cl_2	-37
$SbCl_5 \cdot Me_2O + Me_2O$	CH_2Cl_2	0
$SbCl_5 \cdot Et_2O + Et_2O$	CH_2Cl_2	-10
$SbCl_5 \cdot Me_2CO + Me_2CO$	CH_2Cl_2	-8.5
$SbCl_5 \cdot (Me_2N)Cl_2PO + (Me_2N)Cl_2PO$	CH_2Cl_2	1.1
$[Tb(DMF)_8]^{3+} + DMF$	CD_3NO_2	-38
$[Dy(DMF)_8]^{3+} + DMF$	CD_3NO_2	-37 to -33
$[Ho(DMF)_8]^{3+} + DMF$	CD_3NO_2	-39 to -34
$[Er(DMF)_8]^{3+} + DMF$	CD_3NO_2	-37 to -32
$[Tm(DMF)_8]^{3+} + DMF$	CD_3NO_2	-18
$[Yb(DMF)_8]^{3+} + DMF$	CD_3NO_2	-38 to -33

(continued)

P (kbar)	No. of data	$\Delta V^{\ddagger}$ (cm^3 mol^{-1})	$\Delta\beta^{\ddagger}$ (cm^3 mol^{-1} kbar^{-1})	$\Delta\bar{V}$ (cm^3 mol^{-1})	Ref.	Remarks
1.5	~12	+7.3 ± 0.3			46	
1.5	9	−9.6 ± 0.5			47	
2.0	8	+5.9 ± 0.4			49	
2.0	12	+13.1 ± 1.0	+5.5 ± 0.9		36	
1.2	9	+7.9 ± 1.6	+2.6 ± 2.7		36	
		+20.7 ± 0.3			9	
2.0	10	−7.1 ± 0.5			50	
2.0	10	−8.4 ± 0.3			50	
2.0	10	−7.2 ± 0.6			50	
2.0	10	−5.4 ± 0.1			50	
		−22.8 ± 1.1	−3.9 ± 1.1		9	
		−20.0 ± 1.7	−5.9 ± 1.0		9	
		+24.7 ± 1.7	+4.1 ± 1.8		9	
		+18.2 ± 0.9	+3.2 ± 0.6		9	
		+27.2 ± 1.4	+5.9 ± 1.0		9	
		+30.0 ± 1.5	+3.8 ± 1.4		9	
		+28.1 ± 2.0	+8.1 ± 1.9		9	
		+23.0 ± 0.6			9	
2.0		+5.2 ± 0.2			51	
2.0		+6.1 ± 0.2			51	
2.0		+5.2 ± 0.5			51	
2.0		+5.4 ± 0.3			51	
2.0		+7.4 ± 0.3			51	
2.0		+11.8 ± 0.4	+2.0 ± 0.9		51	

Table 15.2. Volumes of Activation

Reaction: Substitution	Solvent	T (°C)
$TiO^{2+} + H_2O_2 \rightarrow [Ti(O_2)]^{2+} + H_2O$	H_2O	25
$[TiO(nta)(H_2O)]^- + H_2O_2 \rightarrow [Ti(O_2)(nta)(H_2O)]^- + H_2O$	H_2O	25
$[TiO(tpypH_4)]^{4+} + H_2O_2 \rightarrow [Ti(O_2)(tpypH_4)]^{4+} + H_2O$	H_2O	25
$[VO_2(nta)]^{2+} + H_2O_2 \rightarrow [VO(O_2)(nta)]^{2-} + H_2O$	H_2O	25
$[VO_2(nta)]^{2-} + H_2O_2 \xrightarrow{H^+} [VO(O_2)(nta)]^{2-} + H_2O$	H_2O	25
$[V(H_2O)_6]^{2+} + SCN^- \rightarrow [V(H_2O)_5SCN]^+ + H_2O$	H_2O	25
$[V(H_2O)_5SCN]^+ + H_2O \rightarrow [V(H_2O)_6]^{2+} + SCN^-$	H_2O	25
$[Cr(tpps)(H_2O)_2]^{3-} + NCS^- \rightarrow [Cr(tpps)(H_2O)NCS]^{4-} + H_2O$	H_2O	15
$[Cr(tpps)(H_2O)NCS]^{4-} + H_2O \rightarrow [Cr(tpps)(H_2O)_2]^{3-} + NCS^-$	H_2O	15
$[Cr(NH_3)_5Br]^{2+} + H_2O \rightarrow [Cr(NH_3)_5H_2O]^{3+} + Br^-$	H_2O	25
	H_2O	25
	H_2O	25
	H_2O	25
$Cr_2O_7^{2-} + B \rightarrow BCrO_3 + CrO_4^{2-}$		
$B = OH^-$	H_2O	25
$B = NH_3$	H_2O	25
$B = H_2O$	H_2O	25
$B = 2,6$-lutidine	H_2O	25
$[Mn(H_2O)_6]^{2+} + bipy \rightarrow [Mn(bipy)(H_2O)_4]^{2+} + 2H_2O$	H_2O	21
$[Mn(bipy)(H_2O)_4]^{2+} + 2H_2O \rightarrow [Mn(H_2O)_6]^{2+} + bipy$	H_2O	21
$Fe^{3+} + SCN^- \rightarrow [Fe(SCN)]^{2+}$	H_2O	25
	H_2O	20
	DMSO	25
$[Fe(SCN)]^{2+} \rightarrow Fe^{3+} + SCN^-$	H_2O	25
	DMSO	25
$FeOH^{2+} + SCN^- \rightarrow [Fe(SCN)]^{2+} + OH^-$	H_2O	25
	H_2O	20
$Fe^{3+} + Br^- \rightarrow Fe^{3+}, Br^-$	H_2O	25
$Fe^{3+}, Br^- \rightarrow [FeBr]^{2+}$	H_2O	25
$[FeBr]^2 \rightarrow Fe^{3+}, Br^-$	H_2O	25
$Fe^{3+} + Br^- \rightarrow [FeBr]^{2+}$	H_2O	25
$Fe^{3+} + Hipt \rightarrow [Fe(ipt)]^{2+} + H^+$	H_2O	25
	DMSO	25
	DMF	25
$FeOH^{2+} + Hipt \rightarrow [Fe(ipt)]^{2+} + H_2O$	H_2O	25
$FeOH^{2+} + H_2ipt^+ \rightarrow [Fe(ipt)]^{2+} + H_3O^+$	H_2O	25
$Fe^{3+} + Hahx \rightarrow [Fe(ahx)]^{2+} + H^+$	H_2O	25
	DMSO	25
	DMF	25
$FeOH^{2+} + Hahx \rightarrow [Fe(ahx)]^{2+} + H_2O$	H_2O	25
$[Fe(4\text{-Me-phen})_3]^{2+} + CN^- \rightarrow$	33% $MeOH/H_2O$	25
	H_2O	25
$[Fe(4\text{-Me-phen})_3]^{2+} + S_2O_8^{2-} \rightarrow$	H_2O	25
$[Fe(fz)_3]^{4-} + OH^- \rightarrow$	H_2O	25

for Substitution Reactions

P (kbar)	No. of data	$\Delta V^{\ddagger}$ (cm^3 mol^{-1})	$\Delta\beta^{\ddagger}$ (cm^3 mol^{-1} kbar^{-1})	$\Delta\bar{V}$ (cm^3 mol^{-1})	Ref.	Remarks
1.2	6	-6.0 ± 0.4			52	$\mu = 3.0\ m$ (ClO$_4^-$)
1.2	6	-19 ± 2			52	$\mu = 1.0\ m$ (ClO$_4^-$)
1.2	6	-3.3 ± 0.2			52, 53	$\mu = 1.0\ m$ (ClO$_4^-$)
1.5	~10	-3.4 ± 0.5			17	$\mu = 1.0\ m$ (ClO$_4^-$)
1.5	~10	$+1.5 \pm 0.5$			17	$\mu = 1.0\ m$ (ClO$_4^-$)
1.5	9	-2.1 ± 0.8			19	$\mu = 0.5\ M$ (ClO$_4^-$)
1.5	9	-11.5 ± 0.9			19	$\mu = 0.5\ M$ (ClO$_4^-$)
1.0	5	$+7.4 \pm 0.1$			54	[H$^+$] = 0.1 M, $\mu = 1.0\ M$
1.0	5	$+8.2 \pm 0.4$			54	[H$^+$] = 0.1 M, $\mu = 1.0\ M$
2.0		-9.3 ± 2.0			55	0.1 M HClO$_4$
2.0		-7			55	0.1 M HClO$_4$, NaPES
2.0		-4			55	0.1 M HClO$_4$, (C$_4$H$_9$)$_4$NPES
2.0		-9			55	0.1 M HClO$_4$, NaPSS
1.75	17	-17.9 ± 0.6			56	$\mu = 0.1\ M$ (KNO$_3$)
1.3	13	-19.2 ± 0.9			56	$\mu = 0.1\ M$ (KNO$_3$)
1.75	10	-24.9 ± 0.9			56	$\mu = 0.1\ M$ (KNO$_3$)
1.75	10	-26.9 ± 0.7			56	$\mu = 0.1\ M$ (KNO$_3$)
2.0	5	-1.2 ± 0.2		$+3.0 \pm 0.4^a$	57	$\mu = 0.3\ M$ (ClO$_4^-$)
2.0	5	-4.1 ± 0.4			57	$\mu = 0.3\ M$ (ClO$_4^-$)
1.5	~8	-6.1 ± 1.0			58	$\mu = 1.5\ m$ (ClO$_4^-$)
2.0	5	$+6.7 \pm 0.4$		$+8.3 \pm 0.3^a$	59	$\mu = 0.2\ M$ (NO$_3^-$)
2.0	6	$+3.3 \pm 0.6$			60	$\mu = 0.2\ m$ (ClO$_4^-$)
2.0	6	-15.0 ± 1.2			60	$\mu = 0.2\ m$ (ClO$_4^-$)
2.0	6	-8.5 ± 1.4			60	$\mu = 0.2\ m$ (ClO$_4^-$)
1.5	~8	$+8.5 \pm 1.2$			58	$\mu = 1.5\ m$ (ClO$_4^-$)
2.0	5	~0			59	$\mu = 0.2\ M$ (NO$_3^-$)
2.8				$+11 \pm 1^c$	61	$\mu = 0.2\ M$ (ClO$_4^-$)
2.8	5	-19 ± 4			61	$\mu = 0.2\ M$ (ClO$_4^-$)
2.8	5	-2 ± 4			61	$\mu = 0.2\ M$ (ClO$_4^-$)
2.8	5			$+8.1 \pm 0.2^a$	61	$\mu = 0.2\ M$ (ClO$_4^-$)
1.5	~10	-8.7 ± 0.8			62	$\mu = 1.0\ m$ (ClO$_4^-$)
2.0	6	$+10.9 \pm 1.7$			60	$\mu = 0.2\ m$ (ClO$_4^-$)
2.0	6	$+5.0 \pm 0.4$			60	$\mu = 0.2\ m$ (ClO$_4^-$)
1.5	~10	$+4.1 \pm 0.6$			62	$\mu = 1.0\ m$ (ClO$_4^-$)
1.5	~10	-10.0 ± 1.2			62	$\mu = 1.0\ m$ (ClO$_4^-$)
1.5	~8	-10.0 ± 1.4			62	$\mu = 1.0\ m$ (ClO$_4^-$)
2.0	6	$+3.0 \pm 0.3$			60	$\mu = 0.2\ m$ (ClO$_4^-$)
2.0	6	-0.8 ± 0.2			60	$\mu = 0.2\ m$ (ClO$_4^-$)
1.5	~8	$+7.7 \pm 0.6$			62	$\mu = 1.0\ m$ (ClO$_4^-$)
1.4	2	$+13$			63, 64	
1.4	2	$+10$			63, 64	
1.4	2	~0			63, 65	
		-2			65	

continued

Table 15.2.

Reaction: Substitution	Solvent	T (°C)
$[Fe(bsb)_3]^{2+} + OH^- \rightarrow$	H_2O	25
	85% $MeOH/H_2O$	25
$[Fe(hxsb)]^{2+} + OH^- \rightarrow$	H_2O	25
	80% $MeOH/H_2O$	25
$[Fe(bipy)_3]^{2+} + H_2O \rightarrow$	H_2O	30
	H_2O	35
	D_2O	35
	D_2O	35
$Co^{2+} + pada \rightarrow [Co(pada)]^{2+}$	H_2O	25
	DMSO	50
	DMSO	50
	DMF	50
$Co^{2+} + Hmtpp \rightarrow [Co(tmpp)]^+ + H^+$	H_2O	25
$[Co(PPh_3)_2Br_2] + PPh_3 \rightarrow$ exchange of PPh_3	$CDCl_3$	30
$cis\text{-}[Co(en)_2(OH_2)_2]^{3+} + H_2C_2O_4 \rightarrow [Co(en)_2C_2O_4]^+ + 2H_3O^+$	H_2O	70
	H_2O	60
$[Co(NH_3)_5OH]^{2+} + SO_3^{2-} \rightarrow [Co(NH_3)_5SO_3]^+ + OH^-$	H_2O	35
$[Co(NH_3)_5SO_3]^+ + SO_3^{2-} \rightarrow trans\text{-}[Co(NH_3)_4(SO_3)_2]^- + NH_3$	H_2O	25
$[Co(tpps)(H_2O)_2]^{3-} + NCS^- \rightarrow [Co(tpps)(H_2O)NCS]^{4-} + H_2O$	H_2O	20
$[Co(tmpp)(H_2O)_2]^{5+} + SCN^- \rightarrow [Co(tmpp)(H_2O)SCN]^{4+} + H_2O$	H_2O	25
$[Co(NH_3)_5L]^{3+} + H_2O \rightarrow [Co(NH_3)_5H_2O]^{3+} + L$		
$L = OHCH_3$	H_2O	38
$L = OHCH_2CH_3$	H_2O	38
$L = OHCH(CH_3)_2$	H_2O	25
$L = OC(NH_2)_2$	H_2O	34
$L = OC(NH_2)(NHCH_3)$	H_2O	35
$L = OC(NHCH_3)_2$	H_2O	34
$L = OCH(NH_2)$	H_2O	48
$L = OCH(NHCH_3)$	H_2O	48
$L = OCH(N(CH_3)_2)$	H_2O	49
$L = DMSO$	H_2O	41
$[Co(NH_3)_5Br]^{2+} + H_2O \rightarrow [Co(NH_3)_5H_2O]^{3+} + Br^-$	H_2O	25
	H_2O	25
$trans\text{-}[Co(NH_3)_4(CN)DMSO]^{2+} + H_2O$		
$\rightarrow trans\text{-}[Co(NH_3)_4(CN)H_2O]^{2+} + DMSO$	H_2O	20
$trans\text{-}[Co(en)_2(NO_2)Cl]^+ + H_2O$		
$\rightarrow trans\text{-}[Co(en)_2(NO_2)H_2O]^{2+} + Cl^-$	H_2O	10
	H_2O	15
	H_2O	20
$cis\text{-}[Co(en)_2(NO_2)Cl]^+ + H_2O$		
$\rightarrow cis\text{-}[Co(en)_2(NO_2)H_2O]^{2+} + Cl^-$	H_2O	30
$cis\text{-}[Co(bipy)_2(NO_2)Cl]^+ + H_2O$		
$\rightarrow cis\text{-}[Co(bipy)_2(NO_2)H_2O]^{2+} + Cl^-$	H_2O	30
$cis\text{-}[Co(bipy)_2(NO_2)Br]^+ + H_2O$		
$\rightarrow cis\text{-}[Co(bipy)_2(NO_2)H_2O]^{2+} + Br^-$	H_2O	30

(continued)

P (kbar)	No. of data	$\Delta V^{\ddagger}$ (cm^3 mol^{-1})	$\Delta \beta^{\ddagger}$ (cm^3 mol^{-1} kbar^{-1})	$\Delta \bar{V}$ (cm^3 mol^{-1})	Ref.	Remarks
		+11			66	
		+29			66	
		+13			66	
		+5			66	
2.1	6	+12.3 ± 0.5			67	1.0 M HCl
1.4	5	+14.8 + 0.7			67	0.01 M HCl
1.4	5	+13.3 ± 0.4			67	1.0 M DCl
1.4	5	+16.3 + 0.5			67	0.01 M DCl
2.0	5	+11.2 ± 2.5			68	ClO$_4^-$ salt
2.0	6	+8.3 ± 0.5			68	ClO$_4^-$ salt
2.0	6	+6.8 ± 0.5			68	BPh$_4^-$ salt
2.0	6	+10.1 ± 0.8			68	BPh$_4^-$ salt
2.0	9	+8.0 ± 0.3			69	0.28 M Ca(NO$_3$)$_2$
2.6	7	−12.1 ± 0.6	−3.3 ± 0.5		70	
1.5	4	+1.1 ± 0.2			71	2 M HClO$_4$
1.5	4	+3.5 ± 0.8			71	2 M HNO$_3$
1.2	6	+18.6 ± 1.5			72	pH = 9.4, μ = 0.5 M
1.0	10	+13.7 ± 0.7			72	pH = 9.4, μ = 0.5 M
1.0	5	+15.4 ± 0.6			54	[H$^+$] = 0.1 M, μ = 1.0 M
1.2	6	±14 ± 4			73	[H$^+$] = 0.1 m, μ = 2.2 m
1.7	4	+2.2 ± 0.2		+1.6 (25°)[b]	74, 75	0.01 M CF$_3$SO$_3$H
1.7	4	+2.9 ± 0.3		+2.2 (25°)[b]	74, 75	0.01 M CF$_3$SO$_3$H
1.7	4	+3.8 ± 0.2		+2.9 (25°)[b]	74, 75	0.01 M CF$_3$SO$_3$H
1.7	4	+1.3 ± 0.5		+1.2 (25°)[b]	74, 76	$\mu \sim 0$
1.7	4	+0.3 ± 0.3		−1.6 (25°)[b]	74	$\mu \sim 0$
1.7	4	+1.5 ± 0.3		−0.6 (25°)[b]	74	$\mu \sim 0$
1.7	4	+1.1 ± 0.3		+0.5 (25°)[b]	74	0.01 M CF$_3$SO$_3$H
1.7	4	+1.7 ± 0.3		+1.5 (25°)[b]	74	0.01 M CF$_3$SO$_3$H
1.7	4	+2.6 ± 0.4		+2.9 (25°)[b]	74	0.01 M CF$_3$SO$_3$H
1.7	4	+2.0 ± 0.4		+3.1 (25°)[b]	74	0.01 M CF$_3$SO$_3$H
2.0		−8.7			77	0.01 M HClO$_4$
2.0		+12			77	0.01 M HClO$_4$, NaPES
1.5	7	+0.6 ± 0.3			78	μ = 0.1 M (ClO$_4^-$)
2.0	6	−0.3			79	$\mu \sim 0$
2.0	6	+0.1		−10.4[b]	79	$\mu \sim 0$
2.0	6	+0.3			79	$\mu \sim 0$
2.0	6	+0.9		−9.3 (25°)[b]	79	$\mu \sim 0$
2.0	6	+2.9		−9.1 (25°)[b]	79	$\mu \sim 0$
2.0	5	+11.3		−6.7 (25°)[b]	79	$\mu \sim 0$

continued

Table 15.2.

Reaction: Substitution	Solvent	T (°C)
cis-[Co(phen)$_2$(NO$_2$)Br]$^+$ + H$_2$O		
→ cis-[Co(phen)$_2$(NO$_2$)H$_2$O]$^{2+}$ + Br$^-$	H$_2$O	35
[Co(NH$_3$)$_4$(NH$_2$)Cl]$^+$ + H$_2$O → [Co(NH$_3$)$_5$OH]$^{2+}$ + Cl$^-$	H$_2$O	35
[Co(NH$_3$)$_4$(NH$_2$)Br]$^+$ + H$_2$O → [Co(NH$_3$)$_5$OH]$^{2+}$ + Br$^-$	H$_2$O	25
trans-[Co(pn)$_2$Cl$_2$]$^+$ + H$_2$O		
→ trans-[Co(pn)$_2$(Cl)H$_2$O]$^{2+}$ + Cl$^-$	H$_2$O	31
	H$_2$O	40
	H$_2$O	50
	H$_2$O	61
trans-[Co(dtcd)(N$_3$)Cl]$^+$ + H$_2$O		
→ trans-[Co(dtcd)(N$_3$)H$_2$O]$^{2+}$ + Cl$^-$	H$_2$O	25
trans-[Co(dtcd)(N$_3$)Br]$^+$ + H$_2$O → trans-[Co(dtcd)(N$_3$)H$_2$O]$^{2+}$ + Br$^-$	H$_2$O	20
[Co(tren)Cl$_2$]$^+$ + H$_2$O → [Co(tren)(Cl)H$_2$O]$^{2+}$ + Cl$^-$	H$_2$O	14
cis-β-[Co(trien)Cl$_2$]$^+$ + H$_2$O → [Co(trien)(Cl)H$_2$O]$^{2+}$ + Cl$^-$	H$_2$O	25
[Co(NH$_3$)$_5$SO$_4$]$^+$ + H$_2$O → [Co(NH$_3$)$_5$H$_2$O]$^{3+}$ + SO$_4^{2-}$	H$_2$O	35
	H$_2$O	55
[Co(NH$_3$)$_5$SO$_4$]$^+$ + H$_2$O $\xrightarrow{H^+}$ [Co(NH$_3$)$_5$H$_2$O]$^{3+}$ + SO$_4^{2-}$	H$_2$O	35
	H$_2$O	55
[Co(III)X]$^+$ + H$_2$O → [Co(H$_2$O)]$^{2+}$ + X$^-$		
CoX$^+$ = trans-[Co(NH$_3$)$_4$Cl$_2$]$^+$	H$_2$O	25
	H$_2$O	10
CoX$^+$ = trans-R,S-[Co(2,3,2-tet)Cl$_2$]$^+$	H$_2$O	50
	H$_2$O	50
	H$_2$O	30
CoX$^+$ = trans-[Co(en)$_2$Cl$_2$]$^+$	H$_2$O	25
	H$_2$O	30
CoX$^+$ = cis-[Co(en)$_2$Cl$_2$]$^+$	H$_2$O	30
CoX$^+$ = trans-[Co(Meen)$_2$Cl$_2$]$^+$	H$_2$O	30
CoX$^+$ = trans-[Co(Eten)$_2$Cl$_2$]$^+$	H$_2$O	30
CoX$^+$ = trans-[Co(Pren)$_2$Cl$_2$]$^+$	H$_2$O	30
CoX$^+$ = trans-[Co(3,2,3-tet)Cl$_2$]$^+$	H$_2$O	30
CoX$^+$ = trans-[Co(cyclam)Cl$_2$]$^+$	H$_2$O	50
CoX$^+$ = trans-RR,SS-[Co(2,3,2-tet)Cl$_2$]$^+$	H$_2$O	25
CoX$^+$ = trans-R,S-[Co(2,3,2-tet)Cl$_2$]$^+$	H$_2$O	25
CoX$^+$ = trans-[Co(en)$_2$Br$_2$]$^+$	H$_2$O	30

(continued)

P (kbar)	No. of data	$\Delta V^{\ddagger}$ (cm³ mol⁻¹)	$\Delta \beta^{\ddagger}$ (cm³ mol⁻¹ kbar⁻¹)	$\Delta \bar{V}$ (cm³ mol⁻¹)	Ref.	Remarks
2.0	6	+3.3		−12.5 (30°)[b]	79	$\mu \sim 0$
		+15.0		−9.4 (30°)[b]	79	$\mu \sim 0$
		+13.6		−8.6 (30°)[b]	79	$\mu \sim 0$
2.0	5	+14.4			80	pH = 2.5
2.0	5	+10.5			80	pH = 2.5
2.0	5	+6.3			80	pH = 2.5
2.0	5	+1.5			80	pH = 2.5
1.4	5	+8.3 ± 0.5			81	0.1 M HNO₃
1.4	4	+5.3 ± 0.7			81	0.1 M HNO₃
1.4	5	+7.3 ± 0.4			82	
1.4	5	+3.0 ± 0.6			82	
1.5	6	−18.3 ± 0.4		−26 ± 3 (25°)[b]	83	$\mu = 1.0$ m (ClO₄⁻)
1.0	5	−19.7 ± 0.8			83	$\mu = 1.0$ m (ClO₄⁻)
1.5	6	−3.5 ± 0.6			83	$\mu = 1.0$ m (ClO₄⁻)
1.0	5	−3.9 ± 0.5			83	$\mu = 1.0$ m (ClO₄⁻)
1.5	7	−1.3 ± 0.3		−15.5 ± 2.1[b]	4, 78	$\mu \sim 0$
1.5	7	−1.7 ± 0.7			4, 78	$\mu = 0.1$ M (ClO₄⁻)
1.5	7	−1.3 ± 0.4		−13.1 ± 2.1 (25°)[b]	4, 78	$\mu \sim 0$
1.5	7	−1.9 ± 0.4			4, 78	$\mu = 1.0$ M (ClO₄⁻)
1.5	7	−1.0 ± 0.4			4, 78	$\mu = 0.1$ M (ClO₄⁻)
1.5	7	−1.7 ± 1.1		−13.9 ± 1.7[b]	4, 78	$\mu \sim 0$
1.5	7	−1.1 ± 0.9			4, 78	$\mu = 0.1$ M (ClO₄⁻)
1.5	7	−0.3 ± 0.4		−14.2 ± 1.8 (25°)[b]	4, 78	$\mu = 0.1$ M (ClO₄⁻)
1.5	7	−3.1 ± 0.5		−13.1 ± 0.8 (25°)[b]	4, 78	$\mu = 0.1$ M (ClO₄⁻)
1.5	7	−0.3 ± 0.9		−11.7 ± 1.8 (25°)[b]	4, 78	$\mu = 0.1$ M (ClO₄⁻)
1.5	7	+0.3 ± 0.6		−14.9 ± 1.6 (25°)[b]	4, 78	$\mu = 0.1$ M (ClO₄⁻)
1.5	7	−2.8 ± 1.5		−14.9 ± 2.7 (25°)[b]	4, 78	$\mu = 0.1$ M (ClO₄⁻)
1.5	7	−2.0 ± 0.9		−14.3 ± 1.2 (25°)[b]	4, 78	0.1 M HNO₃
1.5	7	+1.0 ± 0.4			4, 78	0.01 M HNO₃
1.5	7	−0.8 ± 1.7			4, 78	0.01 M HNO₃
1.5	7	+1.4 ± 0.6		−14.5 ± 1.7 (25°)[b]	78	$\mu = 0.1$ M (ClO₄⁻)

continued

Table 15.2.

Reaction: Substitution	Solvent	T (°C)
$CoX^+ = trans\text{-}[Co(2,3,2\text{-tet})Br_2]^+$	H_2O	30
$CoX^+ = trans\text{-}[Co(NH_3)_4(CN)Cl]^+$	H_2O	40
$CoX^+ = trans\text{-}[Co(NH_3)_4(CN)Br]^+$	H_2O	40
$CoX^+ = trans\text{-}[Co(NH_3)_4(CN)I]^+$	H_2O	30
$CoX^+ = trans\text{-}[Co(NH_3)_4(CN)N_3]^+$	H_2O	40
$[Co(dmg)_2(Cl)(urea)] + H_2O \rightarrow [Co(dmg)_2(urea)H_2O]^+ + Cl^-$	H_2O	
$[Co(NH_3)_5SO_3]^+ + H_2O \xrightarrow{H^+} trans\text{-}[Co(NH_3)_4(SO_3)H_2O]^+ + NH_4^+$	H_2O	25
$[Co(en)_2CO_3]^+ + H_2O \rightarrow cis\text{-}[Co(en)_2(H_2O)OCO_2]^+$	H_2O	25
$[Co(en)_2CO_3]^+ + H_3O^+ \rightarrow cis\text{-}[Co(en)_2(H_2O)OCO_2H]^{2+}$	H_2O	25
$\alpha\text{-}[Co(edda)CO_3]^- + H_2O \rightarrow \alpha\text{-}[Co(edda)(H_2O)OCO_2]^-$	H_2O	25
$\alpha\text{-}[Co(edda)CO_3]^- + H_3O^+ \rightarrow \alpha\text{-}[Co(edda)(H_2O)OCO_2H]$	H_2O	25
$\beta\text{-}[Co(edda)CO_3]^- + H_2O \rightarrow \beta\text{-}[Co(edda)(H_2O)OCO_2]^-$	H_2O	25
$\beta\text{-}[Co(edda)CO_3]^- + H_3O^+ \rightarrow \beta\text{-}[Co(edda)(H_2O)OCO_2H]$	H_2O	25
$[Co(nta)CO_3]^{2-} + H_2O \rightarrow [Co(nta)(H_2O)OCO_2]^{2-}$	H_2O	25
$[Co(nta)CO_3]^{2-} + H_3O^+ \rightarrow [Co(nta)(H_2O)OCO_2H]^-$	H_2O	25
$\alpha\text{-}[Co(edda)(NH_3)_2]^+ + OH^- \rightarrow \alpha\text{-}[Co(edda)(NH_3)OH] + H_2O$	H_2O	66
$\beta\text{-}[Co(edda)(NH_3)_2]^+ + OH^- \rightarrow \beta\text{-}[Co(edda)(NH_3)OH] + H_2O$	H_2O	43
$\alpha\text{-}[Co(edda)(NO_2)_2]^- + OH^- \rightarrow \alpha\text{-}[Co(edda)(NO_2)OH]^- + NO_2^-$	H_2O	66
$Ni^{2+} + pada \rightarrow [Ni(pada)]^{2+}$	DMSO	50
	DMSO	50
	DMF	35
	DMF	25
	DMF	35
	DMF	50
$Ni^{2+} + suc^{2-} \rightarrow [Ni(suc)]$	H_2O	10
$[Ni(suc)] \rightarrow Ni^{2+} + suc^{2-}$	H_2O	10
$Ni^{2+} + male^{2-} \rightarrow [Ni(male)]$	H_2O	10
$[Ni(male)] \rightarrow Ni^{2+} + male^{2-}$	H_2O	10
$Ni^{2+} + mal^{2-} \rightarrow [Ni(mal)]$	H_2O	20
$Ni^{2+} + gly^{2-} \rightarrow [Ni(gly)]$	H_2O	20
$[Ni(gly)] \rightarrow Ni^{2+} + gly^{2-}$	H_2O	20
$Ni^{2+} + lac^{2-} \rightarrow [Ni(lac)]$	H_2O	20
$[Ni(lac)] \rightarrow Ni^{2+} + lac^{2-}$	H_2O	20
$Ni^{2+} + isoq \rightarrow [Ni(isoq)]^{2+}$	H_2O	25
	DMF	25
	CH_3CN	25
	CH_3OH	25
	C_2H_5OH	25
$[Ni(isoq)]^{2+} \rightarrow Ni^{2+} + isoq$	H_2O	25
	DMF	25
	CH_3OH	25
	C_2H_5OH	25

(continued)

P (kbar)	No. of data	$\Delta V^{\ddagger}$ (cm^3 mol^{-1})	$\Delta\beta^{\ddagger}$ (cm^3 mol^{-1} kbar^{-1})	$\Delta\bar{V}$ (cm^3 mol^{-1})	Ref.	Remarks
1.5	7	$+3.4 \pm 0.6$		-13.1 ± 2.3 $(25°)^b$	78	$\mu = 0.1\ M\ (ClO_4^-)$
1.5	7	-7.3 ± 0.4		-12.5 ± 0.9 $(25°)^b$	78	$\mu = 0.1\ M\ (ClO_4^-)$
1.5	7	-6.2 ± 0.6		-12.3 ± 1.6 $(25°)^b$	78	$\mu = 0.1\ M\ (ClO_4^-)$
1.5	7	-3.3 ± 0.9		-10.8 ± 3.5 $(25°)^b$	78	$\mu = 0.1\ M\ (ClO_4^-)$
1.5	7	$+5.7 \pm 0.2$			78	$0.1\ M\ HNO_3$
		$+3.5 \pm 0.3$			76	$\mu \sim 0$
1.0	5	$+6.0 \pm 1.2$			72	$[H^+] = 1.5\ M, \mu = 2\ M$
1.5	4	-6.1 ± 0.3			84	$pH = 5, \mu = 0.5\ M$
1.0	5	-7.1 ± 0.9			84	$[H^+] = 0.2\ M, \mu = 2\ M$
1.5	4	-14.0 ± 0.9			84	$pH = 5, \mu = 0.5\ M$
1.0	5	-4.2 ± 0.4			84	$[H^+] = 0.05\ M, \mu = 2\ M$
1.5	4	-14.5 ± 2.5			84	$pH = 5, \mu = 0.5\ M$
1.0	5	-3.8 ± 0.4			84	$[H^+] = 0.05\ M, \mu = 2\ M$
1.5	4	-9.4 ± 1.1			84	$pH = 5, \mu = 0.5\ M$
1.0	5	$+0.2 \pm 1.2$			84	$[H^+] = 0.05\ M, \mu = 2\ M$
1.5	4	$+16.6$			85	$pH = 9.4$, buffer
1.5	4	$+22.3$			85	$pH = 9.5$, buffer
1.5	4	$+11.9$			85	$pH = 9.9$, buffer
2.0	6	$+11.3 \pm 1.0$			68	ClO_4^- salt
2.0	5	$+12.4 \pm 0.8$			68	BPh_4^- salt
2.0	6	$+11.5 \pm 1.5$			68	ClO_4^- salt
2.0	5	$+9.2 \pm 1.4$			68	BPh_4^- salt
2.0	5	$+6.4 \pm 1.3$			68	BPh_4^- salt
2.0	5	$+9.5 \pm 0.4$			68	BPh_4^- salt
1.0	6	$+14.1 \pm 0.8$		$+11.1 \pm 0.9^a$	86	$pH \sim 6.8, \mu \sim 0$
1.0	6	$+3.0 \pm 0.9$			86	$pH \sim 6.8, \mu \sim 0$
1.0	6	$+13.5 \pm 0.6$		$+14.9 \pm 0.4^a$	86	$pH \sim 7.3, \mu \sim 0$
1.0	6	-1.4 ± 0.4			86	$pH \sim 7.3, \mu \sim 0$
1.0	6	$+16.0 \pm 2.1$			21	$\mu \sim 0$
1.0	6	$+14.7 \pm 0.5$		$+17.3 \pm 1.0^a$	87	$pH \sim 6, \mu \sim 0$
1.0	6	-2.6 ± 0.7			87	$pH \sim 6, \mu \sim 0$
1.0	6	$+13.6 \pm 0.4$		$+17.6 \pm 0.9^a$	87	$pH \sim 6, \mu \sim 0$
1.0	6	-4.1 ± 0.6			87	$pH \sim 6, \mu \sim 0$
2.0	9	$+7.4 \pm 1.3$			88	$\mu = 0.1\ M\ (ClO_4^-)$
2.0	9	$+9.3 \pm 0.3$			88	$\mu = 0.1\ M\ (ClO_4^-)$
2.0	9	$+9.4 \pm 0.1$			88	$\mu = 0.1\ M\ (ClO_4^-)$
2.0	9	$+12.8 \pm 0.6$			88	$\mu = 0.1\ M\ (ClO_4^-)$
2.0	9	$+12.6 \pm 0.5$		$+1.1 \pm 0.1^b$	88	$\mu = 0.1\ M\ (ClO_4^-)$
2.0	9	$+8.9 \pm 0.8$			88	$\mu = 0.1\ M\ (ClO_4^-)$
2.0	9	$+12.2 \pm 0.3$			88	$\mu = 0.1\ M\ (ClO_4^-)$
2.0	9	$+9.9 \pm 0.5$			88	$\mu = 0.1\ M\ (ClO_4^-)$
2.9	9	$+15.7 \pm 1.1$			88	$\mu = 0.1\ M\ (ClO_4^-)$

continued

Table 15.2.

Reaction: Substitution	Solvent	T (°C)
$[Ni(OAc)]_2 + pan \rightarrow [Ni(pan)] + 2OAc^-$	HAc	25
$[Ni(tmc)]^{2+} + 2H_2O \rightarrow [Ni(tmc)(H_2O)_2]^{2+}$	D_2O	−3 to 87
$[Mo(bipy)(CO)_4] + CN^- \rightarrow$	DMSO	25
	CH_3OH	25
$[Ru(NH_3)_5Cl]^{2+} + H_2O \rightarrow [Ru(NH_3)_5H_2O]^{3+} + Cl^-$	H_2O	60
$[Ru(NH_3)_5H_2O]^{3+} + Cl^- \rightarrow [Ru(NH_3)_5Cl]^{2+} + H_2O$	H_2O	60
$[Rh(tpps)(H_2O)_2]^{3-} + NCS^- \rightarrow [Rh(tpps)(H_2O)NCS]^{4-} + H_2O$	H_2O	15
$[Pd(1,1,7,7\text{-}Me_4dien)H_2O]^{2+} + X^-$		
$\rightarrow [Pd(1,1,7,7\text{-}Me_4dien)X]^+ + H_2O$		
X = Cl	H_2O	25
X = Br	H_2O	25
X = I	H_2O	25
$[Pd(1,1,4,7,7\text{-}Me_5dien)H_2O]^{2+} + X^-$		
$\rightarrow [Pd(1,1,4,7,7\text{-}Me_5dien)X]^+ + H_2O$		
X = Cl	H_2O	25
X = Br	H_2O	25
X = I	H_2O	25
X = N_3	H_2O	25
$[Pd(1,1,4\text{-}Et_3dien)H_2O]^{2+} + Cl^-$		
$\rightarrow [Pd(1,1,4\text{-}Et_3dien)Cl]^+ + H_2O$	H_2O	25
$[Pd(1,1,7,7\text{-}Et_4dien)H_2O]^{2+} + Cl^-$		
$\rightarrow [Pd(1,1,7,7\text{-}Et_4dien)Cl]^+ + H_2O$	H_2O	25
$[Pd(L)X]^+ + H_2O \rightarrow [Pd(L)H_2O]^{2+} + X^-$		
L = dien; X = Cl	H_2O	25
L = 1,4,7-Et_3dien; X = Cl	H_2O	25
L = 1,1,4-Et_3dien; X = Cl	H_2O	25
L = 1,1,7,7-Et_4dien; X = Cl	H_2O	25
L = 1,1,7,7-Et_4dien; X = SCN	H_2O	30
L = 4-Me-1,1,7,7-Et_4dien; X = Cl	H_2O	30
L = 4-Me-1,1,7,7,-Et_4dien; X = I	H_2O	30
L = 4-Me-1,1,7,7,-Et_4dien; X = SCN	H_2O	30
L = 1,1,4,7,7-Et_5dien; X = Cl	H_2O	25
L = 1,1,4,7,7-Et_5dien; X = I	H_2O	30
$[Pd(L)X]^+ + Y^- \rightarrow [Pd(L)Y]^+ + X^-$		
L = dien; X = Cl; Y = I	H_2O	25
L = 1,4,7-Et_3dien; X = Cl; Y = I	H_2O	25
L = 1,1,4-Et_3dien; X = Cl; Y = I	H_2O	25
L = 1,1,7,7-Et_4dien; X = Cl; Y = OH	H_2O	25

(continued)

P (kbar)	No. of data	$\Delta V^{\ddagger}$ ($cm^3\,mol^{-1}$)	$\Delta \beta^{\ddagger}$ ($cm^3\,mol^{-1}\,kbar^{-1}$)	$\Delta \bar{V}$ ($cm^3\,mol^{-1}$)	Ref.	Remarks
0.6	11	$+15.5 \pm 2.6$			89	
1.4	9			-10.0 ± 0.1^a	90	
1.4	3	-9			63, 91	
1.4	3	$+4$			63, 91	
2.9	18	-30.2 ± 2.4		-9 ± 1 $(25°)^b$	92	$\mu \sim 0.01\ M$
				-10 ± 3 $(60°)^b$	92	
1.2	6	-20 ± 1.4			92	$\mu = 0.11\ M$
1.0	5	$+8.8 \pm 0.4$			54	$[H^+] = 0.1\ M,\ \mu = 1\ M$
1.0	5	-7.1 ± 0.2			93	$\mu = 0.1\ M\ (ClO_4^-)$
1.0	5	-7.6 ± 0.3			93	$\mu = 0.1\ M\ (ClO_4^-)$
1.0	5	-9.3 ± 0.8			93	$\mu = 0.1\ M\ (ClO_4^-)$
1.0	5	-4.9 ± 0.4			93	$\mu = 0.1\ M\ (ClO_4^-)$
1.0	5	-7.3 ± 0.4			93	$\mu = 0.1\ M\ (ClO_4^-)$
1.0	5	-9.9 ± 1.2			93	$\mu = 0.1\ M\ (ClO_4^-)$
1.0	5	-11.7 ± 1.2			93	$\mu = 0.1\ M\ (ClO_4^-)$
1.0	5	-2.7 ± 0.2			16	$\mu = 0.1\ M\ (ClO_4^-)$
1.0	5	-3.0 ± 0.2			16	$\mu = 0.1\ M\ (ClO_4^-)$
1.0	3	-12.2 ± 0.8			94	$\mu = 0.1\ M\ (ClO_4^-)$
	5	-10.0 ± 0.6			94	$\mu = 0.1\ M\ (ClO_4^-)$
1.0	5	-10.8 ± 0.7			94	$\mu = 0.1\ M\ (ClO_4^-)$
	5	-10.8 ± 1.0			94	$\mu = 0.1\ M\ (ClO_4^-)$
1.0	5	-14.2 ± 0.6			94	$\mu = 0.1\ M\ (ClO_4^-)$
	5	-14.5 ± 1.2			94	$\mu = 0.1\ M\ (ClO_4^-)$
1.0	5	-13.0 ± 0.6			94	$\mu = 0.1\ M\ (ClO_4^-)$
1.5	5	-10.6 ± 0.4			95	$\mu = 0.5\ M$
1.2	7	-14.3 ± 0.6			94	$\mu = 0.05\ M$
1.5	7	-7.8 ± 0.2			94	$\mu = 0.05\ M$
1.5	5	-10.5 ± 0.6			95	$\mu = 0.5\ M$
1.5	7	-12.8 ± 0.8			94	$\mu = 0.05\ M$
1.2	6	-8.2 ± 0.3			94	$\mu = 0.05\ M$
1.0	5	-10.3 ± 1.0			94	$\mu = 0.1\ M\ (ClO_4^-)$
1.0	5	-11.1 ± 0.8			94	$\mu = 0.1\ M\ (ClO_4^-)$
1.0	5	-11.3 ± 1.3			94	$\mu = 0.1\ M\ (ClO_4^-)$
1.0	5	$+6.4 \pm 0.9$			94	$\mu = 0.1\ M\ (ClO_4^-)$

continued

Table 15.2.

Reaction: Substitution	Solvent	T (°C)
cis-[Pt(PEt$_3$)$_2$(2,4,6-Me$_3$Ph)Br] + S		
$\rightarrow$ *cis*-[Pt(PEt$_3$)$_2$(2,4,6-Me$_3$Ph)S]$^+$ + Br$^-$		
S = C$_2$H$_5$OH	C$_2$H$_5$OH	30
S = DMSO	DMSO	30
cis-[Pt(PEt$_3$)$_2$(2,4,6-Me$_3$Ph)Br] + tu		
$\rightarrow$ *cis*-[Pt(PEt$_3$)$_2$(2,4,6-Me$_3$Ph)tu]$^+$ + Br$^-$	C$_2$H$_5$OH	30
	DMSO	30
trans-[Pt(PEt$_3$)$_2$(2,4,6-Me$_3$Ph)Br] + DMSO		
$\rightarrow$ *trans*-[Pt(PEt$_3$)$_2$(2,4,6-Me$_3$Ph)DMSO]$^+$ + Br$^-$	DMSO	35
trans-[Pt(PEt$_3$)$_2$(2,4,6-Me$_3$Ph)Br] + tu		
$\rightarrow$ *trans*-[Pt(PEt$_3$)$_2$(2,4,6-Me$_3$Ph)tu]$^+$ + Br$^-$	C$_2$H$_5$OH	15
	DMSO	35
[Nd(DMF)$_8$]$^{3+}$ + DMF $\rightarrow$ [Nd(DMF)$_9$]$^{3+}$	DMF	-42 and -17
[Nd(tmp)$_8$]$^{3+}$ + tmp $\rightarrow$ [Nd(tmp)$_9$]$^{3+}$	tmp	25

[a] $\Delta \bar{V}$ determined from the pressure dependence of the equilibrium constant.
[b] $\Delta \bar{V}$ determined from dilatometric or partial molar volume (density) measurements.
[c] Theoretical predicted value.

(continued)

P (kbar)	No. of data	$\Delta V^{\ddagger}$ (cm^3 mol^{-1})	$\Delta\beta^{\ddagger}$ (cm^3 mol^{-1} kbar^{-1})	$\Delta\bar{V}$ (cm^3 mol^{-1})	Ref.	Remarks
1.0	5	-13.0 ± 1.0			96	
1.0	5	-13.5 ± 1.5			96	
1.0	5	-13.9 ± 1.0			96	
1.0	5	-20 ± 3			96	
1.0	5	-6 ± 8			96	
1.0	5	$-9.7 + 0.6$			96	
1.0	5	-12.3 ± 1.8			96	
				$-9.8 + 1.1^a$	97	
				-23.8 ± 1.5^a	97	

Table 15.3. Volumes of Activation

Reaction: Isomerization	Solvent	T (°C)
Racemization		
$(-)_{589}$-Ge(acac)$_3^-$	TCE	60
	PC	60
	CH$_3$CN	50/60
	DMF	50
$(-)_{546}$-[Co(pyrdtc)$_3$]	C$_2$H$_5$OH	43
	DMF	43
	CH$_3$CN	41
	Toluene	55
$(-)_{546}$-[Co(Ph$_2$-dtc)$_3$]	DMF	50
		70
	CHCl$_3$	55
	C$_6$H$_5$Cl	63
	CCl$_4$	69
	Acetone	50
$trans \rightarrow cis$-[Co(en)$_2$(X)H$_2$O]$^{n+}$		
X = H$_2$O	H$_2$O	46
X = H$_2$O	D$_2$O	48
X = H$_2$O	D$_2$O	46
X = CH$_3$COO$^-$	D$_2$O	45
X = CH$_3$COO$^-$	D$_2$O	45
X = SeO$_3$H$^-$	D$_2$O	40
$cis \rightarrow trans$-[Pt(PEt$_3$)$_2$(C$_6$H$_5$)X]		
X = Cl	CH$_3$OH	30
X = Br	CH$_3$OH	30
X = I	CH$_3$OH	30
[Co(NH$_3$)$_5$ONO]$^{2+}$ → [Co(NH$_3$)$_5$NO$_2$]$^{2+}$	H$_2$O	20
	H$_2$O	25
	MFA	35
	DMSO	36
	SFL	54
	NH$_3$	5
cis-[Co(en)$_2$(ONO)$_2$]$^+$ → cis-[Co(en)$_2$(NO$_2$)ONO]$^+$	H$_2$O	25
cis-[Co(en)$_2$(NO$_2$)ONO]$^+$ → cis-[Co(en)$_2$(NO$_2$)$_2$]$^+$	H$_2$O	35
	H$_2$O	35
$trans$-[Co(en)$_2$(ONO)$_2$]$^+$ → $trans$-[Co(en)$_2$(NO$_2$)$_2$]$^+$	H$_2$O	35
cis-[Co(en)$_2$(ONO)$_2$]$^+$ $\xrightarrow{\text{OH}^-}$ cis-[Co(en)$_2$(NO$_2$)$_2$]$^+$	H$_2$O	25
$trans$-[Co(en)$_2$(ONO)$_2$]$^+$ $\xrightarrow{\text{OH}^-}$ $trans$-[Co(en)$_2$(NO$_2$)$_2$]$^+$	H$_2$O	25
[Pd(1,1,7,7,-Et$_4$dien)SCN]$^+$ → [Pd(1,1,7,7-Et$_4$dien)NCS]$^+$	H$_2$O	30
	DMF	30
[Pd(4-Me-1,1,7,7-Et$_4$dien)SCN]$^+$ → [Pd(4-Me-1,1,7,7-Et$_4$dien)NCS]$^+$	H$_2$O	30
Inversion about S		
$trans$-[Pd($\overline{\text{SCH}_2\text{CMe}_2\text{CH}_2}$)$_2Br_2$]	CHCl$_3$	18
$trans$-[Pd($\overline{\text{SCH}_2\text{CH}_2\text{CH}_2}$)$_2Cl_2$]	CHCl$_3$	26
$trans$-[Pd($\overline{\text{S(CH}_2)_5}$)$_2Cl_2$]	CHCl$_3$	-3
$trans$-[Pd($\overline{\text{SCH}_2\text{CMe}_2\text{CH}_2}$)$_2Cl_2$]	CH$_2$Cl$_2$	

P (kbar)	No. of data	$\Delta V^{\ddagger}$ (cm^3 mol^{-1})	$\Delta \beta^{\ddagger}$ (cm^3 mol^{-1} kbar^{-1})	$\Delta \bar{V}$ (cm^3 mol^{-1})	Ref.	Remarks
2.8	7	$+15.1 \pm 0.8$			98	
2.8	7	$+5.4 \pm 0.3$			98	
2.8	4	$0.0 + 0.2$			98	
2.8	7	-4.1 ± 0.6			98	
1.4	5	$+9.8 \pm 0.5$	$+4.0 \pm 0.9$		99	
1.4	5	$+5.2 \pm 0.7$	$+2.2 \pm 1.0$		99	
1.4	5	$+5.4 \pm 0.5$	$+2.5 \pm 0.8$		99	
1.4	5	$+7.8 \pm 0.6$	$+3.1 \pm 0.9$		99	
2.6	3	-2.0			99	
2.6	2	-5.7			99	
2.6	2	-6.6			99	
2.6	2	-9.3			99	
2.6	2	-5.2			99	
2.6	2	-6.8			99	
1.4	5	$+12.4 \pm 0.4$	$+5 \pm 0.5$		100	0.05 M HClO$_4$, $\mu = 1\ M$
1.4	5	$+10.6 \pm 0.2$			100	0.05 M DClO$_4$
1.4	5	$+11.9 \pm 0.2$	$+4.3 \pm 0.3$		100	0.05 M DClO$_4$, $\mu = 1\ M$
1.4	5	$+3.7 + 0.3$			100	0.05 M DClO$_4$
1.4	5	$+2.5 \pm 0.3$			100	0.05 M DClO$_4$, $\mu = 1\ M$
1.4	5	0.0 ± 0.4			100	pD = 1.4, 0.5 M Na$_2$SeO$_3$
1.5	6	$+6.4 \pm 0.4$			101	$\mu = 0.01\ M$ (ClO$_4^-$)
1.5	6	$+5.2 \pm 0.3$			101	$\mu = 0.01\ M$ (ClO$_4^-$)
1.5	6	$+7.2 \pm 0.8$			101	$\mu = 0.01\ M$ (ClO$_4^-$)
1.7	7	$+27 \pm 1.4$	$+5 \pm 2$		102	0.1 M NaOH
1.4	4	-6.5 ± 0.2			103	$\mu \sim 0$
1.4	5	-5.7 ± 0.5			103	$\mu \sim 0$
1.4	5	-3.6 ± 0.3			103	$\mu \sim 0$
1.4	5	-6.8 ± 0.7			103	$\mu \sim 0$
4.0	7	-16 ± 2			26	0.001 M NH$_4$NO$_3$, $\mu = 0.2\ M$ (NO$_3^-$)
1.0	6	-5.6 ± 0.6			104	$\mu = 0.5\ M$ (ClO$_4^-$)
1.5	4	-6.9 ± 0.6			104	$\mu = 0.5\ M$ (ClO$_4^-$)
1.5	4	-3.4 ± 0.2			104	$\mu = 0.5\ M$ (ClO$_4^-$)
1.7	8	-3.6 ± 0.4			104	$\mu = 0.5\ M$ (ClO$_4^-$)
1.0	6	$+19.7 \pm 1.1$			105	$\mu = 0.5\ M$ (ClO$_4^-$)
1.0	6	$+13.6 \pm 1.2$			105	$\mu = 0.5\ M$ (ClO$_4^-$)
1.5	6	-10.1 ± 0.3			95	$\mu = 0.1\ M$
2.0	8	-9.5 ± 0.5			95	$\mu = 0.1\ M$
1.5	6	-10.8 ± 0.3			95	$\mu = 0.5\ M$
2.2		0 ± 2			106	
2.2		0 ± 2			106	
2.2		0 ± 2			106	
2.2		Small positive			106	

Table 15.4. *Volumes of Activation*

Reaction: Electron transfer	Solvent	T (°C)
$[VO(nta)H_2O]^- + H_2O_2 \rightarrow [VO(O_2)(nta)]^{2-} + H_3O^+$	H_2O	25
$[VO(pda)H_2O]^- + H_2O_2 \rightarrow [VO(O_2)(pda)]^{2-} + H_3O^+$	H_2O	25
$[Co(NH_3)_5N_3]^{2+} + Fe^{2+} \rightarrow$	H_2O	40
	DMSO	40
cis-$[Co(en)_2Cl_2]^+ + Fe^{2+} \rightarrow$	DMSO	30
$[Co(NH_3)_5H_2O]^{3+} + [Fe(CN)_6]^{4-} \rightarrow$	H_2O	25
$[Co(NH_3)_5OSO_2]^+ \xrightarrow{H^+} Co^{2+} + 5NH_4^+ + SO_3$	H_2O	25
	H_2O	14.4
$[Mo_2O_4(edta)]^{2-} + [(NH_3)_5Co(\mu\text{-}O_2^-)Co(NH_3)_5]^{5+} \rightarrow$	H_2O	41

* $\Delta \bar{V}$ for ion-pair formation.

Table 15.5. *Volumes of Activation*

Reaction: Miscellaneous	Solvent	T (°C)
Addition		
$[Co(NH_3)_5OH]^{2+} + NO^+ \rightarrow [Co(NH_3)_5ONO]^{2+} + H^+$	H_2O	25
$[Co(NH_3)_5OH]^{2+} + CO_2 \rightarrow [Co(NH_3)_5OCO_2]^+ + H^+$	H_2O	25
$[Rh(NH_3)_5OH]^{2+} + CO_2 \rightarrow [Rh(NH_3)_5OCO_2]^+ + H^+$	H_2O	25
$[Ir(NH_3)_5OH]^{2+} + CO_2 \rightarrow [Ir(NH_3)_5OCO_2]^+ + H^+$	H_2O	25
$[Ir(P(OPh)_3)_2(CO)Cl] + HCl \rightarrow$	Toluene	25
$[Ir(cod)(phen)]^+ + O_2 \rightarrow [Ir(cod)(phen)O_2]^+$	CH_3OH	40
$[Ir(cod)(phen)I] + O_2 \rightarrow [Ir(cod)(phen)O_2]^+ + I^-$	CH_3OH	25
Elimination		
$[Co(NH_3)_5OCO_2H]^{2+} \rightarrow [Co(NH_3)_5OH]^{2+} + CO_2$	H_2O	25
cis-$[Co(en)_2(H_2O)OCO_2H]^{2+} \rightarrow$ *cis*-$[Co(en)_2(H_2O)OH]^{2+} + CO_2$	H_2O	25
α-$[Co(edda)(H_2O)OCO_2H] \rightarrow \alpha$-$[Co(edda)(H_2O)OH] + CO_2$	H_2O	25
β-$[Co(edda)(H_2O)OCO_2H] \rightarrow \beta$-$[Co(edda)(H_2O)OH] + CO_2$	H_2O	25
$[Co(nta)(H_2O)OCO_2H] \rightarrow [Co(nta)(H_2O)OH] + CO_2$	H_2O	25
$[Rh(NH_3)_5OCO_2H]^{2+} \rightarrow [Rh(NH_3)_5OH]^{2+} + CO_2$	H_2O	25
$[Ir(NH_3)_5OCO_2H]^{2+} \rightarrow [Ir(NH_3)_5OH]^{2+} + CO_2$	H_2O	25

for Electron Transfer Reactions

P (kbar)	No. of data	$\Delta V^{\ddagger}$ ($cm^3\ mol^{-1}$)	$\Delta \beta^{\ddagger}$ ($cm^3\ mol^{-1}\ kbar^{-1}$)	$\Delta \bar{V}$ ($cm^3\ mol^{-1}$)	Ref.	Remarks
1.2	6	-10.5 ± 0.3			107	$\mu = 1\ M\ (ClO_4^-)$
1.2	6	-13.0 ± 0.5			107	$\mu = 1\ M\ (ClO_4^-)$
1.5	4	$+12.1 \pm 0.5$			108	$[H^+] = 0.1\ M, \mu = 0.3\ M$
1.5	4	$+6.5 \pm 0.2$			108	$[H^+] = 0.1\ M, \mu = 0.3\ M$
1.2	4	-9.2 ± 5.5		$+10.9 \pm 5.6^*$	109	$[H^+] = 0.02\ M, \mu = 0.3\ M$
1.0	5	$+26.5 \pm 2.4$		$-15.4 \pm 7.9^*$	110	pH = 4.7, $\mu = 0.05\ M$
1.0	5	$+34.4 \pm 2.9$			111	pH = 6.2, $\mu = 1\ M$
1.0	5	$+34.6 \pm 0.5$			111	pH = 6.2, $\mu = 1\ M$
2.9	10	$+12.1 \pm 1.9$	-6.2 ± 2.2	$+23.7 \pm 3.0^*$	112	0.1 M $HClO_4$

for Addition and Elimination Reactions

P (kbar)	No. of data	$\Delta V^{\ddagger}$ ($cm^3\ mol^{-1}$)	$\Delta \beta^{\ddagger}$ ($cm^3\ mol^{-1}\ kbar^{-1}$)	$\Delta \bar{V}$ ($cm^3\ mol^{-1}$)	Ref.	Remarks
1.5	7	-1.8 ± 0.7			113	pH = 3.7, $\mu = 2\ M$
1.0	5	-10.1 ± 0.6			114	pH = 8.5, $\mu = 0.5\ M$
1.0	5	-4.7 ± 0.8			114	pH = 8.5, $\mu = 0.5\ M$
1.0	5	-4.0 ± 1.0			114	pH = 8.5, $\mu = 0.5\ M$
1.0	5	-20.5 ± 1.4			115	
1.5	4	-31.1 ± 1.7			116	0.01 M LiCl
0.7	4	-44.4 ± 1.6			116	0.5 M LiI
1.0	5	$+6.8 \pm 0.3$			114	$[H^+] = 0.1\ M, \mu = 0.5\ M$
1.0	5	-1.1 ± 1.2			84	$[H^+] = 2\ M, \mu = 3\ M$
1.0	5	-0.7 ± 2.4			84	$[H^+] = 1\ M, \mu = 2\ M$
1.0	5	$+0.3 \pm 1.4$			84	$[H^+] = 2\ M, \mu = 2\ M$
1.0	5	-1.5 ± 1.7			84	$[H^+] = 2\ M, \mu = 3\ M$
1.0	5	$+5.2 \pm 0.3$			114	$[H^+] = 0.1\ M, \mu = 0.5\ M$
1.0	5	$+2.5 \pm 0.4$			114	$[H^+] = 0.1\ M, \mu = 0.5\ M$

Table 15.6. Volumes of Activation

Reaction: Photochemical	Solvent	T (°C)
$[Cr(NH_3)_5Cl]^{2+} + H_2O \rightarrow$ *cis*-$[Cr(NH_3)_4(Cl)H_2O]^{2+} + NH_3$	H_2O	20
$\rightarrow [Cr(NH_3)_5H_2O]^{3+} + Cl^-$	H_2O	20
$[Cr(NH_3)_5Br]^{2+} + H_2O \rightarrow$ *cis*-$[Cr(NH_3)_4(Br)H_2O]^{2+} + NH_3$	H_2O	5
$\rightarrow [Cr(NH_3)_5H_2O]^{3+} + Br^-$	H_2O	5
$[Cr(NH_3)_5NCS]^{2+} + H_2O \rightarrow$ *cis*-$[Cr(NH_3)_4(NCS)H_2O]^{2+} + NH_3$	H_2O	15
$\rightarrow [Cr(NH_3)_5H_2O]^{3+} + NCS^-$	H_2O	15
$[Cr(NH_3)_6]^{3+} + H_2O \rightarrow [Cr(NH_3)_5H_2O]^{3+} + NH_3$	H_2O	15
$[Cr(NCS)_6]^{3-} + H_2O \rightarrow [Cr(NCS)_5H_2O]^{2-} + NCS^-$	H_2O	15
$[Cr(CN)_6]^{3-} + H_2O \rightarrow [Cr(CN)_5H_2O]^{2-} + CN^-$	H_2O	15
$[Co(NH_3)_5Br]^{2+} + H_2O \rightarrow [Co(NH_3)_5H_2O]^{3+} + Br^-$	H_2O	20
$\rightarrow Co^{2+} + 5NH_3 + Br$	H_2O	20
	H_2O	20
$[Co(CN)_6]^{3-} + H_2O \rightarrow [Co(CN)_5H_2O]^{2-} + CN^-$	H_2O	15
$[Ru(bipy)_3^{2+}]^* + Q \rightarrow [Ru(bipy)_3]^{3+} + Q^-$		
$Q = [Fe(CN)_6]^{3-}$	H_2O	25
$Q = [(en)_2Co(\mu\text{-}NH_2)(\mu\text{-}O_2)Co(en)_2]^{4+}$	H_2O	25
$Q = [(NH_3)_5Co(\mu\text{-}O_2)Co(NH_3)_5]^{5+}$	H_2O	25
$Q = [(CN)_5Co(\mu\text{-}O_2)Co(CN)_5]^{5-}$	H_2O	25
$Q = Tl^{3+}$	H_2O	25
$[Ru(bipy)_3^{2+}]^* + Q \rightarrow [Ru(bipy)_3]^+ + Q^+$		
$Q = [Fe(CN)_6]^{4-}$	H_2O	25
$Q = [Mo(CN)_6]^{4-}$	H_2O	25
$Q = [Os(CN)_6]^{4-}$	H_2O	25
$Q = [Ir(Cl)_6]^{3-}$	H_2O	25
$Q = [Eu(aq)]^{2+}$	H_2O	25
$Q = [Fe(H_2O)_6]^{2+}$	H_2O	25
$[Ru(bipy)_3^{2+}]^* + Q \rightarrow [Ru(bipy)_3]^{2+} + Q^*$		
$Q = [Co(NH_3)_5H_2O]^{3+}$	H_2O	25
cis-$[Rh(NH_3)_4Cl_2]^+ \rightarrow$ *trans*-$[Rh(NH_3)_4(H_2O)Cl]^{2+} + Cl^-$	H_2O	25
cis-$[Rh(NH_3)_4(H_2O)Cl]^{2+} \rightarrow$ *trans*-$[Rh(NH_3)_4(H_2O)Cl]^{2+}$	H_2O	25
cis-$[Rh(NH_3)_4Br_2]^+ \rightarrow$ *trans*-$[Rh(NH_3)_4(H_2O)Br]^{2+} + Br^-$	H_2O	25
cis-$[Rh(NH_3)_4Br_2]^+ \rightarrow$ "$[Rh(NH_3)_3(H_2O)Br_2]^+$" $+ NH_3$	H_2O	25
cis-$[Rh(NH_3)_4(H_2O)Br]^{2+} \rightarrow$ *trans*-$[Rh(NH_3)_4(H_2O)Br]^{2+}$	H_2O	25
trans-$[Rh(NH_3)_4(OH)Cl]^+ \rightarrow$ *cis*-$[Rh(NH_3)_4(OH)_2]^+ + Cl^-$	H_2O	25
trans-$[Rh(NH_3)_4Cl_2]^+ \rightarrow$ *trans*-$[Rh(NH_3)_4(H_2O)Cl]^{2+} + Cl^-$	H_2O	25
trans-$[Rh(NH_3)_4Br_2]^+ \rightarrow$ *trans*-$[Rh(NH_3)_4(H_2O)Br]^{2+} + Br^-$	H_2O	25
$[Rh(NH_3)_5Cl]^{2+} + H_2O \rightarrow [Rh(NH_3)_5H_2O]^{3+} + Cl^-$	H_2O	25
$\rightarrow$ *trans*-$[Rh(NH_3)_4(H_2O)Cl]^{2+} + NH_3$	H_2O	25
$[Rh(ND_3)_5Cl]^{2+} + D_2O \rightarrow [Rh(ND_3)_5D_2O]^{3+} + Cl^-$	D_2O	25
$\rightarrow$ *trans*-$[Rh(ND_3)_4(D_2O)Cl]^{2+} + ND_3$	D_2O	25
$[Rh(NH_3)_5Br]^{2+} + H_2O \rightarrow [Rh(NH_3)_5H_2O]^{3+} + Br^-$	H_2O	25
$\rightarrow$ *trans*-$[Rh(NH_3)_4(H_2O)Br]^{2+} + NH_3$	H_2O	25
$[Rh(ND_3)_5Br]^{2+} + D_2O \rightarrow [Rh(ND_3)_5D_2O]^{3+} + Br^-$	D_2O	25
$\rightarrow$ *trans*-$[Rh(ND_3)_4(D_2O)Br]^{2+} + ND_3$	D_2O	25

for Photochemical Reactions

P (kbar)	No. of data	$\Delta V^{\ddagger}$ (cm^3 mol^{-1})	$\Delta\beta^{\ddagger}$ (cm^3 mol^{-1} kbar^{-1})	$\Delta\bar{V}$ (cm^3 mol^{-1})	Ref.	Remarks
1.5	4	-9.4 ± 0.4			117	$[H^+] = 0.01\ M,\ \mu = 0.5\ M$
1.5	4	-13.0 ± 0.5			117	$[H^+] = 0.01\ M,\ \mu = 0.5\ M$
1.5	4	-10.2 ± 0.1			117	$[H^+] = 0.01\ M,\ \mu = 0.5\ M$
1.5	4	-12.2 ± 0.3			117	$[H^+] = 0.01\ M,\ \mu = 0.5\ M$
1.5	4	-11.4 ± 0.1			117	$[H^+] = 0.01\ M,\ \mu = 0.5\ M$
1.5	4	-9.8 ± 0.2			117	$[H^+] = 0.01\ M,\ \mu = 0.5\ M$
1.0	5	-12.6 ± 0.5			118	$[H^+] = 0.01\ M,\ \mu = 0.01\ M$
1.0	5	$+2.9 \pm 0.6$			118	$[H^+] = 0.1\ M,\ \mu = 0.5\ M$
1.5	7	$+3.0 \pm 0.2$			118	pH $= 8.9,\ \mu = 0.3\ M$
2.8	2	-0.4 ± 3.2			119	pH $= 4$, buffer
2.8	2	$+6.0 \pm 0.6$			119	pH $= 4$, buffer
2.8	2	$+4.8 \pm 0.3$			119	pH $= 4$, buffer
1.5	7	$+2.0 \pm 0.2$			118	pH $= 6,\ \mu \sim 0$
3.0			0.0 ± 0.1		120	0.1 M HCl, $\mu = 0.25\ M$
3.0			0 to $+1.0$		120	0.05 M H$_2$SO$_4$, $\mu = 0.25\ M$
3.0			$+1 \pm 2$		120	0.05 M H$_2$SO$_4$, $\mu = 0.25\ M$
3.0			0 to $+1.2$		120	0.1 M HCl, $\mu = 0.25\ M$
3.0			$+0.2 \pm 0.1$		120	4.75 M HClO$_4$
3.0				0.0 ± 0.5	120	pH $= 5.2,\ \mu = 0.5\ M$
3.0				$+24.7 \pm 0.6$	120	0.1 M HCl, $\mu = 0.25\ M$
3.0				$+6.8 \pm 2.0$	120	0.1 M HCl, $\mu = 0.25\ M$
3.0				$+1.1 \pm 0.5$	120	0.1 M HCl, $\mu = 0.25\ M$
3.0				-11.0 ± 1.0	120	0.5 M HClO$_4$
3.0				-0.6 ± 0.6	120	0.05 M H$_2$SO$_4$, $\mu = 0.25\ M$
3.0				-2.6 ± 0.6	120	0.05 M H$_2$SO$_4$, $\mu = 0.25\ M$
2.0	5	-3.5 ± 0.3			121	10^{-2}–$10^{-3}\ M$ HClO$_4$
2.0	5	0.0 ± 0.4			121	$10^{-3}\ M$ HClO$_4$
2.0	5	-2.3 ± 0.3			121	$10^{-3}\ M$ HClO$_4$
2.0	5	$+9.3 \pm 0.8$			121	$10^{-3}\ M$ HClO$_4$
2.0	5	-1.0 ± 0.4			121	$10^{-3}\ M$ HClO$_4$
2.0	5	-8.8 ± 0.7			121	0.1 M NaOH
2.0	5	$+2.8 \pm 0.6$			121	$10^{-3}\ M$ HClO$_4$
2.0	5	$+2.9 \pm 0.7$			121	$10^{-3}\ M$ HClO$_4$
2.0	9	-8.6 ± 1.6			25	10^{-2}–$10^{-3}\ M$ HClO$_4$
2.0	9	$+9.3 \pm 1.9$			25	10^{-2}–$10^{-3}\ M$ HClO$_4$
2.0	9	-7.7 ± 1.6			25	10^{-2}–$10^{-3}\ M$ DClO$_4$
2.0	9	$+6.0 \pm 2.2$			25	10^{-2}–$10^{-3}\ M$ DClO$_4$
2.0	9	-6.8 ± 1.6			25	10^{-2}–$10^{-3}\ M$ HClO$_4$
2.0	9	$+8.1 \pm 1.2$			25	10^{-2}–$10^{-3}\ M$ HClO$_4$
2.0	8	-5.3 ± 1.8			25	10^{-2}–$10^{-3}\ M$ DClO$_4$
2.0	8	-7.5 ± 1.1			25	10^{-2}–$10^{-3}\ M$ DClO$_4$

continued

Table 15.6.

Reaction: Photochemical	Solvent	T (°C)
Phosphorescence lifetime		
[Cr(bpy)₃]³⁺	H_2O	
[Cr(en)₃]³⁺	H_2O	
[Ru(bipy)₃]²⁺	H_2O	
[Rh(ND₃)₅Cl]²⁺	D_2O	25
[Rh(ND₃)₅Br]²⁺	D_2O	25
Nonradiative deactivation		
[Rh(ND₃)₅Cl]²⁺	D_2O	25
[Rh(ND₃)₅Br]²⁺	D_2O	25

(continued)

P (kbar)	No. of data	$\Delta V^{\ddagger}$ (cm^3 mol^{-1})	$\Delta \beta^{\ddagger}$ (cm^3 mol^{-1} kbar^{-1})	$\Delta \bar{V}$ (cm^3 mol^{-1})	Ref.	Remarks
2.3		-0.7 ± 0.2			24	
2.3		-0.9 ± 0.2			24	
2.3		-1.6 ± 1.0			24	
2.8	6	$+3.5 \pm 1.1$			25	10^{-3} M DCl
2.0	6	-4.1 ± 0.6			25	10^{-3} M DCl
2.0		-2.6 ± 1.0			25	
2.0		$+2.5 \pm 1.2$			25	

References

References for Chapter 1

1. S. J. Lippard, ed., *Progr. Inorg. Chem.*, **30**, (1983).
2. I. I. Creaser, R. J. Geue, J. MacB. Harrowfield, A. J. Herlt, A. M. Sargeson, M. R. Snow, and J. Springborg, *J. Am. Chem. Soc.*, **104**, 6016 (1982).
3. R. V. Dubs, L. R. Gahan, and A. M. Sargeson, *Inorg. Chem.*, **22**, 2523 (1983).
4. A. Hammershøi and A. M. Sargeson, *Inorg. Chem.*, **22**, 3554 (1983).
4a. I. I. Creaser, A. M. Sargeson, and A. W. Zanella, *Inorg. Chem.*, **22**, 4022 (1983).
5. J. F. Endicott, G. R. Brubaker, T. Ramasami, K. Kumar, K. Dwarakanath, J. Cassel, and D. Johnson, *Inorg. Chem.*, **22**, 3754 (1983).
6. D. Geselowitz, *Inorg. Chem.*, **20**, 4457 (1981).
7. B. S. Brunschwig, C. Creutz, D. H. Macartney, T. K. Sham, and N. Sutin, *Discussions Faraday Soc.*, **74**, 113 (1982).
8. D. J. Szalda, C. Creutz, D. Mahajan, and N. Sutin, *Inorg. Chem.*, **22**, 2372 (1983).
9. F. Moattar, J. R. Walton, and L. E. Bennett, *Inorg. Chem.*, **22**, 550 (1982).
10. H. Cohen, S. Efrima, D. Mayerstein, M. Nutkovich, and K. Wieghardt, *Inorg. Chem.*, **22**, 688 (1983).
11. K. W. Frese, Jr., *J. Phys. Chem.* **85**, 3911 (1981).
12. C.-W. Lee and F. C. Anson, *J. Phys. Chem.*, **87**, 3362 (1983).
13. P. Delahay, *Chem. Phys. Lett.*, **89**, 149 (1982).
14. M. A. de Araujo and H. L. Hodges, *Inorg. Chem.*, **21**, 3167 (1982).
15. K. L. Rollick and J. K. Kochi, *J. Am. Chem. Soc.*, **104**, 1319 (1982).
16. F. Wilkinson and C. Tsiamis, *J. Am. Chem. Soc.*, **105**, 767 (1983).
17. B. W. Carlson and L. L. Miller, *J. Am. Chem. Soc.*, **105**, 7453 (1983).
17a. E. M. Kosower and D. Huppert, *Chem. Phys. Lett.*, **96**, 433 (1983).
18. D. F. Calef and P. G. Wolynes, *J. Phys. Chem.*, **87**, 3387 (1983).
19. D. F. Calef and P. G. Wolynes, *J. Chem. Phys.*, **78**, 470 (1983).
20. L. Salem, *Electrons in Chemical Reactions: First Principles*, Wiley, New York (1982).
21. N. Sutin, *Progr. Inorg. Chem.*, **30**, 440 (1983).
22. N. Sutin, *Acc. Chem. Res.*, **15**, 275 (1982).
23. K. Y. Wong and P. N. Schatz, *Progr. Inorg. Chem.*, **30**, 370 (1983).
24. S. F. Fischer, *Chem. Phys. Lett.*, **91**, 367 (1982).
25. A. M. Kuznetsov, *Chem. Phys. Lett.*, **91**, 34 (1982).

26. J. P. Dahl and J. Ulstrup, *Chem. Phys. Lett.*, **93**, 564 (1982).
27. J. Logan and M. D. Newton, *J. Chem. Phys.*, **78**, 4086 (1983).
28. D. H. Macartney and N. Sutin, *Inorg. Chem.*, **22**, 3530 (1983).
29. J. T. Hupp and M. J. Weaver, *Inorg. Chem.*, **22**, 2557 (1983).
30. E. L. Yee, J. T. Hupp, and M. J. Weaver, *Inorg. Chem.*, **22**, 3465 (1983).
31. J. F. Endicott and T. Ramasami, *J. Am. Chem. Soc.*, **104**, 5252 (1982).
32. L. I. Trakhtenberg, V. L. Klochikhin, and S. Ya. Pshezhetsky, *Chem. Phys.*, **69**, 121 (1982).
33. P. H. Cribb, S. Nordholm, and N. S. Hush, *Chem. Phys.*, **69**, 259 (1962).
34. V. I. Goldanskiĭ, *Pure Appl. Chem.*, **55**, 11 (1983).
35. J. R. Winkler, D. G. Nocera, K. M. Yocom, E. Bordignon, and H. B. Gray, *J. Am. Chem. Soc.*, **104**, 5798 (1982).
36. K. F. Freed, *Chem. Phys. Lett.*, **97**, 489 (1983).
37. R. J. Klingler and J. K. Kochi, *J. Am. Chem. Soc.*, **104**, 4186 (1982).
38. J. V. Caspar, E. M. Kober, B. P. Sullivan, and T. J. Meyer, *J. Am. Chem. Soc.*, **104**, 630 (1982).
39. B. T. Reagor, D. F. Kelley, D. H. Huchital, and P. M. Rentzepis, *J. Am. Chem. Soc.*, **104**, 7400 (1982).
40. E. Buhks, M. Bixon, and J. Jortner, *J. Phys. Chem.*, **85**, 3763 (1981).
40a. J. Ulstrup and J. Jortner, *J. Chem. Phys.*, **63**, 4358 (1975).
41. T. Guarr, E. Buhks, and G. McLendon, *J. Am. Chem. Soc.*, **105**, 3763 (1983).
42. A. I. Fiksel, V. N. Parmon, and K. I. Zamaraev, *Chem. Phys.*, **69**, 135 (1982).
43. L. J. Root and M. J. Ondrechen, *Chem. Phys. Lett.*, **93**, 421 (1982).
44. S. Larsson, *Chem. Phys. Lett.*, **90**, 136 (1982).
45. K. Kumar, F. P. Rotzinger, and J. F. Endicott, *J. Am. Chem. Soc.*, **105**, 7064 (1983).
46. K. A. Norton, Jr., and J. K. Hurst, *J. Am. Chem. Soc.*, **104**, 5960 (1982).
47. L. T. Calcaterra, G. L. Closs, and J. R. Miller, *J. Am. Chem. Soc.*, **105**, 670 (1983).
48. C. A. Stein, N. A. Lewis, and G. Seitz, *J. Am. Chem. Soc.*, **104**, 2596 (1982).
49. N. M. Kostić, R. Margalit, C.-M. Che, and H. B. Gray, *J. Am. Chem. Soc.*, **105**, 7765 (1983).
50. J. R. Winkler, D. G. Nocera, K. M. Yocom, E. Bordignon, and H. B. Gray, *J. Am. Chem. Soc.*, **104**, 5798 (1982).
51. S. S. Isied, G. Worosila, and S. J. Atherton, *J. Am. Chem. Soc.*, **104**, 7659 (1982).
52. A. M. Chang and R. H. Austin, *J. Chem. Phys.* **77**, 5272 (1982).
53. J. L. McGourty, N. V. Blough, and B. M. Hoffman, *J. Am. Chem. Soc.*, **105**, 4470 (1983).
54. A. Siemiarczuk, A. R. McIntosh, T.-F. Ho, M. J. Stillman, K. J. Roach, A. C. Weedon, J. R. Bolton, and J. S. Connolly, *J. Am. Chem. Soc.*, **105**, 7224 (1983).
55. A. R. McIntosh, A. Siemiarczuk, J. R. Bolton, M. J. Stillman, T.-F. Ho, and A. C. Weedon, *J. Am. Chem. Soc.*, **105**, 7216 (1983).
56. A. M. Kuznetsov and J. Ulstrup, *Chem. Phys. Lett.*, **97**, 285 (1983).
57. J. R. Miller, J. A. Peeples, M. J. Schmitt, and G. L. Closs, *J. Am. Chem. Soc.*, **104**, 6488 (1982).
58. S. Strauch, G. McLendon, M. McGuire, and T. Guarr, *J. Phys. Chem.*, **87**, 3579 (1983).
59. J. R. Miller, K. W. Hartmann, and S. Abrash, *J. Am. Chem. Soc.*, **104**, 4296 (1982).
60. J. N. Murrell, *Chem. Phys. Lett.*, **93**, 521 (1982).
61. F. A. Houle, S. L. Anderson, D. Gerlich, T. Turner, and Y. T. Lee, *J. Chem. Phys.*, **77**, 748 (1982).
62. H. S. W. Massey and E. H. S. Burhop, *Electronic and Ionic Impact Phenomena*, Oxford, University Press, (1952), p. 470. (Cf. R. D. Cannon, *Electron Transfer Reactions*, London (Butterworths, 1980), pp. 18 & 33.)
63. S. B. Sears and A. E. DePristo, *J. Chem. Phys.*, **77**, 290 (1982).
64. A. E. DePristo and S. B. Sears, *J. Chem. Phys.*, **77**, 298 (1982).

65. C. H. Becker, *J. Chem. Phys.*, **76**, 5928 (1982).
66. A. E. DePristo, *J. Chem. Phys.*, **78**, 1237 (1983).
67. S. B. Sears and A. E. DePristo, *J. Chem. Phys.*, **77**, 290 (1982).
68. A. E. DePristo and S. B. Sears, *J. Chem. Phys.*, **77**, 298 (1982).
69. A. E. DePristo, *J. Chem. Phys.*, **79**, 1741 (1983).
70. C.-Y. Lee and A. E. DePristo, *J. Am. Chem. Soc.*, **105**, 6775 (1983).
71. G. C. Shields and R. F. Moran, *Chem. Phys. Lett.*, **101**, 287 (1983).
72. E. E. Ferguson, *Chem. Phys. Lett.*, **99**, 89 (1983).
73. K. H. Bowen, G. W. Liesegang, R. A. Sanders, and D. R. Herschbach, *J. Phys. Chem.*, **87**, 557 (1983).
74. C. Creutz, *Progr. Inorg. Chem.*, **30**, 2 (1983).
75. U. Fürholz, H.-B. Bürgi, F. E. Wagner, A. Stebler, J. H. Ammeter, E. Krausz, R. J. H. Clark, M. J. Stead, and A. Ludi, *J. Am. Chem. Soc.*, **106**, 121 (1984).
76. C. Creutz and H. Taube, *J. Am. Chem. Soc.*, **95**, 1086 (1973).
77. S. S. Isied and H. Taube, *Inorg. Chem.*, **15**, 3070 (1976).
78. P. Day, *Int. Rev. Phys. Chem.*, **1**, 149 (1981).
79. R. H. Magnuson, P. A. Lay, and H. Taube, *J. Am. Chem. Soc.*, **105**, 2507 (1983).
80. D. E. Richardson and H. Taube, *J. Am. Chem. Soc.*, **105**, 40 (1983).
81. R. C. Long and D. N. Hendrickson, *J. Am. Chem. Soc.*, **105**, 1513 (1983).
82. E. M. Kober, K. A. Goldsby, D. N. S. Narayana, and T. J. Meyer, *J. Am. Chem. Soc.*, **105**, 4303 (1983).
83. J.-J. Girerd, *J. Chem. Phys.*, **79**, 1766 (1983).
84. C. Sanchez, J. Livage, J. P. Launay, M. Fournier, and Y. Jeannin, *J. Am. Chem. Soc.*, **104**, 3194 (1982).
85. S. P. Harmalker, M. A. Leparulo, and M. T. Pope, *J. Am. Chem. Soc.*, **105**, 4286 (1983).
86. J. P. Launay and F. Babonneau, *Chem. Phys.*, **67**, 295 (1982).
86a. S. A. Borshch, I. N. Kotov, and I. B. Bersuker, *Chem. Phys. Lett.*, **89**, 381 (1982).
87. R. D. Cannon, L. Montri, D. B. Brown, K. M. Marshall, and C. M. Elliott, *J. Am. Chem. Soc.*, **106**, 2591 (1984).
88. D. P. Rillema, R. W. Callahan, and K. B. Mack, *Inorg. Chem.*, **21**, 2589 (1982).
89. S. I. Amer, T. P. Dasgupta, and P. M. Henry, *Inorg. Chem.*, **22**, 1970 (1983).
90. K. J. Pfenning, L. Lee, H. D. Wohlers, and J. D. Petersen, *Inorg. Chem.*, **21**, 2477 (1982).
91. K. J. Moore, L. Lee, G. A. Mabbott, and J. D. Petersen, *Inorg. Chem.*, **22**, 1108 (1983).
92. K. Prassides and P. Day, *J. Chem. Soc., Faraday Trans. 2*, **80**, 85 (1984).
93. K. Ichikawa and W. W. Warren, *Phys. Rev. B* **20**, 900 (1979).
94. W. W. Warren, G. Schönherr, and F. Hensel, *Chem. Phys. Lett.*, **96**, 505 (1983).
95. A. R. Siedle, M. C. Etter, M. E. Jones, G. Filipovich, H. E. Mishmash, and W. Bahmet, *Inorg. Chem.*, **21**, 2624 (1982).
96. P. E. Fanwick and J. L. Huckaby, *Inorg. Chem.* **21**, 3067 (1982).
97. A. L. Beauchamp, D. Layek, and T. Theophanides, *Acta Crystallogr., Sect. B* **38**, 1901 (1982).
98. M. Tanaka, I. Tsujikawa, K. Toriumi, and T. Ito, *Acta Crystallogr., Sect. B* **38**, 2793 (1982).
99. M. Yamashita, H. Ito, K. Toriumi, and T. Ito, *Inorg. Chem.*, **22**, 1566 (1983).
100. C.-M. Che, F. H. Herbstein, W. P. Schaefer, R. E. Marsh, and H. B. Gray, *J. Am. Chem. Soc.*, **105**, 4604 (1983).
101. S. Ahmad, R. J. H. Clark, and M. Kurmoo, *J. Chem. Soc., Dalton Trans.*, 1371 (1982).
102. R. J. H. Clark, M. Kurmoo, and D. N. Mountney, *J. Chem. Soc., Dalton Trans.*, 1851 (1982).
103. R. J. H. Clark, M. Kurmoo, A. M. R. Galas, and M. B. Hursthouse, *J. Chem. Soc., Dalton Trans.*, 2505 (1982).
104. R. J. H. Clark and M. Kurmoo, *J. Chem. Soc., Dalton Trans.*, 2545 (1982).

105. M. Bhaduri, *J. Chem. Phys.*, **77**, 1400 (1982).
106. P. G. Pickup and R. W. Murray, *J. Am. Chem. Soc.*, **105**, 4510 (1983).
107. J. M. Williams, *Adv. Inorg. Chem. Radiochem.*, **26**, 235 (1983).
108. J. Martinsen, L. J. Pace, T. E. Phillips, B. M. Hoffman, and J. A. Ibers, *J. Am. Chem. Soc.*, **104**, 83 (1982).
109. T. K. Sham, *J. Chem. Phys.*, **79**, 1116 (1983).
110. P. Bernhard, H.-B. Burgi, J. Hauser, H. Lehmann, and A. Ludi, *Inorg. Chem.*, **21**, 3936 (1982).
111. A. Kojima, K. Okazaki, S. Ooi, and K. Saito, *Inorg. Chem.*, **22**, 1168 (1983).
112. P. Delahay, *Chem. Phys. Lett.*, **90**, 425 (1982).
113. P. Delahay, *Chem. Phys. Lett.*, **96**, 613 (1983).
114. S. U. M. Khan and J. O'M. Bockris, *Chem. Phys. Lett.*, **99**, 83 (1983).
115. P. Delahay, *Chem. Phys. Lett.*, **99**, 87 (1983).
116. S. U. M. Khan and J. O'M. Bockris, *J. Phys. Chem.*, **87**, 4012 (1983).
117. N. Dowling and P. M. Henry, *Inorg. Chem.*, **21**, 4088 (1982).
118. H. Krentzien and H. Taube, *Inorg. Chem.*, **21**, 4001 (1982).
119. J. E. Sutton, H. Krentzien, and H. Taube, *Inorg. Chem.*, **21**, 2842 (1982).
120. R. R. Ruminski and J. D. Petersen, *Inorg. Chem.*, **21**, 3706 (1982).
121. D. N. Marks, W. O. Siegl, and R. R. Gagné, *Inorg. Chem.*, **21**, 3140 (1982).
122. P. Blanc, C. Madic, and J. P. Launay, *Inorg. Chem.*, **21**, 2923 (1982).

References for Chapter 2

1. D. Geselowitz and H. Taube, in: *Advances in Inorganic and Bioinorganic Mechanisms* (A. G. Sykes, ed.), Vol. 1, p. 391, Academic Press, London (1982).
2. F. Armstrong, in: *Advances in Inorganic and Bioinorganic Mechanisms* (A. G. Sykes, ed.), Vol. 1, p. 65, Academic Press, London (1982).
3. C. Creutz, *Prog. Inorg. Chem.*, **30**, 1 (1983).
4. J. F. Endicott, K. Kumar, T. Ramasami, and F. P. Rotzinger, *Prog. Inorg. Chem.*, **30**, 141 (1983).
5. A. Haim, *Prog. Inorg. Chem.*, **30**, 273 (1983).
6. T. J. Meyer, *Prog. Inorg. Chem.*, **30**, 389 (1983).
7. N. Sutin, *Prog. Inorg. Chem.*, **30**, 441 (1983).
8. D. B. Rorabacher and J. F. Endicott (eds.), Mechanistic Aspects of Inorganic Reactions, American Chemical Society Symposium Series 198, Washington, D.C. (1982).
9. *Faraday Discuss. Chem. Soc.* No. 74, (1983).
10. N. Sutin, *Acc. Chem. Res.* **15**, 275 (1982).
11. I. I. Creaser, R. J. Geue, J. M. Harrowfield, A. J. Herlt, A. M. Sargeson, M. R. Snow, and J. Springborg, *J. Am. Chem. Soc.*, **104**, 6016 (1982).
12. J. R. Winkler, D. G. Nocera, K. M. Yocom, E. Bordignon, and H. B. Gray, *J. Am. Chem. Soc.*, **104**, 5798 (1982).
13. S. S. Isied, G. Worosila, and S. J. Atherton, *J. Am. Chem. Soc.*, **104**, 7659 (1982).
14. N. M. Kostic, R. Margalit, C.-M. Che, and H. B. Gray, *J. Am. Chem. Soc.*, **105**, 7765 (1983).
15. J. E. Sutton, H. Krentzien, and H. Taube, *Inorg. Chem.*, **21**, 2842 (1982).
16. D. E. Richardson, J. P. Sen, J. D. Bohr, and H. Taube, *Inorg. Chem.*, **21**, 3136 (1982).
17. H. Krentzien and H. Taube, *Inorg. Chem.*, **21**, 4001 (1982).
18. D. P. Rillema, R. W. Callahan, and K. B. Mack, *Inorg. Chem.*, **21**, 2589 (1982).
19. C. A. Stein, N. A. Lewis, G. Seitz, and A. D. Baker, *Inorg. Chem.*, **22**, 1124 (1983).
20. C. A. Stein, N. A. Lewis, and G. Seitz, *J. Am. Chem. Soc.*, **104**, 2596 (1982).

21. N. Dowling and P. M. Henry, *Inorg. Chem.*, **21**, 4088 (1982).
22. B. Anderes and D. K. Lavallee, *Inorg. Chem.*, **22**, 2665 (1983).
23. B. T. Reagor, D. F. Kelley, D. H. Huchital, and P. M. Remtzepis, *J. Am. Chem. Soc.*, **104**, 7400 (1982).
24. K. A. Norton and J. K. Hurst, *J. Am. Chem. Soc.*, **104**, 5960 (1982).
25. H. Cohen, E. S. Gould, D. Meyerstein, M. Nutkovich, and C. A. Radlowski, *Inorg. Chem.*, **22**, 1374 (1983).
26. H. A. Boucher, G. A. Lawrance, A. M. Sargeson, and D. F. Sangster, *Inorg. Chem.*, **22**, 3482 (1983).
27. P. Blanc, C. Madic, and J. P. Launay, *Inorg. Chem.*, **21**, 2923 (1982).
28. N. Aoi, G. Matsubayashi, and T. Tanaka, *J. Chem. Soc., Dalton Trans.* 1059 (1983).
29. H. Cohen, S. Efrima, D. Meyerstein, M. Nutkovich, and K. Weighardt, *Inorg. Chem.*, **22**, 688 (1983).
30. J. E. Earley, R. N. Bose, and B. H. Berrie, *Inorg. Chem.*, **22**, 1836 (1983).
31. R. A. Kelley and J. E. Earley, *Inorg. Chim. Acta*, **76**, L167 (1983).
32. R. N. Bose and J. E. Earley, *J. Chem. Soc., Chem. Commun.*, 59 (1983).
33. M. S. Ram, A. H. Martin, and E. S. Gould, *Inorg. Chem.*, **22**, 1103 (1983).
34. R. J. Balahura and A. J. Johnston, *Inorg. Chem.*, **22**, 3309 (1983).
35. G. Ali and N. A. Lewis, *J. Chem. Soc., Chem. Commun.*, 715 (1982).
36. R. J. Balahura, W. L. Purcell, M. E. Victoriano, M. L. Lieberman, V. M. Loyola, W. Fleming, and J. W. Fronabarger, *Inorg. Chem.*, **22**, 3602 (1983).
37. R. J. Balahura, W. C. Kupferschmidt, and W. L. Purcell, *Inorg. Chem.*, **22**, 1456 (1983).
38. J. L. Laird and R. B. Jordan, *Inorg. Chem.*, **21**, 4127 (1982).
39. N. Rajasekar, V. S. Srinivasan, A. N. Singh, and E. S. Gould, *Inorg. Chem.*, **21**, 3245 (1982).
40. V. S. Srinivasan, A. N. Singh, C. A. Radlowski, and E. S. Gould, *Inorg. Chem.*, **21**, 1240 (1982).
41. V. S. Srinivasan and E. S. Gould, *Inorg. Chem.*, **21**, 3854 (1982).
42. V. S. Srinivasan, A. N. Singh, K. Weighardt, N. Rajasekar, and E. S. Gould, *Inorg. Chem.*, **21**, 2531 (1982).
43. V. S. Srinivasan, N. Rajasekar, A. N. Singh, C. A. Radlowski, J. C.-K. Heh, and E. S. Gould, *Inorg. Chem.*, **21**, 2824 (1982).
44. A. M. Bond, G. A. Lawrance, P. A. Lay, and A. M. Sargeson, *Inorg. Chem.*, **22**, 2010 (1983).
45. L. R. Gahan, T. W. Hambley, A. M. Sargeson, and M. R. Snow, *Inorg. Chem.*, **21**, 2699 (1982).
46. R. V. Dubs, L. R. Gahan, and A. M. Sargeson, *Inorg. Chem.*, **22**, 2523 (1983).
47. A. Hammershoi and A. M. Sargeson, *Inorg. Chem.*, **22**, 3554 (1983).
48. I. I. Creaser, A. M. Sargeson, and A. W. Zanella, *Inorg. Chem.*, **22**, 4022 (1983).
49. J. F. Endicott, G. R. Brubaker, T. Ramasami, K. Kumar, K. Dwarakanath, J. Cassel, and D. Johnson, *Inorg. Chem.*, **22**, 3754 (1983).
50. D. Geselowitz, *Inorg. Chem.*, **20**, 4457 (1981).
51. J. F. Endicott, B. Durham, M. D. Glick, T. J. Anderson, J. M. Kuszaj, W. G. Schmonsees, and K. P. Balakrishnan, *J. Am. Chem. Soc.*, **103**, 1431 (1981).
52. V. Houlding, T. Geiger, V. Kolle, and M. Grätzel, *J. Chem. Soc., Chem. Commun.*, 681 (1982).
53. M. A. R. Scandola, F. Scandola, A. Indelli, and V. Balzani, *Inorg. Chim. Acta*, **76**, L67 (1983).
54. P. A. Lay, A. W. H. Mau, W. H. F. Sasse, I. I. Creaser, L. R. Gahan, and A. M. Sargeson, *Inorg. Chem.*, **22**, 2347 (1983).
55. J. F. Endicott and T. Ramasami, *J. Am. Chem. Soc.*, **104**, 5252 (1982).
56. H. A. Boucher, G. A. Lawrance, P. A. Lay, A. M. Sargeson, A. M. Bond, D. F. Sangster, and J. C. Sullivan, *J. Am. Chem. Soc.*, **105**, 4652 (1983).

57. T. Ramasami, J. F. Endicott, and G. R. Brubaker, *J. Phys. Chem.*, **87**, 5057 (1983).
58. E. L. Yee, J. T. Hupp, and M. J. Weaver, *Inorg. Chem.*, **22**, 3465 (1983).
59. G. Wada, T. Inatani, and S. Ichimura, *Bull. Chem. Soc. Jpn.*, **55**, 3441 (1982).
60. J. T. Hupp and M. J. Weaver, *Inorg. Chem.*, **22**, 2557 (1983).
61. S. U. M. Khan and T. O'M. Bockris, *J. Phys. Chem.*, **87**, 4012 (1983).
62. R. Schmid and L. Han, *Inorg. Chem. Acta*, **69**, 127 (1983).
63. R. Schmid, R. W. Soukup, M. K. Arasteh, and V. Gutman, *Inorg. Chim. Acta*, **73**, 21 (1983).
64. J. A. M. Ahmad and W. C. E. Higginson, *J. Chem. Soc.*, **87**, 1449 (1983).
65. J. M. Anast, A. W. Hamburg, and D. W. Margerum, *Inorg. Chem.*, **22**, 2139 (1983).
66. C. A. Koval and D. W. Margerum, *Inorg. Chem.*, **20**, 2311 (1981).
67. L. L. Diaddario, W. R. Robinson, and D. W. Margerum, *Inorg. Chem.*, **22**, 1021 (1983).
68. J. M. Anast and D. W. Margerum, *Inorg. Chem.*, **21**, 3494 (1982).
69. C. K. Murray and D. W. Margerum, *Inorg. Chem.*, **22**, 463 (1983).
70. F. P. Rotzinger, K. Kumar, and J. F. Endicott, *Inorg. Chem.*, **21**, 4111 (1982).
71. K. Kumar, F. P. Rotzinger, and J. F. Endicott, *J. Am. Chem. Soc.*, **105**, 7064 (1983).
72. D. H. Macartney and N. Sutin, *Inorg. Chem.*, **22**, 3530 (1983).
73. D. H. Macartney and A. McAuley, *Can. J. Chem.*, **60**, 2625 (1982).
74. D. H. Macartney and A. McAuley, *Can. J. Chem.*, **61**, 103 (1983).
75. D. H. Macartney and A. McAuley, *Inorg. Chem.*, **22**, 2062 (1983).
76. A. G. Lappin and M. C. M. Laranjeira, *J. Chem. Soc., Dalton Trans.*, 1861 (1982).
77. A. G. Lappin, M. C. M. Laranjeira, and R. D. Peacock, *Inorg. Chem.*, **22**, 786 (1983).
78. J. Korvenranta, H. Saarinen, and M. Nasakkala, *Inorg. Chem.*, **21**, 4296 (1982).
79. A. McAuley and K. F. Preston, *Inorg. Chem.*, **22**, 2111 (1983).
80. E. O. Schlemper and R. K. Murman, *Inorg. Chem.*, **22**, 1077 (1983).
81. G. Neogi, S. Acharya, R. K. Panda, and D. Ramaswamy, *J. Chem. Soc., Dalton Trans.*, 1233 (1983).
82. R. Sahu, G. Neogi, S. Acharya, and R. K. Panda, *Int. J. Chem. Kinet.*, **15**, 823 (1983).
83. C. K. Murray and D. W. Margerum, *Inorg. Chem.*, **21**, 3501 (1982).
84. L. Fabbrizzi, A. Perotti, and A. Poggi, *Inorg. Chem.*, **22**, 1411 (1983).
85. K. Weighardt, W. Schmidt, W. Herrmann, and H. J. Kuppers, *Inorg. Chem.*, **22**, 2953 (1983).
86. M. J. van der Merwe, J. C. A. Boeyens, and R. D. Hancock, *Inorg. Chem.*, **22**, 3489 (1983).
87. L. Fabbrizzi, H. Cohen, and D. Meyerstein, *J. Chem. Soc., Dalton Trans.*, 2125 (1983).
88. L. Fabbrizzi, A. Poggi, and P. Zanello, *J. Chem. Soc., Dalton Trans.*, 2191 (1983).
89. K. M. Davies, *Inorg. Chem.*, **22**, 615 (1983).
90. M. A. deAraiyo and H. L. Hodges, *Inorg. Chem.*, **21**, 3167 (1982).
91. C.-W. Lee and F. C. Anson, *J. Phys. Chem.*, **87**, 3360 (1983).
92. G. S. Yoneda, G. L. Blackmer, and R. A. Holwerda, *Inorg. Chem.*, **16**, 3376 (1977).
93. M. A. Augustin and J. K. Yandell, *Inorg. Chem.*, **18**, 577 (1979).
94. G. Daramola, J. F. Ojo, O. Olubujide, and F. Oriaifo, *J. Chem. Soc., Dalton Trans.*, 2137 (1982).
95. N. J. Curtis, G. A. Lawrance, and A. M. Sargeson, *Aust. J. Chem.*, **36**, 1327 (1983).
96. P. Bernhard, H.-B. Burgi, J. Hauser, H. Lehmann, and A. Ludi, *Inorg. Chem.*, **21**, 3936 (1982).
97. F. Moattor, J. R. Walton, and L. E. Bennett, *Inorg. Chem.*, **22**, 550 (1983).
98. B. Banaś and J. Mrozinski, *Acta Chim. (Hungary)*, **113**, 225 (1983).
99. S. Fukuzumi, N., Nishizawa, and T. Tanaka, *Bull. Chem. Soc. Jpn.*, **55**, 3482 (1982).
100. R. B. Ali, K. Sarawek, A. Wright, and R. D. Cannon, *Inorg. Chem.*, **22**, 351 (1983).
101. M. Chou, C. Creutz, D. Mahajan, N. Sutin, and A. P. Zipp, *Inorg. Chem.*, **21**, 3989 (1982).
102. H. A. Schwarz and C. Creutz, *Inorg. Chem.*, **22**, 707 (1983).

103. C. Creutz, A. D. Keller, N. Sutin, and A. P. Zipp, *J. Am. Chem. Soc.*, **104**, 3618 (1982).
104. D. J. Szalda, C. Creutz, D. Mahajan, and N. Sutin, *Inorg. Chem.*, **22**, 2372 (1983).
105. K. W. Hicks and M. A. Hurless, *Inorg. Chim. Acta*, **74**, 229 (1983).
106. H. Hennig, A. Rehorek, D. Rehorek, P. Thomas, and D. Bazold, *Inorg. Chim. Acta*, **77**, L11 (1983).
107. J. G. Leipoldt, C. R. Dennis, and E. C. Grobler, *Inorg. Chim. Acta*, **77**, L45 (1983).
108. K. W. Hicks, *Inorg. Chim. Acta*, **76**, L115 (1983).
109. R. M. C. Goncalves, H. Kellawi, and D. R. Rosseinsky, *J. Chem. Soc., Dalton Trans.*, 991 (1983).
110. K. Sato, T. Ohsaka, H. Matsuda, and N. Oyama, *Bull. Chem. Soc. Jpn.*, **56**, 1863 (1983).
111. K. Haya, I. Uchida, and S. Toshima, *J. Phys. Chem.*, **87**, 105 (1983).
112. N. Oyama, T. Ohsaka, M. Kaneko, K. Sato, and H. Matsuda, *J. Am. Chem. Soc.*, **105**, 6003 (1983).
113. J. S. Facci, R. H. Schmehl, and R. W. Murray, *J. Am. Chem. Soc.*, **104**, 4959 (1983).
114. M. Sano, H. Kashiwagi, and H. Yamatera, *Inorg. Chem.*, **21**, 3837 (1982).
115. R. van Eldik, *Inorg. Chem.*, **22**, 353 (1983).
116. R. van Eldik and H. Kelm, *Inorg. Chim. Acta*, **60**, 177 (1982).
117. M. Kanesato, M. Ebihara, Y. Sasaki, and K. Saito, *J. Am. Chem. Soc.*, **105**, 5711 (1983).
118. F. B. Veno, Y. Sasaki, T. Ito, and K. Saito, *J. Chem. Soc., Chem. Commun.*, 328 (1982).
119. R. van Eldik and H. Kelm, *Inorg. Chim. Acta*, **73**, 91 (1983).
120. S. Wherland, *Inorg. Chem.*, **22**, 2349 (1983).
121. S. Kondo, Y. Sasaki, and K. Saito, *Inorg. Chem.*, **20**, 429 (1981).
122. U. Sakaguchi, I. Yamamoto, S. Izumoto, and H. Yoneda, *Bull. Chem. Soc. Jpn.*, **56**, 153 (1983).
123. D. A. Geselowitz and H. Taube, *J. Am. Chem. Soc.*, **102**, 4525 (1980).
124. D. T. Richens and A. G. Sykes, *Comments Inorg. Chem.*, **1**, 141 (1981).
125. D. T. Richens and A. G. Sykes, *J. Chem. Soc., Chem. Commun.*, 616 (1983).
126. R. McAllister, K. W. Hicks, M. A. Hurless, S. T. Pittenger, and R. W. Gedridge, *Inorg. Chem.*, **21**, 4098 (1982).
127. A. W. Maverick, J. S. Najdzionek, D. MacKenzie, D. G. Nocera, and H. B. Gray, *J. Am. Chem. Soc.*, **105**, 1878 (1983).
128. A. Peloso, *J. Chem. Soc., Dalton Trans.*, 1285 (1983).
129. N. Rajasekar and E. S. Gould, *Inorg. Chem.*, **22**, 3798 (1983).
130. S. K. Chapman, I. Sanemasa, A. D. Watson, and A. G. Sykes, *J. Chem. Soc., Dalton Trans.*, 1949 (1983).
131. M. Goldberg and I. Pecht, *Biochemistry*, **15**, 4197 (1976).
132. S. K. Chapman, A. D. Watson, and A. G. Sykes, *J. Chem. Soc., Dalton Trans.*, 2543 (1983).
133. S. K. Chapman, J. D. Sinclair-Day, A. G. Sykes, S.-C. Tam, and R. J. P. Williams, *J. Chem. Soc., Chem. Commun.*, 1152 (1983).
134. S. K. Chapman, I. Sanemasa, and A. G. Sykes, *J. Chem. Soc., Dalton Trans.*, 2549 (1983).
135. D. Beoku-Betts, S. K. Chapman, C. V. Knox, and A. G. Sykes, *J. Chem. Soc., Chem. Commun.*, 1150 (1983).
136. O. Farver, Y. Shahak, and I. Pecht, *Biochemistry*, **21**, 1885 (1982).
137. K. O. Burkey and E. L. Gross, *Biochemistry*, **21**, 5886 (1982).
138. O. Farver, Y. Blatt, and I. Pecht, *Biochemistry*, **21**, 3556 (1982).
139. A. F. Corin, R. Bersohn, and P. E. Cole, *Biochemistry*, **22**, 2032 (1983).
140. A. G. Mauk, E. Bordignon, and H. B. Gray, *J. Am. Chem. Soc.*, **104**, 7654 (1982).
141. M. J. Sisley, M. G. Segal, C. S. Stanley, I. K. Adzamil, and A. G. Sykes, *J. Am. Chem. Soc.*, **105**, 225 (1983).
142. S. Dahlin, B. Reinhammar, and M. T. Wilson, *Inorg. Chim. Acta*, **79**, 016 (1983).

143. *Inorg. Chim. Acta*, **79**, (1983).
144. H. A. O. Hill and N. J. Walton, *J. Am. Chem. Soc.*, **104**, 6515 (1982).
145. A. K. Chung, R. M. Weiss, A. Warshel, and T. Takano, *J. Phys. Chem.*, **87**, 1683 (1983).
146. M. R. Mauk, L. S. Reid, and A. G. Mauk, *Biochemistry*, **21**, 1843 (1982).
147. L. S. Reid, V. T. Taniguchi, H. B. Gray, and A. G. Mauk, *J. Am. Chem. Soc.*, **104**, 7516 (1982).
148. S. K. Chapman, M. Davies, C. P. J. Vuik, and A. G. Sykes, *J. Chem. Soc., Chem. Commun.*, 868 (1983).
149. T.-M. Chan, E. L. Ulrich, and J. L. Markley, *Biochemistry*, **22**, 6002 (1983).
150. F. A. Armstrong, P. C. Harrington, and R. G. Wilkins, *J. Inorg. Biochem.*, **18**, 83 (1983).
151. P. C. Harrington and R. G. Wilkins, *J. Inorg. Biochem.*, **19**, 339 (1983).
152. J. Hirose, M. Ueoka, T. Tsuchiga, M. Nakagawa, M. Noji, and Y. Kidani, *Chem. Lett.*, 1429 (1983).
153. J. L. McGourty, N. V. Blough, and B. M. Hoffman, *J. Am. Chem. Soc.*, **105**, 4470 (1983).

References for Chapter 3

1. G. Davies and A. M. El-Sayed, in: *Copper Coordination Chemistry: Biochemical and Inorganic Perspectives* (K. D. Karlin and J. Zubieta, eds.), Adenine Press, Guilderland, New York (1983), pp. 281–309.
2. G. V. Buxton, *NATO Adv. Study Inst. Ser., Ser. C*, **86**, 267 (1982).
3. M. J. Blandamer and J. Burgess, *Pure Appl. Chem.*, **54**, 2285 (1982).
4. E. A. Deutsch, M. J. Root, and D. L. Nosco, *Adv. Inorg. Bioinorg. Mech.*, **1**, 269 (1982).
5. R. A. Henderson, G. J. Leigh, and C. J. Pickett, *Adv. Inorg. Chem. Radiochem.*, **27**, 198 (1983).
6. J. H. Espenson, *Adv. Inorg. Bioinorg. Mech.*, **1**, 1 (1982).
7. J. H. Espenson, *Progr. Inorg. Chem.*, **30**, 189 (1983).
8. A. O. Allen and B. H. J. Bielski, in: *Superoxide Dismutase* (L. Oberley, ed.), Vol. 1, p. 125, CRC Press (1982).
9. G. Cohen and R. A. Greenwald, *Oxy Radicals and Their Scavenger Systems, Volume I. Molecular Aspects* (G. Cohen and R. A. Greenwald, eds.), Elsevier Biomedical, New York (1983).
10. A. B. Ross and P. Neta, *Rate Constants for Reactions of Aliphatic Carbon-Centred Radicals in Aqueous Solution*, National Standards Reference Data Series (U.S. National Bureau of Standards), NSRDS-NBS 70, Notre Dame Radiation Laboratory (1982).
11. G. L. Hug, *Optical Spectra of Nonmetallic Inorganic Transient Species in Aqueous Solution*, National Standards Reference Data Series (U.S. National Bureau of Standards), U.S. NBS 69, Notre Dame Radiation Laboratory (1981).
12. L. Eberson, *Adv. Phys. Org. Chem.*, **18**, 79 (1982).
13. W. K. Wilmarth, D. M. Stanbury, J. E. Byrd, H. N. Po, and C.-P. Chua, *Coord. Chem. Rev.*, **51**, 155 (1983).
14. W. K. Wilmarth, N. Schwartz, and C. R. Giuliano, *Coord. Chem. Rev.*, **51**, 243 (1983).
15. J. Hoigné and H. Bader, *Water Res.*, **17**, 185 (1983).
16. J. Hoigné and H. Bader, *Vom Wasser*, **59**, 253 (1982).
17. M. Orban and I. R. Epstein, *J. Am. Chem. Soc.*, **104**, 5918 (1982).
18. J. T. Kummer, *Inorg. Chim. Acta*, **76**, L291 (1983).
19. P. N. Balasubramanian and E. S. Gould, *Inorg. Chem.*, **22**, 2635 (1983).
20. N. Rajasekar, R. Subramanian, and E. S. Gould, *Inorg. Chem.*, **21**, 4110 (1982).

21. K. K. Sengupta and B. Basu, *Transition Met. Chem.*, **8,** 6 (1983).
22. K. K. Sengupta and B. Basu, *Transition Met. Chem.*, **8,** 3 (1983).
23. S. S. Gupta and Y. K. Gupta, *J. Chem. Soc., Dalton Trans.*, 547 (1983).
24. P. Keswani, M. R. Goyal, and Y. K. Gupta, *J. Chem. Soc., Dalton Trans.*, 1831 (1983).
25. P. Arselli and E. Mentasti, *J. Chem. Soc., Dalton Trans.*, 689 (1983).
26. D. F. C. Morris and T. J. Ritter, *J. Chem. Soc., Dalton Trans.*, 216 (1980).
27. K. K. Sengupta and P. K. Sen, *Inorg. Chem.*, **18,** 979 (1979).
28. A. K. Banerjee, A. Misra, A. K. Basak, and D. Banerjea, *Transition Met. Chem.*, **7,** 239 (1982).
29. S. Acharaya, G. Neogi, and R. K. Panda, *Int. J. Chem. Kinet.*, **15,** 867 (1983).
30. S. Acharya, G. Neogi, R. K. Panda, and D. Ramaswamy, *Bull. Chem. Soc. Jpn.*, **56,** 2814 (1983).
31. S. Acharya, G. Neogi, R. K. Panda, and D. Ramaswamy, *Int. J. Chem. Kinet.*, **14,** 1253 (1982).
32. E. R. Burrows and D. H. Rosenblatt, *J. Org. Chem.*, **48,** 992 (1983).
33. M. Puutio and P. O. I. Virtanen, *Acta Chem. Scand., Ser. A.*, **36,** 689 (1982).
34. R. V. Vrath, B. Sethuram, and T. N. Rao, *Indian J. Chem.*, **21A,** 414 (1982).
35. C. P. Murthy, B. Sethuram, and T. N. Rao, *Monatsh. Chem.*, **113,** 941 (1982).
36. S. Goldstein and G. Czapski, *J. Am. Chem. Soc.*, **105,** 7276 (1983).
37. F. R. Hopf, M. M. Rogic, and J. F. Wolf, *J. Phys. Chem.*, **87,** 4681 (1983).
38. G. Davies and M. A. El-Sayed, *Inorg. Chem.*, **22,** 1257 (1983).
39. W. Zamudio, A.-M. Garcia, and E. Spodine, *Transition Met. Chem.*, **8,** 69 (1983).
40. I. I. Creaser, R. J. Geue, J. M. Harrowfield, A. J. Herlt, A. M. Sargeson, M. R. Snow, and J. Springborg, *J. Am. Chem. Soc.*, **104,** 6016 (1982).
41. A. Bakač, J. H. Espenson, I. I. Creaser, and A. M. Sargeson, *J. Am. Chem. Soc.*, **105,** 7624 (1983).
42. L. I. Simandi, C. R. Savage, Z. A. Schelly, and S. Nemeth, *Inorg. Chem.*, **21,** 2765 (1982).
43. S. Nemeth and L. I. Simandi, *Inorg. Chem.*, **22,** 3151 (1983).
44. R. R. Durand, Jr., C. S. Bencosme, J. P. Collman, and F. C. Anson, *J. Am. Chem. Soc.*, **105,** 2710 (1983).
45. Y. L. Mest, M. L'Her, J. Courtot-Coupez, J. P. Collman, E. R. Evitt, and C. S. Bencosme, *J. Chem. Soc., Chem. Commun.*, 1286 (1983).
46. J. R. Hamon and D. Astruc, *J. Am. Chem. Soc.*, **105,** 5951 (1983).
47. N. Herron, W. P. Schammel, S. C. Jackels, J. J. Grzybowski, L. L. Zimmer, and D. H. Busch, *Inorg. Chem.*, **22,** 1433 (1983).
48. D. P. Riley and D. H. Busch, *Inorg. Chem.*, **22,** 4141 (1983).
49. P. A. Forshey and T. Kuwana, *Inorg. Chem.*, **22,** 699 (1983).
50. F. C. Frederick, W. M. Coleman, and L. T. Taylor, *Inorg. Chem.*, **22,** 792 (1983).
51. C. Ercolani, G. Rossi, F. Monacelli, and M. Verzino, *Inorg. Chim. Acta*, **73,** 95 (1983).
52. D. M. Stanbury, D. Gaswick, G. M. Brown, and H. Taube, *Inorg. Chem.*, **22,** 1975 (1983).
53. D. P. Riley, *Inorg. Chem.*, **22,** 1965 (1983).
54. P. R. Bontchev, H. Mueller, J. Mattusch, M. I. Mitewa, and K. S. Kabassanov, *Dokl. Bolg. Akad. Nauk*, **35,** 1515 (1982).
55. D.-H. Chin, D. T. Sawyer, W. P. Schaefer, and C. J. Simmons, *Inorg. Chem.*, **22,** 752 (1983).
56. A. Bakač and J. H. Espenson, *Inorg. Chem.*, **22,** 779 (1983).
57. M. J. Nicol, *S. Afr. J. Chem.*, **35,** 77 (1982).
58. P. K. Bhattacharyya, R. D. Saini, and P. B. Ruikar, *Int. J. Chem. Kinet.*, **14,** 1219 (1982).
59. J. A. Gilbert, S. W. Gersten, and T. J. Meyer, *J. Am. Chem. Soc.*, **104,** 6872 (1982).
60. P. Banerjee and M. P. Pujari, *Indian J. Chem.*, **22A,** 289 (1983).
61. M. P. Pujari and P. Banerjee, *Transition Met. Chem.*, **8,** 91 (1983).

62. M. Otto and G. Werner, *Anal. Chim. Acta*, **147**, 255 (1983).

63. G. W. Winston, L. Berl, and A. I. Cederbaum, in: *Oxy Radicals and Their Scavenger Systems. Volume I. Molecular Aspects* (G. Cohen and R. A. Greenwald, eds.), Elsevier Biomedical, New York (1983), p. 145.

64. J. H. Espenson and J. D. Melton, *Inorg. Chem.*, **22**, 2779 (1982).

65. C. Giannotti, C. Fontaine, A. Chiaroni, and C. Riche, *J. Organomet. Chem.*, **113**, 57 (1976).

66. T. G. Hoon and C. Y. Mok, *Inorg. Chim. Acta*, **62**, 231 (1982).

67. B. S. Brunschwig, M. H. Chou, C. Creutz, P. Ghosh, and N. Sutin, *J. Am. Chem. Soc.*, **105**, 4832 (1983).

68. G. Nord, B. Pedersen, and E. Bjergbakke, *J. Am. Chem. Soc.*, **105**, 1913 (1983).

69. N. Serpone and F. Bolletta, *Inorg. Chim. Acta*, **75**, 189 (1983).

70. R. C. Thompson, *Inorg. Chem.*, **22**, 584 (1983).

71. P. L. Freund and M. Spiro, *J. Chem. Soc., Faraday Trans. 1*, **79**, 491 (1983).

72. P. P. Tejeda, J. R. Velasco, and F. S. Burgos, *J. Chem. Soc., Dalton Trans.*, 2679 (1983).

73. W. K. Wilmarth, Y.-T. Fanchiang, and J. E. Byrd, *Coord. Chem. Rev.*, **51**, 141 (1983).

74. J. Readman and D. A. House, *Polyhedron*, **1**, 611 (1982).

75. M. J. Blandamer, J. Burgess, S. J. Hamshere, and C. White, *Can. J. Chem.*, **61**, 1361 (1983).

76. P. S. Radhakrishnamurti, C. Janardhana, and G. K. Mohanty, *Bull. Chem. Soc. Jpn.*, **55**, 3617 (1982).

77. M. P. Pujari and P. Banerjee, *J. Chem. Soc., Dalton Trans.*, 1015 (1983).

78. W. Geiseler, *J. Phys. Chem.*, **86**, 4394 (1982).

79. T. Yoshida and Y. Ushiki, *Bull. Chem. Soc. Jpn.*, **55**, 1772 (1982).

80. H. D. Foersterling, H. J. Lamberz, H. Schreiber, and W. Zittlau, *Acta Chim. Acad. Sci. Hung.*, **110**, 251 (1982).

81. Z. Noszticzius and A. Feller, *Acta Chim. Acad. Sci. Hung.*, **110**, 261 (1982).

82. A. M. Zhabotinskii, *Acta Chim. Acad. Sci. Hung.*, **110**, 283 (1982).

83. R. Turcsanyi, *Acta Chim. Acad. Sci. Hung.*, **110**, 305 (1982).

84. A. M. Zhabotinskii, A. M. Zaikin, and A. B. Rovinskii, *React. Kinet. Catal. Lett.*, **20**, 29 (1982).

85. M. Alamgir, M. Orban, and I. R. Epstein, *J. Phys. Chem.*, **87**, 3725 (1983).

86. M. Orban and I. R. Epstein, *J. Phys. Chem.*, **87**, 3212 (1983).

87. A. M. Kjaer and J. Ulstrup, *Inorg. Chem.*, **21**, 3490 (1982).

88. Y. Sulfab, *Polyhedron*, **2**, 679 (1983).

89. N. A. Al-Jallal and Y. Sulfab, *Transition Met. Chem.*, **8**, 51 (1983).

90. M. A. Hussein, A. A. Abdel-Khalek, and Y. Sulfab, *J. Chem. Soc., Dalton Trans.*, 317 (1983).

91. M. A. Hussein and Y. Sulfab, *Transition Met. Chem.*, **7**, 181 (1982).

92. A. Adeymo, A. Valiotti, and P. Hambright, *Inorg. Chim. Acta Lett.*, **64**, L251 (1983).

93. G. D. Jones, M. G. Jones, M. T. Wilson, M. Brunori, A. Colosimo, and P. Sarti, *Biochem. J.*, **109**, 175 (1983).

94. S. A. Kazmi, A. L. Shorter, and J. V. McArdle, *J. Inorg. Biochem.*, **17**, 269 (1982).

95. R. J. Balahura, G. Ferguson, B. L. Ruhl, and R. G. Wilkins, *Inorg. Chem.*, **22**, 3990 (1983).

96. G. Neogi, S. Acharya, R. K. Panda, and D. Ramaswamy, *Int. J. Chem. Kinet.*, **15**, 945 (1983).

97. G. Neogi, S. Acharya, R. K. Panda, and D. Ramaswamy, *Int. J. Chem. Kinet.*, **15**, 881 (1983).

98. P. Banerjee and M. P. Pujari, *Indian J. Chem.*, **214A**, 1123 (1982).

99. R. K. Nandwana and S. K. Solanti, *Cienc. Cult. (Sao Paolo)*, **35**, 979 (1983) [*Chem. Abstr.* 99:146840q].

100. M. Cyfert, *Inorg. Chim. Acta*, **73**, 135 (1983).

101. C. R. Dennis, S. S. Basson, and J. G. Leipoldt, *Polyhedron*, **2**, 1357 (1983).
102. V. Khyarsing and A. P. Filippov, *Kinet. Catal.*, **24**, 250 (1983) [transl. from *Kinetika i Kataliz*, **24**, 302 (1983)].
103. R. van Eldik, J. von Jouanne, and H. Kelm, *Inorg. Chem.*, **21**, 2818 (1982).
104. R. van Eldik, *Inorg. Chem.*, **22**, 353 (1983).
105. K. C. Koshy and G. M. Harris, *Inorg. Chem.*, **22**, 2947 (1983).
106. C. A. Ayoko and M. A. Olaltunji, *Polyhedron*, **2**, 577 (1983).
107. C. Baiocchi, E. Mentasti, and P. Arselli, *Transition Met. Chem.*, **8**, 40 (1983).
108. H. K. Baek and R. A. Holwerda, *Inorg. Chem.*, **22**, 3452 (1983).
109. M. Y. Hazen, J. Silver, and M. T. Wilson, *Inorg. Chim. Acta*, **78**, 1 (1983).
110. K. K. Sengupta, B. Basu, S. Sengupta, and S. Nandi, *Polyhedron*, **2**, 983 (1983).
111. A. K. Indrayan, S. K. Mishra, and Y. K. Gupta, *Indian J. Chem.*, **22A**, 756 (1983).
112. L. J. Kirschenbaum and J. D. Rush, *Inorg. Chem.*, **22**, 3304 (1983).
113. K. J. M. Rao and P. V. K. Rao, *Indian J. Chem.*, **21A**, 1066 (1982).
114. G. R. Buettner, T. P. Doherty, and L. K. Patterson, *FEBS Lett.*, **158**, 143 (1983).
115. C. Bull, G. J. McClune, and J. A. Fee, *J. Am. Chem. Soc.*, **105**, 5290 (1983).
116. J. Butler, W. H. Koppenol, and E. Margoliash, *J. Biol. Chem.*, **257**, 10747 (1982).
117. W. J. Albery, A. W. Foulds, and J. R. Darwent, *J. Photochem.*, **19**, 37 (1982).
118. P. N. Balasubramanian and E. S. Gould, *Inorg. Chem.*, **22**, 1100 (1983).
119. A. Bakač, R. J. Blau, and J. H. Espenson, *Inorg. Chem.*, **22**, 3789 (1983).
120. R. J. Field, N. V. Raghavan, and J. G. Brummer, *J. Phys. Chem.*, **86**, 2443 (1982).
121. L. A. Lednicky and D. M. Stanbury, *J. Am. Chem. Soc.*, **105**, 3098 (1983).
122. S. B. McCullen and T. L. Brown, *J. Am. Chem. Soc.*, **104**, 7496 (1982).
123. Y. Sorek, H. Cohen, W. A. Mulac, K. H. Schmidt, and D. Meyerstein, *Inorg. Chem.*, **22**, 3040 (1983).
124. L. S. Ryvkina and V. M. Berdnikov, *React. Kinet. Catal. Lett.*, **21**, 409 (1982).
125. R. F. Pasternack and D. Pysnik, in: *Oxy Radicals and Their Scavenger Systems. Volume I. Molecular Aspects* (G. Cohen and R. A. Greenwald, eds.), Elsevier Biomedical, New York (1983), p. 151.
126. F. Gotz and E. Lengfelder, in: *Oxy Radicals and Their Scavenger Systems. Volume I. Molecular Aspects* (G. Cohen and R. A. Greenwald eds.), Elsevier Biomedical, New York (1983), p. 228.
127. T. Imamura, K. Hasegawa, and M. Fujimoto, *Chem. Lett.*, 705 (1983).
128. S. E. Jones, G. S. Srivatsa, D. T. Sawyer, T. G. Traylor, and T. C. Mincey, *Inorg. Chem.*, **22**, 3903 (1983).
129. T. Ozawa and T. Kwan, *Chem. Pharm. Bull.*, **31**, 2864 (1983).
130. H. Arzoumanian, R. Lai, J. Metzger, J.-F. Petrignani, and P. Tordo, *Inorg. Chem.*, **22**, 2097 (1983).
131. M. B. Mooiman and J. Pratt, *J. Chem. Soc., Chem. Commun.*, 33 (1981).
132. J. Halpern, *Inorg. Chim. Acta*, **77**, L105 (1983).
133. M. B. Mooiman, J. M. Pratt, and E. H. R. von Barsewisch, *S. Afr. J. Chem.*, **35**, 175 (1982).
134. M. B. Mooiman and J. M. Pratt, *S. Afr. J. Chem.*, **35**, 171 (1982).
135. M. A. Bogorodskaya, I. A. Potapov, M. B. Rozenkevich, and Yu. A. Sakharovskii, *Kinet. Catal.*, **24**, 617 (1983).
136. G. Kalatzis, J. Konstantatos, E. Vrachnou-Astra, and D. Katakis, *J. Am. Chem. Soc.*, **105**, 2897 (1983).
137. S. S. Miller and A. Haim, *J. Am. Chem. Soc.*, **105**, 5624 (1983).
138. R. H. Hill and R. J. Puddephatt, *J. Am. Chem. Soc.*, **105**, 5797 (1983).
139. A. S. Berenblyum, A. G. Knizhnik, S. L. Mund, and I. I. Moiseev, *J. Organomet. Chem.*, **234**, 219 (1982).

140. A. S. Berenblyum, A. P. Aseeva, L. I. Lakhman, and I. I. Moiseev, *J. Organomet. Chem.*, **234**, 237 (1982).
141. F. Ungváry and L. Markó, *Organometallics*, **2**, 1608 (1983).
142. J. H. Espenson and M. S. McDowell, *Organometallics*, **1**, 1514 (1982).
143. T. Funabiki, H. Nakamura, and S. Yoshida, *J. Organomet. Chem.*, **243**, 95 (1983).
144. E. C. Ashby, R. N. DePriest, A. Tuncay, and S. Srivastava, *Tetrahedron Lett.*, **23**, 5251 (1982).
145. K. L. Brown and R. Legates, *J. Organomet. Chem.*, **233**, 259 (1982).
146. L. Eberson, *Acta Chem. Scand., Ser. B*, **36**, 533 (1982).
147. J. C. Brodovitch, A. McAuley, and T. Oswald, *Inorg. Chem.*, **21**, 3442 (1982).
148. A. McAuley, T. Oswald, and R. I. Haines, *Can. J. Chem.*, **61**, 1120 (1983).
149. A. Vlcek, Jr., J. Klima, and A. A. Vlcek, *Inorg. Chim. Acta*, **69**, 191 (1983).
150. M. Bhattacharjee and M. K. Mahanti, *React. Kinet. Catal. Lett.*, **21**, 449 (1982).
151. M. Bhattacharjee and M. K. Mahanti, *Gazz. Chim. Ital.*, **113**, 101 (1983).
152. K. C. Gupta and V. D. Misra, *J. Indian Chem. Soc.*, **60**, 258 (1983).
153. K. K. Sengupta, S. Maiti, T. Samanta, S. Nandi, and A. Barnerjee, *Transition Met. Chem.*, **7**, 274 (1982).
154. P. O. I. Virtanen and S. Karppinen, *Finn. Chem. Lett.*, 55 (1983). [*Chem. Abstr.* 99:59547f].
155. K. Behari, N. N. Pandey, R. K. Khanna, and R. S. Shukla, *Z. Phys. Chem. (Leipzig)*, **263**, 734 (1982).
156. R. R. Nagori, M. Mehta, R. N. Mehrotra, *Indian J. Chem.*, **21A**, 41 (1982).
157. K. Zaw, M. Lautens, and P. M. Henty, *Organometallics*, **2**, 197 (1983).
158. P. M. Reddy, V. Jagannadham, B. Sethuram, and T. N. Rao, *React. Kinet. Catal. Lett.*, **19**, 243 (1982).
159. S. B. Hanna and J. T. Fenton, *Int. J. Chem. Kinet.*, **15**, 925 (1983).
160. G. V. Rao, K. C. Rajanna, and P. K. Saiprakash, *Cienc. Cult. (Sao Paolo)*, **35**, 639 (1983). [*Chem. Abstr.* 99:182277v].
161. P. C. Dash, D. P. Das, B. K. Mohanty, R. K. Samal, and M. C. Nayak, *J. Macromol. Sci., Chem.*, **A17**, 1357 (1982).
162. W. D. Drury and J. M. De Korte, *Inorg. Chem.*, **22**, 121 (1983).
163. S. Baral and P. Neta, *J. Phys. Chem.*, **87**, 1502 (1983).
164. J. H. Espenson, P. Connolly, D. Meyerstein, and H. Cohen, *Inorg. Chem.*, **22**, 1009 (1983).
165. H. Cohen, E. S. Gould, D. Meyerstein, M. Nutkovich, and C. A. Radlowski, *Inorg. Chem.*, **22**, 1374 (1983).
166. R. C. McHatton and J. H. Espenson, *Inorg. Chem.*, **22**, 784 (1983).
167. J.-T. Chen and J. H. Espenson, *Inorg. Chem.*, **22**, 1651 (1983).
168. J. H. Espenson, M. Shimura, and A. Bakač, *Inorg. Chem.*, **21**, 2537 (1982). (en = 1,2-diaminoethane; tn = 1,3-diaminopropane; chxn = *trans*-1,2-diaminocyclohexane.)
169. Y. Sorek, H. Cohen, W. A. Mulac, K. H. Schmidt, and D. Meyerstein, *Inorg. Chem.*, **22**, 3040 (1983).
170. D. Brault and P. Neta, *J. Phys. Chem.*, **86**, 3405 (1982).
171. G. Levey and T. W. Ebbesen, *J. Phys. Chem.*, **87**, 829 (1983).
172. J. Lusztyk, B. Maillard, D. A. Lindsay, and K. U. Ingold, *J. Am. Chem. Soc.*, **105**, 3578 (1983).
173. A. N. Singh, P. N. Balasubramanian, and E. S. Gould, *J. Am. Chem. Soc.*, **22**, 655 (1983).
174. W. A. Mulac, H. Cohen, and D. Meyerstein, *Inorg. Chem.*, **21**, 4016 (1982).
175. J. W. van Leeuwen, C. van Dijk, H. J. Grande, and C. Veeger, *Eur. J. Biochem.*, **127**, 631 (1982).
176. M. Rougee, T. Ebbesen, F. Ghetti, and R. V. Bensasson, *J. Phys. Chem.*, **86**, 4404 (1982).
177. M. Pieniazek and E. Ratajczak, *J. Organomet. Chem.*, **238**, 289 (1982).

178. W. A. Mulac and D. Meyerstein, *J. Am. Chem. Soc.*, **104**, 4124 (1982).
179. E. P. Petryaev, O. I. Shadyro, and N. I. Kovalenko, *Khim. Vys. Energ.*, **16**, 520 (1982). [*Chem. Abstr.* 98:81386d].
180. R. G. Finke, B. L. Smith, B. J. Mayer, and A. A. Molinero, *Inorg. Chem.*, **22**, 3677 (1983).
181. F. Tischler and J. I. Morrow, *Inorg. Chem.*, **22**, 2286 (1983).
182. S. B. Hanna and J. T. Fenton, *Int. J. Chem. Kinet.*, **15**, 925 (1983).
183. T. J. Jones and R. M. Noyes, *J. Phys. Chem.*, **87**, 4686 (1983).
184. V. S. Srinivasan, *Inorg. Chem.*, **21**, 4328 (1982).
185. M. Zielinski, *J. Radioanal. Chem.*, **80**, 237 (1983).
186. R. K. Nanda and N. K. Mohanty, *Indian J. Chem.*, **21A**, 522 (1982).
187. A. L. Jain and K. K. Banerji, *J. Chem. Res. (S)*, 60 (1983).
188. R. R. Rao and E. V. Sundaram, *J. Indian Chem. Soc.*, **59**, 704 (1982).
189. P. M. Reddy, V. Jagannadham, B. Sethuram, and T. N. Rao, *Indian J. Chem.*, **21A**, 483 (1982).
190. E. S. Rudakov and L. K. Volkova, *Kinet. Catal.*, **24**, 458 (1983).
191. L. K. Volkova, Yu. V. Geletii, G. V. Lyubimova, E. S. Rudakov, V. P. Tret'yakov, and A. E. Shilov, *Bull. Acad. Sci. USSR, Div. Chem. Sci.*, **31**, 1473 (1983) (transl. from *Izv. Akad. Nauk SSR, Ser. Khim.* 1654 (1982)).
192. A. K. Bhattacharjee and M. K. Mahanti, *Gazz. Chim. Ital.*, **113**, 1 (1983).
193. E. Baciocchi, C. Rol, and G. V. Sebastiani, *J. Chem. Res.(S)*, 232 (1983).
194. R. Sugimoto, H. Suzuki, Y. Moro-Oka, and T. Ikawa, *Chem. Lett.*, 1863 (1982).
195. D. G. Lee and K. C. Brown, *J. Am. Chem. Soc.*, **104**, 5076 (1982).
196. N. Miyaura and J. K. Kochi, *J. Am. Chem. Soc.*, **105**, 2368 (1983).
197. L. Eberson, *J. Am. Chem. Soc.*, **105**, 3192 (1983).
198. E. Baciocchi, L. Eberson and C. Rol, *J. Org. Chem.*, **47**, 5106 (1982).

References for Chapter 4

1. J. C. Lockhart, *Adv. Inorg. Bioinorg. Mech.* **1**, 217 (1982).
2. R. J. P. Corriu and C. Guerin, *Adv. Organomet. Chem.*, **20**, 265 (1982).
3. F. K. Cartledge, *J. Organomet. Chem. Library*, **13**, 284 (1982).
4. J. C. Martin, *Science*, **221**, 509 (1983).
5. *Comprehensive Organometallic Chemistry. The Synthesis, Reactions and Structures of Organometallic Compounds* (G. Wilkinson, ed.), Pergamon Press, New York (1982).
6. C. G. Salentine, *Inorg. Chem.*, **22**, 3921 (1983).
7. J. Emsley and J. S. Lucas, *J. Chem. Soc., Dalton Trans.*, 1811 (1983).
8. C. J. Adams and I. E. Clark, *Polyhedron*, **2**, 673 (1983).
9. L. Babcock and R. Pizer, *Inorg. Chem.*, **22**, 174 (1983).
10. J. G. Dawber and D. H. Matusin, *J. Chem. Soc., Faraday Trans., 1*, **78**, 2521 (1982).
11. J. Emri and B. Györi, *Polyhedron*, **1**, 673 (1982).
12. J. Emri and B. Györi, *Polyhedron*, **2**, 1273 (1983).
13. K. Niedenzu and H. Nöth, *Chem. Ber.*, **116**, 1132 (1983).
14. H. Nöth and R. Staudigl, *Chem. Ber.*, **115**, 3011 (1982).
15. A. J. Boulton and C. S. Prado, *J. Chem. Soc., Chem. Commun.*, 1008 (1982).
16. D. S. Matteson and D. Majumdar, *Organometallics*, **2**, 1529 (1983).
17. D. S. Matteson, R. Ray, R. R. Rocks, and D. J. Tsai, *Organometallics*, **2**, 1536 (1983).
18. D. S. Matteson and G. Erdik, *Organometallics*, **2**, 1083 (1983).
19. D. J. S. Tsai, P.K. Jesthi, and D. S. Matteson, *Organometallics*, **2**, 1543 (1983).

20. K. K. Wang and H. C. Brown, *J. Am. Chem. Soc.*, **104**, 7148 (1982).

21. D. J. Nelson and H. C. Brown, *J. Am. Chem. Soc.*, **104**, 4907 (1982).

22. D. J. Nelson, C. D. Blue, and H. C. Brown, *J. Am. Chem. Soc.*, **104**, 4913 (1982).

23. H. C. Brown, K. K. Wang, and J. Chandrasekharan, *J. Am. Chem. Soc.*, **105**, 2340 (1983).

24. H. C. Brown, J. Chandrasekharan, and K. K. Wang, *J. Org. Chem.*, **48**, 3689 (1983).

25. H. C. Brown, J. Chandrasekharan, and K. K. Wang, *J. Org. Chem.*, **48**, 2901 (1983).

26. H. C. Brown and N. C. Hébert, *J. Organomet. Chem.*, **255**, 135 (1983).

27. H. C. Brown and J. Chandrasekharan, *Organometallics*, **2**, 1261 (1983).

28. T. Clark, D. Wilhelm, and P. von-R. Schleyer, *J. Chem. Soc., Chem. Commun.*, 606 (1983).

29. R. Chadha and N. K. Ray, *J. Phys. Chem.*, **86**, 3293 (1982).

30. R. Chadha and N. K. Ray, *Theor. Chim. Acta*, **60**, 573 (1982).

31. O. Eisenstein, H. B. Schlegel, and M. M. Kayser, *J. Org. Chem.*, **47**, 2886 (1982).

32. B. F. Spielvogel, A. T. McPhail, J. A. Knight, C. G. Moreland, C. L. Gatchell, and K. W. Morse, *Polyhedron*, **2**, 1345 (1983).

33. H. C. Brown, D. J. Nelson, and C. G. Scouten, *J. Org. Chem.*, **48**, 641 (1983).

34. H. C. Brown and J. Chandrasekharan, *J. Org. Chem.*, **48**, 644 (1983).

35. H. C. Brown and J. Chandrasekharan, *J. Org. Chem.*, **48**, 5080 (1983).

36. H. C. Brown, S. Narasimhan, and V. Somayaji, *J. Org. Chem.*, **48**, 3091 (1983).

37. S. Krishnamurthy and H. C. Brown, *J. Org. Chem.*, **48**, 3085 (1983).

38. S. E. Wood and B. Rickborn, *J. Org. Chem.*, **48**, 555 (1983).

39. J. L. Torregrosa, M. Baboulenc, V. Speziale and A. Lattes, *Compt. Rend. Ser. II*, **297**, 891 (1983).

40. J. A. Baban, J. C. Brand, and B. P. Roberts, *J. Chem. Soc., Chem. Commun.*, 315 (1983).

41. M. C. R. Symons, T. Chen, and C. Glidewell, *J. Chem. Soc., Chem. Commun.*, 326 (1983).

42. G. B. Jacobsen, J. H. Morris, and D. Reed, *J. Chem. Res. (S)*, 319 (1982).

43. J. R. M. Giles, V. P. J. Marti, and B. P. Roberts, *J. Chem. Soc., Chem. Commun.*, 696 (1983).

44. J. A. Baban and B. P. Roberts, *J. Chem. Soc., Chem. Commun.*, 1224 (1983).

45. E. H. Wong, M. G. Gatter, and R. M. Kabbani, *Inorg. Chem.*, **21**, 4022 (1982).

46. M. A. Beckett and J. D. Kennedy, *J. Chem. Soc., Chem. Commun.*, 575 (1983).

47. J. A. Heppert and D. F. Gaines, *Inorg. Chem.*, **22**, 3155 (1983).

48. B. Oh and T. Onak, *Inorg. Chem.*, **21**, 3150 (1982).

49. R. Baxter, R. J. M. Sands, and J. W. Wilson, *J. Chem. Res. (S)*, 94 (1983).

50. C. Brown, R. H. Cragg, T. J. Miller, and D. O'N. Smith, *J. Organomet. Chem.*, **244**, 209 (1983).

51. R. H. Cragg and T. J. Miller, *J. Organomet. Chem.*, **255**, 143 (1983).

52. C. K. Narula and H. Nöth, *Z. Naturforsch. B*, **38**, 1161 (1983).

53. H. Klusik and A. Berndt, *Angew. Chem. Int. Ed.*, **22**, 877 (1983).

54. P. Paetzold and R. Truppat, *Chem. Ber.*, **116**, 1531 (1983).

55. P. Paetzold and C. von Plotho, *Chem. Ber.*, **115**, 2819 (1982).

56. P. Paetzold, C. von Plotho, E. Niecke, and R. Rüger, *Chem. Ber.*, **116**, 1678 (1983).

57. M. N. Schuchman and C. von Sonntag, *Z. Naturforsch. B*, **37**, 1184 (1982).

58. U. Spitzer, R. van Eldik, and H. Kelm, *Inorg. Chem.*, **21**, 2821 (1982).

59. Y. Pocker and T. L. Deits, *J. Am. Chem. Soc.*, **105**, 980 (1983).

60. Y. Pocker, T. L. Deits, C. T. O. Fong, and C. H. Miao, *Inorg. Chem. Acta*, **79**, 37 (1983).

61. S. Lindskog, *Inorg. Chim. Acta*, **79**, 36 (1983).

62. I. Bertini and C. Luchinat, *Inorg. Chim. Acta*, **79**, 36 (1983).

63. D. N. Silverman and C. K. Tu, *Inorg. Chim. Acta*, **79**, 38 (1983).

64. S. H. Koenig and R. D. Brown, III, *Inorg. Chim. Acta*, **79**, 38 (1983).

65. D. E. Penny and T. J. Ritter, *J. Chem. Soc., Faraday Trans. I*, **79**, 2103 (1983).

66. D. Barth, C. Tondre, and J. J. Delpuech, *Int. J. Chem. Kinet.*, **15**, 1147 (1983).

67. E. A. Castro, R. Cortés, J. G. Santos, and J. C. Vega, *J. Org. Chem.*, **47**, 3774 (1982).
68. T. Yoshimura, K. Asada, and S. Oae, *Bull. Chem. Soc. Jpn.*, **55**, 3000 (1982).
69. R. J. Millican, M. Angelopoulos, A. Bose, B. Riegel, D. Robinson, and C. K. Wagner, *J. Am. Chem. Soc.*, **105**, 3622 (1983).
70. P. L. Fabre, J. Devynck, and B. Trémillion, *Bull. Soc. Chim. Fr. Part 1*, 5 (1983).
71. G. K. Surya Prakash, A. Husain, and G. A. Olah, *Angew. Chem. Int. Ed.*, **22**, 50 (1983).
72. J. F. McGarrity and D. P. Cox, *J. Am. Chem. Soc.*, **105**, 3961 (1983).
73. R. K. Harris and C. T. G. Knight, *J. Chem. Soc., Faraday Trans. 2*, **79**, 1525 (1983).
74. R. K. Harris and C. T. G. Knight, *J. Chem. Soc., Faraday Trans. 2*, **79**, 1539 (1983).
75. G. Boxhoorn, O. Sudmeijer, and P. H. G. van Kasteren, *J. Chem. Soc., Chem. Commun.*, 1416 (1983).
76. N. H. Ray and R. J. Plaisted, *J. Chem. Soc., Dalton Trans.*, 475 (1983).
77. J. C. Martin and W. H. Stevenson, III, *Phosphorus Sulfur*, **18**, 81 (1983).
78. R. J. P. Corriu, M. Poirer, and G. Royo, *J. Organomet. Chem.*, **233**, 165 (1982).
79. B. J. Helmer, R. West, R. J. P. Corriu, M. Poirer, G. Royo, and A. de Saxce, *J. Organomet. Chem.*, **251**, 295 (1983).
80. T. Inoue, *Inorg. Chem.*, **22**, 2435 (1983).
81. A. R. Bassindale and T. Stout, *J. Organomet. Chem.*, **238**, C41 (1982).
82. S. Kozuka and I. Naribayashi, *Mem. Fac. Eng. Osaka City Univ.*, **22**, 115 (1981).
83. R. Damrauer, C. H. de Puy, and V. M. Bierbaum, *Organometallics*, **1**, 1553 (1982).
84. K. Ito and T. Ibaraki, *Bull. Chem. Soc. Jpn.*, **56**, 295 (1983).
85. K. Ito and T. Ibaraki, *Bull. Chem. Soc. Jpn.*, **55**, 2973 (1982).
86. R. J. P. Corriu, S. Ould-Kada, and G. F. Lanneau, *J. Organomet. Chem.* **248**, 23 (1983).
87. R. J. P. Corriu, S. Ould-Kada, and G. F. Lanneau, *J. Organomet. Chem.* **248**, 39 (1983).
88. G. Cerveau, E. Colomer, and R. J. P. Corriu, *J. Organomet. Chem.*, **236**, 33 (1982).
89. N. M. K. El-Durini and R. A. Jackson, *J. Chem. Soc., Perkin Trans. 2*, 1275 (1983).
90. F. K. Cartledge, *Organometallics*, **2**, 425 (1983).
91. C. E. Peishoff and W. L. Jorgensen, *J. Org. Chem.*, **48**, 1970 (1983).
92. C. Eaborn, *J. Organomet. Chem.*, **239**, 93 (1982).
93. S. A. I. Al-Shali and C. Eaborn, *J. Organomet. Chem.*, **246**, C34 (1983).
94. C. Eaborn, Y. Y. El-Kaddar, and P. D. Lickiss, *J. Chem. Soc., Chem. Commun.*, 1450 (1983).
95. R. Damrauer, C. Eaborn, D. A. R. Happer, and A. I. Mansour, *J. Chem. Soc., Chem. Commun.*, 348 (1983).
96. Z. H. Arube, J. Chojnowski, C. Eaborn, and W. A. Stańczyk, *J. Chem. Soc., Chem. Commun.*, 493 (1983).
97. C. E. Eaborn and D. E. Reed, *J. Chem. Soc., Chem. Commun.*, 495 (1983).
98. A. I. Al-Wassil, C. Eaborn, and A. K. Saxena, *J. Chem. Soc., Chem. Commun.*, 974 (1983).
99. K. Tamao, M. Akita, and M. Kumada, *J. Organomet. Chem.*, **254**, 13 (1983).
100. B. J. Helmer and R. West, *Organometallics*, **1**, 1463 (1982).
101. C. Eaborn and A. I. Masour, *J. Organomet. Chem.*, **254**, 273 (1983).
102. J. B. Lambert and W. J. Schulz, Jr., *J. Am. Chem. Soc.*, **105**, 1671 (1983).
103. G. A. Olah and L. D. Field, *Organometallics*, **1**, 1485 (1982).
104. P. R. Jones, R. A. Pierce, and A. H. B. Cheng, *Organometallics*, **2**, 12 (1983).
105. G. A. Olah, A. L. Berrier, L. D. Field, and G. K. Surya Prakash, *J. Am. Chem. Soc.*, **104**, 1349 (1982).
106. C. Chatgilialoglu, K. U. Ingold, J. Lusztyk, A. S. Nazran, and J. C. Scaiano, *Organometallics*, **2**, 1332 (1983).
107. M. A. Ring, H. E. O'Neal, S. F. Rickborn, and B. A. Sawrey, *Organometallics*, **2**, 1891 (1983).
108. C. Chatgilialoglu, K. U. Ingold, and J. C. Scaiano, *J. Am. Chem. Soc.*, **104**, 5123 (1982).

109. C. Chatgilialoglu, K. U. Ingold, and J. C. Scaiano, *J. Am. Chem. Soc.*, **105**, 3292 (1983).
110. T. Dohmaru and Y. Nagata, *Bull. Chem. Soc. Jpn.*, **56**, 1847 (1983).
111. T. Dohmaru and Y. Nagata, *J. Phys. Chem.*, **86**, 4522 (1982).
112. T. Dohmaru and Y. Nagata, *Bull. Chem. Soc. Jpn.*, **56**, 2387 (1983).
113. C. Chatgilialoglu, K. U. Ingold, and J. C. Scaiano, *J. Am. Chem. Soc.*, **104**, 5119 (1982).
114. O. W. Webster, W. R. Hertler, D. Y. Sogah, W. B. Farnham, and T. V. Rajanbabu, *J. Am. Chem. Soc.*, **105**, 5706 (1983).
115. T. J. Barton, *Petrarch Register and Review of Organosilicon Compounds S-5 Edition*, 26, Petrarch Systems, Inc., USA (1982).
116. M. Gielen, *Top. Stereochem.*, **12**, (*Top. Inorg. Organomet. Stereochem.*), 37 (1981).
117. P. Jutzi, H. Saleske, D. Bühl, and H. Grohe, *J. Organomet. Chem.*, **252**, 29 (1983).
118. Y. A. Alexandrov and B. I. Tarunin, *J. Organomet. Chem.*, **238**, 125 (1982).
119. M. B. Taraban, T. V. Leshina, K. M. Salikhov, R. Z. Sagdeev, Y. N. Molin, O. I. Margorskaya, and N. S. Vyazankin, *J. Organomet. Chem.*, **256**, 31 (1983).
120. H. Sakurai, Y. Nakadaira, and H. Tobita, *Chem. Lett.*, 1855 (1982).
121. M. Schriewer and W. P. Neumann, *J. Am. Chem. Soc.*, **105**, 897 (1983).
122. H. Strehlow, I. Wagner, and P. Hildebrandt, *Ber. Bunsenges. Phys. Chem.*, **87**, 516 (1983).
123. D. S. Ross, K. F. Kuhlmann, and R. Malhotra, *J. Am. Chem. Soc.*, **105**, 4299 (1983).
124. N. C. Marziano and M. Sampoli, *J. Chem. Soc., Chem. Commun.*, 523 (1983).
125. A. H. Clemens, J. H. Ridd, and J. P. B. Sandall, *J. Chem. Soc., Chem. Commun.*, 343 (1983).
126. P. Helsby and J. H. Ridd, *J. Chem. Soc., Perkin Trans. 2*, 1191 (1983).
127. D. S. Ross, R. Malhotra, and W. C. Ogier, *J. Chem. Soc., Chem. Commun.*, 1353 (1982).
128. P. Helsby and J. H. Ridd, *J. Chem. Soc., Perkin Trans. 2*, 311 (1983).
129. F. Al-Omran and J. H. Ridd, *J. Chem. Soc., Perkin Trans. 2*, 1185 (1983).
130. B. N. Maya and G. Stedman, *J. Chem. Soc., Dalton Trans.*, 257 (1983).
131. A. Mertens, K. Lammertsma, M. Arvanaghi, and G. A. Olah, *J. Am. Chem. Soc.*, **105**, 5657 (1983).
132. R. E. Robertson, K. M. Koshy, A. Annessa, J. N. Ong, J. M. W. Scott, and M. J. Blandamer, *Can. J. Chem.*, **60**, 1780 (1982).
133. C. A. F. Johnson and C. M. Mechie, *J. Chem. Res. (S)*, 212 (1983).
134. W. A. Pryor, J. W. Lightsey, and D. F. Church, *J. Am. Chem. Soc.*, **104**, 6685 (1982).
135. E. C. Tuazon, W. P. L. Carter, R. V. Brown, A. M. Winer, and J. N. Pitts, Jr., *J. Phys. Chem.*, **87**, 1600 (1983).
136. G. P. Panigrahi and R. Nayak, *Int. J. Chem. Kinet.*, **15**, 989 (1983).
137. P. Keswani and Y. K. Gupta, *Indian J. Chem.*, **22A**, 763 (1983).
138. P. G. Bowers and R. M. Noyes, *J. Am. Chem. Soc.*, **105**, 2572 (1983).
139. S. B. Oblath, S. S. Markowitz, T. Novakov, and S. G. Chang, *J. Phys. Chem.*, **86**, 4853 (1982).
140. S. B. Oblath, S. S. Markowitz, T. Novakov, and S. G. Chang, *Inorg. Chem.*, **22**, 579 (1983).
141. A. Citterio, A. Gentile, F. Minisci, M. Serravalle, and S. Ventura, *J. Chem. Soc., Chem. Commun.*, 916 (1983).
142. P. N. Balasubramanian and E. S. Gould, *Inorg. Chem.*, **22**, 1100 (1983).
143. L. Castedo, R. Riguera, and M. P. Vázquez, *J. Chem. Soc., Chem. Commun.*, 301 (1983).
144. L. R. Dix and D. L. H. Williams, *J. Chem. Res. (S)*, 190 (1982).
145. S. E. Aldred, D. L. H. Williams, and M. Garley, *J. Chem. Soc., Perkin Trans. 2*, 777 (1982).
146. M. P. Doyle, J. W. Terpstra, R. A. Pickering, and D. M. LePoire, *J. Org. Chem.*, **48**, 3379 (1983).
147. R. P. Müller and J. R. Huber, *J. Phys. Chem.*, **87**, 2460 (1983).
148. K. A. Joergensen, M. T. M. El-Wassimy, and S. O. Lawesson, *Acta. Chem. Scand. B*, **37**, 785 (1983).

149. B. C. Challis and J. A. Challis, in: *The Chemistry of Functional Groups, Supplement F, The Chemistry of Amino, Nitroso and Nitro Compounds and Their Derivatives* (S. Patai, ed.), p. 1135, Wiley, New York (1982).
150. A. Meyer, D. L. H. Williams, R. Bonett, and S. L. Ooi, *J. Chem. Soc., Perkin Trans. 2,* 1383 (1982).
151. R. N. Leoppky and W. Tomasik, *J. Org. Chem.,* **48**, 2751 (1983).
152. V. G. Koshechko and A. N. Inozemtsev, *Zh. Obshch. Khim. SSSR* **53**, 2119 (1982); *Chem. Abstr.,* **100**, 67915d.
153. R. E. Lyle, W. G. Krueger, and V. E. Gunn, *J. Org. Chem.* **48**, 3574 (1983).
154. J. Cassado, A. Castro, J. R. Leis, M. A. L. Quintela, and M. Mosquera, *Monatsh. Chem.,* **114**, 639 (1983).
155. J. Cassado, A. Castro, M. A. L. Quintela, M. Mosquera, and M. F. R. Prieto, *Monatsh. Chem.,* **114**, 647 (1983).
156. R. N. Loeppky, J. R. Outram, W. Tomasik, and J. M. Faulconer, *Tetrahedron Lett.,* **24**, 4271 (1983).
157. J. W. Lown and S. M. S. Chauhan, *J. Org. Chem.,* **48**, 507 (1983).
158. T. A. Meyer and D. L. H. Williams, *J. Chem. Soc., Chem. Commun.,* 1067 (1983).
159. F. T. Bonner, J. Kada, and K. G. Phelan, *Inorg. Chem.,* **22**, 1389 (1983).
160. K. G. Phelan and G. Stedman, *J. Chem. Soc., Dalton Trans.,* 1603 (1982).
161. D. A. Nissen and D. E. Meeker, *Inorg. Chem.,* **22**, 716 (1983).
162. R. J. Gowland and G. Stedman, *J. Chem. Soc., Chem. Commun.,* 1038 (1983).
163. M. T. Beck and G. Rábai, *J. Chem. Soc., Dalton Trans.,* 1687 (1982).
164. M. Orbán and I. R. Epstein, *J. Am. Chem. Soc.,* **104**, 5918 (1982).
165. M. N. Hughes, J. R. Lusty, and H. L. Wallis, *J. Chem. Soc., Dalton Trans.,* 261 (1983).
166. M. N. Hughes and P. E. Wimbledon, *Inorg. Chim. Acta,* **65**, L129 (1982).
167. M. N. Hughes, H. L. Wallis, and P. E. Wimbledon, *J. Chem. Soc., Dalton Trans.,* 2181 (1982).
168. P. Keswani, M. R. Goyal, and Y. K. Gupta, *J. Chem. Soc., DaltonTrans.,* 1831 (1983).
169. V. C. Hand and D. W. Margerum, *Inorg. Chem.,* **22**, 1449 (1983).
170. M. P. Snyder and D. W. Margerum, *Inorg. Chem.* **21**, 2545 (1982).
171. R. A. Isaac and J. C. Morris, *Environ. Sci. Technol.,* **17**, 738 (1983).
172. D. S. Mahadevappa, K. S. Rangappa, N. M. M. Gowda, and B. T. Gowda, *Int. J. Chem. Kinet.,* **14**, 1183 (1982).
173. B. T. Gowda and D. S. Mahadevappa, *J. Chem. Soc., Perkin Trans. 2,* 323 (1983).
174. D. S. Mahadevappa, M. S. Ahmed, and N. M. M. Gowda, *Int. J. Chem. Kinet.,* **15**, 775 (1983).
175. V. C. Hand, M. P. Snyder, and D. W. Margerum, *J. Am. Chem. Soc.,* **105**, 4022 (1983).
176. D. S. Mahadevappa, S. Ananda, A. S. A. Murthy, and K. S. Rangappa, *React. Kinet. Catal. Lett.* **23**, 181 (1983).
177. W. D. Stanbro and M. J. Lenkevich, *Int. J. Chem. Kinet.,* **15**, 1321 (1983).
178. J. E. Wajon and J. C. Morris, *Inorg. Chem.,* **21**, 4258 (1982).
179. R. Kerbachi, R. Minkwitz, and U. Engelhardt, *Inorg. Chim. Acta,* **65**, L103 (1982).
180. C. A. Ogle, S. W. Martin, M. P. Dziobak, M. W. Urban, and G. D. Mendenhall, *J. Org. Chem.,* **48**, 3728 (1983).
181. T. Minato, S. Yamabe, and H. Oda, *Can. J. Chem.,* **60**, 2740 (1982).
182. K. Schlosser and S. Steenken, *J. Am. Chem. Soc.* **105**, 1504 (1983).
183. A. S. Nazran and D. Griller, *J. Am. Chem. Soc.,* **105**, 1970 (1983).
184. P. S. Skell, R. L. Tlumak, and S. Seshadri, *J. Am. Chem. Soc.,* **105**, 5125 (1983).
185. T. Huston, I. C. Hisatune, and J. Heicklen, *Can. J. Chem.,* **61**, 2070 (1983).
186. A. J. Frank and M. Grätzel, *Inorg. Chem.,* **21**, 3834 (1982).

187. M. Watanabe, *Bull. Chem. Soc. Jpn.*, **55**, 3766 (1982).
188. G. Kura, *Bull. Chem. Soc. Jpn.*, **56**, 3769 (1983).
189. M. Tsuhako, A. Nakahama, S. Ohashi, H. Nariai, and I. Motooka, *Bull. Chem. Soc. Jpn.*, **56**, 1372 (1983).
190. J. Emsley and S. Niazi, *Polyhedron*, **2**, 375 (1983).
191. J. P. Richard and P. A. Frey, *J. Am. Chem. Soc.*, **105**, 6605 (1983).
192. R. D. Sammons, H.-T. Ho, and P. A. Frey, *J. Am. Chem. Soc.*, **104**, 5841 (1982).
193. S. Meyerson, E. S. Kuhn, F. Ramirez, and J. F. Marecek, *J. Am. Chem. Soc.*, **104**, 7231 (1982).
194. M. A. Reynolds, N. J. Oppenheimer, and G. L. Kenyon, *J. Am. Chem. Soc.*, **105**, 6603 (1983).
195. P. M. Cullis, *J. Am. Chem. Soc.*, **105**, 7783 (1983).
196. B. T. Khan and P. N. Rao, *Inorg. Chim. Acta*, **67**, 79 (1982).
197. J. A. Gerlt, M. A. Reynolds, P. C. Demou, and G. L. Kenyon, *J. Am. Chem. Soc.*, **105**, 6469 (1983).
198. M. A. Reynolds, J. A. Gerlt, P. C. Demou, N. J. Oppenheimer, and G. L. Kenyon, *J. Am. Chem. Soc.*, **105**, 6475 (1983).
199. K. S. Dhathathreyan, S. S. Krishnamurthy, A. R. V. Murthy, R. A. Shaw, and M. Woods, *J. Chem. Soc., Dalton Trans.*, 1549 (1982).
200. D. I. Loewus and F. Eckstein, *J. Am. Chem. Soc.*, **105**, 3286 (1983).
201. T. A. Modro and B. P. Rijkmans, *J. Org. Chem.*, **47**, 3208 (1982).
202. J. Mollin, J. Láznička, and F. Kašpárek, *Collect. Czech. Chem. Commun.*, **48**, 232 (1983).
203. D. G. Gorenstein and K. Taira, *J. Am. Chem. Soc.*, **104**, 6130 (1982).
204. J. J. C. van Lier, L. J. M. van de Ven, J. W. de Haan, and H. M. Buck, *J. Phys. Chem.*, **87**, 3501 (1983).
205. K. W. Y. Abell and A. J. Kirby, *J. Chem. Soc., Perkin Trans. 2*, 1171 (1983).
206. N. E. Jacobsen and P. A. Bartlett, *J. Am. Chem. Soc.*, **105**, 1613 (1983).
207. N. E. Jacobsen and P. A. Bartlett, *J. Am. Chem. Soc.*, **105**, 1619 (1983).
208. V. Mizrahi and T. A. Modro, *J. Org. Chem.*, **48**, 3030 (1983).
209. G. McGall and R. A. McLelland, *J. Chem. Soc., Chem. Commun.*, 1222 (1982).
210. A. F. Janzen, A. E. Lemire, R. K. Marat, and A Queen, *Can. J. Chem.*, **61**, 2264 (1983).
211. P. J. J. M. van Ool and H. M. Buck, *Recl. Trav. Chim. Pays-Bas*, **102**, 215 (1983).
212. R. G. Cavell, S. Pirakitigoon, and L. V. Griend, *Inorg. Chem.*, **22**, 1378 (1983).
213. R. G. Cavell and L. V. Griend, *Inorg. Chem.*, **22**, 2066 (1983).
214. R. J. P. Corriu, G. F. Lanneau, and D. LeClerq, *Phosphorus Sulfur*, **18**, 197 (1983).
215. M. J. P. Hargen, *J. Chem. Soc., Perkin Trans. 1*, 2127 (1983).
216. N. Bourne and A. Williams, *J. Am. Chem. Soc.*, **105**, 3357 (1983).
217. M. T. Skoog and W. P. Jencks, *J. Am. Chem. Soc.*, **105**, 3356 (1983).
218. K. B. Dillon and A. W. G. Platt, *J. Chem. Soc., Chem. Commun.*, 1089 (1983).
219. K. B. Dillon and A. W. G. Platt, *J. Chem. Soc., Dalton Trans.*, 1199 (1982).
220. K. B. Dillon and A. W. G. Platt, *J. Chem. Soc., Dalton Trans.*, 1159 (1983).
221. R. M. K. Deng, K. B. Dillon, and A. W. G. Platt, *Phosphorus Sulfur* **18**, 93 (1983).
222. M. Anpo, K. U. Ingold, and J. K. S. Wan, *J. Phys. Chem.*, **87**, 1674 (1983).
223. M. Anpo, R. Sutcliffe, and K. U. Ingold, *J. Am. Chem. Soc.*, **105**, 3580 (1983).
224. R. F. Hudson, C. Brown, and A. Maron, *Chem. Ber.*, **115**, 2560 (1982).
225. A. K. Gupta, K. S. Gupta, and Y. K. Gupta, *J. Chem. Soc., Dalton Trans.*, 1845 (1982).
226. S. C. Dharma Rao, A. K. Panda, and S. N. Mahapatro, *J. Chem. Soc., Perkin Trans. 2*, 769 (1983).
227. O. Dahl, *Phosphorus Sulfur*, **18**, 201 (1983).
228. A. Skowrońska, E. Krawczyk, and J. Burski, *Phosphorus Sulfur*, **18**, 233 (1983).

229. C. Srinivasan and K. Pitchumani, *Int. J. Chem. Kinet.*, **14**, 1315 (1982).
230. A. Heesing and H. Steinkamp, *Chem. Ber.*, **115**, 2854 (1982).
231. J. G. Dawber, J. C. Tebby, and A. A. C. Waite, *J. Chem. Soc., Perkin Trans. 2*, 1923 (1983).
232. S. Bracher, J. I. G. Cadogan, I. Gosney, and S. Yaslak, *J. Chem. Soc., Chem. Commun.*, 857 (1983).
233. J. I. G. Cadogan, A. H. Cowley, I. Gosney, M. Pakulski, and S. Yaslak, *J. Chem. Soc., Chem. Commun.*, 1408 (1983).
234. J. I. G. Cadogan, *Phosphorus Sulfur*, **18**, 229 (1983).
235. C. D. Baer, J. O. Edwards, P. H. Rieger, and C. M. Silva, *Inorg. Chem.*, **22**, 1402 (1983).
236. L. J. Kirschenbaum and J. D. Rush, *Inorg. Chem.*, **22**, 3304 (1983).
237. I. N. Lisichkin, *Deposited Doc.* 1982, *VINITI* 273; *Chem. Abstr.* **98**, 41422f.
238. S. U. Choi and Y. I. Pae, *Bull. Korean Chem. Soc.*, **3**, 144 (1982).
239. S. U. Choi, Y. I. Pae, and S. H. Rhyu, *Bull. Korean Chem. Soc.*, **3**, 55 (1982).
240. B. A. Arbuzov, V. V. Klochkov, A. V. Aganov, Yu. M. Mareev, Yu. Yu. Samitov, and V. S. Vinogradova, *Izv. Akad. Nauk SSSR, Ser. Khim.*, 529 (1983); *Chem. Abstr.*, **98**, 205072v.
241. B. Maillard, K. U. Ingold, and J. C. Scaiano, *J. Am. Chem. Soc.*, **105**, 5095 (1983).
242. E. Niki, S. Yokoi, J. Tsuchiya, and Y. Kamiya, *J. Am. Chem. Soc.*, **105**, 1498 (1983).
243. D. M. Storch, C. J. Dymek, Jr., and L. P. Davis, *J. Am. Chem. Soc.*, **105**, 1765 (1982).
244. K. Fuke, M. Ueda, and M. Itoh, *J. Am. Chem. Soc.*, **105**, 1091 (1983).
245. E. L. Clennan and M. E. Mehrsheikh-Mohammadi, *J. Am. Chem. Soc.*, **105**, 5932 (1983).
246. A. A. Gorman, I. R. Gould, and I. Hamblett, *J. Am. Chem. Soc.*, **104**, 7098 (1982).
247. C. W. Jefford, S. Kohmoto, J. Boukouvalas, and U. Burger, *J. Am. Chem. Soc.*, **105**, 6498 (1983).
248. S. L. Wilson and G. B. Schuster, *J. Am. Chem. Soc.*, **105**, 679 (1983).
249. J. R. Hurst and G. B. Schuster, *J. Am. Chem. Soc.*, **104**, 6854 (1982).
250. K. Sehested, J. Holcman, and E. J. Hart, *J. Phys. Chem.*, **87**, 1951 (1983).
251. W. A. Pryor, D. Giamalva, and D. F. Church, *J. Am. Chem. Soc.*, **105**, 6858 (1983).
252. J. W. Agopovich and C. W. Gillies, *J. Am. Chem. Soc.*, **105**, 5047 (1983).
253. J.-I. Choe and R. L. Kuczkowski, *J. Am. Chem. Soc.*, **105**, 4839 (1983).
254. J.-I. Choe, M. Srinivasan, and R. L.Kuczkowski, *J. Am. Chem. Soc.*, **105**, 4703 1983).
255. L. Andrews and C. K. Kohlmiller, *J. Phys. Chem.*, **86**, 4548 (1982).
256. W. A. Pryor, D. G. Prier, and D. F. Church, *J. Am. Chem. Soc.*, **105**, 2883 (1983).
257. W. A. Pryor, C. K. Govindan, and D. F. Church, *J. Am. Chem. Soc.*, **104**, 7563 (1982).
258. W. A. Pryor, N. Onto, and D. F. Church, *J. Am. Chem. Soc.*, **105**, 3614 (1983).
259. E. Niki, Y. Yamamoto, T. Saito, K. Nagano, S. Yokoi, and Y. Kamiya, *Bull. Chem. Soc. Jpn.*, **56**, 223 (1983).
260. K. Namba and S. Nakayama, *Bull. Chem. Soc. Jpn.*, **55**, 3339 (1982).
261. K. Makino, M. M. Mossoba, and P. Riesz, *J. Phys. Chem.*, **87**, 1369 (1983).
262. D. T. Sawyer, T. S. Calderwood, K. Yamaguchi, and C. T. Angelis, *Inorg. Chem.*, **22**, 2577 (1983).
263. I. B. Afanas'ev, N. S. Kuprianova, and N. I. Polozova, *Int. J. Chem. Kinet.*, **15**, 1045 (1983).
264. P. A. Narayana, D. Suryanarayana, and L. Kevan, *J. Am. Chem. Soc.*, **104**, 3552 (1982).
265. I. B. Afanas'ev and N. S. Kuprianova, *Int. J. Chem. Kinet.*, **15**, 1057 (1983).
266. C. Bull, G. J. McClune, and J. A. Fee, *J. Am. Chem. Soc.*, **105**, 5290 (1983).
267. N. Getoff and M. Prucha, *Z. Naturforsch. A*, **38**, 589 (1983).
268. O. S. Fedorova and V. M. Berdnikov, *React. Kinet. Catal. Lett.* **23**, 73 (1983).
269. I. B. Afanas'ev and N. S. Kuprianova, *Int. J. Chem. Kinet.* **15**, 1063 (1983).
270. D. T. Sawyer, G. Chiericato, Jr., and T. Tsuchiya, *J. Am. Chem. Soc.*, **104**, 6273 (1982).
271. D. E. Cabelli and B. H. J. Bleiski, *J. Phys. Chem.*, **87**, 1809 (1983).

272. Z. M. Galbaćs and L. J. Csányi, *J. Chem. Soc., Dalton Trans.*, 2353 (1983).
273. O. U. Špalek, J. Balej, and I. Paseka, *J. Chem. Soc., Faraday Trans. 1*, **78**, 2349 (1982).
274. A. K. Gawley and W. J. Hood, *J. Chem. Soc., Faraday Trans 1*, **78**, 2815 (1982).
275. H. W. Richter and W. H. Waddell, *J. Am. Chem. Soc.*, **104**, 4630 (1982).
276. G. Levey and T. W. Ebbeson, *J. Phys. Chem.*, **87**, 829 (1983).
277. J. A. Gilbert, S. W. Gersten, and T. J. Meyer, *J. Am. Chem. Soc.*, **104**, 6872 (1982).
278. T. C. Bruice, *J. Chem. Soc., Chem. Commun.*, 14 (1983).
279. H. Mimoun, *Angew. Chem. Int. Ed.*, **21**, 734 (1982).
280. Y. Sawaki, H. Inoue, and Y. Ogata, *Bull. Chem. Soc. Jpn.*, **56**, 1133 (1983).
281. E. Buncel, H. Wilson, and C. Chuaqui, *J. Am. Chem. Soc.*, **104**, 4896 (1982).
282. E. Buncel, C. Chuaqui, and H. Wilson, *Int. J. Chem. Kinet.*, **14**, 823 (1982).
283. H. G. Hertz, H. Versmold, and C. Yoon, *Ber. Bunsenges. Phys. Chem.*, **87**, 577 (1983).
284. J. V. McArdle and M. R. Hoffmann, *J. Phys. Chem.*, **87**, 5425 (1983).
285. A. Huss, Jr., P. K. Lim, and C. A. Eckert, *J. Phys. Chem.*, **86**, 4224 (1982).
286. A. Huss, Jr., P. K. Lim, and C. A. Eckert, *J. Phys. Chem.*, **86**, 4229 (1982).
287. A. Huss, Jr., P. K. Lim, and C. A. Eckert, *J. Phys. Chem.*, **86**, 4233 (1982).
288. F. C. Laman, C. L. Gardner, and D. T. Fouchard, *J. Phys. Chem.*, **86**, 3130 (1982).
289. V. D. Parker, *Acta Chem. Scand. A* **37**, 423 (1983).
290. T. Ozawa and T. Kwan, *Polyhedron*, **2**, 1019 (1983).
291. K. C. Koshy and G. M. Harris, *Inorg. Chem.*, **22**, 2947 (1983).
292. U. Spitzer and R. van Eldik, *Inorg. Chem.*, **21**, 4008 (1982).
293. A. Hopkins, N. Bourne, and A. Williams, *J. Am. Chem. Soc.*, **105**, 3358 (1983).
294. A. Hopkins and A. Williams, *J. Chem. Soc., Chem. Commun.*, 37 (1983).
295. C. W. Perkins and J. C. Martin, *J. Am. Chem. Soc.*, **105**, 1377 (1983).
296. R. J. Cremlyn, P. H. Gore, A. O. O. Ikejiani, and D. F. C. Morris, *J. Chem. Res. (S)*, 194 (1982).
297. C. J. Garnett, A. J. Lambie, W. H. Beck, and M. Liler, *J. Chem. Soc., Faraday Trans. 1*, **79**, 953 (1983).
298. C. J. Garnett, A. J. Lambie, W. H. Beck, and M. Liler, *J. Chem. Soc., Faraday Trans. 1*, **79**, 965 (1983).
299. D. N. Harpp and R. A. Smith, *J. Am. Chem. Soc.*, **104**, 6045 (1982).
300. Z. Wu, T. B. Back, R. Ahmad, R. Yamdagni, and D. A. Armstrong, *J. Phys. Chem.*, **86**, 4417 (1982).
301. G. M. Whitesides, J. Houk, and M. A. K. Patterson, *J. Org. Chem.*, **48**, 112 (1983).
302. B. Boduszek and J. L. Kice, *J. Org. Chem.*, **47**, 3199 (1982).
303. B. Boduszek and J. L. Kice, *J. Org. Chem.*, **48**, 995 (1983).
304. S. Oae and H. Togo, *Bull. Chem. Soc. Jpn.*, **56**, 3802 (1983).
305. S. Oae and H. Togo, *Bull. Chem. Soc. Jpn.*, **56**, 3813 (1983).
306. S. Oae and H. Togo, *Bull. Chem. Soc. Jpn.*, **56**, 3818 (1983).
307. B. Meyer, K. Ward, K. Koshlap, and L. Peter, *Inorg. Chem.*, **22**, 2345 (1983).
308. T. Takata and S. Oae, *Bull. Chem. Soc. Jpn.*, **55**, 3937 (1982).
309. F. Akiyama, *Bull. Chem. Soc. Jpn.*, **56**, 2657 (1983).
310. G. Lowe and S. J. Salamone, *J. Chem. Soc., Chem. Commun.*, 1392 (1983).
311. C. Srinivasan and K. Pitchumani, *Int. J. Chem. Kinet.*, **14**, 789 (1982).
312. C. Srinivasan, P. Kuthalingam, and N. Arumgam, *Int. J. Chem. Kinet.*, **14**, 1139 (1982).
313. S. B. Jonnalagadda, V. Choudary, and A. K. Bhattacharya, *J. Chem. Soc., Perkin Trans. 2*, 849 (1983).
314. J. K. Rasmussen, S. M. Heilmann, P. E. Toren, A. V. Pocius, and T. A. Kotnour, *J. Am. Chem. Soc.*, **105**, 6845 (1983).
315. M. Kimura, A. Kobayashi, and K. Boku, *Bull. Chem. Soc. Jpn.*, **55**, 2068 (1982).

316. C. Walling, G. M. El-Taliawi, and C. Zhao, *J. Org. Chem.*, **48**, 4914 (1983).
317. C. Walling, C. Zhao, and G. M. El-Taliawi, *J. Org. Chem.*, **48**, 4910 (1983).
318. M. D. G. Constante and F. Ferranti, *Gazz. Chim. Ital.*, **112**, 173 (1982).
319. P. C. M. van Noort, H. P. W. Vermeeren, and R. Louw, *Recl. Trav. Chim. Pays-Bas*, **102**, 312 (1983).
320. S. Oae, T. Takata, and Y. H. Kim, *Bull. Chem. Soc. Jpn.*, **55**, 2484 (1982).
321. F. Freeman and C. N. Angeletakis, *J. Am. Chem. Soc.*, **105**, 4039 (1983).
322. F. Freeman, C. N. Angeletakis, and T. J. Maricich, *J. Org. Chem.*, **47**, 3403 (1982).
323. M. C. Hovey, *J. Am. Chem. Soc.*, **104**, 4196 (1982).
324. C. Gu and C. S. Foote, *J. Am. Chem. Soc.*, **104**, 6060 (1982).
325. J.-J. Liang, C.-L. Lu, M. L. Kacher, and C. S. Foote, *J. Am. Chem. Soc.*, **105**, 4717 (1983).
326. T. Akasaka and W. Ando, *J. Chem. Soc., Chem. Commun.*, 1203 (1983).
327. W. Ando, H. Miyazaki, and T. Akasaka, *Tetrahedron Lett.*, **23**, 2655 (1982).
328. C. Chatgilialoglu, L. Lunazzi, and K. U. Ingold, *J. Org. Chem.*, **48**, 3588 (1983).
329. J. L. Kice and H. Slebocka-Tilk, *J. Am. Chem. Soc.*, **104**, 7123 (1982).
330. G. H. Schmidt and D. G. Garratt, *J. Org. Chem.*, **48**, 4169 (1983).
331. J. E. Bäckvall, J. Bergman, and L. Engman, *J. Org. Chem.*, **48**, 3918 (1983).
332. F. Bigolo, E. Leporati, M. A. Pellinghelli, G. Crisponi, P. Deplano, and E. F. Trogu, *J. Chem. Soc., Dalton Trans.*, 1763 (1983).
333. R. T. Mahdi and J. D. Miller, *J. Chem. Soc., Dalton Trans.*, 1071 (1983).
334. O. Ito, *J. Am. Chem. Soc.*, **105**, 850 (1983).
335. M. S. Subhani, *Rev. Roum. Chim.*, **28**, 23 (1983).
336. W. Tötsch and F. Sladky, *Z. Naturforsch. B*, **38**, 1025 (1983).
337. J. M. Miller and J. H. Clark, *J. Chem. Soc., Chem. Commun.*, 1318 (1982).
338. T. Sugawara, M. Yudasaka, Y. Yokohama, T. Fujiyama, and H. Iwamura, *J. Phys. Chem.*, **86**, 2705 (1982).
339. J. M. Miller, R. K. Kanippayoor, and J. H. Clark, *J. Chem. Soc., Dalton Trans.*, 683 (1983).
340. J. Emsley, J. Lucas, R. J. Parker, and R. E. Overill, *Polyhedron*, **2**, 19 (1983).
341. *Chem. Eng. News*, **61**, 18 (1983).
342. A. A. Vigalok, G. G. Petrova, S. G. Lukashina, and I. V. Vigalok, *Russian J. Inorg. Chem.*, **27**, 1091 (1982).
343. D. N. Kevill and N. S. Posselt, *J. Chem. Res. (S)*, 264 (1983).
344. P. S. Radhakrishnamurti, C. Janardhana, G. P. Behera, and G. K. Mohanty, *Int. J. Chem. Kinet.*, **14**, 801 (1982).
345. P. S. Radhakrishnamurti, S. C. Pali, and B. R. Dev, *Int. J. Chem. Kinet.*, **14**, 1267 (1982).
346. M. A. Brusa and A. J. Colussi, *Int. J. Chem. Kinet.*, **14**, 1211 (1982).
347. C. S. Reddy and E. V. Sundaram, *Int. J. Chem. Kinet.*, **15**, 307 (1983).
348. S. P. Singh and R. K. Prasad, *J. Indian Chem. Soc.*, **60**, 350 (1983).
349. M. P. Pujari and P. Banerjee, *J. Chem. Soc., Dalton Trans.*, 1015 (1983).
350. A. M. Kjaer and J. Ulstrup, *Inorg. Chem.*, **21**, 3490 (1982).
351. S. C. Negi and K. K. Banerji, *J. Org. Chem.*, **48**, 3329 (1983).
352. B. Shah and K. K. Banerji, *J. Chem. Soc., Perkin Trans. 2*, 33 (1983).
353. C. Walling, G. M. El-Taliawi, and C. Zhao, *J. Am. Chem. Soc.*, **105**, 5119 (1983).
354. D. D. Tanner, T. C.-S. Ruo, H. Takiguchi, A. Guillaume, D. W. Reed, B. P. Setiloane, S. L. Tan, and C. P. Meintzer, *J. Org. Chem.*, **48**, 2743 (1983).
355. P. S. Radhakrishnamurti, C. Janardhana, and G. K. Mohanty, *Bull. Chem. Soc. Jpn.*, **55**, 3617 (1982).
356. M. A. Hussein, A. A. Abdel-Khalek, and Y. Sulfab, *J. Chem. Soc., Dalton Trans.*, 317 (1983).
357. M. S. Subhani, *Rev. Roum. Chim.*, **28**, 281 (1983).

358. M. S. Subhani, *Rev. Roum. Chim.*, **28**, 397 (1983).
359. B. C. Schardt and C. L. Hill, *Inorg. Chem.*, **22**, 1563 (1983).
360. D. D. Tanner, D. W. Reed, and B. P. Setiloane, *J. Am. Chem. Soc.*, **104**, 3917 (1982).
361. J. Muchová and V. Holba, *Collect. Czech. Chem. Commun.* **48**, 1158 (1983).
362. O. Gurel and D. Gurel, *Top. Curr. Chem.* **118** 1, 75 (1983).
363. A. M. Zhabotinskii, *Kem. Kozl.* **57**, 23 (1982).
364. W. Geiseler, *Ber. Bunsenges. Phys. Chem.*, **86**, 721 (1982).
365. I. R. Epstein, K. Kustin, P. De Kepper, and M. Orbán, *Sci. Am.*, **248**, 96 (1983).
366. F. D'alba and S. Di Lorenzo, *J. Chem. Soc., Faraday Trans. 1*, **79**, 39 (1983).
367. J. J. Tyson, *J. Phys. Chem.*, **86**, 3006 (1982).
368. P. De Kepper and K. Bar-Eli, *J. Phys. Chem.*, **87**, 480 (1983).
369. Z. Noszticzus, E. Noszticzius, and Z. A. Schelly, *J. Phys. Chem.*, **87**, 510 (1983).
370. O. Decroly and A. Goldbeter, *Proc. Natl. Acad. Sci. U.S.A.*, **79**, 6917 (1982).
371. M. Orbán and I. R. Epstein, *J. Phys. Chem.*, **86**, 3907 (1982).
372. M. Alamgir and I. R. Epstein, *J. Am. Chem. Soc.*, **105**, 2500 (1983).
373. M. Orbán, C. Dateo, P. De Kepper, and I. R. Epstein, *J. Am. Chem. Soc.*, **104**, 5911 (1982).
374. M. Orbán, P. De Kepper, and I. R. Epstein, *J. Phys. Chem.*, **86**, 431 (1982).
375. I. R. Epstein, C. E. Dateo, P. De Kepper, K. Kustin, and M. Orbán, *Springer Ser. Synergetics*, **12**, 188 (1981).
376. M. Alamgir, M. Orbán, and I. R. Epstein, *J. Phys. Chem.*, **87**, 3725 (1983).
377. M. Orbán and I. R. Epstein, *Springer Ser. Synergetics*, **12**, 197 (1982).
378. W. Geiseler, *J. Phys. Chem.*, **86**, 4394 (1982).
379. L. Adamčíková and P. Ševčík, *Int. J. Chem. Kinet.*, **14**, 735 (1982).
380. K. Bar-Eli and W. Geiseler, *J. Phys. Chem.*, **87**, 3769 (1983).
381. K. Bar-Eli and W. Geiseler, *Kem. Kozl.*, **57**, 269 (1982).
382. M. Alamgir, P. De Kepper, M. Orbán, and I. R. Epstein, *J. Am. Chem. Soc.*, **105**, 2641 (1983).
383. J. Rinzel and G. B. Ermentrout, *J. Phys. Chem.*, **86**, 2954 (1982).
384. M. Orbán and I. R. Epstein, *J. Phys. Chem.*, **87**, 3212 (1983).
385. A. M. Dallison, D. R. J. Macer, and G. A. Rodley, *Inorg. Chim. Acta*, **76**, L219 (1983).
386. L. Treindl and A. Olexová, *Electrochim. Acta*, **28**, 1495 (1983).
387. N. Ganapathsubramanian and R. M. Noyes, *J. Phys. Chem.*, **86**, 5158 (1982).
388. E. Kórös and E. Koch, *Thermochim. Acta*, **71**, 287 (1983).
389. E. Kórös and M. Varga, *React. Kinet. Catal. Lett.*, **21**, 521 (1982).
390. Z. Varadi and M. T. Beck, *React. Kinet. Catal. Lett.*, **21**, 527 (1982).
391. H. Ritschel and H. R. Weigt, *Z. Physik. Chem. (Leipzig)*, **264**, 593 (1983).
392. E. Kórös and M. Varga, *J. Phys. Chem.*, **86**, 4839 (1982).
393. N. Ganapathisubramanian and R. M. Noyes, *J. Phys. Chem.*, **86**, 5155 (1982).
394. K. Yoshikawa, *Bull. Chem. Soc. Jpn.*, **55**, 2042 (1982).
395. H. D. Foersterling, H. H. Lambertz, H. Schreiber, and W. Zittlau, *Kem. Kozl.*, **57**, 283 (1982).
396. H. D. Foesterling, H. J. Lambertz, and H. Schreiber, *Z. Naturforsch A* **38**, 483 (1983).
397. N. I. Butkovskaya, I. I. Morozov, V. L. Tal'rose, and E. S. Vasilev, *Chem. Phys.*, **79**, 21 (1983).
398. F. Bolletta and V. Balzani, *J. Am. Chem. Soc.*, **104**, 4250 (1982).
399. I. Tkáč and L. Treindl, *Collect. Czech. Chem. Commun.*, **48**, 13 (1983).
400. L. Adamčíková and O. Knappová, *Collect. Czech. Chem. Commun.*, **48**, 2335 (1983).
401. L. Adamčíková and P. Ševčík, *Collect. Czech. Chem. Commun.*, **47**, 2333 (1982).
402. L. Treindl and A. Nagy, *Collect. Czech. Chem. Commun.*, **48**, 3229 (1983).
403. A. Tockstein and M. Handlířová, *Collect. Czech. Chem. Commun.*, **47**, 2454 (1982).

404. D. Janjic, *Chimia*, **36**, 123 (1982).
405. R. Ramaswamy and S. Ramanathan, *Proc. Indian Acad. Sci.* (*Chem. Sci.*), **92**, 221 (1983).
406. V. J. Farage and D. Janjic, *Inorg. Chim. Acta*, **65**, L33 (1982).
407. M. A. Brusa and A. J. Colussi, *Int. J. Chem. Kinet.*, **15**, 1335 (1983).
408. T. Tlaczala and A. Bartecki, *Z. Phys. Chem.* (*Leipzig*), **264**, 507 (1983).
409. L. F. Salter and J. G. Sheppard, *Int. J. Chem. Kinet.*, **14**, 815 (1982).
410. L. Treindl and V. Dorovský, *Collect. Czech. Chem. Commun.*, **47**, 2831 (1982).
411. R. P. Rastogi, *Indian J. Chem.*, **22A**, 830 (1983).
412. D. Edelson, *J. Phys. Chem.*, **87**, 1204 (1983).
413. D. O. Cooke, *Int. J. Chem. Kinet.*, **14**, 1047 (1982).
414. S. D. Furrow, *J. Phys. Chem.*, **86**, 3089 (1982).
415. A. Hanna, A. Saul, and K. Showalter, *J. Am. Chem. Soc.*, **104**, 3838 (1982).
416. N. Ganapathisubramanian and K. Showalter, *J. Phys. Chem.*, **87**, 1098 (1983).

References for Chapter 5

1. P. S. Pregosin, *Coord. Chem. Rev.*, **44**, 247 (1982).
2. M. J. Blandamer and J. Burgess, *Pure Appl. Chem.*, **54**, 2285 (1982).
3. J. A. S. Howell and P. M. Burkinshaw, *Chem. Rev.*, **83**, 557 (1983).
4. E. C. Constable, *Polyhedron*, **2**, 551 (1983).
5. F. R. Hartley and J. A. Davies, *Rev. Inorg. Chem.*, **4**, 27 (1982).
6. J. U. Mondal and D. M. Blake, *Coord. Chem. Rev.*, **47**, 205 (1982).
7. J. A. Ibers, L. J. Pace, J. Martinsen, and B. M. Hoffman, *Struct. Bonding*, **50**, 1 (1982).
8. (a) S. Haghighi, C. A. McAuliffe, and M. E. Friedman, *Rev. Inorg. Chem.*, **3**, 291 (1981).
   (b) A. T. M. Marcelis and J. Reedijk, *Recl. Trav. Chim. Pay-Bas*, **102**, 121 (1983).
9. M. Cusumano, A. Gianetto, G. Guglielmo, and V. Ricevuto, *J. Chem. Soc., Dalton Trans.*, 2445 (1982).
10. M. J. Blandamer, J. Burgess, and P. P. Duce, *Transition Met. Chem.*, **8**, 308 (1983).
11. R. Gosling and M. L. Tobe, *Inorg. Chem.*, **22**, 1235 (1983).
12. G. Annibale, L. Cattalini, L. Canovese, G. Michelon, G. Marangoni, and M. L. Tobe, *Inorg. Chem.*, **22**, 975 (1983).
13. (a) L. I. Elding and O. Gröning, *Inorg. Chem. Acta*, **31**, 243 (1978).
    (b) J. A. Davis, *Adv. Inorg. Chem. Radiochem.*, **24**, 115 (1981).
14. J. J. Pesek and W. R. Mason, *Inorg. Chem.*, **22**, 2958 (1983).
15. J. K. Beattie, *Inorg. Chim. Acta*, **76**, L69 (1983).
16. H. G. Gray and R. J. Olcott, *Inorg. Chem.*, **1**, 481 (1962).
17. (a) M. Kotowski, D. A. Palmer, and H. Kelm, *Inorg. Chim. Acta*, **44**, L113 (1980).
    (b) R. Romeo and M. Cusumano, *Inorg. Chim. Acta*, **49**, 167 (1981).
18. (a) W. H. Baddley and F. Basolo, *J. Am. Chem. Soc.*, **88**, 2944 (1966).
    (b) J. B. Goddard and F. Basolo, *Inorg. Chem.*, **7**, 936 (1968).
    (c) R. Roulet and H. B. Gray, *Inorg. Chem.*, **9**, 2101 (1972).
19. L. Canovese, M. Cusumano, and A. Giannetto, *J. Chem. Soc., Dalton Trans.*, 195 (1983).
20. E. L. J. Breet, R. van Eldik, and H. Kelm, *Polyhedron*, **2**, 1181 (1983).
21. W. E. Hill, D. M. A. Minahan, G. G. Taylor, and C. A. McAuliffe, *J. Am. Chem. Soc.*, **104**, 6001 (1982).
22. W. E. Hill, D. M. A. Minahan, C. A. McAuliffe, and K. L. Minton, *Inorg. Chim. Acta*, **74**, 9 (1983).

23. J. C. Briggs, C. A. McAuliffe, W. E. Hill, D. M. A. Minahan, J. G. Taylor, and G. Dyer, *Inorg. Chem.*, **21**, 4204 (1982).
24. W. E. Hill, D. M. A. Minahan, and C. A. McAuliffe, *Inorg. Chem.*, **22**, 3382 (1983).
25. L. I. Elding, B. Kellenberger, and L. M. Venanzi, *Helv. Chim. Acta*, **66**, 1677 (1983).
26. E. L. J. Breet and R. van Eldik, *Inorg. Chim. Acta*, **76**, L301 (1983).
27. G. Annibale, L. Canovese, L. Cattalini, G. Marangoni, G. Michelon, and M. L. Tobe, *J. Chem. Soc., Dalton Trans.*, 775 (1983).
28. U. Belluco, L. Cattalini, F. Basolo, R. G. Pearson, and A. Turco, *J. Am. Chem. Soc.*, **87**, 241 (1969).
29. L. Mochida and J. C. Bailar, *Inorg. Chem.*, **22**, 1834 (1983).
30. G. W. Watt and W. A. Cude, *J. Am. Chem. Soc.*, **90**, 6382 (1968).
31. M. Kretschmer, P. S. Pregosin, and H. Rüegger, *J. Organomet. Chem.*, **241**, 87 (1983).
32. S. J. S. Kerrison and P. J. Sadler, *J. Chem. Soc., Dalton Trans.*, 2363 (1982).
33. G. Al-Takhin, H. Skinner, and A. A. Zaki, *J. Chem. Soc., Dalton Trans.*, 2323 (1983).
34. G. Al-Takhin, G. Pilcher, J. Bickerton and A. A. Zaki, *J. Chem. Soc., Dalton Trans.*, 2657 (1983).
35. S. N. Bhattacharya and C. V. Senoff, *Inorg. Chem.*, **22**, 1607 (1983).
36. G. Ferguson, R. McCrindle, A. J. McAlees, M. Parvez, and D. K. Stephenson, *J. Chem. Soc., Dalton Trans.*, 1865 (1983).
37. R. McCrindle, G. Ferguson, A. J. McAlees, M. Parves, and D. K. Stephenson, *J. Chem. Soc., Dalton Trans.*, 1291 (1982).
38. L. Helm, P. Meier, A. E. Merbach, and P. A. Tregloan, *Inorg. Chim. Acta*, **73**, 1 (1983).
39. A. G. Constable, C. R. Langrick, B. Shabanzadeh, and B. L. Shaw, *Inorg. Chim. Acta*, **65**, L151 (1982).
40. J. J. MacDougall, J. H. Nelson, and F. Mathey, *Inorg. Chem.*, **21**, 2145 (1982).
41. L. E. Erickson, T. A. Ferrett, and L. F. Buhse, *Inorg. Chem.*, **22**, 1461 (1983).
42. I. Ugi, D. Marquarding, H. Klusacek, P. Gillespie, and F. Ramirez, *Acc. Chem. Res.*, **4**, 288 (1971).
43. L. E. Erickson and D. C. Brower, *Inorg. Chem.*, **21**, 838 (1982).
44. J. A. Davies and V. Uma, *J. Electroanal. Chem.*, **158**, 13 (1983).
45. M. J. Blandamer, J. Burgess, and R. Romeo, *Inorg. Chim. Acta*, **65**, L179 (1982).
46. R. S. Paonessa and W. C. Trogler, *J. Am. Chem. Soc.*, **104**, 1138 (1982).
47. Yu K. Grishkin, V. A. Roznyatovsky, Yu A. Ustynyuk, S. N. Titova, G. A. Domrachev, and G. A. Razuvaev, *Polyhedron*, **2**, 895 (1983).
48. R. A. Jones, T. C. Wright, J. L. Atwood, and W. E. Hunter, *Organometallics*, **2**, 470 (1983).
49. F. R. Hartley, S. G. Murray, and D. M. Potter, *J. Organometal. Chem.*, **254**, 119 (1983).
50. D. Bandyopadhyay, P. Bandyopadhyay, A. Chakravorty, F. A. Cotton, and L. R. Falvello, *Inorg. Chem.*, **22**, 1315 (1983).
51. M. Kubota, R. K. Rothrock, M. R. Kernan, and R. B. Haven, *Inorg. Chem.*, **21**, 2491 (1982).
52. G. Bodenhausen, J. A. Deli, C. Anklin, and P. S. Pregosin, *Inorg. Chim. Acta*, **77**, 417 (1983).
53. C. A. Bignozzi, C. Bartocci, C. Chiorboli, and V. Carassiti, *Inorg. Chim. Acta*, **70**, 87 (1983).
54. Y.-T. Fanchiang, G. T. Bratt, and H. P. C. Hogenkamp, *J. Chem. Soc., Dalton Trans.*, 1929 (1983).
55. R. W. Hay and A. K. Basak, *J. Chem. Soc., Dalton Trans.*, 1819 (1982).
56. M. Chou, C. Creutz, D. Mahajan, N. Sutin, and A. P. Zipp, *Inorg. Chem.*, **21**, 3989 (1982).
57. R. L. Batstone-Cunningham, N. W. Dodgen, J. P. Hunt, and D. M. Roundhill, *J. Chem. Soc., Dalton Trans.*, 1473 (1983).
58. K. Chandrasekhar and H.-B. Burgi, *J. Am. Chem. Soc.*, **105**, 7081 (1983).

References for Chapter 6

1. P. Moore, this series, Vol. 2.
2. L. F. Larkworthy, *Coord. Chem. Rev.*, **45**, 105 (1982).
3. J. H. Espenson, M. Shimura, and A. Bakac, *Inorg. Chem.*, **21**, 2537 (1982).
4. W. A. Mulac, H. Cohen, and D. Meyerstein, *Inorg. Chem.*, **21**, 4016 (1982).
5. A. Bakac and J. H. Espenson, *Inorg. Chem.*, **21**, 2537 (1982).
6. J. H. Espenson, P. Connolly, D. Meyerstein, and H. Cohen, *Inorg. Chem.*, **22**, 1009 (1983).
7. H. Ogino, M. Shimura, and N. Tanaka, *J. Chem. Soc., Chem. Commun.*, 1063 (1983).
8. H. Ogino, M. Shimura, and N. Tanaka, *Inorg. Chem.*, **21**, 126 (1982).
9. V. Gold and D. L. Wood, *J. Chem. Soc., Chem. Commun.*, **97**, 45104f (1982).
10. F. Basolo, P. Moore, and R. G. Pearson, *Inorg. Chem.*, **5**, 223 (1966).
11. S. H. McClaugherty and C. M. Grisham, *Inorg. Chem.*, **21**, 4133 (1982).
12. L. Kolosci and K. Schug, *Inorg. Chem.*, **22**, 3053 (1983).
13. N. E. Dixon, G. A. Lawrence, P. A.Lay, and A. M. Sargeson, *Inorg. Chem.*, **22**, 846 (1983).
14. D. Yang and D. A. House, *Polyhedron*, **2**, 1267 (1983).
15. A. D. Kirk and S. G. Glover, *Inorg. Chem.*, **22**, 176 (1983).
16. N. Ise, T. Okubo, and Y. Yamamura, *J. Phys. Chem.*, **86**, 1694 (1982).
17. N. Ise, T. Okubo, and Y. Yamamura, *Ber. Bunsenges. Phys. Chem.*, **86**, 922 (1982).
18. A. O. Adeniran, G. J. Baker, G. J. Bennett, M. J. Blandamer, J. Burgess, N. K. Dhammi, and P. P. Duce, *Transition Met. Chem.*, **7**, 183 (1982).
19. J. U. Hwang, J. L. Chung, and S. O. Bek, *J. Chem. Soc., Chem. Commun.*, **28**, 95 (1984).
20. J. W. Vaughn, *Inorg. Chem.*, **22**, 844 (1983).
21. Y. N. Shevchenko and V. V. Sachok, *Zh. Neorg. Khim.*, **28**, 1973 (1983).
22. D. A. House and D. Yang, *Inorg. Chim. Acta*, **74**, 179 (1983).
23. D. A. House and D. Yang, *Inorg Chem.*, **21**, 2999 (1982).
24. D. A. House and O. Nor, *Inorg. Chim. Acta*, **72**, 195 (1983).
25. D. Yang and D. A. House, *Inorg. Chim. Acta*, **64**, L167 (1982).
26. H. R. Maecke, B. F. Mentzen, J. P. Puaux, and A. W. Adamson, *Inorg. Chem.*, **21**, 3080 (1982).
27. Z. Chen, M. Cimolino, and A. W. Adamson, *Inorg. Chem.*, **22**, 3035 (1983).
28. P. Kita, *Polish J. Chem.*, **56**, 913 (1982).
29. S. Das, R. N. Banerjee, and D. Banerjea, *J. Coord. Chem.*, **13**, 123 (1984).
30. A. K. Basak and C. Chatterjee, *Bull. Chem. Soc. Jpn.*, **56**, 318 (1983).
31. A. Dhur, S. Das, R. N. Banerjee, and D. Banerjea, *Transition Met. Chem.*, **7**, 125 (1982).
32. B. Chakravorty, S. Modak, and B. Nandi, *Transition Met. Chem.*, **7**, 113 (1982).
33. A. A. Koroleva and E. G. Timofeeva, *Koord. Khim.*, **8**, 314 (1982).
34. N. Nakano, Y. Narusawa, and M. Tsuchiya, *J. Inorg. Nucl. Chem.*, **43**, 3011 (1981).
35. R. F. Johnston and R. A. Holwerda, *Inorg. Chem.*, **22**, 2942 (1983).
36. B. K. Niogy and G. S. De, *Chem. Abstr.* **99**, 219437a (1983).
37. M. Bhattacharya and G. S. De, *Indian J. Chem.*, **21A**, 898 (1982).
38. I. A. Khan, M. Shahid, and Kabir-Ud-Din, *Indian. J. Chem.*, **22A**, 382 1983).
39. F. Capitan, E. J. Alonso, and M. C. Valencia, *Chem. Abstr.*, **96**, 1304802 (1982).
40. M. Bhattacharya and G. S. De, *Indian. J. Chem.*, **22A**, 626 (1983).
41. E. Garcia-Espana, J. Moratal, and J. Faus, *J. Coord. Chem.*, **12**, 41 (1982).
42. N. F. Kosenko and I. I. Batishcheva, *Chem. Abstr.* **99**, 2011830 (1983).
43. W. V. Malik, S. P. Srivastava, T. K. Thallum, and V. K. Gupta, *Acta Chim. Acad. Sci. Hung.*, **109**, 345 (1982).
44. M. K. Chaidhuri and N. Ray, *Synth. React. Inorg. Met.-Org. Chem.*, **12**, 879 (1982).

45. R. J. Balahura, W. L. Purcell, M. E. Vicotriano, M. L. Lieberman, V. M. Loyola, W. Fleming, and J. W. Fronabarger, *Inorg. Chem.*, **22**, 3602 (1983).
46. J. L. Laird and R. B. Jordon, *Inorg. Chem.*, **21**, 4127 (1982).
47. R. J. Balahura, W. Kupferschmidt, and W. L. Purcell, *Inorg. Chem.*, **22**, 1456 (1983).
48. R. J. Balahura and A. W. Johnston, *Inorg. Chem.*, **22**, 3309 (1983).
49. H. Stunzi and W. Marty, *Inorg. Chem.*, **22**, 2145 (1983).
50. I. Ostrich and A. J. Leffler, *Inorg. Chem.*, **22**, 921 (1983).
51. M. Martinez and M. Ferrer, *Inorg. Chim. Acta*, **69**, 123 (1983).
52. W. A. Wickramasinghe, P. H. Bird, M. A. Jamieson, N. Serpone, and M. Maestri, *Inorg. Chim. Acta*, **64**, (1982).
53. Y. N. Shevchenko and N. I. Yashina, *Chem. Abstr.*, **98**, 171871n (1983).
54. Y. N. Shevchenko, V. V. Sachok, and V. Y. Dudarev, *Chem. Abstr.*, **97**, 103249v (1982).
55. D. A. House, *Inorg. Chim. Acta*, **60**, 145 (1982).
56. A. L. Hale, W. Levason, and F. P. McCullough, *Inorg. Chem.*, **21** (1982).
57. T. Laier, B. Nielson, and J. Springborg, *Acta Chem. Scand., Ser. A*, **36**, 91 (1982).
58. H. R. Fischer, J. Glerup, D. J. Hodgson, and E. Pedersen, *Inorg. Chem.*, **21**, 3063 (1982).
59. S. J. Cline, D. J. Hodgson, S. Kallesoe, S. Larsen, and E. Pedersen, *Inorg. Chem.*, **22**, 637 (1983).
60. D. J. Listen, K. S. Murray, and B. O. West, *J. Chem. Soc., Chem. Commun.*, 1109 (1982).
61. P. Leupin, A. G. Sykes, and K. Wieghardt, *Inorg. Chem.*, **22**, 1253 (1983).
62. V. A. Raman, T. S. P. Rao, N. R. Anipindi, and M. N. Sastri, *Polish J. Chem.*, **55**, 659 (1981).
63. S. Saric and I. J. Gal, *Chem. Abstr.*, **98**, 150242t (1983).
64. R. Vasantha, B. U. Nair, T. Ramasami, and D. Ramaswamy, *Chem. Abstr.*, **97**, 155276w (1982).
65. N. A. P. Kane-Maguire, J. A. Bennett, and P. K. Miller, *Inorg. Chim. Acta*, **76**, L123 (1983).
66. J. Eriksen and O. Moensted, *Acta Chem. Scand., Ser. A*, **37**, 579 (1983).
67. J. G. Leipoldt, R. Van Eldik, and H. Kelm, *Inorg. Chem.*, **22**, 4146 (1983).
68. T. W. Kallen and E. J. Senko, *Inorg. Chem.*, **22**, 2924 (1983).
69. F. Bolletta, M. Maestri, L. Moggi, M. A. Jamieson, N. Serpone, M. S. Henry, and M. Z. Hoffman, *Inorg. Chem.*, **22**, 2502 (1983).
70. J. Lillie and W. L. Waltz, *Inorg. Chem.*, **22**, 1473 (1983).
71. M. T. Gandolfi, M. Maestri, D. Sandrini, and V. Balzani, *Inorg. Chem.*, **22**, 3435 (1983).
72. A. D. Kirk and C. Namasivayam, *Inorg. Chem.*, **22**, 2961 (1983).
73. P. Riccieri and E. Zinato, *Inorg. Chem.* **22**, 2305 (1983).
74. N. A. P. Kane-Maguire, W. S. Crippen, and P. K. Miller, *Inorg. Chem.*, **22**, 696 (1983).
75. N. A. P. Kane-Maguire, M. M. Allen, J. M. Vaught, J. S. Hallock, and A. L. Heatherington, *Inorg. Chem.*, **22**, 3851 (1983).
76. N. J. Curtis, G. A. Lawrence, and A. M. Sargeson, *Aust. J. Chem.*, **36**, 1495 (1983).
77. H. Kanno, T. Shimotori, S. Utsuno, and J. Fujita, *Chem. Lett.*, 939 (1983).
78. P. Kita, *Polish J Chem.*, **56**, 919 (1982).
79. P. Kita and L. Chamarczuk, *Polish J. Chem.*, **55**, 2251 (1981).
80. M. N. Bishnu, B. Chakravorti, R. N. Banerjee, and D. Banerjea, *J. Coord. Chem.*, **13**, 63 (1983).
81. A. Uehara, Y. Nishiyama, and R. Tsuchiya, *Inorg. Chem.*, **21**, 2422 (1982).
82. A. Uehara, Y. Nishiyama, and R. Tsuchiya, *Inorg. Chem.*, **22**, 2864 (1983).
83. M. Monfort, M. Serra, A. Escuer, and J. Ribas, *Thermochim. Acta*, **69**, 397 (1983).
84. M. Serra, A. Escuer, J. Ribas, M. D. Baro, and J. Casabo, *Thermochim. Acta* **56**, 183 (1982).
85. M. Serra, A. Escuer, J. Ribas, and M. D. Baro, *Thermochim. Acta*, **64**, 237 (1983).
86. M. Corbella, M. Serra, L. Martinez-Sarrion, and J. Ribas, *Thermochim. Acta*, **57**, 283 (1982).
87. Y. N. Shevchenko and N. I. Yashina, *Zh. Neorg. Khim.*, **27**, 1441 (1982).

88. G. Liptay, C. Varhelyi, E. Brandt-Petrik, and J. Zsako, *Chem. Abstr.* **97**, 137705f (1982).
89. P. O'Brien, *Polyhedron*, **2**, 233 (1983).
90. B. M. L. Bhatia, *Thermochim. Acta*, **58**, 367 (1982).
91. N. Rajasekar, R. Subramaniam, and E. S. Gould, *Inorg. Chem.*, **22**, 971 (1983).
92. K. Ohtsuka, *Chem. Abstr.* **100**, 40426h (1984); **97**, 151462f (1982); **97**, 28157v (1982).
93. N. A. Al-Jallal and Y. Sulfab, *Transition Met. Chem.*, **8**, 51 (1983).
94. P. Russev, P. R. Bontchev, K. Kobassanov, and A. Molinovski, *Inorg. Chim. Acta*, **70**, 179 (1983).
95. A. Okuma, N. Takeuchi, S. Tsuji, and N. Okazaki, *Inorg. Chim. Acta*, **74**, 77 (1983).

References for Chapter 7

1. R. W. Hay, *Coord. Chem. Rev.* **57**, 1 (1984).
2. M. L. Tobe, in: *Advances in Inorganic and Bioinorganic Mechanisms* (A. G. Sykes, ed.), Vol. II, Academic Press, London (1983).
3. N. E. Dixon and A. M. Sargeson, in: *Zinc Enzymes* (T. Spiro, ed.), Wiley, New York (1983).
4. G. A. Lawrance, *Inorg. Chem.*, **21**, 3687 (1982).
5. W. G. Jackson and C. M. Begbie, *Inorg. Chim. Acta*, **61**, 167 (1982).
6. D. A. House and O. Nor, *Inorg. Chim. Acta*, **72**, 195 (1983).
7. C.-K. Poon, T.-C. Lau, C.-L. Wong, and Y.-P. Kan, *J. Chem. Soc., Dalton Trans.*, 1641 (1983).
8. Y. Kitamura, *Bull. Chem. Soc. Jpn.*, **55**, 3625 (1982).
9. S. Das, R. N. Banerjee, S. G. Bhattacharya, and D. Banerjea, *Transition Met. Chem.*, **8**, 241 (1983).
10. J. N. Cooper, J. G. Bentsen, T. M. Handel, K. M. Strohmaier, W. A. Porter, B. G. Johnson, A. M. Carr, D. A. Farnath, and S. L. Appleton, *Inorg. Chem.*, **22**, 3060 (1983).
11. B. Chakravarty, S. Modak, P. K. Das, and A. K. Sil, *J. Inorg. Nucl. Chem.*, **43**, 3253 (1981).
12. R. W. Hay and A. K. Basak, *Inorg. Chim. Acta*, **69**, 1 (1983).
13. R. W. Hay, unpublished observations.
14. S. Das, S. Bhattacharya, R. N. Banerjee, and D. Banerjea, *Transition Met. Chem.*, **7**, 249 (1982).
15. M. Ida, M. Ando, and H. Yamatera, *Bull. Chem. Soc. Jpn.*, **55**, 1441 (1982).
16. W. L. Reynolds, M. Glavas, and E. Dzelilovic, *Inorg. Chem.*, **22**, 1946 (1983).
17. D. A. Buckingham, C. R. Clark, and W. S. Webley, *Inorg. Chem.*, **21**, 3353 (1982).
18. D. A. House and W. Marty, *Inorg. Chim. Acta*, **65**, L47 (1982).
19. P. M. Coddington and K. E. Hyde, *Inorg. Chem.*, **22**, 2211 (1983).
20. R. Van Eldik, U. Spitzer, and H. Kelm, *Inorg. Chim. Acta*, **74**, 149 (1983).
21. D. Smith, M. F. Amira, P. B. Abdullah, and C. B. Monk, *J. Chem. Soc., Dalton Trans.*, 337 (1983).
22. P. B. Abdullah and C. B. Monk, *J. Chem. Soc., Dalton Trans.*, 1175 (1983).
23. W. G. Jackson, C. M. Begbie, and M. L. Randall, *Inorg. Chim. Acta*, **70**, 7 (1983).
24. W. G. Jackson, M. L. Randall, A. M. Sargeson, and W. Marty, *Inorg. Chem.*, **22**, 1013 (1983).
25. W. G. Jackson, D. P. Fairlie, and M. L. Randall, *Inorg. Chim. Acta*, **70**, 197 (1983).
26. A. R. Gainsford and D. A. House, *Inorg. Chim. Acta*, **74**, 205 (1982).
27. M. Humanes, C. Chatterjee and M. L. Tobe (to be published).
28. G. Bombieri, E. Forseltini, A. Del Pra, M. L. Tobe, C. Chatterjee, and C. J. Cooksey, *Inorg. Chim. Acta*, **75**, 93 (1983).

29. E. C. Niederhoffer, C. J. Raleigh, A. E. Martell, P. Rudolf, and A. Clearfield, *Cryst. Struct. Commun.*, **11**, 1163 (1982).
30. G. Bombieri, E. Forsellini, A. Del Pra, M. L. Tobe, and C. Chatterjee, *Inorg. Chim. Acta*, **68**, 205 (1983).
31. D. A. Buckingham, J. D. Edwards, and G. M. McLaughlin, *Inorg. Chem.*, **21**, 2770 (1982).
32. See, for example, P. C. Jain and E. C. Lingafelter, *J. Am. Chem. Soc.*, **89**, 724 (1967); M. di Vaira and P. L. Orioli, *Acta Crystallogr., Sect. B* **24**, 595 (1968); M. di Vaira and P. L. Orioli, *Inorg. Chem.*, **6**, 490 (1967).
33. D. A. Buckingham, C. R. Clark, and W. S. Webley, *Aust. J. Chem.*, **33**, 263 (1980).
34. D. A. Buckingham, P. J. Cresswell, and A. M. Sargeson, *Inorg. Chem.*, **14**, 1485 (1975).
35. B. Chakravarty and P. K. Das, *Transition Met. Chem.*, **7**, 340 (1982).
36. B. Chakravarty and P. K. Das, *Transition Met. Chem.*, **8**, 146 (1983).
37. M. N. Bishnu, B. Chakravarty, R. N. Banerjee, and D. Banerjea, *J. Coord. Chem.*, **13**, 63 (1983).
38. T. P. Dasgupta and G. Sadler, *Inorg. Chim. Acta*, **78**, L173 (1983).
39. R. W. Hay and A. K. Basak, *Inorg. Chim. Acta*, **73**, 179 (1983).
40. C. A. Andrade and H. Taube, *J. Am. Chem. Soc.*, **86**, 1328 (1964).
41. S. Sheel, D. T. Meloon, and G. M. Harris, *Inorg. Chem.*, **1**, 170 (1962).
42. R. B. Jordan and H. Taube, *J. Am. Chem. Soc.*, **88**, 4406 (1966).
43. R. W. Hay and A. K. Basak, *Inorg. Chim. Acta*, **71**, 73 (1983).
44. L. S. Bark, M. B. Davies, and S. R. Dawson, *Polyhedron*, **1**, 203 (1982).
45. R. W. Hay, R. Bembi, W. T. Moodie, and P. R. Norman, *J. Chem. Soc., Dalton Trans.*, 2131 (1982).
46. For a discussion, see R. A. Henderson and M. L. Tobe, *Inorg. Chem.*, **16**, 2576 (1977).
47. L. R. Gahan, G. A. Lawrance, and A. M. Sargeson, *Aust. J. Chem.*, **35**, 1119 (1982).
48. N. K. Mohanty and R. K. Nanda, *Transition Met. Chem.*, **7**, 202 (1982).
49. B. F. Anderson, J. D. Bell, D. A. Buckingham, P. J. Cresswell, G. J. Gainsford, L. G. Marzilli, G. B. Robertson, and A. M. Sargeson, *Inorg. Chem.*, **16**, 3233 (1977).
50. A. M. Sargeson, *Pure Appl. Chem.*, **33**, 527 (1973).
51. D. A. House, *Coord. Chem. Rev.*, **23**, 223 (1977) and references therein.
52. R. Van Eldik, A. C. Dash, and G. M. Harris, *Inorg. Chim. Acta*, **77**, L143 (1983).
53. A. C. Dash and G. M. Harris, *Inorg. Chem.*, **20**, 4011 (1981).
54. S. Balt, H. J. Gamelkoorn, and W. E. Renkema, *J. Chem. Soc., Dalton Trans.* 2415 (1983).
55. S. Balt, J. Brennan, and W. de Kieviet, *J. Inorg. Nucl. Chem.*, **41**, 331 (1979).
56. See Reference 59.
57. R. Dreos Garlatti, G. Tauzher, and G. Costa, *Inorg. Chim. Acta*, **71**, 9 (1983).
58. W. K. Wilmarth, K. R. Ashley, J. C. Harmon, J. Fredericks, and A. L. Crumbliss, *Coord. Chem. Rev.*, **51**, 225 (1983).
59. S. Balt, H. J. C. Gamelkoorn, H. J. A. M. Kuipers, and W. E. Renkema, *Inorg. Chem.*, **22**, 3072 (1983).
60. I. M. Sidahmed and C. F. Wells, *J. Chem. Soc., Dalton Trans.*, 1035 (1983).
61. C. N. Elgy and C. F. Wells, *J. Chem. Soc., Dalton Trans.*, 1617 (1983).
62. K. Tsukahara, H. Oshita, Y. Emoto, and Y. Yamamoto, *Bull. Chem. Soc. Jpn.*, **55**, 2107 (1982).
63. R. Dreos Garlatti, G. Tauzher, and G. Costa, *Inorg. Chim. Acta*, **70**, 83 (1983).
64. A. C. Dash, R. K. Nanda, and N. Ray, *J. Coord. Chem.*, **12**, 91 (1982).
65. B. Chakravarty and P. K. Das, *Transition Met. Chem.*, **8**, 165 (1983).
66. W. K. Wilmarth, J. E. Byrd, H. N. Po, H. K. Wilcox, and P. H. Tewari, *Coord. Chem. Rev.*, **51**, 181 (1983).
67. W. K. Wilmarth, J. E. Byrd, and H. N. Po, *Coord. Chem. Rev.*, **51**, 209 (1983).

68. A. Haim, *Inorg. Chem.*, **21**, 2887 (1982).
69. A. Haim and W. K. Wilmarth, *Inorg. Chem.*, **1**, 573 (1962).
70. M. G. Burnett and M. W. Gilfillan, *J. Chem. Soc., Dalton Trans.*, 1578 (1981).
71. F. Basolo, B. D. Stone, and R. G. Pearson, *J. Am. Chem. Soc.*, **75**, 819 (1953).
72. J. N. Cooper, C. A. Pennell, and B. C. Johnson, *Inorg. Chem.*, **22**, 1956 (1983).
73. M. C. Ghosh and P. Banerjee, *Bull. Chem. Soc. Jpn.*, **56**, 2871 (1983).
74. S. Balt and A. M. van Herk, *Transition Met. Chem.*, **8**, 152 (1983).
75. G. Tauzher, R. Dreos Garlatti, and G. Costa, *Inorg. Chim. Acta* **75**, 145 (1983).
76. P. O'Brien, *Polyhedron*, **2**, 233 (1983).
77. J. A. Chambers, T. J. Goodwin, M. D. W. Mulqii, P. A. Williams, and R. S. Vagg, *Inorg. Chim. Acta*, **75**, 241 (1983).
78. W. Rindermann and R. van Eldik, *Inorg. Chim. Acta*, **68**, 35 (1983).
79. K. Miyoshi, N. Katoda, and H. Yoneda, *Inorg. Chem.*, **22**, 1839 (1983).
80. S. Balt, H. J. A. M. Kuipers, and W. E. Renkema, *J. Chem. Soc., Dalton Trans.*, 1739 (1983).
81. W. Rindermann, R. van Eldik, and H. Kelm, *Inorg. Chim. Acta*, **61**, 173 (1982).
82. W. G. Jackson, G. A. Lawrance, P. A. Lay, and A. M. Sargeson, *Aust. J. Chem.*, **35**, 1561 (1982).
83. R. Tsuchiya, A. Uehara, and Y. Muramatsu, *Bull. Chem. Soc. Jpn.*, **55**, 3770 (1982).
84. W. L. Purcell, *Inorg. Chem.*, **22**, 1205 (1983).
85. M. F. Hoq, C. R. Johnson, S. Paden, and R. E. Shepherd, *Inorg. Chem.*, **22**, 2693 (1983).
86. P. Deplano and E. F. Trogu, *Inorg. Chim. Acta*, **61**, 265 (1982).
87. P. Deplano and E. F. Trogu, *Inorg. Chim. Acta*, **61**, 261 (1982).
88. P. Deplano and E. F. Trogu, *Inorg. Chim. Acta*, **68**, 153 (1983).
89. C. H. Langford and H. B. Gray, *Ligand Substitution Processes*, Benjamin, New York (1965).
90. M. H. M. Abou-El-Wafe and M. G. Burnett, *J. Chem. Soc., Chem. Comm.*, 833 (1983).
91. G. Chiavon and C. Paradisi, *Inorg. Chim. Acta*, **81**, 219 (1984).
92. E. Campi, C. Paradisi, G. Schiavon, and M. L. Tobe, *J. Chem. Soc., Chem. Comm.*, 682 (1966).
93. See, for example, R. W. Hay and B. Jeragh, *Transition Met. Chem.*, **4**, 289 (1979); R. W. Hay and B. Jeragh, *J. Chem. Soc., Dalton Trans.*, 1343 (1979); R. W. Hay and B. Jeragh, *Transition Met. Chem.*, **5**, 252 (1980).
94. See, for example, G. M. Harris and K. E. Hyde, *Inorg. Chem.*, **17**, 1892 (1978); K. E. Hyde, G. H. Fairchild, and G. M. Harris, *Inorg. Chem.*, **15**, 2631 (1976).
95. See, for example, T. P. Dasgupta and G. M. Harris, *J. Am. Chem. Soc.*, **93**, 91 (1971); T. P. Dasgupta and G. M. Harris, *Inorg. Chem.*, **13**, 1275 (1974); G. M. Harris and K. E. Hyde, *Inorg. Chem.*, **17**, 1892 (1978).
96. R. van Eldik and G. M. Harris, *Inorg. Chim. Acta*, **70**, 147 (1983).
97. K. Koshy and T. P. Dasgupta, *Inorg. Chim. Acta*, **61**, 207 (1982).
98. J. F. Glenister, K. E. Hyde, and G. Davies, *Inorg. Chem.*, **21**, 2331 (1982).
99. U. Spitzer and R. van Eldik, *Inorg. Chem.*, **21**, 4008 (1982).
100. R. van Eldik and G. M. Harris, *Inorg. Chem.*, **19**, 880 (1980).
101. R. van Eldik, *Inorg. Chim. Acta*, **42**, 49 (1980).
102. R. van Eldik, J. von Jouanne, and H. Kelm, *Inorg. Chem.*, **21**, 2818 (1982).
103. See for example, R. van Eldik, D. A. Palmer, H. Kelm, and G. M. Harris, *Inorg. Chem.*, **19**, 3679 (1980) and literature cited therein.
104. W. G. Jackson, G. A. Lawrance, P. A. Lay, and A. M. Sargeson, *J. Chem. Soc., Chem. Commun.*, 70 (1982).
105. R. van Eldik and G. M. Harris, *Inorg. Chim. Acta*, **65**, L125 (1982).
106. U. Spitzer, R. van Eldik, and H. Kelm, *Inorg. Chem.*, **21**, 2821 (1982).
107. A. C. Dash and G. M. Harris, *Inorg. Chem.*, **21**, 2336 (1982).

108. A. C. Dash, *Inorg. Chem.*, **22**, 837 (1983).
109. P. Natarajan and A. Radhakrishnan, *J. Chem. Soc., Dalton Trans.*, 2403 (1982).
110. K. Angermann, R. Schmidt, R. van Eldik, H. Kelm, and F. Wasgestian, *Inorg. Chem.*, **21**, 1175 (1982).
111. T. G. Hoon and C. Y. Mok, *Inorg. Chim. Acta*, **62**, 231 (1982).
112. B. Kräutler and R. Stepanek, *Helv. Chim. Acta*, **66**, 1493 (1983).
113. G. Ferraudi, P. A. Grutsch, and C. Kutal, *Inorg. Chim. Acta*, **59**, 249 (1982).
114. J. L. Reed, *Inorg. Chem.*, **21**, 2829 (1982).
115. R. L. de la Vega, W. R. Ellis, Jr., and W. L. Purcell, *Inorg. Chim. Acta*, **68**, 97 (1983).
116. P. J. Lawson, M. G. McCarthy, and A. M. Sargeson, *J. Am. Chem. Soc.*, **104**, 6710 (1982).
117. N. E. Dixon and A. M. Sargeson, *J. Am. Chem. Soc.*, **104**, 6716 (1982).
118. See for example, C. R. Clark, R. F. Tasker, D. A. Buckingham, D. R. Knighton, D. R. K. Harding, and W. S. Hancock, *J. Am. Chem. Soc.*, **103**, 7023 (1981).
119. S. S. Isied, A. Vassilian, and J. M. Lyon, *J. Am. Chem. Soc.*, **104**, 3910 (1982).
120. P. D. Ford, K. B. Nolan, and D. C. Povey, *Inorg. Chim. Acta*, **61**, 189 (1982).
121. A. Pasini and L. Casella, *J. Inorg. Nucl. Chem.*, **36**, 2133 (1974).
122. A. Miyanaga, U. Sakaguchi, Y. Morimoto, Y. Kushi, and H. Yoneda, *Inorg. Chem.*, **21**, 1387 (1982).
123. L. R. Solujić and M. B. Ćelap, *Inorg. Chim. Acta*, **67**, 103 (1982).
124. A. C. Dash, N. Mullick, and R. K. Nanda, *Transition Met. Chem.*, **7**, 204 (1982).
125. S. H. McClaugherty and C. M. Grisham, *Inorg. Chem.*, **21**, 4133 (1982).
126. P. R. Norman and R. D. Cornelius, *J. Am. Chem. Soc.*, **104**, 2356 (1982).
127. R. D. Cornelius and P. R. Norman, *Inorg. Chim. Acta*, **65**, L193 (1982).
128. R. W. Hay and R. Bembi, *Inorg. Chim. Acta*, **78**, 143 (1983).
129. B. Anderson, R. M. Milburn, J. MacB. Harrowfield, G. B. Robertson, and A. M. Sargeson, *J. Am. Chem. Soc.*, **99**, 2652 (1977).
130. D. R. Jones, L. F. Lindoy, A. M. Sargeson, and M. R. Snow, *Inorg. Chem.*, **21**, 4155 (1982).
131. J. H. Dimmit and J. H. Weber, *Inorg. Chem.*, **21**, 1554 (1982).
132. Y.-T. Fanchiang, *Inorg. Chem.*, **21**, 2344 (1982).
133. F. T. T. Ng, G. L. Rempel, and J. Halpern, *Inorg. Chim. Acta*, **77**, L165 (1983).
134. I. Ya. Levitin, A. L. Sigan, R. M. Bodnar, R. G. Gasanov, and M. E. Vol'pin, *Inorg. Chim. Acta*, **76**, L169 (1983).
135. O. Vollárova and J. Benko, *J. Chem. Soc., Dalton Trans.*, 2359 (1983).

References for Chapter 8

1. K. Wieghardt, M. Hahn, W. Swiridoff, and J. Weiss, *Angew. Chem. Int. Ed.*, **22**, 491 (1983).
2. B. V. DePamphilis, A. G. Jones, and A. Davison, *Inorg. Chem.*, **22**, 2292 (1983).
3. M. J. Abrams, A. Davison, A. G. Jones, C. E. Costello, and H. Pang, *Inorg. Chem.*, **22**, 2798 (1983).
4. K. Libson, B. L. Barnett, and E. Deutsch, *Inorg. Chem.*, **22**, 1695 (1983).
5. J. Burgess and R. D. Peacock, in: *Gmelin Handbuch der Anorganische Chemie, 8 Auflage: Technetium Supplement Volume 2* (H. K. Kugler and C. Keller, eds.), Chap. 13, Springer-Verlag, Berlin (1983).
6. H. E. Toma and J. M. Malin, *Inorg. Chem.*, **12**, 1039 (1973).
7. P. J. Morando and M. E. Blesa, *J. Chem. Soc., Dalton Trans.*, 2147 (1982).
8. N. D. Lis de Katz and N. E. Katz, *Mh. Chem.*, **113**, 745 (1982).

9. J. A. Olabe and L. A. Gentil, *Transition Met. Chem.*, **8**, 65 (1983).
10. D. B. Soria, M. del V. Hidalgo, and N. E. Katz, *J. Chem. Soc., Dalton Trans.*, 1555 (1982).
11. H. E. Toma, A. A. Batista, and H. B. Gray, *J. Am. Chem. Soc.*, **104**, 7509 (1982).
12. J. A. Salas, M. Katz, and N. E. Katz, *J. Solution Chem.*, **12**, 115 (1983).
13. G. C. Pedrosa, J. A. Salas, M. Katz, and N. E. Katz, *J. Coord. Chem.*, **12**, 145 (1983).
14. N. Y. M. Iha and H. E. Toma, *Ann. Acad. Bras. Cienc.*, **54**, 491 (1982); *Chem. Abstr.* **98**, 136578a (1983).
15. M. J. Blandamer, J. Burgess, K. W. Morcom, and R. Sherry, *Transition Met. Chem.*, **8**, 354 (1983).
16. A. L. Coelho, H. E. Toma, and J. M. Malin, *Inorg. Chem.*, **22**, 2703 (1983).
17. H. E. Toma, J. M. Malin, and E. Giesbrecht, *Inorg. Chem.*, **12**, 2084 (1973).
18. H. E. Toma and M. S. Takasugi, *Polyhedron*, **1**, 429 (1982).
19. H. E. Toma, E. Giesbrecht, and R. L. E. Rojas, *Can. J. Chem.*, **61**, 2520 (1983).
20. P. J. Morando, V. I. E. Bruyère, M. A. Blesa, and J. A. Olabe, *Transition Met. Chem.*, **8**, 99 (1983).
21. P. J. Morando, V. I. E. Bruyère, M. A. Blesa, and A. Esteban, *Thermochim. Acta*, **62**, 249 (1983).
22. H. E. Toma, E. Giesbrecht, and R. L. E. Rojas, *Quim. Nova*, **6**, 72 (1983); *Chem. Abstr.*, **99**, 94413f (1983).
23. H. E. Toma, N. M. Moroi, and N. Y. M. Iha, *Ann. Acad. Bras. Cienc.*, **54**, 315 (1982).
24. A. P. Szecsy and A. Haim, *J. Am. Chem. Soc.*, **104**, 3063 (1982).
25. H. E. Toma and A. A. Batista, *J. Inorg. Biochem.*, **20**, 53 (1984).
26. J. A. Olabe and H. O. Zerga, *Inorg. Chem.*, **22**, 4156 (1983).
27. H. E. Toma and N. Y. M. Iha, *Inorg. Chem.*, **21**, 3573 (1982).
28. V. Gáspár and M. T. Beck, *Polyhedron*, **2**, 387 (1983).
29. V. Gáspár and M. T. Beck, *Magy. Kem. Foly.*, **88**, 433 (1982); *Chem. Abstr.*, **98**, 25405x (1983).
30. K. J. Moore, L. Lee, J. E. Figard, J. A. Gelroth, A. J. Stinson, H. D. Wohlers, and J. D. Petersen, *J. Am. Chem. Soc.*, **105**, 2274 (1983).
31. E. Hejmo, T. Jarzynowski, E. Porcel-Ortega, T. Senkowski, and Z. Stasicka, *Proc. 9th Conf. Coord. Chem.*, 99 (1983); *Chem. Abstr.*, **99**, 113622s (1983).
32. A. P. Szecsy, S. S. Miller, and A. Haim, *Inorg. Chim. Acta*, **28**, 189 (1978).
33. J. Casado, M. A. Lopez Quintela, M. Mosquera, M. F. Rodríguez Prieto, and J. Vázquez Tato, *Ber. Bunsenges. Phys. Chem.*, **87**, 1208 (1983).
34. A. M. da C. Ferreira and H. E. Toma, *J. Chem. Soc., Dalton Trans.*, 2051 (1983).
35. *Coord. Chem. Rev.*, **51**(2) (1983)—see pp. 120–220.
36. C. R. Johnson and R. E. Shepherd, *Inorg. Chem.*, **22**, 1117 (1983).
37. J. O. Edwards, *Coord. Chem. Rev.*, **8**, 87 (1972).
38. J. Burgess and R. H. Prince, *J. Chem. Soc. A*, 2111 (1970).
39. A. M. Kjaer and J. Ulstrup, *Inorg. Chem.*, **21**, 3490 (1982).
40. N. F. Ashford, M. J. Blandamer, J. Burgess, D. Laycock, M. Waters, P. Wellings, R. Woodhead, and F. M. Mekhail, *J. Chem. Soc., Dalton Trans.*, 869 (1979).
41. S. Tachiyashiki and H. Yamatera, *Polyhedron*, **2**, 9 (1983).
42. L. Johansson, *Chem. Scripta*, **9**, 30 (1976).
43. G. I. Gromova, A. K. Pyartman, and V. E. Mironov, *Russ. J. Inorg. Chem.*, **23**, 1875 (1978).
44. M. J. Blandamer, J. Burgess, P. Wellings, and M. V. Twigg, *Transition Met. Chem.*, **6**, 129 (1981).
45. M. J. Blandamer, J. Burgess, and B. Clark, *J. Chem. Soc., Chem. Commun.*, 659 (1983).
46. P. O'Brien, *Polyhedron*, **2**, 233 (1983).
47. M. J. Blandamer and J. Burgess, *Pure Appl. Chem.*, **54**, 2285 (1982).

48. M. J. Blandamer, J. Burgess, T. Digman, P. P. Duce, J. P. McCann, R. H. Reynolds, and D. M. Sweeney, *Transition Met. Chem.*, **8**, 148 (1983).
49. V. D. Gusev, V. A. Shormanov, and G. A. Krestov, *Russ. J. Phys. Chem.*, **56**, 1530 (1982).
50. J.-M. Lucie, D. R. Stranks, and J. Burgess, *J. Chem. Soc., Dalton Trans.*, 245 (1975).
51. K. Yoshitani, M. Yamamoto, and Y. Yamamoto, *Bull. Chem. Soc. Jpn.*, **56**, 1978 (1983).
52. G. A. Lawrence, D. R. Stranks, and S. Suvachittanont, *Inorg. Chem.*, **18**, 82 (1979).
53. J. Burgess and C. D. Hubbard, *J. Chem. Soc., Chem. Commun.*, 1482 (1983).
54. D. Sengupta, A. Pal, and S. C. Lahiri, *J. Chem. Soc., Dalton Trans.*, 2685 (1983).
55. F. M. Van Meter and H. M. Neumann, *J. Am. Chem. Soc.*, **98**, 1382 (1976).
56. L. L. Costanzo, S. Giuffrida, G. de Guidi, and S. Pistara, *Inorg. Chim. Acta*, **73**, 165 (1983).
57. N. Serpone, G. Ponterini, M. A. Jamieson, F. Bolletta, and M. Maestri, *Coord. Chem. Rev.*, **50**, 209 (1983).
58. R. D. Gillard, *Coord. Chem. Rev.*, **50**, 303 (1983).
59. E. C. Constable, *Polyhedron*, **2**, 551 (1983).
60. M. J. Blandamer, J. Burgess, P. P. Duce, K. S. Payne, R. Sherry, P. Wellings, and M. V. Twigg, *Transition Met. Chem.*, **9**, 163 (1984).
61. E. Buncel, M. R. Crampton, M. J. Strauss, and F. Terrier, *Electron Deficient Aromatic- and Heteroaromatic-Base Interactions: The Chemistry of Anionic Sigma Complexes*, Elsevier, Amsterdam (1984).
62. F. Terrier, *Chem. Rev.*, **82**, 77 (1982).
63. G. A. Artamkina, M. P. Egorov, and I. P. Beletskaya, *Chem. Rev.*, **82**, 427 (1982).
64. R. Schmid and L. Han, *Inorg. Chim. Acta*, **69**, 127 (1983).
65. N. Serpone and F. Bolletta, *Inorg. Chim. Acta*, **75**, 189 (1983).
66. G. Nord, B. Pedersen, and E. Bjergbakke, *J. Am. Chem. Soc.*, **105**, 1913 (1983).
67. P. Bandyopadhyay, D. Bandyopadhyay, A. Chakravorty, F. A. Cotton, L. R. Falvello, and S. Han, *J. Am. Chem. Soc.*, **105**, 6327 (1983).
68. G. Nord, A. C. Hazell, R. G. Hazell, and O. Farver, *Inorg. Chem.*, **22**, 3429 (1983); P. J. Spellane, R. J. Watts, and C. J. Curtis, *Inorg. Chem.*, **22**, 4060 (1983).
69. L. Chassot and A. von Zelewsky, *Helv. Chim. Acta*, **66**, 2443 (1983).
70. F. Pomposo and D. V. Stynes, *Inorg. Chem.*, **22**, 569 (1983).
71. N. K. Kildahl, K. J. Balkus, and M. J. Flynn, *Inorg. Chem.*, **22**, 589 (1983).
72. J. S. Olson, R. E. McKinnie, M. P. Mims, and D. K. White, *J. Am. Chem. Soc.*, **105**, 1522 (1983).
73. D. P. Riley and D. H. Busch, *Inorg. Chem.*, **22**, 4141 (1983).
74. K. Sriraman, B. S. R. Sarma, N. R. Sastry, and K. Kalides, *Z. Phys. Chem. (Leipzig)*, **264**, 155 (1983).
75. D. Brault and P. Neta, *J. Phys. Chem.*, **87**, 3320 (1983).
76. M. M. Doeff and D. A. Sweigart, *Inorg. Chem.*, **21**, 3699 (1982).
77. H. C. Bajaj and P. C. Nigam, *Transition Met. Chem.*, **7**, 190 (1982).
78. H. C. Bajaj and P. C. Nigam, *Transition Met. Chem.*, **8**, 105 (1983).
79. S. E. Ronco and P. J. Aymonino, *Inorg. Chim. Acta*, **77**, L31 (1983).
80. T. Sekine and K. Inaba, *Chem. Lett.*, 1669 (1983).
81. V. M. Nekipelov, V. A. Ivanchenko, and K. I. Zamaraev, *Kinet. Katal.*, **24**, 591 (1983); *Chem. Abstr.*, **99**, 94428q (1983).
82. L. Dozsa, J. E. Sutton, and H. Taube, *Inorg. Chem.*, **21**, 3997 (1982).
83. A. Yeh, N. Scott, and H. Taube, *Inorg. Chem.*, **21**, 2542 (1982).
84. C.-K. Poon, T.-C. Lau, C.-L. Wong, and Y.-P. Kan, *J. Chem. Soc., Dalton Trans.*, 1641 (1983).
85. C.-K. Poon, S.-S. Kwong, C.-M. Che, and Y.-P. Kan, *J. Chem. Soc., Dalton Trans.*, 1457 (1982).

86. Y. Ilan and H. Taube, *Inorg. Chem.*, **22**, 1655 (1983).
87. F. Pomposo, D. Carruthers, and D. V. Stynes, *Inorg. Chem.*, **21**, 4245 (1982).
88. S. Goswami, A. R. Chakravarty, and A. Chakravorty, *Inorg. Chem.*, **21**, 2737 (1982).
89. S. Goswami, A. R. Chakravarty, and A. Chakravorty, *Inorg. Chem.*, **22**, 602 (1983).
90. A. R. Chakravarty, A. Chakravorty, F. A. Cotton, L. R. Falvello, B. K. Ghosh, and M. Tomas, *Inorg. Chem.*, **22**, 1892 (1983).
91. S. C. Grocott and S. B. Wild, *Inorg. Chem.*, **21**, 3526, 3535 (1982).
92. G. A. Heath, A. J. Lindsay, and T. A. Stephenson, *J. Chem. Soc., Dalton Trans.*, 2429 (1982).
93. D. E. Richardson, J. P. Sen, J. D. Buhr, and H. Taube, *Inorg. Chem.*, **21**, 3136 (1982).
94. C. M. Elson, J. Gulens, I. J. Itzkovitch, and J. A. Page, *J. Chem. Soc., Chem. Commun.*, 875 (1970).
95. A. Khair, *Dhaka University Studies B*, **31**, 35 (1983); *Chem. Abstr.*, **99**, 186370t (1983).
96. F. Bottomley and M. Mukaida, *J. Chem. Soc., Dalton Trans.*, 1933 (1982).
97. M. F. Tweedle and H. Taube, *Inorg. Chem.*, **21**, 3361 (1982).
98. K. Kalyanasundaram, *Coord. Chem. Rev.*, **46**, 159 (1982).
99. R. Fasano and P. E. Hoggard, *Inorg. Chem.*, **22**, 566 (1983).
100. E. M. Kober and T. J. Meyer, *Inorg. Chem.*, **21**, 3967 (1982).
101. B. Durham, J. V. Caspar, J. K. Nagle, and T. J. Meyer, *J. Am. Chem. Soc.*, **104**, 4803 (1982).
102. J. V. Caspar and T. J. Meyer, *J. Am. Chem. Soc.*, **105**, 5583 (1983).
103. T. K. Foreman, J. B. S. Bonilha, and D. G. Whitten, *J. Phys. Chem.*, **86**, 3436 (1982).
104. J.-G. Xu and G. B. Porter, *Can. J. Chem.*, **60**, 2856 (1982).
105. J. V. Caspar and T. J. Meyer, *Inorg. Chem.*, **22**, 2444 (1983).
106. A. Sugimori, H. Uchida, T. Akiyama, M. Mukaida, and K. Shimizu, *Chem. Lett.*, 1135 (1982).
107. D. R. Prasad and G. Ferraudi, *Inorg. Chem.*, **21**, 4241 (1982).
108. F. Barigelletti, A. Juris, V. Balzani, P. Belser, and A. von Zelewsky, *Inorg. Chem.*, **22**, 3335 (1983).
109. P. J. Steel, F. Lahousse, D. Lerner, and C. Marzin, *Inorg. Chem.*, **22**, 1488 (1983).
110. R. R. Ruminski and J. D. Petersen, *Inorg. Chem.*, **21**, 3706 (1982).
111. B. P. Sullivan and T. J. Meyer, *Inorg. Chem.*, **21**, 1037 (1982).
112. S. F. Clark and J. D. Petersen, *Inorg. Chem.*, **22**, 620 (1983).
113. J. M. Calvert and T. J. Meyer, *Inorg. Chem.*, **21**, 3978 (1982).
114. M. I. Cruz, H. Nijs, J. J. Fripiat, and H. Van Damme, *J. Chim. Phys.*, **79**, 753 (1982).
115. R. S. Vagg and P. A. Williams, *Inorg. Chem.*, **22**, 355 (1983).
116. C. P. Guengerich and K. Schug, *Inorg. Chem.*, **22**, 181 (1983).
117. F. Bottomley, W. V. F. Brooks, D. E. Paéz, P. S. White, and M. Mukaida, *J. Chem. Soc., Dalton Trans.*, 2465 (1983).
118. C. Bartocci, C. A. Bignozzi, F. Scandola, R. Rumin, and P. Courtot, *Inorg. Chim. Acta*, **76**, L119 (1983).
119. D. J. Gulliver, A. L. Hale, W. Levason, and S. G. Murray, *Inorg. Chim. Acta*, **69**, 25 (1983).
120. N. E. Dixon, G. A. Lawrance, P. A. Lay, and A. M. Sargeson, *Inorg. Chem.*, **22**, 846 (1983).
121. Y. Ilan and H. Taube, *Inorg. Chem.*, **22**, 3144 (1983).
122. C.-K. Poon and T.-C. Lau, *Inorg. Chem.*, **22**, 1664 (1983).
123. C.-K. Poon, T.-C. Lau, and C.-M. Che, *Inorg. Chem.*, **22**, 3893 (1983).
124. A. Ceulemans, D. Beyens, and L. G. Vanquickenborne, *Inorg. Chem.*, **22**, 1113 (1983).
125. L. G. Vanquickenborne and A. Ceulemans, *Coord. Chem. Rev.*, **48**, 157 (1983).
126. A. Girandeau, P. Lemoine, M. Gross, J. Rosé, and P. Braunstein, *Inorg. Chim. Acta*, **62**, 117 (1982).
127. R. Schöllhorn, R. Steffen, and K. Wagner, *Angew. Chem. Int. Ed.*, **22**, 555 (1983).

128. J. Gonzales-Velasco, I. Rubinstein, R. J. Crutchley, A. B. P. Lever, and A. J. Bard, *Inorg. Chem.*, **22**, 822 (1983).
129. J. G. Gaudiello, P. R. Sharp, and A. J. Bard, *J. Am. Chem. Soc.*, **104**, 6373 (1982).
130. R. G. Egdell, J. B. Goodenough, A Hamnett, and C. G. Naish, *J. Chem. Soc., Faraday Trans. 1*, **79**, 893 (1983).
131. W. D'Olieslager, L. Heerman, and M. Clarysse, *Polyhedron*, **2**, 1107 (1983).
132. H. H. Fricke and W. Preetz, *Z. Naturforsch. B*, **38**, 917 (1983).
133. A. W. Zanella and P. C. Ford, *Inorg. Chem.*, **14**, 42, 700 (1975).
134. B. Anderes and D. K. Lavallee, *Inorg. Chem.*, **22**, 3724 (1983).
135. H. Schulz and W. Preetz, *Z. Anorg. Allg. Chem.*, **490**, 55 (1982).
136. F. Cristiani, F. A. Devillanova, A. Diaz, G. Verani, and L. Antolini, *Transition Met. Chem.*, **8**, 236 (1983).
137. J. E. Armstrong and R. A. Walton, *Inorg. Chem.*, **22**, 1545 (1983).
138. N. J. Curtis, N. E. Dixon, and A. M. Sargeson, *J. Am. Chem. Soc.*, **105**, 5347 (1983).
139. L. H. Skibsted and P. C. Ford, *Inorg. Chem.*, **22**, 2749 (1983).
140. B. Chakravarty and B. Nandi, *J. Chin. Chem. Soc.* (*Taipei*), **29**, 205 (1982); *Chem. Abstr.*, **97**, 203742u (1982).
141. D. A. Palmer and R. van Eldik, *Chem. Rev.*, **83**, 651 (1983).
142. C. Creutz, A. D. Keller, N. Sutin, and A. P. Zipp, *J. Am. Chem. Soc.*, **104**, 3618 (1982).
143. M. N. Bishnu, B. Chakravarti, R. N. Banerjee, and D. Banerjea, *J. Coord. Chem.*, **13**, 63 (1983).
144. A. O. Adeniran, G. J. Baker, G. J. Bennett, M. J. Blandamer, J. Burgess, N. K. Dhammi, and P. P. Duce, *Transition Met. Chem.*, **7**, 183 (1982).
145. M. H. Abraham, T. Hill, H. C. Ling, R. A. Schulz, and R. A. C. Watt, *J. Chem. Soc., Faraday Trans. 1*, **80**, 489 (1984).
146. C. Tissier, *Compt. Rend.*, **286C**, 35 (1978).
147. M. J. Blandamer, J. Burgess, and P. P. Duce, *Transition Met. Chem.*, **8**, 184 (1983).
148. U. Spitzer, R. van Eldik, and H. Kelm, *Inorg. Chem.*, **21**, 2821 (1982).
149. J. G. Leipoldt, R. van Eldik, and H. Kelm, *Inorg. Chem.*, **22**, 4146 (1983).
150. T. W. Swaddle, *J. Chem. Soc., Chem. Commun.*, 832 (1982).
151. T. W. Swaddle and M. K. S. Mak, *Can. J. Chem.*, **61**, 473 (1983).
152. M. E. F. Sheridan, M.-J. Jun, and C.-F. Liu, *Inorg. Chim. Acta*, **76**, L109 (1983).
153. V. M. Savostina, O. A. Shpigun, T. V. Chebrikova, O. D. Choporova, and A. V. Garmash, *Russ. J. Inorg. Chem.*, **27**, 550 (1982).
154. A. Peloso, *J. Chem. Soc., Dalton Trans.*, 2141 (1982).
155. W. Weber, R. van Eldik, H. Kelm, J. Dibenedetto, Y. Ducommun, H. Offen, and P. C. Ford, *Inorg. Chem.*, **22**, 623 (1983).
156. M. J. Camara and J. H. Lunsford, *Inorg. Chem.*, **22**, 2498 (1983).
157. J. S. Svendsen and L. H. Skibsted, *Acta Chem. Scand.*, **37A**, 443 (1983).
158. L. Mønsted and L. H. Skibsted, *Acta Chem. Scand.*, **37A**, 663 (1983).
159. D. A. Sexton, P. C. Ford, and D. Magde, *J. Phys. Chem.*, **87**, 197 (1983).
160. L. H. Skibsted, W. Weber, R. van Eldik, H. Kelm, and P. C. Ford, *Inorg. Chem.*, **22**, 541 (1983).
161. M. M. Muir, L. B. Zinner, L. A. Paguaga, and L. M. Torres, *Inorg. Chem.*, **21**, 3448 (1982).
162. K. Howland and L. H. Skibsted, *Acta Chem. Scand.*, **37A**, 647 (1983).
163. D. B. Miller, P. K. Miller, and N. A. P. Kane-Maguire, *Inorg. Chem.*, **22**, 3831 (1983).
164. S. Muralidharan, G. Ferraudi, and K. Schmatz, *Inorg. Chem.*, **21**, 2961 (1982).
165. D. R. Prasad and G. Ferraudi, *Inorg. Chem.*, **21**, 2967 (1982).
166. S. Muralidharan and G. Ferraudi, *J. Phys. Chem.*, **87**, 4877 (1983).
167. Xiucen Yang and C. Kutal, *J. Am. Chem. Soc.*, **105**, 6038 (1983).

168. P. C. Ford, D. Wink, and J. Dibenedetto, *Progr. Inorg. Chem.*, **30**, 213 (1983).
169. G. Rio and F. Larèze, *Bull. Soc. Chim. Fr.*, I-433 (1982).
170. C.-K. Poon, T.-W. Tang, and C.-M. Che, *J. Chem. Soc., Dalton Trans.*, 1647 (1983).
171. J. E. Fergusson and D. A. Rankin, *Aust. J. Chem.*, **36**, 863, 871 (1983).
172. N. L. Kovalenko, N. Ya. Rogin, and G. D. Mal'chikov, *Russ. J. Inorg. Chem.*, **27**, 553 (1982).
173. A. V. Garmash, *Chem. Abstr.*, **98**, 209145u (1983).
174. I. I. Alekseeva, G. N. Latysheva, and L. E. Romanovskaya, *Russ. J. Inorg. Chem.*, **27**, 705 (1982).
175. K. C. Koshy and G. M. Harris, *Inorg. Chem.*, **22**, 2947 (1983).
176. T. G. Appleton, J. R. Hall, N. S. Ham, F. W. Hess, and M. A. Williams, *Aust. J. Chem.*, **36**, 673 (1983).
177. W. K. Wilmarth, Y.-T. Fanchiang, and J. E. Byrd, *Coord. Chem. Rev.*, **51**, 141 (1983).
178. R. Kuroda, S. Neidle, I. M. Ismail, and P. J. Sadler, *Inorg. Chem.*, **22**, 3620 (1983).
179. M. Kretschmer and L. Heck, *Z. Anorg. Allg. Chem.*, **490**, 215 (1982).
180. V. V. Vasil'ev, I. A. Konovalova, and G. A. Shagisultanova, *Koord. Khim.*, **9**, 1110 (1983); *Chem. Abstr.*, **99**, 166850w (1983).
181. E. W. Abel, K. G. Orrell, and A. W. G. Platt, *J. Chem. Soc., Dalton Trans.*, 2345 (1983).

References for Chapter 9

1. B. I. Gislason and H. Strehlow, *Aust. J. Chem.*, **36**, 1941 (1983).
2. T. W. Swaddle, *J. Chem. Soc., Chem. Comm.*, 832 (1982).
3. Y. Ducommun, D. Zbinden, and A. E. Merbach, *Helv. Chim. Acta*, **65**, 1385 (1982).
4. A. Hioki, S. Funahashi, and M. Tanaka, *Inorg. Chem.*, **22**, 749 (1983).
5. F. L. Dickert and S. W. Hellman , *Ber. Bunsenges. Phys. Chem.*, **87**, 513 (1983).
6. C. H. McAteer and P. Moore, *J. Chem. Soc., Dalton Trans.*, 353 (1983).
7. H. W. Dodgen and J. P. Hunt, *Inorg. Chem.*, **22**, 1146 (1983).
8. A. Hioki, S. Funahashi, and M. Tanaka, *Inorg. Chim. Acta*, **76**, L151 (1983).
9. K. Ishihara, S. Funahashi, and M. Tanaka, *Inorg. Chem.*, **22**, 2564 (1983).
10. P. D. Chattopadhyay and B. Kratochvil, *Can. J. Chem.*, **61**, 1842 (1983).
11. R. L. Batstone-Cunningham, H. W. Dodgen, and J. P. Hunt, *Inorg. Chem.*, **21**, 3831 (1982).
12. L. L. Rusnak, E. S. Yang, and R. B. Jordan, *Inorg. Chem.*, **17**, 1810 (1978).
13. R. G. Wilkins, *Comments Inorg. Chem.*, **2**, 187 (1983).
14. J. E. Finholt, P. Leupin, and A. G. Sykes, *Inorg. Chem.*, **22**, 3315 (1983).
15. R. Doss, R. van Eldik, and H. Kelm, *Ber. Bunsenges. Phys. Chem.*, **86**, 925 (1982).
16. S. Funahashi, K. Ishihara, and M. Tanaka, *Inorg. Chem.*, **22**, 2070 (1983).
17. K. Ishihara, S. Funahashi, and M. Tanaka, *Inorg. Chem.*, **22**, 194 (1983).
18. K. Bridger, R. C. Patel, and E. Matijevic, *J. Phys. Chem.*, **87**, 1192 (1983).
19. D. L. Pisaniello, L. Helm, P. Meier, and A. E. Merbach, *J. Am. Chem. Soc.*, **105**, 4528 (1983).
20. D. L. Pisaniello, L. Helm, D. Zbinden, and A. E. Merbach, *Helv. Chim. Acta* **66**, 1872 (1983).
21. S. F. Lincoln, A. M. Hounslow, and A. J. Jones, *Aust. J. Chem.*, **35**, 2393 (1982).
22. H. Doine, Y. Ikeda, H. Tomiyasu, and H. Fukutomi, *Bull. Chem. Soc. Jpn.*, **56**, 1989 (1983).
23. B. G. Cox, C. Guminski, P. Firman, and H. Schneider, *J. Phys. Chem.*, **87**, 1357 (1983).
24. E. Schmidt and A. I. Popov, *J. Am. Chem. Soc.*, **105**, 1873 (1983).
25. C. C. Chen and S. Petrucci, *J. Phys. Chem.*, **86**, 2601 (1982).
26. B. G. Cox, P. Firman, and H. Schneider, *Inorg. Chim. Acta*, **69**, 161 (1983).

27. S. Funahashi, Y. Yamaguchi, K. Ishihara, and M. Tanaka, *J. Chem. Soc., Chem. Comm.*, 976 (1982).
28. N. L. Gottke and C. D. Hubbard, *Inorg. Chim. Acta*, **68**, 219 (1983).
29. M. J. Hynes and M. T. O'Shea, *Inorg. Chim. Acta*, **73**, 201 (1983).
30. S. Chopra and R. B. Jordan, *Inorg. Chem.*, **22**, 1708 (1983).
31. R. L. Reeves, G. S. Calabrese, and S. A. Harkaway, *Inorg. Chem.*, **22**, 3076 (1983).
32. S. A. Bajue, T. S. Dasgupta, and G. C. Lalor, *Polyhedron*, **2**, 431 (1983).
33. I. Ando, S. Nishijima, K. Ujimoto, and H. Kurihara, *Bull. Chem. Soc. Jpn.*, **55**, 2881 (1982).
34. I. Ando, K. Yoshizumi, K. Ito, K. Ujimoto, and H. Kurihara, *Bull. Chem. Soc. Jpn.*, **56**, 1368 (1983).
35. T. Inoue, K. Sugahara, K. Kojima, and R. Shimozawa, *Inorg. Chem.*, **22**, 3977 (1983).
36. T. Inoue, K. Kojima, and R. Shimozawa, *Inorg. Chem.*, **22**, 3972 (1983).
37. R. A. Read and D. W. Margerum, *Inorg. Chem.*, **22**, 3447 (1983).
38. C. N. Elgy and C. F. Wells, *J. Chem. Soc., Faraday Trans. 1*, **79**, 2439 (1983).
39. P. D. I. Fletcher, J. R. Hicks, and V. C. Reinsborough, *Can. J. Chem.*, **61**, 1594 (1983).
40. P. D. I. Fletcher and B. H. Robinson, *J. Chem. Soc., Faraday Trans. 1*, **79**, 1959 (1983).
41. J. Garcia-Rosas and H. Schneider, *Inorg. Chim. Acta*, **70**, 183 (1983).
42. T. Okubo and A. Enokida, *J. Chem. Soc., Faraday Trans. 1*, **79**, 1639 (1983).
43. M. Harada, M. Mori, M. Adachi, and W. Eguchi, *J. Chem. Eng. Jpn.*, **16**, 187 (1983); *Chem. Abstr.*, **99**, 44123r (1983).
44. T. S. Shoupe and C. D. Hubbard, *Polyhedron*, **1**, 361 (1982).
45. G. M. Cole, D. W. Doll, and S. L. Holt, *J. Am. Chem. Soc.*, **105**, 4477 (1983).
46. D. C. Weatherburn, *Aust. J. Chem.*, **36**, 433 (1983).
47. Y. Funaki, S. Harada, K. Okumiya, and T. Yasunaga, *J. Am. Chem. Soc.*, **104**, 5325 (1982).
48. J. Oakes and E. G. Smith, *J. Chem. Soc., Faraday Trans. 1*, **79**, 543 (1983).
49. S. P. Kasprzyk and R. G. Wilkins, *Inorg. Chem.*, **21**, 3349 (1982).
50. M. J. Hynes and M. T. O'Shea, *J. Chem. Soc., Dalton Trans.*, 331 (1983).
51. B. G. Cox, J. Garcia-Rosas, and H. Schneider, *Nouv. J. Chim.*, **6**, 397 (1982); *Chem Abstr.*, **98**, 41405c (1983).
52. M. Tanaka, *Pure Appl. Chem.*, **55**, 151 (1983).
53. G. Schneider and M. Zeppezauer, *J. Inorg. Biochem.*, **18**, 59 (1983).
54. M. K. S. Mak and C. H. Langford, *Inorg. Chim. Acta*, **70**, 237 (1983).
55. B. Perlmutter-Hayman, F. Secco, and M. Venturini, *J. Chem. Soc., Dalton Trans.*, 1945 (1982).
56. T. Sekine and K. Inaba, *Bull. Chem. Soc. Jpn.*, **55**, 3773 (1982).
57. G. Laurenczy, L. Radics, and E. Brucher, *Inorg. Chim. Acta*, **75**, 219 (1983).
58. M. C. Gennaro, P. Mirti, and C. Casalino, *Polyhedron*, **2**, 13 (1983).
59. J. T. Carr and P. A. Tregloan, *Aust. J. Chem.*, **36**, 843 (1983).
60. K. Ishihara, S. Funahashi, and M. Tanaka, *Inorg. Chem.*, **22**, 3589 (1983).
61. R. K. Panda, S. Acharya, G. Neogi, and D. Ramaswamy, *J. Chem. Soc., Dalton Trans.*, 1225 (1983).
62. G. Neogi, S. Acharya, R. K. Panda, and D. Ramaswamy, *Int. J. Chem. Kin.*, **15**, 521 (1983).
63. H. L. Hodges and M. A. de Araugo. *Inorg. Chem.*, **21**, 3236 (1982).
64. R. Yamada, K. Tamura, S. Harada, and T. Yasunaga, *Bull. Chem. Soc. Jpn.*, **55**, 3413 (1982).
65. D. P. Parr, C. Rhodes, and R. Nakon, *Inorg. Chim. Acta*, **80**, L11 (1983).
66. C. R. Krishnamoorthy and R. Nakon, *Inorg. Chim. Acta*, **80**, L33 (1983).
67. R. J. Pell, H. W. Dodgen, and J. P. Hunt, *Inorg. Chem.*, **22**, 529 (1983).
68. D. N. Hague and K. G. Moodley, *S. Afr. J. Chem.*, **36**, 10 (1983); *Chem. Abstr.*, **98**, 132966c (1983).

69. R. van Eldik, E. L. J. Breet, M. Kotowski, D. A. Palmer, and H. Kelm, *Ber. Bunsenges. Phys. Chem.*, **87**, 904 (1983).
70. J. J. Pesek and W. R. Mason, *Inorg. Chem.*, **22**, 2958 (1983).
71. E. J. Rose, P. N. Venkatasubramanian, J. C. Swartz, R. D. Jones, F. Basolo, and B. M. Hoffman, *Proc. Natl. Acad. Sci. U.S.A.*, **79**, 5742 (1982).
72. J. P. Collman, J. I. Brauman, B. L. Iverson, J. L. Sessler, R. M. Morris, and Q. H. Gibson, *J. Am. Chem. Soc.*, **105**, 3052 (1983).
73. N. K. Kildahl, K. J. Balkus, and M. J. Flynn, *Inorg. Chem.*, **22**, 589 (1983).
74. L. Helm, P. Meier, A. E. Merbach, and P. A. Tregloan, *Inorg. Chim. Acta*, **73**, 1 (1983).
75. J. Sachinidis and M. W. Grant, *Aust. J. Chem.*, **36**, 2019 (1983).
76. M. Tubino and A. E. Merbach, *Inorg. Chim. Acta*, **71**, 149 (1983).
77. S. F. Lincoln, J. H. Coates, B. G. Doddridge, A. M. Hounslow, and D. L. Pisaniello, *Inorg. Chem.*, **22**, 2869 (1983).
78. S. F. Lincoln, A. M. Hounslow, and J. H. Coates, *Inorg. Chim. Acta*, **77**, L7 (1983).
79. Y. Ikeda, H. Tomiyasu, and H. Fukutomi, *Bull. Chem. Soc. Jpn.*, **56**, 1060 (1983).

References for Chapter 10

1. J. A. S. Howell and P. M. Burkinshaw, *Chem. Rev.*, **83**, 557 (1983).
2. D. J. Darensbourg, *Adv. Organomet. Chem.*, **21**, 113 (1982).
3. J. D. Atwood, M. J. Wovkulich, and D. C. Sonnenberger, *Acc. Chem. Res.*, **16**, 350 (1983).
4. A. Poë, *Chem. Brit.*, 997 (1983).
5. Q. Z. Shi, T. G. Richmond, W. C. Trogler, and F. Basolo, *J. Am. Chem. Soc.*, **104**, 4032 (1982).
6. Q. Z. Shi, T. G. Richmond, W. C. Trogler, and F. Basolo, *J. Am. Chem. Soc.*, **106**, 71 (1984).
7. T. G. Richmond, Q. Z. Shi, W. C. Trogler, and F. Basolo, *J. Chem. Soc., Chem. Commun.*, 650 (1983).
8. T. G. Richmond, Q. Z. Shi, W. C. Trogler, and F. Basolo, *J. Am. Chem. Soc.*, **106**, 76 (1984).
9. S. B. McCullen, H. W. Walker, and T. L. Brown, *J. Am. Chem. Soc.*, **104**, 4007 (1982).
10. A. F. Hepp and M. S. Wrighton, *J. Am. Chem. Soc.*, **105**, 5934 (1983).
11. H. Yesaka, T. Kobayashi, K. Yasufuku, and S. Nagakura, *J. Am. Chem. Soc.*, **105**, 6249 (1983).
12. J. W. Hershberger, R. J. Klingler, and J. K. Kochi, *J. Am. Chem. Soc.*, **105**, 61 (1983).
13. J. W. Hershberger, C. Amatore, and J. K. Kochi, *J. Organomet. Chem.*, **250**, 345 (1983).
14. R. S. Herrick, D. M. Leazer, and J. L. Templeton, *Organometallics*, **2**, 834 (1983).
15. G. R. Dobson and K. J. Asali, *Inorg. Chem.*, **22**, 1835 (1983).
16. G. R. Dobson, S. E. Mansour, D. E. Halverson, and E. S. Erikson, *J. Am. Chem. Soc.*, **105**, 5505 (1983).
17. R. J. Kazlauskas and M. S. Wrighton, *J. Am. Chem. Soc.*, **104**, 5784 (1982).
18. D. J. Darensbourg, A. Rokicki, and R. Kudaroski, *Organometallics*, **1**, 1161 (1982).
19. M. J. Wovkulich and J. D. Atwood, *Organometallics*, **1**, 1316 (1982).
20. N. J. Coville and M. O. Albers, *Inorg. Chim. Acta*, **65**, L7 (1982).
21. M. O. Albers, N. J. Coville, and E. Singleton, *J. Organomet. Chem.*, **234**, C13 (1982).
22. G. W. Harris, M. O. Albers, J. C. A. Boeyens, and N. J. Coville, *Organometallics*, **2**, 609 (1983).
23. M. G. Peterleitner, M. V. Tolstaya, V. V. Krivykh, L. I. Denisovitch, and M. I. Rybinskaya, *J. Organomet. Chem.*, **254**, 313 (1983).
24. M. E. Rerek, L. N. Ji, and F. Basolo, *J. Chem. Soc., Chem. Commun.*, 1208 (1983).
25. C. P. Casey, J. M. O'Connor, W. D. Jones, and K. J. Haller, *Organometallics*, **2**, 535 (1983).

26. K. A. M. Creber and J. K. S. Wan, *Can. J. Chem.*, **61**, 1017 (1983).
27. K. A. M. Creber and J. K. S. Wan, *Transition Met. Chem.*, **8**, 253 (1983).
28. S. P. Modi and J. D. Atwood, *Inorg. Chem.*, **22**, 26 (1983).
29. K. S. Suslick, J. W. Goodale, P. F. Schubert, and H. H. Wang, *J. Am. Chem. Soc.*, **105**, 5781 (1983).
30. Q. Z. Shi, T. G. Richmond, W. C. Trogler, and F. Basolo, *Organometallics*, **1**, 1033 (1982).
31. B. D. Fabian and J. A. Labinger, *Organometallics*, **2**, 659 (1983).
32. N. J. Coville, M. O. Albers, and E. Singleton, *J. Chem. Soc., Dalton Trans.*, 947 (1983).
33. N. J. Coville, M. O. Albers, and E. Singleton, *J. Am Chem. Soc., Dalton Trans.*, 1389 (1982).
34. M. E. Rerek and F. Basolo, *Organometallics*, **2**, 372 (1983).
35. K. Broadley, N. G. Connelly, and W. E. Geiger, *J. Chem. Soc., Dalton Trans.*, 121 (1983).
36. J. G. Leipoldt, S. S. Basson, J. J. J. Schlebusch, and E. C. Grobler, *Inorg. Chim. Acta*, **62**, 113 (1982).
37. J. J. Turner, M. B. Simpson, M. Poliakoff, and W. B. Maier, *J. Am. Chem. Soc.*, **105**, 3898 (1983).
38. M. Arewgoda, B. H. Robinson, and J. Simpson, *J. Am. Chem. Soc.*, **105**, 1893 (1983).
39. A. Darchen, E. K. Lhadi, and H. Patin, *J. Organomet. Chem.*, **259**, 189 (1983).
40. E. K. Lhadi, C. Mahe, H. Patin, and A. Darchen, *J. Organomet. Chem.*, **246**, C61 (1983).
41. R. W. Wegman and T. L. Brown, *Inorg. Chem.*, **22**, 183 (1983).
42. N. J. Coville, A. M. Stolzenberg, and E. L. Muetterties, *J. Am. Chem. Soc.*, **105**, 2499 (1983).
43. A. M. Stolzenberg and E. L. Muetterties, *J. Am. Chem. Soc.*, **105**, 822 (1983).
44. A. Poe and C. Sekhar, *J. Chem. Soc., Chem. Commun.*, 566 (1983).
45. K. S. Suslick and P. F. Schubert, *J. Am. Chem. Soc.*, **105**, 6042 (1983).
46. J. L. Atwood, I. Bernal, F. Calderazzo, L. G. Canada, R. Poli, R. D. Rogers, C. A. Veracini, and D. Vitali, *Inorg. Chem.*, **22**, 1797 (1983).
47. D. R. Tyler, M. A. Schmidt, and H. B. Gray, *J. Am. Chem. Soc.*, **105**, 6018 (1983).
48. R. H. Magnuson, R. Meirowitz, S. Zulu, and W. P. Giering, *J. Am. Chem. Soc.*, **104**, 5790 (1982).
49. R. H. Magnuson, R. Meirowitz, S. J. Zulu, and W. P. Giering, *Organometallics*, **2**, 460 (1983).
50. D. Miholova and A. A. Vlcek, *J. Organomet. Chem.*, **240**, 413 (1982).
51. S. J. McLain, *J. Am. Chem. Soc.*, **105**, 6355 (1983).
52. J. A. Labinger and J. S. Miller, *J. Am. Chem. Soc.*, **104**, 6856 (1982).
53. J. A. Labinger, J. N. Bonfiglio, D. L. Grimmett, S. T. Masuo, E. Shearin, and J. S. Miller, *Organometallics*, **2**, 733 (1983).
54. D. L. Grimmett, J. A. Labinger, J. N. Bonfiglio, S. T. Masuo, E. Shearin, and J. S. Miller, *J. Am. Chem. Soc.*, **104**, 6858 (1982).
55. D. L. Grimmett, J. A. Labinger, J. N. Bonfiglio, S. T. Masuo, E. Shearin, and J. S. Miller, *Organometallics*, **2**, 1325 (1983).
56. R. B. Hitam, R. Narayanaswamy, and A. J. Rest, *J. Chem. Soc., Dalton Trans.*, 615 (1983).
57. J. D. Cotton and R. D. Markwell, *Inorg. Chim. Acta*, **63**, 13 (1982).
58. J. D. Cotton, H. A. Kimbin, and R. D. Markwell, *J. Organomet. Chem.*, **232**, C75 (1982).
59. K. C. Brinkman, G. D. Vaughn, and J. A. Gladysz, *Organometallics*, **1**, 1056 (1982).
60. D. M. DeSimone, P. J. Desrosiers, and R. P. Hughes, *J. Am. Chem. Soc.*, **104**, 4842 (1982).
61. C. P. Casey and L. M. Baltusis, *J. Am. Chem. Soc.*, **104**, 6347 (1982).
62. H. Brunner, B. Hammer, I. Bernal, and M. Draux, *Organometallics*, **2**, 1595 (1983).
63. T. C. Flood, K. D. Campbell, H. H. Downs, and S. Nakanishi, *Organometallics*, **2**, 1590 (1983).
64. C. R. Jablonski and Y. P. Wang, *Inorg. Chem.*, **21**, 4037 (1982).
65. G. Bellachioma, G. Cardaci, and G. Reichenbach, *J. Chem. Soc., Dalton Trans.*, 2593 (1983).
66. M. Pankowski and M. Bigorgne, *J. Organomet. Chem.*, **251**, 333 (1983).

67. S. Sakaki, K. Kitaura, K. Morokuma, and K. Ohkubo, *J. Am. Chem. Soc.*, **105**, 2280 (1983).
68. D. C. Roe, *J. Am. Chem. Soc.*, **105**, 7770 (1983).
69. H. Scordia, R. Kergoat, M. M. Kubicki, J. E. Buerchais, and P. L'Haridon, *Organometallics*, **2**, 1681 (1983).
70. M. Brookhart, A. R. Pinhas, and A. Lukacs, *Organometallics*, **1**, 1730 (1982).
71. S. Muralidharan and J. H. Espenson, *Inorg. Chem.*, **22**, 2787 (1983).
72. M. D. Seidler and R. G. Bergman, *Organometallics*, **2**, 1897 (1983).
73. W. P. Weiner and R. G. Bergman, *J. Am. Chem. Soc.*, **105**, 3922 (1983).
74. S. G. Slater, R. Lusk, B. F. Schumann, and M. Darensbourg, *Organometallics*, **1**, 1662 (1982).
75. G. R. Clark, T. J. Collins, K. Marsden, and W. R. Roper, *J. Organomet. Chem.*, **259**, 215 (1983).
76. T. R. Gaffney and J. A. Ibers, *Inorg. Chem.*, **21**, 2857 (1982).
77. D. Mansuy, J. P. Battioni, D. Dupre, and E. Sartori, *J. Am. Chem. Soc.*, **104**, 6159 (1982).

References for Chapter 11

1. J. Halpern, *Acc. Chem. Res.* **15**, 238 (1982).
2. A. M. Kashin and I. P. Beletskaya, *Russ. Chem. Rev.* (*Eng. Trans.*), **51**, 503 (1982).
3. *Organometallic Chemistry*, SPR (E. W. Abel and F. G. A. Stone, eds.), Royal Society of Chemistry, London (1981).
4. M. D. Johnson, *Acc. Chem. Res.*, **16**, 343 (1983).
5. (a) J. H. Espenson, in: *Advances in Organic and Bioinorganic Mechanisms* (A. G. Sykes, ed.), Vol. 1, p. 1, Academic Press, London (1982). (b) *J. H. Espenson, Progr. Inorg. Chem.*, **30**, 189 (1983).
6. J. H. Espenson, M. Shimura, and A. Bakac, *Inorg. Chem.*, **21**, 2537 (1982).
7. J.-T. Chen and J. H. Espenson, *Inorg. Chem.*, **22**, 1651 (1983).
8. W. A. Mulac, H. Cohen, and D. Meyerstein, *Inorg. Chem.*, **21**, 4016 (1982).
9. A. Bakac, R. J. Blau, and J. H. Espenson, *Inorg. Chem.*, **22**, 3789 (1983).
10. H. Ogino, M. Shimura, and N. Tanaka, *J. Chem. Soc., Chem. Commun.*, 1063 (1983).
11. J. H. Espenson, P. Connolly, D. Meyerstein, and H. Cohen, *Inorg. Chem.*, **22**, 1009 (1983).
12. M. Green, in: *Mechanisms of Inorganic and Organometallic Reactions* (M. V. Twigg, ed.), Vol. 2, p. 273, Plenum Press, New York (1984).
13. D. Dolphin, *B*12, Wiley-Interscience, New York (1982).
14. J. Halpern, *Pure Appl. Chem.*, **7**, 1059 (1983).
15. F. T. Ng, G. L. Rempel, and J. Halpern, *Inorg. Chim. Acta*, **77**, L165 (1983).
16. M. Green, in: *Mechanisms of Inorganic and Organometallic Reactions* (M. V. Twigg, ed.), Vol. 2, p. 276, Plenum Press, New York (1984).
17. L. Randaccio, N. Bresciani-Pahor, P. J. Toscano, and L. G. Marzilli, *J. Am. Chem. Soc.*, **103**, 6347 (1981).
18. N. Bresciani-Pahor, L. Randaccio, P. J. Toscano, A. C. Sandercock, and L. G. Marzilli, *J. Chem. Soc., Dalton Trans.*, 129 (1982).
19. N. Bresciani-Pahor, L. Randaccio, P. J. Roscano, and L. G. Marzilli, *J. Chem. Soc., Dalton Trans.*, 567 (1982).
20. N. Besciani-Pahor, M. Calligaris, G. Nardin, and L. Randaccio, *J. Chem. Soc., Dalton Trans.*, 2549 (1982).
21. N. Bresciani-Pahor, L. Randaccio, M. Summers, and P. J. Roscano, *Inorg. Chim. Acta*, **68**, 69 (1983).
22. R. Taube, *Pure Appl. Chem.*, **55**, 165 (1983).

23. R. G. Finke, B. L. Smith, B. J. Mayer, and A. A. Molinero, *Inorg. Chem.*, **22**, 3677 (1983).
24. M. Green, in: *Mechanisms of Inorganic and Organometallic Reactions* (M. V. Twigg, ed.), Vol. 2, pp. 278–279, Plenum Press, New York (1984).
25. P. Bougeard, A. Bury, C. J. Cooksey, M. D. Johnson, J. M. Hungerford, and G. M. Lampman, *J. Am. Chem. Soc.*, **104**, 5230 (1982).
26. J. W. Grate and G. N. Schrauzer, *Organometallics*, **1**, 1155 (1982).
27. S. B. Fergusson and M. C. Baird, *Inorg. Chim. Acta* **63**, 41 (1982).
28. I. Ya. Levitin, A. L. Sigar, R. M. Bodnar, R. G. Gasanov, and M. E. Vol'pin, *Inorg. Chim. Acta*, **76**, L169 (1983).
29. W. A. Mulac and D. Meyerstein, *J. Am. Chem. Soc.*, **104**, 4124 (1982).
30. Y. Sorek, H. Cohen, W. A. Mulac, K. H. Schmidt, and D. Meyerstein, *Inorg. Chem.*, **22**, 3040 (1983).
31. K. A. Robinson, E. Itabashi, and H. B. Mark, *Inorg. Chem.*, **21**, 3571 (1982).
32. K. A. Robinson, H. V. Parekh, and H. B. Mark, *Inorg. Chem.* **22**, 458 (1983).
33. M. Green, in: *Mechanisms of Inorganic and Organometallic Reactions* (M. V. Twigg, ed.), Vol. 2, p. 282, Plenum Press, New York (1984).
34. D. N. Ramakrisna Rao and M. C. R. Symons, *J. Organomet. Chem.*, **244**, C43 (1983); *J. Chem. Soc., Chem. Commun.*, 954 (1982) *J. Chem. Soc., Faraday Trans. 1*, **79**, 269 (1983).
35a. Y.-T. Fanchiang, *Inorg. Chem.*, **22**, 1683 (1983).
35b. Y.-T. Fanchiang, *Organometallics*, **2**, 121 (1983).
36. M. Green, in: *Mechanisms of Inorganic and Organometallic Reactions* (M. V. Twigg, ed.), Vol. 2, p. 281, Plenum Press, New York (1984).
37. Y.-T. Fanchiang, *J. Chem. Soc., Chem. Commun.*, 1369 (1982).
38. Y.-T. Fanchiang, J. J. Pignatello, and J. M. Wood, *Organometallics*, **2**, 1748 and 1752 (1983).
39. B. T. Golding, p. 543 of Vol. I of Reference 13.
40. V. Amir-Ebrahimi, R. Hamilton, M. V. Kulkarni, E. A. McIlgorm, T. R. B. Mitchell, and J. J. Rooney, *J. Mol. Catal.*, **22**, 21 (1983).
41. R. G. Finke, W. P. McKenna, D. A. Schiraldi, B. L. Smith, and C. Pierpont, *J. Am. Chem. Soc.*, **103**, 7592 (1983).
42. R. G. Finke and D. A. Schiraldi, *J. Am. Chem. Soc.*, **105**, 7605 (1983).
43. J. J. Russell, H. S. Rzepa, and D. A. Widdowson, *J. Chem. Soc., Chem. Commun.*, 625 (1983).
44. W. H. Hersch and R. G. Bergmann, *J. Am. Chem. Soc.*, **105**, 5846 (1983).
45. K. H. Theopold and R. G. Bergmann, *Organometallics*, **1**, 1571 (1982).
46. G. K. Yang and R. G. Bergmann, *J. Am. Chem. Soc.*, **105**, 6045 (1983).
47. P. L.-K. Hung, G. Sundararajan, and J. San Filippo. *Organometallics*, **1**, 957 (1982).
48. F. J. Landro, J. A. Gurak, J. W. Chinn, and R. J. Lagow, *J. Organomet. Chem.*, **249**, 1 (1983).
49. M. H. Chisholm, L.-S. Tan, and J. C. Huffman, *J. Am. Chem. Soc.*, **104**, 4879 (1982).
50. A. Connor, M. T. Zafarani-Moattar, J. Bickerton, N. I. El Saied, S. Suradi, R. Carson, G. Al Takhin, and H. A. Skinner, *Organometallics*, **1**, 1166 (1982).
51. L. F. Halle, P. B. Armentrout, and J. L. Beauchamp, *Organometallics*, **1**, 963 (1982).
52. L. F. Halle, P. B. Armentrout, and J. L. Beauchamp, *Organometallics*, **2**, 1829 (1983).
53. D. Mansuy, J.-P. Battioni, D. Dupre, and E. Sartori, *J. Am. Chem. Soc.*, **104**, 6159 (1982).
54. S. Okeya, B. F. Taylor, and P. M. Maitlis, *J. Chem. Soc., Chem. Commun.*, 971 (1983).
55. J. W. Labadie and J. K. Stille, *J. Am. Chem. Soc.*, **105**, 6129 (1983). See also F. K. Sheffy and J. K. Stille, *J. Am. Chem. Soc.*, **105**, 7173 (1983).
56. J. K. Jawad, R. J. Puddephatt, and M. A. Stalteri, *Inorg. Chem.*, **17**, 332 (1982).
57. R. J. Puddephatt and M. A. Stalteri, *Organometallics*, **2**, 1400 (1983).
58. R. J. Cross and J. Gemmill, *J. Chem. Soc., Chem. Commun.*, 1343 (1982).

59. R. J. Cross and A. J. McLennan, *J. Organomet. Chem.*, **255**, 113 (1983).
60. C. N. Wilker and R. Hoffmann, *J. Am. Chem. Soc.*, **105**, 5285 (1983).
61. O. A. Reutov, *J. Organomet. Chem.*, **250**, 145 (1983).
62. H. Kurosawa, H. Okada, M. Sato, and T. Hattori, *J. Organomet. Chem.*, **250**, 83 (1983).
63. J. San Filippo and J. Silbermann, *J. Am. Chem. Soc.*, 2831, **104** (1982); K.-W. Lee and J. San Filippo, *Organometallics*, 1496, 1 (1982).
64. T. Migita, K. Nagai, and M. Kosugi, *Bull. Chem. Soc. Jpn.*, **56**, 2480 (1983).
65. V. S. Petrosyan, *J. Organomet. Chem.*, **250**, 157 (1983).
66. P. L. Watson, *J. Chem. Soc., Chem. Commun.*, **276**, 1983.
67. P. L. Watson, *J. Am. Chem. Soc.*, **105**, 6491 (1983).
68. J. W. Bruno, T. J. Marks, and L. R. Morss, *J. Am. Chem. Soc.*, **105**, 6824 (1983).
69. M. Green, in: *Mechanisms of Inorganic and Organometallic Reactions* (M. V. Twigg, ed.), Vol. 2, pp. 277–278, Plenum Press, New York (1984).
70. A. R. Dias, M. S. Salema, and J. A. Martinho Simoes, *Organometallics*, **1**, 971 (1982).
71. M. Green, in: *Mechanisms of Inorganic and Organometallic Reactions* (M. V. Twigg, ed.), Vol. 2, p. 297, Plenum Press, New York (1984).
72. T. H. Tulip and D. L. Thorn, *J. Am. Chem. Soc.*, **103**, 2448 (1981).
73. M. Brookhart and M. L. H. Green, *J. Organomet. Chem.*, **250**, 395 (1983).
74. J. U. Mondal and D. M. Blake, *Coord. Chem. Rev.*, **47**, 205 (1982).
75. J. Halpern, *Acc. Chem. Res.*, **15**, 332 (1982).
76. M. Chanon, *Bull. Soc. Chim. Fr.*, II-198 (1982).
77. J. Halpern, *Inorg. Chim. Acta*, **62**, 31 (1982).
78. G. B. Sergeer, V. V. Sagorsky, and F. Z. Badaer, *J. Org. Chem.*, **243**, 123 (1983).
79. R. B. Hallock, O. T. Beachley, Y.-J. Li, W. M. Sanders, M. R. Churchill, W. E. Hunter and J. L. Atwood, *Inorg. Chem.*, **22**, 3683 (1983).
80. R. S. Grev and H. F. Schaeffer, *J. Chem. Soc., Chem. Commun.*, 785 (1983).
81. E. E. Siefert, S. D. Witt, K.-L. Loh, and Y.-N. Tang, *J. Organomet. Chem.*, **293**, 239 (1982).
82. C. McDade, J. C. Green, and J. E. Bercaw, *Organometallics*, **1**, 1629 (1982).
83. R. P. Planalp, R. A. Andersen, and A. Zalkin, *Organometallics*, **2**, 16 (1983).
84. J. A. Marsella, J. C. Huffman, K. G. Caulton, B. Longato, and J. R. Norton, *J. Am. Chem. Soc.*, **104**, 6360 (1982).
85. J. M. Mayer, C. J. Curtis, and J. E. Bercaw, *J. Am. Chem. Soc.*, **105**, 2651 (1983).
86. R. J. Goddard, R. Hoffmann, and E. D. Jemmis, *J. Am. Chem. Soc.*, **102**, 7667 (1980).
87. K. A. Mahmond, A. J. Rest, and H. G. Alt, *J. Chem. Soc., Chem. Commun.*, 1011 (1983).
88. J. Chetwynd-Talbot, P. Grebenik, and R. N. Perutz, *Inorg. Chem.*, **21**, 3647 (1982).
89. J. L. Graff, T. J. Sobieralski, M. S. Wrighton, and G. L. Geoffroy, *J. Am. Chem. Soc.*, **104**, 7526 (1982).
90. F. A. Cotton and G. N. Mott, *J. Am. Chem. Soc.*, **104**, 5978 (1982).
91. T. E. Bitterwolf, *J. Organomet. Chem.*, **252**, 305 (1983).
92. S. F. Pedersen and R. R. Schrock, *J. Am. Chem. Soc.*, **104**, 7483 (1982).
93. M. Green, in: *Mechanisms of Inorganic and Organometallic Reactions* (M. V. Twigg, ed.), Vol. 2, p. 286, Plenum Press, New York (1984).
94. J. R. Sweet and W. A. G. Graham, *J. Am. Chem. Soc.*, **105**, 305 (1983); *Organometallics*, **2**, 135 (1983).
95. F. G. N. Cloke, A. E. Derome, M. L. H. Green, and D. O'Hare, *J. Chem. Soc., Chem. Commun.*, 1312 (1983).
96. G. A. Ozin and J. G. McCaffrey, *J. Am. Chem. Soc.*, **104**, 7351 (1982); *Inorg. Chem.*, **22**, 1397 (1983).
97. Z. H. Kafafi, R. H. Hange, L. Fredin, W. E. Billups, and J. L. Margraw, *J. Chem. Soc., Chem. Commun.*, 1230 (1983).

98. H. Azizian and R. H. Morris, *Inorg. Chem.*, **22**, 6 (1983).
99. G. Bellachioma and G. Cardaci, *Inorg. Chem.*, **21**, 3232 (1982).
100. M. J. Breen, M. R. Duttera, G. L. Geoffroy, G. C. Novotnak, D. A. Roberts, P. M. Shulman, and G. R. Steinmetz, *Organometallics*, **1**, 1008 (1982).
101. T. R. Gaffney and J. A. Ibers, *Inorg. Chem.*, **21**, 2851 (1982).
102. L. M. Bavaro, P. Montangero, and J. E. Keisher, *J. Am. Chem. Soc.*, **105**, 4977 (1983).
103. Y. Doi, K. Koshizuka, and T. Keii, *Inorg. Chem.*, **21**, 2732 (1982).
104. L. J. Farrugia, M. Green, D. R. Hankey, A. G. Orpen, and F. G. Stone, *J. Chem. Soc., Chem. Commun.*, 310 (1983).
105. L. F. Halle, R. Houriet, M. M. Kappes, R. H. Staley, and J. L. Beachamp, *J. Am. Chem. Soc.*, **104**, 6293 (1982).
106. R. Houriet, L. F. Halle, and J. L. Beauchamp, *Organometallics*, **1**, 1818 (1983).
107. D. B. Jacobson and B. S. Freiser, *J. Am. Chem. Soc.*, **105**, 736 (1983).
108. D. B. Jacobson and B. S. Freiser, *J. Am. Chem. Soc.*, **105**, 5197 (1983).
109. G. B. Byrd and B. S. Freiser, *J. Am. Chem. Soc.*, **104**, 5944 (1982).
110. D. B. Jacobson and B. S. Freiser, *J. Am. Chem. Soc.*, **105**, 7492 (1982).
111. J. Halpern, *Inorg. Chim. Acta* **77**, L105 (1983).
112. M. Green, in: *Mechanisms of Inorganic and Organometallic Reactions* (M. V. Twigg, ed.), Vol. 2, p. 288, Plenum Press, New York (1984).
113. J. H. Espenson and M. S. McDowell, *Organometallics*, **1**, 1514 (1982).
114. A. Puxeddi and G. Costa, *J. Chem. Soc., Dalton Trans.*, 1285 (1982).
115. M. Okabe and M. Tada, *Bull. Chem. Soc. Jpn.*, **55**, 1498 (1982).
116. R. S. Drago, J. G. Miller, M. A. Hoselton, R. D. Farris, and M. J. Desmond, *J. Am. Chem. Soc.*, **105**, 444 (1983).
117. D. Milstein, *Organometallics*, **1**, 1549 (1982).
118. D. Milstein, *J. Am. Chem. Soc.*, **104**, 5227 (1982).
119. F. Delgado, A. Cabrera, and J. Gomez-Lara, *J. Mol. Catal.*, **22**, 83 (1983).
120. Y. Aoyama, T. Yoshida, K.-I. Sakurai, and H. Ogoshi, *J. Chem. Soc., Chem. Commun.*, 478 (1983).
121. W. D. Jones and F. J. Feher, *J. Am. Chem. Soc.*, **104**, 4240 (1982).
122. W. D. Jones and F. J. Feher, *Organometallics*, **2**, 562 (1983).
123. S. Fukuzumi, N. Nishizawa, and T. Tanaka, *Bull. Chem. Soc. Jpn.*, **56**, 709 (1983).
124. J. U. Mondal, K. G. Young and D. M. Blake, *J. Organomet. Chem.*, **240**, 447 (1982).
125. B. B. Wayland, B. A. Woods, and V. M. Minda, *J. Chem. Soc., Chem. Commun.*, 634 (1982).
126. R. G. Finke, G. Gaughan, C. Pierpont, and J. H. Noordik, *Organometallics*, **2**, 1481 (1983).
127. C. E. Johnson, B. J. Fisher, and R. Eisenberg, *J. Am. Chem. Soc.*, **105**, 7772 (1983).
128. D. R. Anton and R. H. Crabtree, *Organometallics*, **2**, 621 (1983).
129. R. H. Crabtree and R. J. Uriarte, *Inorg. Chem.*, **22**, 4152 (1983).
130. J. K. Hoyano, A. D. McMaster, and W. A. G. Graham, *J. Am. Chem. Soc.*, **105**, 7190 (1983).
131. M. J. Wax, J. M. Stryker, J. M. Buchanan, and R. G. Bergmann, *Chem. Eng. News*, **61**(37), 33 (1983).
132. A. H. Janowicz and R. G. Bergman, *J. Am. Chem. Soc.*, **105**, 3929 (1983).
133. T. R. Gaffney and J. A. Ibers, *Inorg. Chem.*, **21**, 2854 (1982).
134. P. Kalck and J.-J. Bonnet, *Organometallics*, **1**, 1211 (1982).
135. M. R. A. Blomberg and P. E. M. Siegbahn, *J. Chem. Phys.*, **78**, 5682 (1983).
136. M. R. A. Blomberg, U. Brandemark, and P. E. M. Siegbahn, *J. Am. Chem. Soc.*, **105**, 5557 (1983).
137. Y. Abe, A. Yamamoto, and T. Yamamoto, *Organometallics*, **2**, 1466 (1983).
138. M. Green, in: *Mechanisms of Inorganic and Organometallic Reactions* (M. V. Twigg, ed.), Vol. 2, p. 293, Plenum Press, New York (1984).
139. M. Green, in: *Mechanisms of Inorganic and Organometallic Reactions* (M. V. Twigg, ed.), Vol. 1, p. 238, Plenum Press, New York (1984).

140. M. Green, in: *Mechanisms of Inorganic and Organometallic Reactions* (M. V. Twigg, ed.), Vol. 2, p. 294, Plenum Press, New York (1984).
141. M. Anton, G. Muller, and J. Sales, *Transition Met. Chem.* **8**, 79 (1983).
142. G. Favaro, A. Morvillo, and A. Turco, *J. Organomet. Chem.*, **241**, 251 (1983).
143. E.-I. Negishi, *Acc. Chem. Res.*, **15**, 340 (1982).
144. M. Green, in: *Mechanisms of Inorganic and Organometallic Reactions* (M. V. Twigg, ed.), Vol. 2, p. 195, Plenum Press, New York (1984).
145. F. Ozawa, A. Yamamoto, T. Ikariya, and R. H. Grubbs, *Organometallics*, **1**, 1481 (1982).
146. T. Hayashi, T. Hagihara, M. Konishi, and M. Kumada, *J. Am. Chem. Soc.*, **105**, 7767 (1983).
147. J. O. Noell and P. J. Hay, *J. Am. Chem. Soc.*, **104**, 4578 (1982).
148. R. S. Paonesssa and W. C. Trogler, *J. Am. Chem. Soc.*, **104**, 1138 (1982).
149. P. K. Monaghan and R. J. Puddephatt, *Inorg. Chim. Acta*, **76**, L237 (1983).
150. G. Ferguson, M. Parvez, P. K. Monaghan, and R. J. Puddephatt, *J. Chem. Soc., Chem. Commun.*, 267 (1983).
151. P. K. Monaghan and R. J. Puddephatt, *Organometallics*, **2**, 1698 (1983).
152. P. K. Monaghan and R. J. Puddephatt, *Inorg. Chim. Acta*, **65**, L59 (1982).
153. A. G. Constable, R. Langrick, B. Shabanzadeh, and B. L. Shaw, *Inorg. Chim. Acta*, **65**, L151 (1982).
154. T. R. Gaffney and J. A. Ibers, *Inorg. Chem.*, **21**, 2860 (1982).
155. A. Flores-Riveros and O. Novaro, *J. Organomet. Chem.*, **235**, 383 (1982).
156. M. Green, in: *Mechanisms of Inorganic and Organometallic Reactions* (M. V. Twigg, ed.), Vol. 2, p. 193, Plenum Press, New York (1984).
157. L. Abis, A. Sen, and J. Halpern, *J. Am. Chem. Soc.*, **100**, 2915 (1978).
158. S. Komiya, Y. Morimoto, A. Yamamoto, and T. Yamamoto, *Organometallics*, **1**, 1528 (1982).
159. M. Green, in: *Mechanisms of Inorganic and Organometallic Reactions* (M. V. Twigg, ed.), Vol. 2, p. 299, Plenum Press, New York (1984).
160. C. Bartocei, A. Maldotti, S. Sostero, and O. Traverso, *J. Organomet. Chem.*, **253**, 253 (1983).
161. U. Belluco, B. Crociani, R. Michelin, and P. Uguagliati, *Pure Appl. Chem.*, **55**, 47 (1983); see also U. Belluco, R. A. Michelin, P. Uguagliati, and R. Crociani, *J. Organomet. Chem.*, **250**, 565 (1983).
162. S. Sostero, O. Traverso, R. Ros, and R. A. Michelin, *J. Organomet. Chem.*, **246**, 325 (1983).
163. R. A. Michelin, S. Faglia, and P. Uguagliati, *Inorg. Chem.*, **22**, 1831 (1983).
164. A. Morvillo and A. Turco, *J. Organomet. Chem.*, **258**, 383 (1983).
165. S. S. M. Ling and R. J. Puddephatt, *J. Chem. Soc., Chem. Commun.*, 412 (1982).
166. S. S. M. Ling, R. J. Puddephatt, L. Manojlovic-Muir, and K. W. Muir, *J. Organomet. Chem.*, **255**, C11 (1983).
167. R. H. Hill, P. de Mayo, and R. J. Puddephatt, *Inorg. Chem.*, **21**, 3642 (1982); see also H. C. Foley, R. H. Morris, T. S. Targos, and G. L. Geoffrey, *J. Am. Chem. Soc.*, **103**, 7337 (1981).
168. M. Green, in: *Mechanisms of Inorganic and Organometallic Reactions* (M. V. Twigg, ed.), Vol. 2, p. 299, Plenum Press, New York (1984).
169. R. H. Hill and R. J. Puddephatt, *Organometallics*, **2**, 1472 (1983).
170. K. A. Azam and R. J. Puddephatt, *Organometallics*, **2**, 1396 (1983); R. H. Hill and R. J. Puddephatt, *J. Am. Chem. Soc.*, **105**, 5707 (1983).
171. V. P. Dyadchenko, *Russ. Chem. Rev.* (*Engl. Trans.*), **51**, 265 (1982).
172. J. W. Bruno, T. J. Marks, and V. W. Day, *J. Am. Chem. Soc.*, **104**, 7357 (1982).
173. M. Green, in: *Mechanisms of Inorganic and Organometallic Reactions* (M. V. Twigg, ed.), Vol. 2, p. 297, Plenum Press, New York (1984).
174. W. J. Evans, D. J. Wink, and D. R. Stanley, *Inorg. Chem.*, **21**, 2564 (1982).

References for Chapter 12

1. W. A. Kiel, G.-Y. Lin, A. G. Constable, F. B. McCormick, C. E. Strouse, O. Eisenstein, and J. A. Gladysz, *J. Am. Chem. Soc.*, **104**, 4865 (1982).
2. W. A. Kiel, G.-Y. Lin, G. S. Bodner, and J. A. Gladysz, *J. Am. Chem. Soc.*, **105**, 4958 (1983).
3. M. Brookhart, D. Timmers, J. R. Tucker, G. D. Williams, G. R. Husk, H. Brunner, and B. Hammer, *J. Am. Chem. Soc.*, **105**, 6721 (1983).
4. L. R. Beanan, Z. A. Rahman, and J. B. Keister, *Organometallics*, **2**, 1062 (1983).
5. N. M. Kostic and R. F. Fenske, *Organometallics*, **1**, 974 (1982).
6. G. N. Mott and A. J. Carty, *Inorg. Chem.*, **22**, 2726 (1983).
7. J. E. Jensen, L. L. Campbell, S. Nakanishi, and T. C. Flood, *J. Organomet. Chem.*, **244**, 61 (1983).
8. O. Eisenstein and R. Hoffmann, *J. Am. Chem. Soc.*, **103**, 4308 (1981), and references therein.
9. L. Maresca and G. Natile, *J. Chem. Soc., Chem. Commun.*, 40 (1983), and references therein.
10. L. Maresca, G. Natile, A.-M. Manotti-Lanfredi, and A. Tiripicchio, *J. Am. Chem. Soc.*, **104**, 7661 (1982).
11. D. L. Reger, K. A. Belmore, E. Mintz, N. G. Charles, E. A. H. Griffith, and E. L. Amma, *Organometallics*, **2**, 101 (1983).
12. J.-E. Bäckvall, R. E. Nordberg, K. Zetterberg, and B. Åkermark, *Organometallics*, **2**, 1625 (1983).
13. J. E. Bäckvall and R. E. Nordberg, *J. Am. Chem. Soc.*, **103**, 4959 (1981), and references therein.
14. J. W. Faller and K.-H. Chao, *J. Am. Chem. Soc.*, **105**, 3893 (1983).
15. D. A. Brown, J. P. Chester, and N. J. Fitzpatrick, *Inorg. Chem.*, **21**, 2723 (1982).
16. S. A. Stanton, S. W. Felman, C. S. Parkhurst, and S. A. Godleski, *J. Am. Chem. Soc.*, **105**, 1964 (1983).
17. J. C. Calabrese, S. D. Ittel, H. S. Choi, S. G. Davis, and D. A. Sweigart, *Organometallics*, **2**, 226 (1983).
18. S. G. Davies, J. Hibberd, and S. J. Simpson, *J. Organomet. Chem.*, **246**, C16 (1983).
19. J. G. Atton and L. A. P. Kane-Maguire, *J. Chem. Soc., Dalton Trans.*, 1491 (1982).
20. J. J. Guy, B. E. Reichert, and G. M. Sheldrick, *Acta Crystallogr., Sect. B*, **32**, 2504 (1976).
21. D. A. Brown, S. K. Chawla, W. K. Glass, and F. M. Hussein, *Inorg. Chem.*, **21**, 2726 (1982).
22. D. J. Evans and L. A. P. Kane-Maguire, *Inorg. Chem. Acta*, **62**, 109 (1982).
23. G. R. John, L. A. P. Kane-Maguire, T. I. Odiaka, and C. Eaborn, *J. Chem. Soc., Dalton Trans.*, 1721 (1983).
24. J. G. Atton and L. A. P. Kane-Maguire, *J. Organomet. Chem.*, **246**, C23 (1983).
25a. E. G. Bryan, A. L. Burrows, B. F. G. Johnson, J. Lewis, and G. M. Schiavon, *J. Organomet. Chem.*, **129**, C19 (1977).
25b. A. L. Burrows, Ph.D. thesis, University of Cambridge (1978).
26. L. A. P. Kane-Maguire, T. I. Odiaka, S. Turgoose, and P. A. Williams, *J. Chem. Soc., Dalton Trans.*, 2489 (1981), and references therein.
27. D. J. Evans and L. A. P. Kane-Maguire, *J. Organomet. Chem.*, **236**, C15 (1982), and references therein.
28. J. A. S. Howell and M. J. Thomas, *J. Chem. Soc., Dalton Trans.*, 1401 (1983).
29. B. M. R. Bandara, A. J. Birch, L. F. Kelly, and T. Chak Khor, *Tetrahedron Letter.*, **24** 2491 (1983).
30. J. A. S. Howell and M. J. Thomas, *J. Organomet. Chem.*, **247**, C21 (1983).
31. Y. K. Chung, H. S. Choi, D. A. Sweigart, and N. G. Connelly, *J. Am. Chem. Soc.*, **104**, 4245 (1982).

32. Y. K. Chung, E. D. Honig, W. T. Robinson, D. A. Sweigart, N. G. Connelly, and S. D. Ittel, *Organometallics*, **2**, 1479 (1983).
33. G. Schiavon and C. Paradisi, *J. Organomet. Chem.*, **243**, 351 (1983).
34. C. Paradisi and G. Schiavon, *J. Organomet. Chem.*, **246**, 197 (1983).
35. Y. K. Chung, E. D. Honig, and D. A. Sweigart, *J. Organomet. Chem.*, **256**, 277 (1983).
36. L. A. P. Kane-Maguire and D. A. Sweigart, *Inorg. Chem.*, **18**, 700 (1979).
37. G. Faraone, F. Cusmano, and R. Pietropaolo, *J. Organomet. Chem.*, **26**, 147 (1971).
38. M. Brookhart, A. R. Pinhas, and A. Lukacs, *Organometallics*, **1**, 1730 (1982).
39. Y. K. Chung, P. G. Williard, and D. A. Sweigart, *Organometallics*, **1**, 1053 (1982).
40. M. F. Semmelhack, J. L. Garcia, D. Cortes, R. Farina, R. Hong, and B. K. Carpenter, *Organometallics*, **2**, 467 (1983).
41. T. A. Albright and B. K. Carpenter, *Inorg. Chem.*, **19**, 3092 (1980).
42. W. R. Jackson, I. D. Rae, M. G. Wong, M. F. Semmelhack, and J. N. Garcia, *J. Chem. Soc., Chem. Commun.*, 1359 (1982).
43. E. P. Kündig, V. Desobry, and D. P. Simmons, *J. Am. Chem. Soc.*, **105**, 6962 (1983).
44. L. N. Nekrasov, L. P. Yur'eva, and S. M. Peregudova, *J. Organomet. Chem.*, **238**, 185 (1982).
45. C. A. Bunton, M. J. Mhala, J. R. Moffatt, and W. E. Watts, *J. Organomet. Chem.*, **253**, C33 (1983).
46. D. W. Clack, M. Monshi, and L. A. P. Kane-Maguire, *J. Organomet. Chem.*, **120**, C25 (1976).
47. C. A. Bunton, K. Lal, and W. E. Watts, *J. Organomet. Chem.*, **247**, C14 (1983).
48. G. R. John, L. A. P. Kane-Maguire, and D. A. Sweigart, *J. Organomet. Chem.*, **120**, C47 (1976).
49. O. Koch, F. Edelmann, and U. Behrens, *Chem. Ber.*, **115**, 1313 (1982).
50. U. Belluco, R. A. Michelin, P. Uguagliati, and B. Crociani, *J. Organomet. Chem.*, **250**, 565 (1983).
51. G. S. Lewandos, J. W. Maki, and J. P. Ginnebaugh, *Organometallics*, **1**, 1700 (1982).
52. A. Wong and J. A. Gladysz, *J. Am. Chem. Soc.*, **104**, 4948 (1982).
53. W. G. Hatton and J. A. Gladysz, *J. Am. Chem. Soc.*, **105**, 6157 (1983).
54. J. H. Merrifield, G.-Y. Lin, W. A. Kiel, and J. A. Gladysz, *J. Am. Chem. Soc.* **105**, 5811 (1983).
55. M. F. Semmelhack and C. Ullenius, *J. Organomet. Chem.*, **235**, C10 (1982).
56. H. Fischer, E. O. Fischer, and R. Cai, *Chem. Ber.*, **115**, 2707 (1982).
57. H. Fischer, E. O. Fischer, R. Cai, and D. Himmelreich, *Chem. Ber.*, **116**, 1009 (1983).
58. H. Fischer, J. Mühlemeier, R. Märkl, and K. H. Dötz, *Chem. Ber.*, **115**, 1355 (1982).

References for Chapter 13

1. H. Azizian, K. R. Dixon, C. Eaborn, A. Pidcock, N. M. Shuaib, and J. Vinaixa, *J. Chem. Soc., Chem. Commun.*, 1020 (1982).
2. Yu. K. Grishin, V. A. Roznyatovsky, Yu. A. Ustynyuk, S. N. Titova, G. A. Domrachev, and G. A. Razuvaev, *Polyhydron*, **2**, 895 (1983).
3. A. Goodacre and K. G. Orrell, *J. Chem. Soc., Dalton Trans.*, 153 (1983).
4. D. R. Anton and R. H. Crabtree, *Organometallics*, **2**, 621 (1983).
5. J. M. Brown, F. M. Dayrit, and D. Lightowler, *J. Chem. Soc., Chem. Commun.*, 414 (1983).
6. M. Kotzian, C. G. Kreiter, G. Michael, and S. Özkar, *Chem. Ber.*, **116**, 3637 (1983).
7. S. Özkar and C. G. Kreiter, *J. Organomet. Chem.*, **256**, 57 (1983).
8. C. G. Kreiter and S. Özkar, *Z. Naturforsch. B*, **38**, 1424 (1983).
9. O. Schneidsteger, G. Huttner, V. Bejenke, and W. Gartzke, *Z. Naturforsch. B*, **38**, 1598 (1983).

10. D. L. Reger, C. A. Swift, and L. Lebioda, *J. Am. Chem. Soc.*, **105**, 5343 (1983).

11. J. A. L. Palmer and S. B. Wild, *Inorg. Chem.*, **22**, 4054 (1983).

12. S. L. Hawthorne, A. H. Bruder, and R. C. Fay, *Inorg. Chem.*, **22**, 3368 (1983).

13. R. C. Fay and A. F. Lindmark, *J. Am. Chem. Soc.*, **105**, 2118 (1983).

14. C. G. Kreiter and U. Koemm, *Z. Naturforsch. B*, **38**, 943 (1983).

15. U. Koemm and C. G. Kreiter, *J. Organomet. Chem.*, **240**, 27 (1982).

16. J. A. K. Howard, P. Mitrprachachon, and A. Roy, *J. Organomet. Chem.*, **235**, 375 (1982).

17. M. Cosandey, M. von Büren, and H.-J. Hansen, *Helv. Chim. Acta*, **66**, 1 (1983).

18. M. R. Burke, J. Takats, F.-W. Grevels, and J. G. A. Reuvers, *J. Am. Chem. Soc.*, **105**, 4092 (1983).

19. M. A. Arthers and S. M. Nelson, *J. Coord. Chem.*, **13**, 29 (1983).

20. D. Parker, *J. Organomet. Chem.*, **240**, 83 (1982).

21. H. Lehmkuhl, J. Grundke, and R. Mynott, *Chem. Ber.*, **116**, 159 (1983).

22. H. Lehmkuhl, S. Pasynkiewicz, R. Benn, and A. Rufinska, *J. Organomet. Chem.*, **240**, C27 (1982).

23. N. M. Boag, R. J. Goodfellow, M. Green, B. Hessner, J. A. K. Howard, and F. G. A. Stone, *J. Chem. Soc., Dalton Trans.*, 2585 (1983).

24. J. M. Brown, P. A. Chaloner, and G. A. Morris, *J. Chem. Soc., Chem. Commun.*, 664 (1983).

25. P. B. Winston, S. J. Nieter Burgmayer, and J. L. Templeton, *Organometallics*, **2**, 167 (1983).

26. R. S. Herrick, S. J. Nieter Burgmayer, and J. L. Templeton, *Inorg. Chem.*, **22**, 3275 (1983).

27. M. H. Chisholm, K. Folting, D. M. Hoffman, J. C. Huffman, and J. Leonelli, *J. Chem. Soc., Chem. Commun.*, 589 (1983).

28. H. G. Alt and H. I. Hayen, *Angew. Chem. Int. Ed. Engl.*, **22**, 1008 (1983).

29. D. F. R. Gilson, G. Gomez, I. S. Butler, and P. J. Fitzpatrick, *Canad. J. Chem.*, **61**, 737 (1983).

30. A. Shaver, A. Eisenberg, K. Yamada, A. J. F. Clark, and S. Farrokyzad, *Inorg. Chem.*, **22**, 4154 (1983).

31. D. F. R. Gilson and G. Gomez, *J. Organomet. Chem.*, **240**, 41 (1982).

32. M. J. McGlinchey, J. L. Fletcher, B. G. Sayer, P. Bougeard, R. Faggiani, G. J. L. Lock, A. D. Bain, C. Rodger, E. P. Kündig, D. Astruc, J.-R. Hamon, P. Le Maux, S. Top, and G. Jaouen, *J. Chem. Soc., Chem., Commun.*, 634 (1983).

33. D. R. Wilson, R. D. Ernst, and T. H. Cymbaluk, *Organometallics*, **2**, 1220 (1983).

34. M. Moll and H.-J. Seibold, *J. Organomet. Chem.*, **248**, 343 (1983).

35. A. Scrivanti, G. Carturan, and B. Crociani, *Organometallics*, **2**, 1612 (1983).

36. A. T. Patton, C. E. Strouse, C. B. Knobler, and J. A. Gladysz, *J. Am. Chem. Soc.*, **105**, 5804 (1983).

37. W. A. Kiel, G.-Y. Lin, A. G. Constable, F. B. McCormick, C. E. Strouse, O. Eisenstein, and J. A. Gladysz, *J. Am. Chem. Soc.*, **104**, 4865 (1982).

38. A. Wong and J. A. Gladysz, *J. Am. Chem. Soc.*, **104**, 4948 (1982).

39. M. Lukszo, S. Himmel, and W. Malisch, *Angew. Chem. Int. Ed. Engl.*, **22**, 416 (1983).

40. J. R. Sweet and W. A. G. Graham, *J. Am. Chem. Soc.*, **105**, 305 (1983); *Organometallics*, **2**, 135 (1983).

41. W. D. Jones and F. J. Feher, *J. Am. Chem. Soc.*, **104**, 4240 (1982).

42. M. J. Hails, B. E. Mann, and C. M. Spencer, *J. Chem. Soc., Chem. Commun.*, 120 (1983).

43. T. A. Albright, P. Hofmann, R. Hoffmann, C. P. Lillya, and P. A. Dobosh, *J. Am. Chem. Soc.*, **105**, 3396 (1983).

44. L. K. K. LiShingMan, J. G. A. Reuvers, J. Takats, and G. Deganello, *Organometallics*, **2**, 28 (1983).

45. H. Lehmkuhl and C. Naydowski, *J. Organomet. Chem.*, **240**, C30 (1982).

46. H. Werner, H.-J. Krans, V. Schubert, K. Ackermann, and P. Hofmann, *J. Organomet. Chem.*, **250**, 517 (1983).
47. C. A. Ghilardi, P. Innocenti, S. Midollini, and A. Orlandini, *J. Organomet. Chem.*, **231**, C78 (1982).
48. M. Y. Dahrensbourg, R. Bau, M. W. Marks, R. R. Burch, J. C. Deaton, and S. Slater, *J. Am. Chem. Soc.*, **104**, 6961 (1982).
49. M. F. Lappert, A. Singh, J. L. Atwood, W. E. Hunter, *J. Chem. Soc., Chem. Commun.*, 206 (1983).
50. J. C. Bommer and K. W. Morse, Inorg. Chim. Acta, **74**, 25 (1983).
51. J. Browning, G. W. Bushnell, K. R. Dixon, and A. Pidcock, *Inorg. Chem.*, **22**, 2226 (1982).
52. E. W. Abel, G. D. King, K. G. Orrell, G. M. Pring, V. Sik, and T. S. Cameron, *Polyhedron*, **2**, 1117 (1983).
52a. E. W. Abel, G. D. King, K. G. Orrell, and V. Sik, *Polyhedron*, **2**, 1363 (1983).
53. P. J. Pogorzelec and D. H. Reid, *J. Chem. Soc., Chem. Commun.*, 289 (1983).
54. S. G. Davies and S. J. Simpson, *J. Organomet. Chem.*, **240**, C48 (1982).
55. A. J. Canty, N. J. Minchin, J. M. Patrick, and A. H. White, *Aust. J. Chem.*, **36**, 1107 (1983).
56. P. Braunstein, D. Matt, and Y. Dusausoy, *Inorg. Chem.*, **22**, 2043 (1983).
57. R. B. A. Pardy, M. J. Taylor, E. C. Constable, J. D. Merch, and J. K. M. Sanders, *J. Organomet. Chem.*, **231**, C25 (1982).
58. M. Brookhart, M. L. H. Green, and R. B. A. Pardy, *J. Chem. Soc., Chem. Commun.*, 691 (1983).
59. M. Brookhart, W. Lamanna, and A. R. Pinhas, *Organometallics*, **2**, 638 (1983).
60. M. Brookhart and A. Lukacs, *Organometallics*, **2**, 649 (1983).
61. R. J. Kazlauskas and M. S. Wrighton, *J. Am. Chem. Soc.*, **104**, 6005 (1982).
62. D. C. Roe, *J. Am. Chem. Soc.*, **105**, 7770 (1983).
63. W. G. Hatton and J. A. Gladysz, *J. Am. Chem. Soc.*, **105**, 6157 (1983).
64. H. W. Turner, R. R. Shrock, J. D. Fellmann, and S. J. Holmes, *J. Am. Chem. Soc.*, **105**, 4942 (1983).
65. S. J. Holmes, D. N. Clark, H. W. Turner, and R. R. Schrock, *J. Am. Chem. Soc.*, **104**, 6322 (1982).
66. M. D. Curtis, N. A. Fotinos, L. Messerle, and A. P. Sattelberger, *Inorg. Chem.*, **22**, 1559 (1983).
67. J.-P. Lecomte, J.-M. Lehn, D. Parker, J. Guilhem, and C. Pascard, *J. Chem. Soc., Chem. Commun.*, 296 (1983).
68. S. Okeya, B. F. Taylor, and P. M. Maitlis, *J. Chem. Soc., Chem. Commun.*, 971 (1983).
69. R. J. Puddephatt, K. A. Azam, R. H. Hill, M. P. Brown, C. D. Nelson, R. P. Moulding, K. R. Seddon, and M. C. Grossel, *J. Am. Chem. Soc.*, **105**, 5642 (1983).
70. K. A. Azam and R. J. Puddephatt, *Organometallics*, **2**, 1396 (1983).
71. M. C. Grossel, R. P. Moulding, and K. R. Seddon, *J. Organomet. Chem.*, **247**, C32 (1983).
72. H. C. Aspinall, A. J. Deeming, and S. Donovan-Mtunzi, *J. Chem. Soc., Dalton Trans.*, 2669 (1983).
73. A. F. Dyke, S. A. R. Knox, M. J. Morris, and P. J. Naish, *J. Chem. Soc., Dalton Trans.*, 1417 (1983).
74. K.-H. Franzreb and C. G. Kreiter, *J. Organomet. Chem.*, **246**, 189 (1983).
74a. P. O. Nubel and T. L. Brown, *Organometallics*, **3**, 29 (1984); *J. Am. Chem. Soc.*, **106**, 644 (1984).
75. R. D. Adams, D. A. Katahira, and L.-W. Yang, *Organometallics*, **1**, 231 (1982).
76. J. A. King and K. P. C. Vollhardt, *J. Am. Chem. Soc.*, **105**, 4846 (1983).
77. T. V. O'Halloran and S. J. Lippard, *J. Am. Chem. Soc.*, **105**, 3341 (1983).

78. C. B. Knobler, H. D. Kaesz, G. Minghetti, A. L. Bandini, G. Banditelli, and F. Bonati, *Inorg. Chem.*, **22**, 2324 (1983).
79. K. H. Theopold and R. G. Bergman, *J. Am. Chem. Soc.*, **105**, 464 (1983).
80. W. J. Laws and R. J. Puddephatt, *J. Chem. Soc., Chem. Commun.*, 1020 (1983).
81. R. E. Colborn, A. F. Dyke, S. A. R. Knox, K. A. Mead, and P. Woodward, *J. Chem. Soc., Dalton Trans.*, 2099 (1983).
82. B. E. Hanson and J. S. Mancini, *Organometallics*, **2**, 126 (1983).
83. D. M. Hoffman, R. Hoffmann, and C. R. Fisel, *J. Am. Chem. Soc.*, **104**, 3858 (1982).
84. M. J. McGlinchey, M. Mlekuz, P. Bougeard, B. G. Sayer, A. Marinetti, J.-Y. Saillard, and G. Jaouen, *Can. J. Chem.*, **61**, 1319 (1983).
85. D. M. Hoffman and R. Hoffmann, *J. Chem. Soc., Dalton Trans.*, 1471 (1982).
86. M. J. Freeman, M. Green, A. G. Orpen, I. D. Slater, and F. G. A. Stone, *J. Chem. Soc., Chem. Commun.*, 1332 (1983).
87. B. T. Heaton, L. Strona, R. Della Pergola, J. L. Vidal, and R. C. Schoening, *J. Chem. Soc., Dalton Trans.*, 1941 (1983).
88. G. Granozzi, E. Tondello, M. Casarin, S. Aime, and D. Osella, *Organometallics*, **2**, 430 (1983).
89. J. K. Kouba, E. L. Muetterties, M. R. Thompson, and V. W. Day, *Organometallics*, **2**, 1065 (1983).
90. J. T. Park, J. R. Shapley, M. R. Churchill, and C. Bueno, *Inorg. Chem.*, **22**, 1579 (1983).
91. R. D. Barr, M. Green, J. A. K. Howard, T. B. Marder, F. G. A. Stone, *J. Chem. Soc., Chem., Commun.*, 759 (1983).
92. B. E. Mann, C. M. Spencer, and A. K. Smith, *J. Organomet. Chem.*, **244**, C17 (1983).
93. S. Aime, D. Osella, R. Gobetto, B. F. G. Johnson, and L. Milone, *Inorg. Chim. Acta*, **68**, 141 (1983).
94. F. Richter, H. Beurich, M. Müller, N. Gärtner, and H. Vahrenkamp, *Chem. Ber.*, **116**, 3774 (1983).
95. A. Ceriotti, G. Longoni, R. Della Pergola, B. T. Heaton, and D. O. Smith, *J. Chem. Soc., Dalton Trans.*, 1433 (1983).
96. J. B. Keister and J. R. Shapley, *Inorg. Chem.*, **21**, 3304 (1982).
97. A. C. Wills, F. W. B. Einstein, R. M. Ramadan, and R. K. Pomeroy, *Organometallics*, **2**, 935 (1983).
98. N. J. Forrow, S. A. R. Knox, M. J. Morris, and A. G. Orpen, *J. Chem. Soc., Chem. Commun.*, 234 (1983).
99. L. Busetto, J. C. Jeffery, R. M. Mills, F. G. A. Stone, M. J. Went, and P. Woodward, *J. Chem. Soc., Dalton Trans.*, 101 (1983).
100. R. J. Goudsmit, B. F. G. Johnson, J. Lewis, P. R. Raithby, and M. J. Rosales, *J. Chem. Soc., Dalton Trans.*, 2257 (1983).
101. S. Aime and A. J. Deeming, *J. Chem. Soc., Dalton Trans.*, 1807 (1983).

References for Chapter 14

1. P. Biloen, *J. Mol. Catal.* **21**, 17 (1983).
2. D. R. Anton and R. H. Crabtree, *Organometallics*, **2**, 855 (1983).
3. R. J. Cross, *Catalysis*, **5**, 366 (1982).
4. J. A. Connor, M. T. Zafarani-Moattar, J. Bickerton, N. I. El Saied, S. Suradi, R. Carson, G. Al Takhin, and H. A. Skinner, *Organometallics*, **1**, 1166 (1982).
5. B. D. Dombek and A. M. Harrison, *J. Am. Chem. Soc.*, **105**, 2485 (1983).

6. K. Kropp, V. Skibbe, G. Erker, and C. Kruger, *J. Am. Chem. Soc.*, **105**, 3353 (1983).
7. P. Leoni and M. Pasquali, *J. Organomet. Chem.*, **255**, C31 (1983).
8. (a) J. Halpern, *Acc. Chem. Red.*, **15**, 238 (1982); (b) J. Halpern, *Acc. Chem. Res.*, **15**, 322 (1982).
9. (a) Z. Dawoodi, M. L. H. Green, V. S. B. Mtetwa, and K. Prout, *J. Chem. Soc., Chem. Commun.*, 802 (1982); (b) M. Brookhart and M. L. H. Green, *J. Organomet. Chem.*, **250**, 395 (1983).
10. P. L. Watson, *J. Am. Chem. Soc.*, **105**, 6491 (1983).
11. (a) J. K. Hoyano, A. D. McMaster, and W. A. G. Graham, *J. Am. Chem. Soc.*, **105**, 7190 (1983); (b) J. K. Hoyano and W. A. G. Graham, *J. Am. Chem. Soc.*, **104**, 3723 (1982); see also A. H. Janowicz and R. G. Bergman, *J. Am. Chem. Soc.*, **105**, 3929 (1983).
12. K. A. Mahmoud, A. J. Rest, and H. G. Alt, *J. Chem. Soc., Chem. Commun.*, 1011 (1983).
13. E. L. Muetterties and M. J. Krausse, *Angew. Chem. Int. Ed. Engl.*, **22**, 135 (1983).
14. P. Pino and G. Consiglio, *Pure Appl. Chem.*, **55**, 1781 (1983); H. Brunner, *Angew. Chem. Int. Ed. Engl.*, **22**, 897 (1983); T. Hayashi and M. Kumada, *Acc. Chem. Res.*, **15**, 395 (1982).
15. K. K. Pandey, *Coord. Chem. Rev.*, **51**, 69 (1983).
16. O. N. Temkin and L. G. Bruk, *Russ. Chem. Rev. (Engl. Transl.)*, **52**, 117 (1983).
17. B. A. Murrer, and M. J. H. Russell, *Catalysis*, **6**, 169 (1983).
18. M. Bianchi, G. Menchi, P. Frediani, U. Matteoli, and F. Piacenti, *J. Organomet. Chem.*, **247**, 89 (1983).
19. H. P. Withers, Jr., and D. Seyferth, *Inorg. Chem.*, **22**, 2931 (1983).
20. P. Pino, F. Oldani, and G. Consiglio, *J. Organomet. Chem.*, **250**, 491 (1983).
21. J. P. Collman, J. A. Belmont and J. I. Brauman, *J. Am. Chem. Soc.*, **105**, 7288 (1983).
22. J. C. Barborak and K. Cann, *Oragnometallics*, **1**, 1726 (1982).
23. Y. Matsui and M. Orchin, *J. Organomet. Chem.*, **246**, 57 (1983).
24. (a) M. J. Mirbach, N. Topalsavoglou, T. N. Phu, M. F. Mirbach, and A. Saus, *Chem. Ber.*, **116**, 1422 (1983); (b) A. Saus, T. N. Phu, M. J. Mirbach, and M. F. Mirbach, *J. Mol. Catal.*, **18**, 117 (1983).
25. (a) J. M. Brown, L. R. Canning, A. G. Kent, and P. J. Sidebottom, *J. Chem. Soc., Chem. Commun.*, 721 (1982); (b) J. M. Brown and A. G. Kent, *J. Chem. Soc., Chem. Commun.*, 723 (1982).
26. P. W. N. M. Van Leeuwen and C. F. Roobeek, *J. Organomet. Chem.*, **258**, 343 (1983).
27. (a) M. Matsumoto and M. Tamura, *J. Mol. Catal.*, **16**, 195 (1982); (b) M. Matsumoto and M. Tamura, *J. Mol. Catal.*, **16**, 209 (1982).
28. A. M. Trzeciak and J. J. Ziółkowski, *J. Mol. Catal.*, **19**, 41 (1983).
29. G. K. Anderson, C. Billard, H. C. Clark, and J. A. Davies, *Inorg. Chem.*, **22**, 439 (1983).
30. G. Cavinato and L. Toniolo, *J. Organomet. Chem.*, **241**, 275 (1983).
31. P. Isnard, B. Denise, R. P. A. Sneeden, J. M. Cognion, and P. Durual, *J. Organomet. Chem.*, **256**, 135 (1983).
32. D. Milstein and J. L. Huckaby, *J. Am. Chem. Soc.*, **104**, 6150 (1982).
33. G. Consiglio and P. Pino, *Top. Curr. Chem.*, **105**, 77 (1982).
34. T. Hayashi, M. Tanaka, I. Ogata, T. Kodama, T. Takahashi, Y. Uchida, and T. Uchida, *Bull. Chem. Soc. Jpn.*, **56**, 1780 (1983).
35. G. Consiglio, P. Pino, L. I. Flowers, and C. V. Pittman, Jr., *J. Chem. Soc., Chem. Commun.*, 612 (1983).
36. J. K. Stille and G. Parrinello, *J. Mol. Catal.*, **21**, 203 (1983).
37. M. J. Chen, H. M. Feder, and J. W. Rathke, *J. Am. Chem. Soc.*, **104**, 7346 (1982); *J. Mol. Catal.*, **17**, 331 (1982).
38. J. T. Martin and M. C. Baird, *Organometallics*, **2**, 1073 (1983).
39. P. B. Francoisse and C. F. Thyrion, *Ind. Eng. Chem. Prod. Res. Dev.* **22**, 542 (1983).

40. M. Röper, H. Loevenich, and J. Korff, *J. Mol. Catal.*, **17**, 315 (1982).
41. D. J. Drury, M. J. Green, D. J. M. Ray, and A. J. Stevenson, *J. Organomet. Chem.*, **236**, C23 (1982).
42. H. des Abbayes, A. Buloup, and G. Tanguy, *Organometallics*, **2**, 1730 (1983).
43. P. Andrianary, G. Jenner, and A. Kiennemann, *J. Organomet. Chem.*, **252**, 209 (1983).
44. J. Gauthier-Lafaye, R. Perron, and Y. Colleville, *J. Mol. Catal.*, **17**, 339 (1982).
45. G. R. Steinmetz and T. H. Larkins, *Organometallics*, **2**, 1879 (1983).
46. M. Hidai, M. Orisaku, M. Ue, Y. Koyasu, T. Kodama, and Y. Uchida, *Organometallics*, **2**, 292 (1983).
47. (a) G. Jenner, H. Kheradmand, A. Kiennemann, and A. Deluzarche, *J. Mol. Catal.*, **18**, 61 (1983); (b) M. Hidai, Y. Koyasu, M. Orisaku, and Y. Uchida, *Bull. Chem. Soc. Jpn.*, **55**, 3951, 1982; (c) G. Braca, G. Sbrana, G. Valentini, and M. Cini, *J. Mol. Catal.*, **17**, 323 (1982); (d) H. Kheradmand, A. Kiennemann, and G. Jenner, *J. Organomet. Chem.*, **251**, 339 (1983).
48. D. Forster and T. C. Singleton, *J. Mol. Catal.*, **17**, 299 (1982).
49. (a) G. Luft and M. Schrod, *J. Mol. Catal.*, **20**, 175 (1983); (b) M. Schrod, G. Luft, and J. Grobe, *J. Mol. Catal.*, **22**, 169 (1983).
50. *J. Mol. Catal.*, **17**, 117–401 (1982). (International Symposium of Catalytic Reactions of One-carbon Molecules (Belgium, June 1982).)
51. R. Ugo and R. Psaro, *J. Mol. Catal.*, **20**, 53 (1983).
52. (a) G. Henrici-Olivé and S. Olivé, *J. Mol. Catal.*, **16**, 111 (1982); (b) *J. Mol. Catal.*, **18**, 367 (1983); (c) *J. Mol. Catal.*, **19**, 397 (1983); (d) D. G. H. Ballard, *J. Mol. Catal.*, **19**, 393 (1983).
53. D. De Montauzon and R. Mathieu, *J. Organomet. Chem.*, **252**, C83 (1983).
54. R. P. Planalp and R. A. Andersen, *J. Am. Chem. Soc.*, **105**, 7774 (1983).
55. H. Kletzin, H. Werner, O. Serhadli, and M. L. Ziegler, *Angew. Chem. Int. Ed. Engl.*, **22**, 46 (1983).
56. L. E. McCandlish, *J. Catal.*, **83**, 362 (1983).
57. K. M. Nicholas, *Organometallics*, **1**, 1713 (1982).
58. B. D. Dombek, *J. Organomet. Chem.*, **250**, 467 (1983).
59. (a) J. F. Knifton, *J. Chem. Soc., Chem. Commun.*, 729 (1983); (b) R. Whyman, *J. Chem. Soc., Chem. Commun.*, 1439 (1983).
60. D. G. Parker, R. Pearce, and D. W. Prest, *J. Chem. Soc., Chem. Commun.*, 1193 (1982).
61. A. S. C. Chan, W. E. Carroll, and D. E. Willis, *J. Mol. Catal.*, **19**, 377 (1983).
62. J. P. Collman, J. I. Brauman, G. Tustin, and G. S. Wann III, *J. Am. Chem. Soc.*, **105**, 3913 (1983).
63. (a) W. A. R. Slegeir, R. S. Sapienza, R. Rayford, and L. Lam, *Organometallics*, **1**, 1728 (1982); (b) D. J. Darensbourg and A. Rokicki, *Organometallics*, **1**, 1685 (1982).
64. P. Yarrow, H. Cohen, C. Ungermann, D. Vandenberg, P. C. Ford, and R. G. Rinker, *J. Mol. Catal.*, **22**, 239 (1983).
65. D. Choudhury and D. J. Cole-Hamilton, *J. Chem. Soc., Dalton Trans.*, 1885 (1982).
66. C. P. Kubiak, C. Woodcook, and R. Eisenberg, *Inorg. Chem.*, **21**, 2119 (1982).
67. R. H. Fish, A. D. Thormodsen, and G. A. Cremer, *J. Am. Chem. Soc.*, **104**, 5235 (1982).
68. M. A. Radhi, G. Pályi, and L. Markó, *J. Mol. Catal.*, **22**, 195 (1983).
69. K. Kaneda, M. Yasumura, and T. Imanaka, *J. Chem. Soc., Chem. Commun.*, 935 (1982).
70. R. A. Sheldon, *J. Mol. Catal.*, **20**, 1 (1983).
71. H. Mimoun, *Angew. Chem. Int. Ed. Engl.*, **21**, 734 (1982).
72. N. S. Enikolopyan, K. A. Bogdanova, and K. A. Askarov, *Russ. Chem. Rev. (Engl. Transl.)* **52**, 13 (1983); (b) I. V. Kozhevnikov and K. I. Matveev, *Russ. Chem. Rev. (Engl. Transl.)*, **51**, 1075 (1982); (c) I. V. Kozhevnikov, *Russ. Chem. Rev. (Engl. Transl.)*, **52**, 138 (1983); (d) K. B. Sharpless, S. S. Woodard, and M. G. Finn, *Pure Appl. Chem.* **55**, 1823 (1983).

73. P. Chaumette, H. Mimoun, L. Saussine, J. Fischer, and A. Mitschler, *J. Organomet. Chem.*, **250**, 291 (1983).
74. O. Bortolini, V. Conte, F. Di Furia, and G. Modena, *J. Mol. Catal.*, **19**, 331 (1983).
75. A. Arcoria, F. P. Ballistreri, G. Tomaselli, F. Di Furia, and G. Modena, *J. Mol. Catal.*, **18**, 177 (1983).
76. A. Herbowski, J. M. Sobczak, and J. J. Ziólkowski, *J. Mol. Catal.*, **19**, 309 (1983).
77. S. M. Kulikov and I. V. Kozhevnikov, *Kinet. Katal.*, **24**, 42 (1983).
78. F. Di Furia, R. Fornasier, and U. Tonellato, *J. Mol. Catal.*, **19**, 81 (1983).
79. (a) J. T. Groves and T. E. Nemo, *J. Am. Chem. Soc.*, **105**, 5786 (1983); (b) J. T. Groves and R. S. Myers, *J. Am. Chem. Soc.*, **105**, 5791 (1983).
80. (a) O. Bortolini, F. D. Furia, and G. Modena, *J. Mol. Catal.*, **16**, 69 (1982); (b) *ibid.*, p. 61.
81. D. P. Riley, *Inorg. Chem.*, **22**, 1965 (1983).
82. L. Carlton, G. Read, and M. Urgelles, *J. Chem. Soc., Chem. Commun.*, **586** (1983).
83. E. D. Nyberg, D. C. Pribich, and R. S. Drago, *J. Am. Chem. Soc.*, **105**, 3538 (1983).
84. H. Bönnemann, W. Nunez, and D. M. M. Rohe, *Helv. Chim. Acta*, **66**, 177 (1983).
85. M. K. Dickson, N. S. Dixit and D. M. Roundhill, *Inorg. Chem.*, **22**, 3130 (1983).
86. M. T. Atlay, M. Preece, G. Strukul, and B. R. James, *Canad. J. Chem.*, **61**, 1332 (1983).
87. W. K. Wan, K. Zaw, and P. M. Henry, *J. Mol. Catal.*, **16**, 81 (1982).
88. O. S. Andell and J. E. Bäckvall, *J. Organomet. Chem.*, **244**, 401 (1983).
89. M. Iriuchijima, *Sekiyu Gakkaishi*, **26**, 8 (1983).
90. J. E. Baeckvall, *Pure Appl. Chem.*, **55**, 1669 (1983); *Acc. Chem. Res.* **16**, 335 (1983).
91. (a) M. G. Volkhonskii, V. A. Likholobov, and Yu. I. Ermakov, *Kinet. Katal.*, **24**, 578 (1983); (b) N. I. Kuznetsova, V. A. Likholobov, M. A. Fedotov, and Y. I. Yermakov, *J. Chem. Soc., Chem. Commun.*, 973 (1982).
92. M. A. Andrews and C. W. F. Cheng, *J. Am. Chem. Soc.*, **104**, 4268 (1982); see also S. E. Diamond, F. Mares, A. Szalkiewicz, D. A. Muccigrosso, and J. P. Solar, *J. Am. Chem., Soc.*, **104**, 4267.
93. A. E. Martell, *Pure Appl. Chem.*, **55**, 125 (1983).
94. S. Tsuruya, K. Kinumi, K. Hagi, and M. Masai, *J. Mol. Catal.*, **22**, 47 (1983).
95. J. M. Brown, L. R. Canning, A. J. Downs, and A. M. Forster, *J. Organomet. Chem.*, **225**, 103 (1983).
96. M. J. Mirbach, T. N. Phu, and A. Saus, *J. Organomet. Chem.*, **236**, 309 (1982).
97. (a) Y. Doi, K. Koshizuka, and T. Keii, *Inorg. Chem.*, **21**, 2732 (1982); (b) Y. Doi, S. Tamura, and K. Koshizuka, *J. Mol. Catal.*, **19**, 213 (1983).
98. M. Airoldi, G. Deganello, G. Dia, and G. Gennaro, *Inorg. Chim. Acta*, **68**, 179 (1983).
99. S. R. Patil, D. N. Sen, and R. V. Chaudhari, *J. Mol. Catal.*, **19**, 233 (1983).
100. J. Halpern, *Inorg. Chim. Acta*, **77**, L105 (1983).
101. R. Lykvist and R. Larsson, *J. Mol. Catal.*, **19**, 1 (1983).
102. J. W. E. Coenen, *Recl. Trav. Chim.*, **102**, 57 (1983).
103. P. Soegaard-Andersen and J. Ulstrup, *Acta Chem. Scan., Ser. A*, **37**, 585 (1983).
104. J. G. Wadkar and R. V. Chaudhari, *J. Mol. Catal.*, **22**, 103 (1983).
105. C. R. Landis and J. Halpern, *J. Organomet. Chem.*, **250**, 485 (1983); see also J. M. Brown and P. A. Chaloner, *J. Chem. Soc., Perkin Trans. 2*, 711 (1982).
106. R. R. Burch, A. J. Shustermann, E. L. Muetterties, R. G. Teller, and J. M. Williams, *J. Am. Chem. Soc.*, **105**, 3546 (1983).
107. O. W. Howarth, C. H. McAteer, P. Moore, and G. E. Morris, *J. Chem. S⟨c., Chem. Commun.*, 745 (1982).
108. R. H. Crabtree and M. W. Davis, *Organometallics*, **2**, 681 (1983); G. Stork and D. E. Kahne, *J. Am. Chem. Soc.*, **105**, 1072 (1983).
109. D. H. Goldsworthy, F. R. Hartley, and S. G. Murray, *J. Mol. Catal.*, **19**, 257 (1983); 269 (1983).

110. R. Bertani, G. Carturan, and A. Scrivanti, *Angew. Chem. Int. Ed. Engl.*, **22**, 246 (1983).
111. R. Wilczynski, W. A. Fordyce, and J. Halpern, *J. Am. Chem. Soc.*, **105**, 2066 (1983).
112. C. R. Landis and J. Halpern, *Organometallics*, **2**, 840 (1983).
113. D. Pieta, A. M. Trzeciak, and J. J. Ziółkowski, *J. Mol. Catal.*, **18**, 193 (1983).
114. (a) S. Töros, L. Kollar, B. Heil, and L. Markó, *J. Organomet. Chem.*, **253**, 375 (1983);
     (b) S. Töros, L. Kollar, B. Heil, and L. Markó, *J. Organomet. Chem.*, **255**, 377 (1983).
115. L. Markó and J. Palágyi, *Transition Met. Chem.*, **8**, 207 (1983).
116. (a) J. Halpern, *Pure Appl. Chem.*, **55**, 99 (1983); (b) E. I. Klabunovskii, *Russ. Chem. Rev.*
     (*Engl. Transl.*), **51**, 1103 (1982); (c) W. S. Knowles, *Acc. Chem. Res.*, **16**, 106 (1983).
117. J. M. Brown, P. A. Chaloner, and G. A. Morris, *J. Chem. Soc., Chem. Commun.*, 664 (1983).
118. J. M. Brown and D. Parker, *Organometallics*, **1**, 950 (1982).
119. (a) E. Ruch and I. Ugi, *Top. Stereochem.*, **4**, 99 (1969); (b) H. Brunner, B. Schönhammer,
     B. Schönhammer, and C. Steinberger, *Chem. Ber.*, **116**, 3527 (1983).
120. (a) D. Baudry, M. Ephritikhine, H. Felkin, and R. Holmes-Smith, *J. Chem. Soc., Chem.
     Commun.*, 788 (1983); (b) D. Baudry, M. Ephritikhine, H. Felkin, and J. Zakrzewski, *J.
     Chem. Soc., Chem. Commun.*, 1235 (1982).
121. G. B. Shul'pin, G. V. Nizova, and A. E. Shilov, *J. Chem. Soc., Chem. Commun.*, 671 (1983).
122. K. Yoshinaga, T. Kito, and K. Ohkubo, *Bull. Chem. Soc. Jpn.*, **56**, 1786 (1983).
123. M. Bianchi, U. Matteoli, P. Frediani, G. Menchi, and F. Piacenti, *J. Organomet. Chem.*,
     240, 59 (1982).
124. S. Rajagopal, S. Vancheesan, J. Rajaram, and J. C. Kuriacose, *J. Mol. Catal.*, **22**, 137 (1983).
125. S. Mathukumaru-Pillai, S. Vancheesan, J. Rajaram, and J. C. Kuriacose, *J. Mol. Catal.*,
     20, 169 (1983).
126. J. Blum, S. Shtelzer, P. Albin, and Y. Sasson, *J. Mol. Catal.*, **16**, 167 (1982).
127. Y. Matsui, T. E. Nalesnik, and M. Orchin, *J. Mol. Catal.*, **19**, 303 (1983).
128. (a) D. Beaupére, L. Nadjo, R. Uzan, and P. Bauer, *J. Mol. Catal.*, **18**, 73 (1983); (b) D.
     Beaupére, L. Nadjo, R. Uzan, and P. Bauer, *J. Mol. Catal.*, **18**, 195 (1983); (c) D. Beaupére,
     L. Nadjo, R. Uzan, and P. Bauer, *J. Mol. Catal.*, **18**, 185 (1983); (d) D. Beaupére, P.
     Bauer, L. Nadjo, and R. Uzan, *J. Organomet. Chem.*, **238**, C12 (1982).
129. S. Goswami, A. R. Chakravarty, and A. Chakravorty, *J. Chem. Soc., Chem. Commun.*,
     1288 (1982).
130. S. Bhaduri and K. R. Sharma, *J. Chem. Soc., Chem. Commun.*, 1412 (1983).
131. H. Lehmkuhl and Y. L. Tsien, *Chem. Ber.*, **116**, 2437 (1983).
132. J. Halpern, *Pure Appl. Chem.*, **55**, 1059 (1983).
133. M. Balci and Y. Sutbeyas, *Tetrahedron Lett.*, **24**, 4135 (1983).
134. D. Milstein, *J. Am. Chem. Soc.*, **104**, 5227 (1982).
135. W. H. Campbell and P. W. Jennings, *Organometallics*, **1**, 1071 (1982).
136. R. L. Pruett and R. T. Kacmarcik, *Organometallics*, **1**, 1693 (1982).
137. W. J. Richter, *J. Mol. Catal.*, **18**, 145 (1983).
138. J. C. Mol., *CHEMTECH*, **13**, 250 (1983).
139. R. Larsson, *J. Mol. Catal.*, **20**, 81 (1983).
140. P. J. Krusic and F. N. Tebbe, *Inorg. Chem.*, **21**, 2900 (1982).
141. J. B. Lee, K. C. Ott, and R. H. Grubbs, *J. Am. Chem. Soc.*, **104**, 7491 (1982).
142. H. Balcar, B. Matyska, and M. Švestka, *J. Mol. Catal.*, **20**, 159 (1983).
143. J. Kress and J. A. Osborn, *J. Am. Chem. Soc.*, **105**, 6346 (1983).
144. J. G. Hamilton, K. J. Ivin, J. J. Rooney, and L. C. Waring, *J. Chem. Soc., Chem. Commun.*,
     159 (1983).
145. C. Larroche, J. P. Laval, A. Lattes, M. Leconte, F. Quignard, and J. M. Basset, *J. Chem.
     Soc., Chem. Commun.*, 220 (1983).
146. B. Tinland, F. Quignard, M. Leconte, and J. M. Basset, *J. Am. Chem. Soc.*, **105**, 2924 (1983).

147. R. Taube and K. Seyferth, *J. Organometallic Chem.*, **249**, 365 (1983).
148. E. Thorn-Csanyi and H. Timm, *Makromol. Chem., Rapid Commun.*, **4**, 435 (1983).
149. M. Petit, A. Mortreux, and F. Petit, *J. Chem. Soc., Chem. Commun.*, 1385 (1982).
150. L. G. McCullough, M. L. Listemann, R. R. Schrock, M. R. Churchill, and J. W. Ziller, *J. Am. Chem. Soc.*, **105**, 6729 (1983).
151. C. W. Jung, J. D. Fellmann, and P. E. Garrou, *Organometallics*, **2**, 1042 (1983).
152. P. L. Watson and D. C. Roe, *J. Am. Chem. Soc.*, **104**, 6471 (1982).
153. J. Soto, M. L. Steigerwald, and R. H. Grubbs, *J. Am. Chem. Soc.*, **104**, 4479 (1982).
154. (a) A. Sen and R. R. Thomas, *Organometallics*, **1**, 1251 (1982). (b) A. Sen and T. W. Lai, *Organometallics*, **2**, 1059 (1983).
155. D. T. Tsou, J. D. Burrington, E. A. Maher, and R. K. Grasselli, *J. Mol. Catal.*, **22**, 29 (1983).
156. M. Peuckert and W. Keim, *Organometallics*, **2**, 594 (1983).
157. W. Keim, A. Behr, and G. Kraus, *J. Organomet. Chem.*, **251**, 377 (1983).
158. M. Peuckert, W. Keim, S. Storp, and R. S. Weber, *J. Mol. Catal.*, **20**, 115 (1983).
159. A. Pruvot, D. Commereuc, and Y. Chauvin, *J. Mol. Catal.*, **22**, 179 (1983).
160. T. C. Clarke, C. S. Yannoni, and T. J. Katz, *J. Am. Chem. Soc.*, **105**, 7787 (1983).
161. P. Q. Adams, D. L. Davies, A. F. Dyke, S. A. R. Knox, K. A. Mead, and P. Woodward, *J. Chem. Soc., Chem. Commun.*, 222 (1983).
162. H. A. Schäfer, R. Marcy, T. Rüping, and H. Singer, *J. Organomet. Chem.*, **240**, 17 (1982).
163. J. Chatt and R. L. Richards, *J. Organomet. Chem.*, **239**, 65 (1982).
164. P. Pelikán, R. Boča, and M. Magovà, *J. Mol. Catal.*, **19**, 243 (1983).
165. P. Sobota and Z. Janas, *J. Organomet. Chem.*, **243**, 35 (1983).
166. W. Hussain, G. J. Leigh, and C. J. Pickett, *J. Chem. Soc., Chem. Commun.*, 747 (1982).

References for Chapter 15

1. T. Asano and W. J. le Noble, *Chem. Rev.*, **78**, 407 (1978).
2. G. A. Lawrance and D. R. Stranks, *Acc. Chem. Res.*, **12**, 403 (1979).
3. R. van Eldik and H. Kelm, *Rev. Phys. Chem. Jpn.*, **50**, 185 (1980).
4. D. A. Palmer and H. Kelm, *Coord. Chem. Rev.*, **36**, 89 (1981).
5. M. J. Blandamer, J. Burgess, R. E. Robertson, and J. M. W. Scott, *Chem. Rev.*, **82**, 259 (1982).
6. W. J. le Noble and H. Kelm, *Angew. Chem. Int. Ed.*, **11**, 841 (1980).
7. M. J. Blandamer and J. Burgess, *Pure Appl. Chem.*, **55**, 55 (1983).
8. M. J. Blandamer and J. Burgess, *Pure Appl. Chem.*, **54**, 2285 (1982).
9. A. E. Merbach, *Pure Appl. Chem.*, **54**, 1479 (1982).
10. T. W. Swaddle, *Rev. Phys. Chem. Jpn.*, **50**, 230 (1980).
11. T. W. Swaddle, in: *Mechanistic Aspects of Inorganic Reactions* (D. B. Rorabacher and J. F. Endicott, eds.), ACS Symposium Series No. 198, p. 39 (1982).
12. T. W. Swaddle, in: *Advanced in Inorganic and Bioinorganic Mechanisms* (A. G. Sykes, ed.), Adacemic Press, p. 95 (1983).
13. M. F. Mirbach, M. J. Mirbach, and A. Saus, *Chem. Rev.*, **82**, 59 (1982).
14. K. Heremans, *Rev. Phys. Chem. Jpn.*, **50**, 259 (1980).
15. K. Heremans, *Rev. Sci. Instrum.*, **51**, 806 (1980).
16. R. van Eldik, D. A. Palmer, R. Schmidt, and H. Kelm, *Inorg. Chim. Acta*, **50**, 131 (1981).
17. S. Funahashi, K. Ishihara, and M. Tanaka, *Inorg. Chem.*, **20**, 51 (1981).
18. K. Ishihara, S. Funahashi, and M. Tanaka, *Rev. Sci. Instrum.*, **53**, 1231 (1982).
19. P. J. Nichols, Y. Ducommun, and A. E. Merbach, *Inorg. Chem.*, **22**, 3993 (1983).

20. R. Doss, R. van Eldik, and H. Kelm, *Rev. Sci. Instrum.*, **53**, 1592 (1982).
21. T. Inoue, K. Kojima, and R. Shimozawa, *Chem. Lett.*, 259 (1981).
22. H. Vanni, W. L. Earl, and A. E. Merbach, *J. Magn. Reson.*, **29**, 11 (1978).
23. J. Jonas, D. L. Hasha, W. J. Lamb, G. A. Hoffman, and T. Eguchi, *J. Magn. Reson.*, **42**, 169 (1981).
24. A. D. Kirk and G. B. Porter, *J. Phys. Chem.*, **84**, 2998 (1980).
25. W. Weber, R. van Eldik, H. Kelm, J. DiBenedetto, Y. Ducommun, H. Offen, and P. C. Ford, *Inorg. Chem.*, **22**, 623 (1983).
26. S. Balt, J. Ph. Musch, W. E. Renkema, and H. Ronde, *J. Phys. E: Sci. Instrum.*, **16**, 829 (1983).
27. T. Asano, *Rev. Phys. Chem. Jpn.*, **49**, 109 (1979).
28. S. Wherland, *Inorg. Chem.*, **22**, 2349 (1983).
29. S. D. Hamann and W. J. le Noble, *J. Chem. Educ.* (in press).
30. C. H. Langford, *Inorg. Chem.*, **18**, 3288 (1979).
31. K. E. Newman and A. E. Merbach, *Inorg. Chem.*, **19**, 2481 (1980).
32. T. W. Swaddle, *Inorg. Chem.*, **19**, 3203 (1980).
33. T. W. Swaddle, *J. Chem. Soc., Chem. Commun.*, 832 (1982).
34. T. W. Swaddle and M. K. S. Mak, *Can. J. Chem.*, **61**, 473 (1983).
35. T. W. Swaddle, *Inorg. Chem.*, **22**, 2663 (1983).
36. C. Ammann, P. Moore, and A. E. Merbach, *Helv. Chim. Acta*, **63**, 268 (1980).
37. Y. Ducommun, L. Helm, A. Hugi, D. Zbinden, and A. E. Merbach, in: *High Pressure in Research and Industry* (C.-M. Backman, T. Johannisson, and L. Tegnér, eds.), Uppsala, Sweden, University of Uppsala, p. 684 (1982).
38. Y. Ducommun, D. Zbinden, and A. E. Merbach, *Helv. Chim. Acta*, **65**, 1385 (1982).
39. Y. Ducommun., K. E. Newman, and A. E. Merbach, *Helv. Chim. Acta*, **62**, 2511 (1979).
40. Y. Ducommun., K. E. Newman, and A. E. Merbach, *Inorg. Chem.*, **19**, 3696 (1980).
41. F. K. Meyer, K. E. Newman, and A. E. Merbach, *J. Am. Chem. Soc.*, **101**, 5588 (1979).
42. M. J. Sisley, Y. Yano, and T. W. Swaddle, *Inorg. Chem.*, **21**, 1141 (1982).
43. T. W. Swaddle and A. E. Merbach, *Inorg. Chem.*, **20**, 4212 (1981).
44. F. K. Meyer, A. R. Monnerat, K. E. Newman, and A. E. Merbach, *Inorg. Chem.*, **21**, 774 (1982).
45. A. Monnerat, P. Moore, K. A. Newman, and A. E. Merbach, *Inorg. Chim. Acta*, **47**, 139 (1981).
46. Y. Yano, M. T. Fairhurst, and T. W. Swaddle, *Inorg. Chem.*, **19**, 3267 (1980).
47. L. Helm, P. Meier, A. E. Merbach, and P. A. Tregloan, *Inorg. Chim. Acta*, **73**, 1 (1983).
48. Y. Ducommun, W. L. Earl, and A. E. Merbach, *Inorg. Chem.*, **18**, 2754 (1979).
49. R. L. Batstone-Cunningham, H. W. Dodgen, and J. P. Hunt, *Inorg. Chem.*, **21**, 3831 (1982).
50. M. Tubino and A. E. Merbach, *Inorg. Chim. Acta*, **71**, 149 (1983).
51. D. L. Pisaniello, L. Helm, P. Meier, and A. E. Merbach, *J. Am. Chem. Soc.*, **105**, 4528 (1983).
52. M. Inamo, S. Funahashi, and M. Tanaka, *Inorg. Chem.*, **22**, 3734 (1983).
53. M. Inamo, S. Funahashi, and M. Tanaka, *Inorg. Chim. Acta*, **76**, L93 (1983).
54. J. G. Leipoldt, R. van Eldik, and H. Kelm, *Inorg. Chem.*, **22**, 4146 (1983).
55. N. Ise, T. Okubo, and Y. Yamamura, *J. Phys. Chem.*, **86**, 1694 (1982).
56. P. Moore, Y. Ducommun., P. J. Nichols, and A. E. Merbach, *Helv. Chim. Acta*, **66**, 2445 (1983).
57. R. Doss and R. van Eldik, *Inorg. Chem.*, **21**, 4108 (1982).
58. S. Funahashi, K. Ishihara, and M. Tanaka, *Inorg. Chem.*, **22**, 2070 (1983).
59. R. Doss, R. van Eldik, and H. Kelm, *Ber. Bunsenges. Phys. Chem.*, **86**, 925 (1982).
60. K. Ishihara, S. Funahashi, and M. Tanaka, *Inorg. Chem.*, **22**, 3589 (1983).
61. B. B. Hasinoff, *Can. J. Chem.*, **57**, 77 (1979).

62. K. Ishihara, S. Funahashi, and M. Tanake, *Inorg. Chem.*, **22**, 194 (1983).
63. J. Burgess, A. J. Duffield, and R. Sherry, *J. Chem. Soc., Chem. Commun.*, 350 (1980).
64. F. M. Mikhail, P. Askalani, J. Burgess, and R. Sherry, *Transition Met. Chem.*, **6**, 51 (1981).
65. J. Burgess and C. D. Hubbard, *Inorg. Chim. Acta*, **64**, L71 (1982).
66. J. Burgess and C. D. Hubbard, *J. Chem. Soc., Chem. Commun.*, 1482 (1983).
67. G. A. Lawrance, D. R. Stranks, and T. R. Sullivan, *Aust. J. Chem.*, **34**, 1763 (1981).
68. E. F. Caldin and R. C. Greenwood, *J. Chem. Soc., Faraday Trans. 1*, **77**, 773 (1981).
69. S. Funahashi, Y. Yamaguchi, K. Ishihara, and M. Tanaka, *J. Chem. Soc., Chem. Commun.*, 976 (1982).
70. F. K. Meyer, W. L. Earl, and A. E. Merbach, *Inorg. Chem.*, **18**, 888 (1979).
71. R. van Eldik, *Inorg. Chim. Acta*, **49**, 5 (1981).
72. U. Spitzer and R. van Eldik, *Inorg. Chem.*, **21**, 4008 (1982).
73. S. Funahashi, M. Inamo, K. Ishihara, and M. Tanaka, *Inorg. Chem.*, **21**, 447 (1982).
74. G. A. Lawrance, *Inorg. Chem.*, **21**, 3687 (1982).
75. G. A. Lawrance, *Inorg. Chim. Acta*, **54**, L225 (1981).
76. G. A. Lawrance and S. Suvachittanont, *Inorg. Chim. Acta*, **44**, L61 (1980).
77. N. Ise, T. Maruno, and T. Okubo, *Proc. R. Soc. London, Ser. A*, **370**, 485 (1980).
78. G. Daffner, D. A. Palmer, and H. Kelm, *Inorg. Chim. Acta*, **61**, 57 (1982).
79. Y. Kitamura, *Bull. Chem. Soc. Jpn.*, **55**, 3625 (1982).
80. S.-O. Oh, G. Roessling, and H. Lentz, *Z. Phys. Chem.*, **125**, 183 (1981).
81. G. A. Lawrance and S. Suvachittanont, *Aust. J. Chem.*, **33**, 273 (1980).
82. G. A. Lawrance, *Inorg. Chim. Acta*, **45**, L275 (1980).
83. M. J. Sisley and T. W. Swaddle, *Inorg. Chem.*, **20**, 2799 (1981).
84. R. van Eldik and U. Spitzer, *Transition Met. Chem.*, **8**, 351 (1983).
85. Y. Kitamura, *Bull. Chem. Soc. Jpn.*, **52**, 3280 (1979).
86. T. Inoue, K. Kojima, and R. Shimozawa, *Inorg. Chem.*, **22**, 3972 (1983).
87. T. Inoue, K. Sugahara, K. Kojima, and R. Shimozawa, *Inorg. Chem.*, **22**, 3977 (1983).
88. K. Ishihara, S. Funahashi, and M. Tanaka, *Inorg. Chem.*, **22**, 2564 (1983).
89. A. Hioki, S. Funahashi, and M. Tanaka, *Inorg. Chim. Acta*, **76**, L151 (1983).
90. A. E. Merbach, P. Moore, and K. E. Newman, *J. Magn. Reson.*, **41**, 30 (1980).
91. M. J. Blandamer, J. Burgess, J. G. Chambers, and A. J. Duffield, *Transition Met. Chem.*, **6**, 156 (1981).
92. M. T. Fairhurst and T. W. Swaddle, *Inorg. Chem.*, **18**, 3241 (1979).
93. E. L. J. Breet, R. van Eldick, and H. Kelm, *Polyhedron*, **2**, 1181 (1983).
94. R. van Eldik, E. L. J. Breet, M. Kotowski, D. A. Palmer, and H. Kelm, *Ber. Bunsenges. Phys. Chem.*, **87**, 904 (1983).
95. M. Mares, D. A. Palmer, and H. Kelm, *Inorg. Chim. Acta*, **60**, 123 (1982).
96. D. A. Palmer, R. van Eldik, and H. Kelm, *Z. Anorg. Allg. Chem.*, **468**, 77 (1980).
97. D. L. Pisaniello, P. J. Nichols, Y. Ducommun, and A. E. Merbach, *Helv. Chim. Acta*, **65**, 1025 (1982).
98. F. B. Ueno, A. Nagasawa, and K. Saito, *Inorg. Chem.*, **20**, 3504 (1981).
99. G. A. Lawrance, M. J. O'Connor, S. Suvachittanont, D. R. Stranks, and P. A. Tregloan, *Inorg. Chem.*, **19**, 3443 (1980).
100. G. A. Lawrance and S. Suvachittanont, *Aust. J. Chem.*, **33**, 1649 (1980).
101. W. J. Louw, R. van Eldik, and H. Kelm, *Inorg. Chem.*, **19**, 2878 (1980).
102. W. G. Jackson, G. A. Lawrance, P. A. Lay, and A. M. Sargeson, *Inorg. Chem.*, **19**, 904 (1980).
103. W. G. Jackson, G. A. Lawrance, P. A. Lay, and A. M. Sargeson, *Aust. J. Chem.*, **35**, 1561 (1982).
104. W. Rindermann, R. van Eldik, and H. Kelm, *Inorg. Chim. Acta*, **61**, 173 (1982).

105. W. Rindermann and R. van Eldik, *Inorg. Chim. Acta*, **68**, 35 (1983).
106. R. L. Batstone-Cunningham, H. W. Dodgen, J. P. Hunt, and D. M. Roundhill, *J. Chem. Soc., Dalton Trans.*, 1473 (1983).
107. S. Funahashi, S. Funada, M. Inamo, R. Kurita, and M. Tanaka, *Inorg. Chem.*, **21**, 2202 (1982).
108. R. van Eldik, *Inorg. Chem.*, **21**, 2501 (1982).
109. R. van Eldik and H. Kelm, *Inorg. Chim. Acta*, **60**, 177 (1982).
110. R. van Eldik and H. Kelm, *Inorg. Chim. Acta*, **73**, 91 (1983).
111. R. van Eldik, *Inorg. Chem.*, **22**, 353 (1983).
112. Y. Sasaki, F. B. Ueno, and K. Saito, *J. Chem. Soc., Chem. Commun.*, 1135 (1981).
113. H. Ghazi-Bajat, R. van Eldik, and H. Kelm, *Inorg. Chim. Acta*, **60**, 81 (1982).
114. U. Spitzer, R. van Eldik, and H. Kelm, *Inorg. Chem.*, **21**, 2821 (1982).
115. M. Walper and H. Kelm, *Z. Phys. Chem. (Neue Folge)*, **113**, 207 (1978).
116. D. J. A. de Waal, T. I. A. Gerber, W. J. Louw, and R. van Eldik, *Inorg. Chem.*, **21**, 2002 (1982).
117. K. Angermann, R. van Eldik, H. Kelm, and F. Wasgestian, *Inorg. Chem.*, **20**, 955 (1981).
118. K. Angermann, R. van Eldik, H. Kelm, and F. Wasgestian, *Inorg. Chim. Acta*, **49**, 247 (1981).
119. A. D. Kirk, C. Namasivayam, G. B. Porter, M. A. Rampi-Scandola, and A. Simmons, *J. Phys. Chem.*, **87**, 3108 (1983).
120. F. B. Ueno, Y. Sasaki, T. Ito, and K. Saito, *J. Chem. Soc., Chem. Comoun.*, 328 (1982).
121. L. H. Skibsted, W. Weber, R. van Eldik, H. Kelm, and P. C. Ford, *Inorg. Chem.*, **22**, 541 (1983).

Author Index

The page on which an author is cited is given first, followed by the reference number(s) in parentheses.

Paéz, D. E., 235 (117)
Page, J. A., 232 (94)
Paguaga, L. A., 245 (161)
Pakulski, M., 126 (233)
Pal, A., 226 (54)
Palágyi, J., 386 (115)
Pali, S. C., 137 (345)
Palmer, D. A., 145 (17a), 205
 (103), 238, 241, 243, 246 (141),
 257 (69), 399 (4), 400 (16),
 409, 411 (78), 415 (16) (94–95),
 417 (96), 419 (95)
Palmer, J. A. L., 346 (11)
Palyi, G., 380 (68)
Panda, A. K., 125 (226)
Panda, R. K., 31 (82), 50 (81–82),
 62 (29–30), 63 (30–31), 73 (96–
 97), 256 (61–62)
Pandey, K. K., 370 (15)
Pandey, N. N., 82 (155)
Pang, H., 216 (3)
Panigrahi, G. P., 111 (136)
Panowski, M., 279 (66)
Paonessa, R. S., 153 (46), 314
 (148)
Paradisi, C., 203 (91–92), 333
 (33–34)
Pardy, R. B. A., 354 (57–58), 356
 (58)
Parekh, H. V., 290 (32)
Park, J. T., 364 (90)
Parker, D. G., 378 (60)
Parker, D., 348 (20), 358 (67),
 387 (118)
Parker, R. J., 137 (340)
Parker, V. D., 130 (289)
Parkhurst, C. S., 326 (16)
Parmon, V. N., 9 (42)
Parr, D. P., 257 (65)
Parrinello, G., 374 (36)
Parvez, M., 149 (36–37), 314
 (150)
Pascard, C., 358 (67)
Paseka, I., 128 (273)
Pasini, A., 210 (121)
Pasquali, M., 370 (7)
Pasternack, R. F., 77 (125)
Pasynkiewicz, S., 348 (22)
Patel, R. C., 253 (18)
Patil, S. R., 384 (99)
Patin, H., 273 (39–40)
Patrick, J. M., 354 (55)
Patterson, L. K., 76, 77 (114)
Patterson, M. A. K., 132 (301)
Patton, A. T., 350 (36)
Payne, K. S., 227 (60)

Peacock, R. D., 23, 33, 50, 54,
 (77), 216 (5)
Pearce, R., 378 (60)
Pearson, R. G., 147 (28), 160
 (10), 199 (71)
Pecht, I., 33 (138), 55 (131) (138),
 56 (136) (138)
Pedersen, B., 69 (68), 227, 229
 (66)
Pedersen, E., 171 (59)
Pedersen, S. F., 305 (92)
Pedrosa, G. C., 216, 217, 218, 222
 (13)
Peeples, J. A., 10 (57)
Peishoff, C. E., 105 (91)
Pelikán, P., 396 (164)
Pell, R. J., 257 (67)
Pellinghelli, M. A., 136 (332)
Peloso, A., 54 (128), 243 (154)
Pennell, C. A., 199 (72)
Penny, D. E., 100 (65)
Peregudova, S. M., 335 (44)
Perkins, C. W., 131 (295)
Perlmutter-Hayman, B., 256 (55)
Perotti, A., 50 (84)
Perron, R., 375 (44)
Perutz, R. N., 305 (88)
Pesek, J. J., 144, 152, (14), 257
 (70)
Peter, L., 133 (307)
Peterleitner, M. G., 268 (23)
Petersen, J. D., 13 (120), 15 (90–
 91), 16 (90), 221 (30), 234 (30)
 (110) (112), 246 (30)
Petit, F., 392 (149)
Petit, M., 392 (149)
Petrignani, J. F., 78 (130)
Petrosyan, V. S., 300 (65)
Petrova, G. G., 137 (342)
Petrucci, S., 254 (25)
Petryaev, E. P., 83 (179)
Peuckert, M., 394 (156) (158)
Pfenning, K. J., 15, 16 (90)
Phelan, K. G., 113 (159), 114
 (160)
Phillips, T. E., 17 (108)
Phu, T. N., 372 (24a) (24b), 384
 (96)
Piacenti, F., 371 (18), 388 (123)
Pickering, R. A., 113 (146)
Pickett, C. J., 59, 77 (5), 396
 (166)
Pickup, P. G., 16 (106)
Pidcock, A., 343 (1), 353 (51)
Pieniazek, M., 83 (177)
Pierce, R. A., 107 (104)

Pierpont, C., 292 (41), 311 (126)
Pieta, D., 386 (113)
Pietropaolo, R., 334 (37)
Pignatello, J. J., 291 (38)
Pilcher, G., 148 (34)
Pinhas, A. R., 280 (70), 334 (38),
 355, 356 (59)
Pino, P., 370 (14), 371 (20), 373
 (33), 374 (35)
Pirakitigoon, S., 123 (212)
Pisaniello, D. L., 253 (19–20),
 258 (77), 405 (51), 417 (97)
Pistara, S., 226 (56)
Pitchumani, K., 125 (229), 134
 (311)
Pittenger, S. T., 44, 54, (126)
Pittman, C. V., 374 (35)
Pitts, J. N., 111 (135)
Pizer, R., 94 (9)
Plaisted, R. J., 102 (76)
Planalp, R. P., 303 (83), 377 (54)
Platt, A. W. G., 124 (218–221),
 249 (181)
Po, H. N., 60, 62, 70, 74, 75 (13),
 199 (66–67)
Pocius, A. V., 134 (314)
Pocker, Y., 100 (59–60)
Poe, A., 263 (4), 274 (44)
Poggi, A., 50 (84) (88)
Pogorzelec, P. J., 353 (53)
Poirer, M., 63 (78), 103 (79)
Poli, R., 275 (46)
Poliakoff, M., 272 (37)
Polozova, N. I., 128 (263)
Pomeroy, R. K., 366 (97)
Pomposo, F., 228 (70), 231 (81)
Ponterini, G., 227 (57)
Poon, C. K., 236 (122–123), 247
 (170)
Pope, M. T., 14 (85)
Popov, A. I., 254 (24)
Porcel-Ortega, E., 221 (31)
Porter, G. B., 233 (104), 400 (24),
 423 (119), 425 (24)
Porter, W. A., 183 (10)
Posselt, N. S., 137 (343)
Potapov, I. A., 79 (135)
Potter, D. M., 154 (49)
Povey, D. C., 209 (120)
Prado, C. S., 95 (15)
Prasad, D. R., 233 (107), 246
 (165)
Prasad, R. K., 138 (348)
Prassides, K., 15 (92)
Pratt, J., 79 (131)

General Subject Index

Acetate, catalysis by, 160
Acid effect on reduction, 24
Acid catalysis (*see also* Catalysis)
  $[Co(BH)_2gly]^{2+}$, 185
  $[Co(NH_3)_5SO_3]^-$ decomposition, 53
  decarboxylation, 186, 203
  heterolysis, 338
  Mo(V) dimer isomerization, 215
Acid halide decarbonylation, 310
Activation parameters (*see also* Volumes of
    activation)
  and crystal field, 270
  effect of solvent, 222
  H_2 elimination, 80
ADP, 119
A-frame molecules, 358
Alanine, Cr(III) complex, 169
Alcohols, oxidation, 82
Alkanes
  activation of, 390
  oxidation of, 88
Alkenes
  hydrogenation, 384
  insertion of, 279
  metathesis, 391
  oxidation, 88, 127
  polymerization, 393
Alkoxide nucleophiles, 331
Alkynes
  metathesis, 391
  polymerization, 395
  reaction with O_3, 127
  rotation of, 362

Amines, disproportionation, 393
Amino acid synthesis, 208
Amino alcohols, Cr(III) complexes, 161
Ammonia solvent, 200
Anation of
  Co(III) complexes, 197
  $[Coen_2(NO_2)(H_2O)]^{2+}$, 199
  Cu(II) five-coordinate, 258
  M(III) porphyrins, 172
  Pd(II) complexes, 145
  $[Pd(triL)(H_2O)]^{2+}$, 146
Antimony, 126
Apicophilicity, scale of, 103
Aquation of
  aquocobalamin, 199
  Co(III) complexes, 181
  $[CoBr_2(cyclam)]^+$, 182
  $[Co(NH_3)_5L]^{n+}$, 186
  Cr(III) amines, 165
  $[Cr(OX)_3]^{3-}$, 167
  Fe(II) diimines, 225
  $[M(NH_3)_5Br]^{2+}$, 162
  Pt(IV) complexes, 248
  Rh(III) complexes, 238
  $[Rh(NH_3)_5X]^{2+}$, photochemical, 244
  $[Rh(NH_3)_4X_2]^+$, 245
  $[Ru(NH_3)_5BzNC]^{2+}$, 231
Arbuzov elimination, 271
Arenes
  addition to, 333
  hydrogenation, 386
  redox potential, 334
Arsenic, 126